Materials Horizons: From Nature to Nanomaterials

Series Editor

Vijay Kumar Thakur, School of Aerospace, Transport and Manufacturing,
Cranfield University, Cranfield, UK

Materials are an indispensable part of human civilization since the inception of life on earth. With the passage of time, innumerable new materials have been explored as well as developed and the search for new innovative materials continues briskly. Keeping in mind the immense perspectives of various classes of materials, this series aims at providing a comprehensive collection of works across the breadth of materials research at cutting-edge interface of materials science with physics, chemistry, biology and engineering.

This series covers a galaxy of materials ranging from natural materials to nanomaterials. Some of the topics include but not limited to: biological materials, biomimetic materials, ceramics, composites, coatings, functional materials, glasses, inorganic materials, inorganic-organic hybrids, metals, membranes, magnetic materials, manufacturing of materials, nanomaterials, organic materials and pigments to name a few. The series provides most timely and comprehensive information on advanced synthesis, processing, characterization, manufacturing and applications in a broad range of interdisciplinary fields in science, engineering and technology.

This series accepts both authored and edited works, including textbooks, monographs, reference works, and professional books. The books in this series will provide a deep insight into the state-of-art of Materials Horizons and serve students, academic, government and industrial scientists involved in all aspects of materials research.

Review Process

The proposal for each volume is reviewed by the following:

1. Responsible (in-house) editor
2. One external subject expert
3. One of the editorial board members.

The chapters in each volume are individually reviewed single blind by expert reviewers and the volume editor.

Sushanta K. Sethi · Hariome Sharan Gupta · Akarsh Verma
Editors

Polymer Composites: From Computational to Experimental Aspects

Springer

Editors
Sushanta K. Sethi
Department of Mechanical,
Materials and Aerospace Engineering
IIT Dharwad, Karnataka, India

Akarsh Verma
University of Petroleum and Energy Studies
Dehradun, Uttarakhand, India

Hariome Sharan Gupta
Department of Polymer and Process
Engineering
Indian Institute of Technology Roorkee
Roorkee, India

ISSN 2524-5384 ISSN 2524-5392 (electronic)
Materials Horizons: From Nature to Nanomaterials
ISBN 978-981-97-0887-1 ISBN 978-981-97-0888-8 (eBook)
https://doi.org/10.1007/978-981-97-0888-8

This Springer imprint is published by the registered company Springer Nature Singapore Pte Ltd.
The registered company address is: 152 Beach Road, #21-01/04 Gateway East, Singapore 189721, Singapore

Contents

Chapter 1
Polymer Composites: Synthesis, Application, and Basic Theoretical Aspects

Nishtha Arora, Sachin Dua, Smruti Vardhan Purohit, Bibek Dash, Manishkumar D. Yadav, Bikash Kumar Jena, and T. Senthilkumar

Abbreviations

Sr. No.	Abbreviations
(1) Polymer matrix composite	PMC
(2) Polystyrene	PS
(3) Fiber-reinforced polymer	FRP
(4) Polycarbonate	PC
(5) Polyethylene	PE
(6) Polypropylene	PP

N. Arora, S. Dua and S. V. Purohit—Contribute equally to the work.

N. Arora · S. Dua · T. Senthilkumar (✉)
Polymeric Materials Area, Chemical and Material Sciences Division, CSIR-Indian Institute of Petroleum, Dehradun 248005, India
e-mail: t.senthilkumar@iip.res.in

S. V. Purohit · B. K. Jena (✉)
Materials Chemistry Department, CSIR-Institute of Minerals and Materials Technology, Bhubaneswar 751013, India
e-mail: bikash@immt.res.in

B. Dash (✉)
Process Engineering and Instrumentation Department, CSIR-Institute of Minerals and Materials Technology, Bhubaneswar 751013, India
e-mail: bibek@immt.res.in

N. Arora · S. Dua · S. V. Purohit · B. Dash · B. K. Jena · T. Senthilkumar
Academy of Scientific and Innovative Research (AcSIR), Ghaziabad 201002, India

M. D. Yadav
Department of Chemical Engineering, Institute of Chemical Technology, Mumbai 400019, India

S. K. Sethi et al. (eds.), *Polymer Composites: From Computational to Experimental Aspects*, Materials Horizons: From Nature to Nanomaterials,
https://doi.org/10.1007/978-981-97-0888-8_1

(7) Polyether Ether Ketone	PEEK
(8) Polybenzimidazole	PBI
(9) Poly(vinylalcohol-leucine)	PVAL
(10) Poly(lactic acid)	PLA
(11) Polycaprolactone	PCL
(12) Multi-walled carbon nanotube	MWCNT
(13) Carbon nanotube	CNT
(14) Single-walled carbon nanotube	SWCNT
(15) Electromagnetic interference	EMI

1 Introduction to Polymeric Composites

1.1 Definition and Overview

Polymers have been fundamental in advanced applications for an extended period due to their versatility and ease of molding to specific needs [1]. Nonetheless, using a single polymer often falls short of meeting the demands of advanced applications. Consequently, polymer composites have gained worldwide attention. A composite material consists of two or more distinct phases—a matrix/continuous phase and a dispersed phase—resulting in bulk properties vastly different from any individual constituent. The matrix phase, primarily continuous, plays a pivotal role by hosting and sharing the load with the dispersed (reinforcing) phase. The reinforcing phase, typically more robust than the matrix, is embedded in a discontinuous form within the matrix. This reinforcing phase contributes significantly to the composite's overall strength, earning it the designation of a reinforcing phase [2]. These composites can incorporate metals, carbon, ceramics, and other polymers as matrices and reinforcements [1]. The exceptional chemical resistance offered by polymers rivals that of metals. Additionally, key properties like mechanical strength, corrosion, and fatigue resistance are notably higher in polymer matrix composites (PMCs) than in metals. The incorporation of reinforcing agents in polymer matrices leads to substantial enhancements in mechanical, electrical, and thermal characteristics. Incorporating polymers into the matrix can be achieved through diverse approaches, primarily involving methods like melt intercalation, solution blending, or in-situ polymerization. By blending fibers and polymers in various combinations, novel materials with outstanding performance characteristics can be tailored to meet diverse needs. Furthermore, in addition to fibers, various fillers, including $CaCO_3$, clay, mica, and glass microspheres, are repurposed to enhance the properties of polymers (Fig. 1) [1]. While the characteristics of the reinforcing elements greatly influence a composite's final properties, the initial material quality and chosen production methods also significantly affect its performance. The objective of composite manufacturing is to create components that exhibit desired traits by amalgamating the strengths of fibers

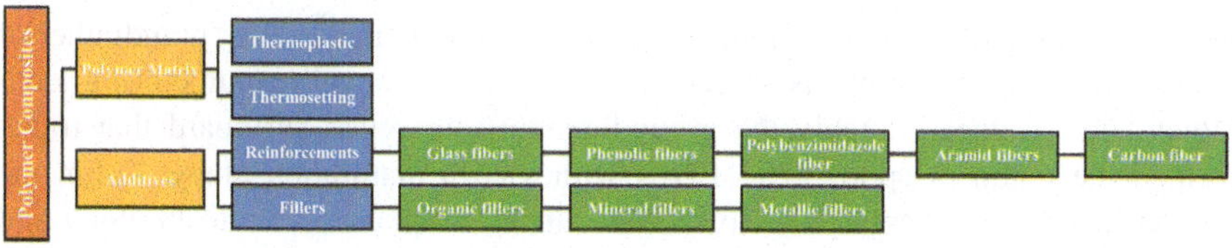

Fig. 1 Schematic representation of components of polymer composites

or particles with the matrix while simultaneously reducing or concealing their individual shortcomings [3]. PMCs can be crafted through various fabrication methods, each influencing the mechanical and physical traits of the final product. The properties of these composites rely on factors like the reinforcement type, volume ratio, how layers are stacked, their orientation, and the total number of layers. The choice of manufacturing techniques is predominantly guided by the type of polymer, be it thermoplastic or thermosetting, owing to their distinct chemical and physical characteristics [4]. When polymer-based composites are combined, they exhibit outstanding qualities such as essential thermal expansion properties, cost-effectiveness, convenient manufacturing, resistance to corrosion and abrasion, fatigue resistance, exceptional specific strength, stiffness, and lightweight nature. PMCs hold immense significance across diverse sectors, including electronics, industry, construction, marine, civil engineering, household, energy, military, sports, communications, medicine, aerospace, and automotive industries due to their exceptional attributes [5]. The chapter extensively explores various facets of synthesizing polymer composites, covering techniques like melt intercalation, solution mixing, and in-situ polymerization while incorporating a wide array of fillers and reinforcements. Additionally, the chapter delves into the detailed significance of these composites across diverse sectors, including construction, healthcare, aerospace, automobiles, electronics, and marine applications.

1.2 Historical Development

The modern age of composite materials commenced with the advent of synthetic plastics, replacing traditional natural resins sourced from plants and animals that were formerly the sole binding agents. In the early 1900s, the development of plastics like vinyl, polystyrene (PS), phenolic, and polyester surpassed the capabilities of natural resins [6]. However, pure plastics lacked adequate strength for certain structural uses, necessitating reinforcement for enhanced strength and stability. In a pivotal moment in 1935, Owens Corning unveiled fiberglass, a glass fiber that, when blended with plastic polymers, yielded an exceptionally robust and lightweight structure, marking the genesis of the fiber-reinforced polymers (FRPs) industry [7]. Following World War II, a burgeoning composites industry emerged as a niche, led by innovators like Brandt Goldsworthy, frequently hailed as the "grandfather of composites." Goldsworthy spearheaded the creation of novel manufacturing techniques and products, notably the groundbreaking fiberglass surfboard that transformed the realm of sports. The 1970s witnessed a maturation phase within the composites sector, characterized by advancements in plastic resin technology and the enhancement of reinforcing fibers. DuPont's development of Kevlar, an aramid fiber known for its exceptional tensile strength, high density, and lightweight properties, stood as a hallmark achievement. Concurrently, carbon fiber emerged during this era and progressively began replacing steel components. Presently, the composites industry remains dynamic, with a significant focus on renewable energy ventures. Notably, the manufacture of wind turbine blades is a leading area of growth, where advanced composite materials continually push the boundaries of size and efficiency. Goldsworthy's pioneering contribution to composites included the invention of pultrusion, a manufacturing process pivotal for producing consistently robust fiberglass-reinforced products. In the contemporary landscape, this process finds extensive application across a diverse spectrum, encompassing tool handles, arrow shafts, ladder rails, railway flooring, pipes, protective covering, and healthcare instruments. In India, the introduction of advanced composites traces back to 1980 when Dr. APJ Abdul Kalam utilized them to craft the nozzle for the Satellite Launch Vehicle (SLV-3). Dr. Kalam stands as a pioneering figure in advancing research in this field and rightfully earns the title of the "father of advanced composites" within the country. PMCs represent a significant composite material category with a broad spectrum of applications. The growth of advanced composites in India can be attributed to the diverse array of raw materials available, various processing technologies, and their adaptability for shaping into intricate forms. However, achieving the optimal composite product in terms of cost-effectiveness hinges on identifying the perfect amalgamation of raw materials and processing techniques, all contingent on its processability [8, 9]. In the present day, composite materials are witnessing a rapid expansion of their usage across industries like automotive, appliances, and consumer goods. Additionally, composites are entering the realm of nanotechnology, showcasing a promising future in this domain. According to market data analysis, the composites market size reached USD 86.56 billion in 2021 and is anticipated to attain USD 150.28 billion by 2028, with an estimated Compound Annual Growth Rate (CAGR) of 8.2% between 2022 and 2028 (Fig. 2) [10].

1.3 Importance and Applications

In today's technologically advanced society, polymers and their composites have emerged as pivotal materials for a wide array of practical applications. Owing to extensive research and expanding knowledge, polymer-based materials have taken the lead and are swiftly displacing other materials in various domains [11]. Constant advancements in polymer-derived materials provide viable alternatives to traditional substances, even in sectors where polymers were historically deemed unsuitable.

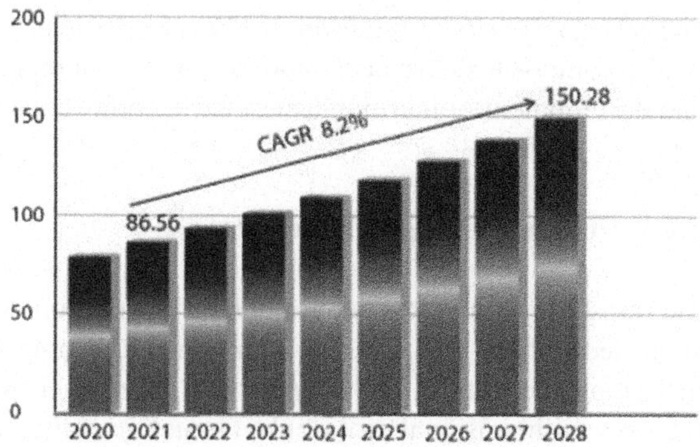

Fig. 2 Global composites market size by 2020 to 2028 (USD Billion). Adapted from ref. [10]

Notably, polymers have now stepped in as substitutes for metals and ceramics, particularly in industries such as construction, aerospace, automotive, and healthcare [12]. The trajectory is clear that this paradigm shift toward polymers will persist, propelled by their inherent characteristics and potential for sustainability [5]. When polymer-based composites are combined, they present exceptional qualities such as precise thermal expansion attributes, cost-effectiveness, effortless manufacturing, resilience to corrosion and abrasion, excellent fatigue resistance, high specific strength, elevated rigidity, and lightweight nature. The significance of PMCs reverberates across a multitude of sectors, including construction, household, energy, defense, industrial, sports, marine, communication, electronics, civil engineering, healthcare, aerospace, and automotive, due to their unparalleled attributes [5, 13].

1.4 Polymer Matrix Materials

Matrix material is a continuous phase that bonds the fibers together [14]. PMCs have garnered significant attention primarily because of their cost-effectiveness and superior specific strength and stiffness when compared to traditional metallic alloys. Furthermore, PMCs provide increased design versatility and enhanced resistance to both corrosion and fatigue [15].

The matrix materials employed in polymer composites play a pivotal role in influencing the composite's overall properties and performance. They contribute to both shape and size stability in the end product and enhance the material's ability to withstand outdoor conditions and maintain its stability. The selection of the matrix material depends upon the particular application and the intended properties sought for the composite. Matrix materials should be coordinately considered and selected from the three aspects of usage performance, processing property, and economic efficiency [14]. PMCs are composed of a polymer matrix, which can be either thermoplastic materials like polycarbonate (PC), polyvinylchloride, nylon, and PS, or thermoset materials such as unsaturated polyester and epoxy. These matrices are combined with a dispersed phase consisting of materials like carbon, ceramics, glass, and other polymers in the form of fibers or particles [16]. Traditionally, thermoset-matrix composites have been more prevalent, but there is currently a rapid evolution and development of thermoplastic-matrix composites [17].

1.4.1 Thermosetting Polymer Matrix

Thermosetting polymers retain their hard and non-softening characteristics when subjected to heat. They consist of a network of polymers with covalent bonds that resist movement at elevated temperatures. These thermosets are in high demand due to their three-dimensional cross-linked structure, which imparts exceptional attributes like temperature and solvent resistance [18]. Thermosetting polymer matrix materials offer a range of merits that make them suitable for diverse applications. This includes high-temperature resistance, excellent dimensional stability, chemical resistance, high strength and stiffness, low creep, excellent adhesion, electrical insulation properties, durable coatings, rigidity and hardness, excellent chemical stability, and more [8, 18]. While thermosetting polymer matrix materials offer many advantages, it's crucial to note that they also have limitations, such as the inability to be remelted and reshaped after curing. This irreversibility can be a drawback in some situations. The choice of matrix material should consider the specific requirements of the application to ensure the best performance. Thermosetting composites find a wide array of applications, ranging from aerospace structures like fuselages, airframes, nose cones, and re-entry surfaces to automotive, railway, and wind turbine components. They are also used in chemical storage tanks, armored panels, civil structures, air conditioning ducts, furniture, and sports equipment [8]. The commonly used thermosetting matrix includes the following:

- Epoxy: Epoxy resins are widely used for their exceptional mechanical properties, adhesion, and resistance to chemicals. They are often used in aerospace, automotive, and construction applications [19].
- Polyester: Polyester resins are known for their ease of processing and are commonly used in boat hulls, automotive parts, and laminates [20, 21].
- Vinyl Ester: Vinyl ester resins offer good corrosion resistance and are suitable for chemical storage tanks and pipes [22, 23].
- Phenolic: Phenolic resins are highly heat-resistant and find use in aircraft components, high-temperature insulation, and brake components [24, 25].

1.4.2 Thermoplastic Polymer Matrix

Thermoplastic polymers consist of long, interconnected chains of molecules and are often referred to as thermosoftening plastics [18]. These composites have carved out a niche for themselves owing to their ease of processing, consistent quality, robust damage resistance, and reduced environmental risks. They are being employed in various sectors, including automotive, aviation, packaging, and refrigeration [8]. In numerous manufacturing procedures, thermoplastic matrices undergo heating and shaping via techniques like molding, extrusion, injection, or thermoforming, followed by a cooling process to uphold the intended structure of the end product. Importantly, this process is reversible, allowing for reshaping if needed [26]. Thermoplastic polymer matrix materials offer various benefits that make them advantageous for various applications. Some of the key merits of thermoplastic polymer matrix materials include recyclability, ease of processing, short processing times, a wide range of material options, chemical resistance, good impact resistance, high strength-to-weight ratio, flexibility and toughness, good electrical insulation properties, suitable for 3d printing, ductility, thermal conductivity control, and many more [27–29]. The choice of matrix material should consider the specific requirements of the application to ensure the best performance. The commonly used thermosetting matrix includes the following:

- Polyethylene (PE): PE-based matrices are used in applications requiring low cost, chemical resistance, and electrical insulation [30, 31].
- Polypropylene (PP): PP matrices are lightweight and have good chemical resistance. They are used in automotive and consumer goods [32, 33].
- Polyamide (Nylon): Nylon is renowned for its exceptional strength and durability, which makes it well-suited for various uses, including gears, bearings, and structural components [34].
- PC: PC matrices offer transparency and impact resistance and are used in applications like eyewear lenses and electrical enclosures [26, 35, 36].
- Polyether Ether Ketone (PEEK): PEEK is a high-performance thermoplastic with exceptional heat and chemical resistance commonly used in aerospace and medical devices [26].

1.5 Reinforcement Materials

The reinforcement serves as the structural backbone, providing mechanical strength in terms of tensile strength and rigidity. The reinforcements can range from elongated particles to continuous strands and are filamentary, either organic or inorganic in nature. This transition from fine particles to continuous fibers is crucial [37]. The predominant choice for reinforcement is E glass fiber, constituting over 95% of applications. Additionally, aramid fiber, like Kevlar, holds considerable interest [38, 39]. In general, the fibers perform effectively in traction in a composite structure (anisotropic). However, compared to metallic structures (which are isotropic), these composites display relatively lower performance in compression and notably weaker shear resistance. Typically, these reinforcements are utilized as fibers or modified variants.

1.5.1 Glass Fibers

At present, glass fibers dominate as the primary choice for reinforcement in advanced composite materials across a wide array of industrial applications [40]. Glass fibers play a pivotal role as the fundamental strengthening component in the widespread application of high-performance composite materials.

1.5.2 Phenolic Fiber

The industrial polymer produced under the trade name KYNOL is recognized for its outstanding thermal performance and commendable dimensional stability.

1.5.3 Polybenzimidazole (PBI) Fiber

PBI stands as a highly efficient fiber known for its cutting-edge properties within polymer composite, useful in the aerospace, automobile, and wind power industries [41]. Referred to as PBI fiber, it is crafted by spinning a poly(2,20-(m-phenylene)-5,50-bibenzimidazole) in dimethylacetamide solvent, resulting in its unique characteristics.

1.5.4 Aramid Fibers

Presently, high-performance polymeric composite, bolstered by aramid fiber reinforcement, demonstrates remarkable economic viability in industrial sectors. These sophisticated materials boast exceptional toughness, outstanding impact resistance, low density, and an extended lifespan [42–44]. Aromatic polyamide fibers, with KEVLAR being the most renowned, are created through a polycondensation methodology involving para-phenylene diamine and terephthalic acid chloride, conducted in N-methylpyrolidone solvent [26].

1.5.5 Carbon Fiber

Carbon fibers play a significant role in crafting composite materials and find extensive applications across various industries, including space construction, aeronautics, aircraft, and automobiles [45, 46]. Renowned for their outstanding mechanical and thermal resistance properties, carbon fibers exhibit remarkably high compressive and tensile strength as well as exceptional rigidity. Derived from polyacrylonitrile and processed through carbonization and subsequent graphitization, carbon fibers are widely recognized as the premier choice in fiber materials.

1.6 Filler Materials

Filler refers to an inert material incorporated into the base polymer, allowing for significant modifications in electrical, mechanical, or thermal properties, enhancing the appearance of surface, or simply reducing the total cost of the resulting composite [26, 47]. Various fillers are employed in polymer composites to modify and enhance their properties. In numerous research endeavors, the proportion of filler ranges typically from 4 to 5% based on weight [48]. The selection of filler relies on the particular application requirements, cost considerations, and the desired properties of the eventual composite material [49].

1.6.1 Organic Fillers

Organic fillers are materials derived from living organisms or carbon-based compounds and used in polymer composites to improve specific properties. It includes natural fibers, cellulose fillers, vegetable fibers, starch-based fillers, chitosan, and many more [48].

1.6.2 Mineral Fillers

Mineral fillers are inorganic substances utilized to improve a range of properties for polymer composites. These are frequently employed because of their abundant natural availability, contributing to cost savings in the production of composite products, and their non-toxic nature [47]. These include carbonates, silica, alumina, clay, talc, and various oxides. Different mineral fillers offer unique advantages and can be tailored to meet the needs of specific applications.

1.6.3 Metallic Fillers

Metallic fillers consist of metals or metallic compounds that are incorporated into polymer composites to augment specific properties. Metal-filled polymer composites offer advantages such as low specific weight, high resistance to corrosion, flexibility, and cost-effective processing methods. These composites provide a wide range of options for modifying their properties [50]. Adding metallic powders such as copper, aluminum, and zinc imparts electrical or thermal conductivity to plastic materials. Iron enhances resistance to abrasion, while lead offers shielding against radiation and absorbs sound within dense materials [26].

Various other additives are also utilized to enhance the physical characteristics, environmental stability, and rheological behavior of the polymers. These include lubricants, plasticizers, and many more. Sabirneeza et al. synthesized poly(vinylalcohol-leucine) (PVAL) composite-based corrosion inhibitor and studied its performance on mild steel in 1 M HCl. The findings revealed that PVAL worked as a mixed-type inhibitor, demonstrating a high inhibition efficiency of 95% at the optimum concentration of 0.6 wt% [52]. A remarkable improvement of 628 times in the wear resistance of epoxy/graphene composite was observed at 5 wt% loadings, rendering it well-suited for applications in electronics, marine, and aerospace sectors [58].

The following table (Table 1) summarizes the significant advancements in polymer properties achieved through composite forms. The upcoming sections emphasize diverse methods for synthesizing polymer composites and explore their wide range of application areas.

2 Methods of Synthesis

2.1 Melt Intercalation

Melt intercalation is the conventional and widely accepted method for producing thermoplastic polymer nanocomposites. This process includes heating the matrix phase to elevated temperatures, introducing the filler, and subsequently blending the material to ensure even dispersion of filler in the matrix, as shown in Fig. 3. One of its advantages is its environmentally friendly nature, as it doesn't involve the use of solvents [61]. It is the preferred method for producing clay/polymer nanocomposites with thermoplastics or elastomeric polymeric matrix. This method aligns well with existing industrial methods like injection molding and extrusion [62]. The primary benefit of this approach is its lack of toxic substances. However, a drawback emerges with the limited dispersion of the filler in the matrix, particularly at higher filler concentrations. This is due to the elevated viscosity of the composites [63]. Maiti et al. described a process of preparation of a mixture of Polycaprolactone (PCL) and multi-walled carbon nanotubes (MWCNTs) through melt blending. Subsequently, they synthesized a nanocomposite of PC and ε-PCL-MWCNT. A PCL-MWCNT masterbatch containing 3.5 wt% of MWCNT was created at 65 °C and 60 rpm using a mixer. This mixture was then further blended with PC at 280 °C and 60 rpm for an additional 10 min. A uniform dispersion of carbon nanotubes (CNTs) at minimal concentrations was observed in scanning electron microscope analysis. No chemical alteration of CNTs was necessary since the percolation threshold was achieved at just 0.14 wt%, indicating the successful formation of an interconnected network even at low CNTs loading [64]. Lee et al. prepared biodegradable nanocomposites by blending a polymer resin and layered silicates via the melt intercalation method. Initially, a blend of organoclay and polymer chips (GREENPOL) was prepared by employing a mechanical oscillator. Subsequently, this mixture was dried at 80 °C, and the nanocomposites were then processed through melt compounding using a twin-screw extruder, varying the organoclay content. With an increase in the clay loading, there was an improvement in the storage modulus, accompanied by the observation of a transition toward solid-like behavior. Moreover, there was an increase in the shear

Table 1 Advancements in polymer properties through composite forms

Polymers Composite	Loading	Remarks	References
Nylon 12/ IF-WS_2	2 wt%	The composites experienced a 27% rise in tensile property and a 28% improvement in bending strength	[51]
PVAL	0.6 wt%	The composite exhibited over 95% efficiency in inhibiting corrosion at an optimum concentration compared to the untreated metal	[52]
Epoxy/ WS_2-PEI	0.25 wt%	The fracture toughness exhibited an 82.98% increase, and the flexural strength saw a rise of 65% when compared to the pure epoxy	[53]
PVP/ $Zn_{1-x}Sn_xS$	1 wt%	The polymeric composite films demonstrate improved refractive index (n) and optical conductivity ($\sigma_{opt.}$) compared to the plain films	[54]
Epoxy/ GBN	5 wt%	The composite exhibited a substantial increase (~140%) in thermal conductivity. It also displayed elevated electrical resistivity of 3.05×10^{12} Ω•cm and minimal dielectric loss with tanδ less than 0.08	[55]
PVDF/ BT@BN	5 wt%	High electric breakdown strength ($E_b \approx 580$ kV/mm) and notable discharged energy density ($U_d \approx 17.6$ J/cm^3), surpassing the PVDF film by 1.76 and 2.8 times, respectively	[56]
PDMS/ CB-GF	8 wt%	The composite displays superior characteristics, with thermal conductivity 222% higher and storage modulus 40% greater than pure PDMS	[57]
Epoxy/ Graphene	5 wt%	The composite displayed a wear resistance 628 times greater than pure epoxy at 10 N, making it suitable for aerospace, electronics, and marine components	[58]
PLA/IL/ APP	3 wt%	The composite achieved notable flame retardancy (limiting oxygen index: 27.2, UL-94: V-0 grade). Furthermore, the elongation at the break rose from 8.5% in PLA to 204.6%	[59]
PU/GN	3 wt%	Nanocomposite films exhibited a 30% decrease in gas permeability, enhancing helium gas barrier properties. Additionally, their improved weather resistance led to reduced photooxidation and carbonyl index	[60]

IF-WS_2—fullerene-like tungsten disulfide, WS_2-PEI—branched polyethyleneimine-functionalized WS_2, PVP—polyvinyl pyrrolidine, GBN—hybrid graphene and boron nitride nanoparticle, PVDF—poly(vinylidene fluoride), BT@BN—$BaTiO_3$ nanoparticles embedded in boron nitride nanosheets, PDMS—polydimethylsiloxane, CB-GF—carbon black-graphene foam, PLA—poly(lactic acid), IL—ionic liquid (tetrabutylphosphonium), APP—ammonium polyphosphate, PU—polyurethane, GN—functionalized-graphene

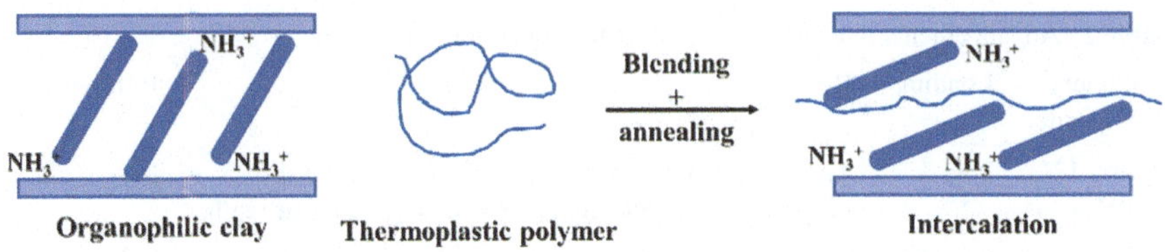

Fig. 3 Schematic representation of melt intercalation process [66]

viscosity of the nanocomposite, accompanied by the observation of shear-thinning behavior [65].

2.2 Solution Mixing/Solution Intercalation/Solution Blending

Solution mixing is one of the simplest techniques for creating polymer composites. The method comprises the following three steps:

- Dispersion of filler.
- Integration of the polymer.
- Elimination of the solvent through distillation or evaporation [63, 67].

In this procedure, nanofillers are initially immersed in a solvent, such as toluene, H_2O, or $CHCl_3$. The polymer and nanofillers are further blended in the solution, resulting in the intercalation of polymer chains and solvent displacement. After the removal of the solvent, the exfoliated structure persists, leading to the formation of polymer nanocomposites, as shown in Fig. 4 [62]. In contrast to melt intercalation, the driving factor in this process is the increase in entropy achieved by the removal of the solvent [61]. In the solution intercalation procedure, typically, high-velocity shear mixing, stirring, or ultrasonication is done to blend the filler suspension in the polymeric solution [68]. The first graphene-based nanocomposite was developed using an ultrasonic-based solution intercalation technique [69]. This method has found widespread application because of its efficiency in distributing nanofillers, regardless of the polymer's polarity, although its success relies on the harmony between the polymer and reinforcements concerning solvent compatibility. Despite the numerous merits of solution mixing, effectively removing solvents poses a considerable challenge, hindering its broad industrial adoption. The drawbacks include the expensive nature of solvents and the environmental impact of their disposal, which can impede scaling up and the ultimate adoption of this process by the industry. Moreover, while solution mixing achieves satisfactory dispersion in its liquid state, intricate details of each element or processing variables (like solvent type and quantity, mixing duration and velocity, sonication, etc.) profoundly impact the final results of the process [68]. Solution mixing can offer advantages at the laboratory scale since it doesn't necessitate costly equipment, and although it entails several steps, the operational procedures

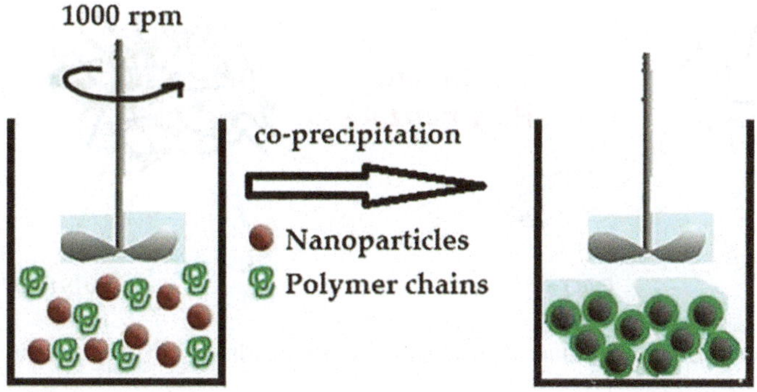

Fig. 4 Schematic diagram for the solution blending process. Reprinted with permission from ref. [71]

are relatively straightforward. This method is amenable to use at moderate temperature conditions and is well-suited for generating numerous small-sized specimens [70].

2.3 *In-Situ Polymerization*

In the method of in-situ polymerization, the typical process entails combining nanomaterial with a pure monomer or a monomer solution, followed by initiating polymerization while the nanomaterials are dispersed in the mixture (Fig. 5) [72]. This approach is exceptionally successful in achieving a consistent dispersion of carbon-based fillers within the matrix, leading to a robust interplay between the matrix and the filler. In-situ polymerization has also been successfully employed in various graphene oxide nanocomposite systems [73]. This method holds particular significance in the creation of composites with polymers that are either insoluble or thermally unstable as matrix materials. This is due to the inability of these matrices to dissolve in solvents suitable for solution mixing or to fuse during melt mixing. In comparison to the previously mentioned methods for preparing nanocomposites, this technique yields the most superior dispersion of the filler [68]. In-situ polymerization offers several distinct advantages. Firstly, it enables the synthesis of nanocomposites using both thermoplastic and thermoset matrices. Additionally, it allows for the grafting of polymers onto the filler surface, typically enhancing the properties of the resulting composite. This method can also achieve partially exfoliated structures because of the effective dispersion and intercalation of fillers within the polymer matrix [61]. The key challenge for achieving well-dispersed particles within the resulting polymer is ensuring effective dispersion of the particles within the monomer, as sedimentation processes in monomers can occur rapidly [73]. Avella et al. developed a novel abrasion-resistant nanocomposite using a poly(methyl methacrylate)

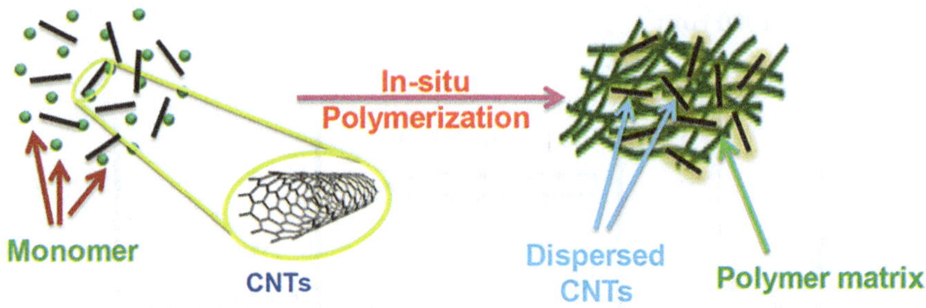

Fig. 5 Schematic route of in-situ polymerization process. Reprinted with permission from ref. [76]

matrix and calcium carbonate ($CaCO_3$) nanopowder through an in-situ polymerization method. They examined how the nanopowders affected the chemical and physical properties of the polymer matrix by conducting analyses of thermal, morphological, and mechanical characteristics [74]. Allen et al. synthesized composite films and fibers of single-walled carbon nanotubes (SWCNT) and conductive polymer through an in-situ polymerization method. The results indicate an improved in-situ polymerization workability of SWCNT/conjugated polymer gels, enabling the creation of transparent electrodes and composite fibers [75].

No single technique meets all requirements perfectly. Consequently, there have been attempts to combine methods, such as employing a blend of solution processing and melt mixing, in-situ polymerization coupled with solvent processing, or in-situ polymerization combined with melt processing.

3 Applications of Polymer Composites

Polymer components are progressively replacing metals in various applications, including structural elements, housings, flexures, and bearings, especially in the automotive industry, in order to reduce weight [77]. Reinforced polymer composites find extensive utility across diverse applications due to their ability to fulfill a broad spectrum of often intricate performance requirements. Most of these materials incorporate some form of glass or carbon fiber reinforcement, which has historically delivered the most substantial enhancements in mechanical properties to the composites [78]. Due to its lightweight nature and exceptional mechanical properties, polymer composites are employed as structural components in imaging devices. The extensive utilization of composite materials across industries is attributed to their favorable strength-to-density and hardness-to-density ratios [79].

3.1 Construction Industry

In recent decades, the construction industry has encountered new challenges due to the increasing demand for infrastructure projects, coupled with the depletion of natural resources like water, limestone, and clay. To address the issue, there is a growing need to explore alternative materials to create innovative construction materials. These materials should not only enhance structural performance but also mitigate problems related to poor durability, ultimately reducing repair costs. Consequently, the construction industry, which is a prominent consumer of composite materials, has witnessed a historical alliance with versatile polymers. Polymer concrete composites represent a groundbreaking strategy for advancing and addressing environmental concerns within the domains of raw material extraction, formulation of mix ratios, and the establishment of construction standards for both the construction and restoration of concrete structures. Polymer concrete, as a building material, is the result of polymerizing a blend of monomers and aggregates. The polymer matrix effectively unites the aggregates, forming a cohesive material. The inception of polymer concrete was conceived with the primary aim of serving as an alternative to traditional cement, particularly tailored for specific applications in the construction industry [80]. Unreinforced polymer composites have found application in various non-load-bearing roles within construction, adorning trimmings, kitchenware, vanities, and cladding [81]. However, the past decade has seen a focused endeavor to introduce reinforced composites into the construction sector for pivotal load-bearing functions. Polymer composites have historically been embraced within the construction realm, gracing them with their utility in non-critical assignments such as baths, vanities, cladding, and decorative elements [82]. Through a synergistic reaction, the polymer and cement interact to form an exclusive microstructure exhibiting distinct properties. The polymer plays a pivotal role in these materials, serving three primary functions as follows:

(1) Acting as a rheological aid by coating individual cement particles, thus diminishing interactions among them.
(2) Filling pores and voids between unreacted cement grains.
(3) Engaging in chemical interactions with the hydration products of cement, resulting in the formation of essential microstructural components known as the interphase regions [83, 84].

In recent times, FRP composite materials have experienced a surge in consideration within the construction sector for structural load-bearing roles. This reinforcement primarily involves the inclusion of fibers or other materials that enhance the material's strength, often in one or more directions. Glass, carbon, basalt, and aramid fibers are regarded as the most effective options as reinforcements in structural engineering. Carbon, glass, and aramid fiber composites are frequently utilized in the construction of building structures. These composites have proven their mettle as a compelling and competitive choice for renovating and fortifying existing civil structures, often replacing conventional steel within reinforced concrete and, albeit

Fig. 6 First FRP bridge in Ginzi, Bulgaria. Reprinted with permission from ref. [89]

to a lesser extent, in the creation of new civil edifices [85]. Like in various other applications of FRP composites, their utilization in bridge construction is on the rise. This increasing use can be attributed to four pivotal advantages as follows:

(1) Exceptional mechanical properties enable weight reduction owing to a favorable strength-to-stiffness-to-density ratio.
(2) Strong resistance to corrosion, enhancing the overall durability of the structure.
(3) Minimal maintenance requirements, leading to cost savings and reduced upkeep efforts.
(4) Capability to shape and mold intricate geometries and forms for bridge components and structures.

These benefits serve as catalysts for the adoption of FRP composites in crafting various components of bridges and even in the creation of large-span structures. The initial efforts to introduce FRP composites into the bridge construction industry date back to the mid-1970s [86]. One of the early examples of a road bridge constructed using FRP composites is the Ginzi Highway bridge (shown in Fig. 6), erected in Bulgaria from 1981 to 1982 [87]. This bridge featured a Glass-FRP slab manufactured as a single element using the hand lay-up technique, and it had a span of 10 m [88].

3.2 *Automobile Applications*

PMCs are extensively utilized in the automotive industry due to their lightweight properties and cost-effectiveness. The mechanical attributes of PMCs play a crucial role in vehicle design, enabling them to fulfill various criteria, such as reducing vehicle weight, thereby enhancing fuel efficiency and reducing exhaust emissions, consequently contributing to a reduction in air pollution. A frequently employed composite material in the automotive sector consists of a polymer matrix reinforced with whiskers or fibers. The automotive industry's first use of PMCs dates back to 1953 when the Chevrolet Corvette introduced a fiberglass body, marking the initial application of PMCs in the automotive field [79]. Over the last decade, European car manufacturers and suppliers have increasingly adopted natural fiber composites with both thermoplastic and thermoset matrices for diverse automotive applications. These include door panels, seat backs, headliners, package trays, dashboards, and interior components. Meanwhile, glass FRPs have demonstrated their ability to satisfy the structural and durability requirements of both interior and exterior automotive components [90]. The automotive industry has witnessed a substantial use of composites, particularly in manufacturing hybrid electric vehicles and electric vehicle batteries. These materials have contributed significantly to reducing vehicle weight and increasing driving range. In addition, sports cars and supercars have integrated polymer composites to enhance their capabilities and achieve significant reductions in emissions, meeting accepted standards. This shift has resulted in reduced overall vehicle weight, expenses, manufacturing intricacy, maintenance requirements, and extended lifespan while also allowing for flexible design with a focus on passenger safety. Polymer composites have demonstrated improved crash performance, a benefit particularly harnessed in Formula 1 racing cars, leading to a considerable reduction in life-threatening accidents associated with racing. Car manufacturers like Porsche and Mercedes-Benz are now utilizing polymer composites in the creation of high-end sports cars, particularly in the construction of passenger compartments [91]. Figure 7 shows the instance of polymer nanocomposites employed within the parts of the car.

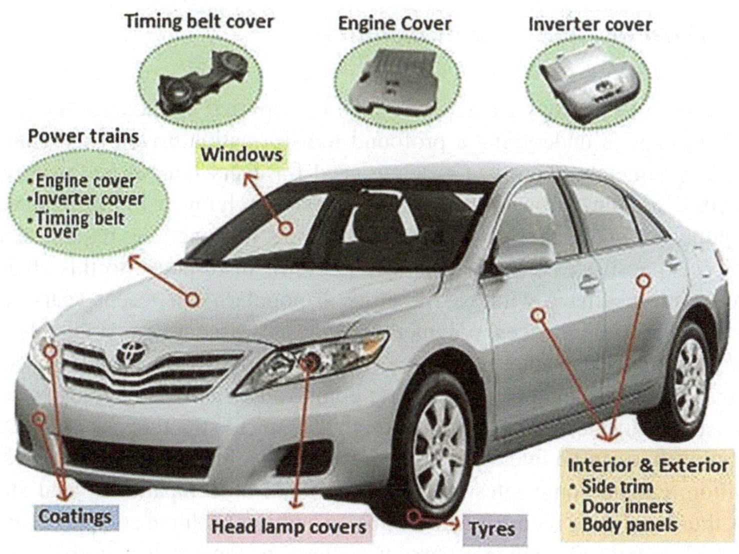

Fig. 7 Application of polymer nanocomposite parts. Reprinted with permission from ref. [92]

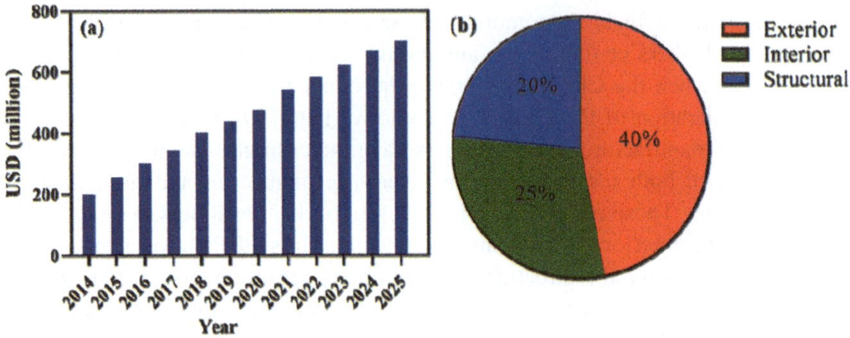

Fig. 8 **a** The market revenue for PMCs in the United States from 2014–2025. **b** The global utilization pattern of PMCs across various automotive components. Reprinted with permission from ref. [93]

Over the past few decades, there has been a consistent and significant increase in the revenue generated from using PMCs in the automotive sector, as depicted in Fig. 8a. An interesting observation from Fig. 8b is that a substantial proportion, around 65%, of these PMCs has found application in both the exterior and interior components of automobiles. According to a market report, the evaluation of the automotive polymer composite industry in 2014 amounted to USD 200 million, and there is a strong indication that this figure is projected to grow to USD 700 million by the year 2025 [93].

3.3 *Electronics and Communications*

As information technology advances rapidly, exemplified by the advent of 5G, the global landscape is undergoing a profound transformation driven by digital technology. Consequently, there is a growing need for everything to be "measurable" and "digital," as conventional rigid electronic devices relying on inorganic materials, semiconductors, and metals do not adequately meet the emerging requirements for flexibility, expansiveness, and wearability [94–96]. In response to this challenge, flexible electronics have witnessed a surge in popularity in recent years. Within this realm of flexible devices, polymeric composites assume a pivotal role owing to their versatile and adaptable features [97–99]. Primarily, in contrast to conventional manufacturing methods, polymeric composites can be strategically tailored using cost-effective techniques. Moreover, the chemical structures of polymers and their composites can be intelligently tailored to suit a broad range of applications. In addition, certain composites exhibit exceptional biocompatibility and stability in vivo, thus bestowing them with the ability to monitor health in complex biological contexts. Consequently, polymer composites have become indispensable materials in the realm of flexible electronics [100, 101]. The diagram presented in Fig. 9 illustrates the application of polypyrrole (PPy)-based composite in the domain of flexible electronics.

The proliferation of highly integrated circuitry in electronic devices has led to the generation of unwanted EM radiation, interfering with neighboring devices and causing malfunctions. Therefore, managing EM radiation has become a significant challenge in various sectors, including electronics, communications, military, and medical instruments. To mitigate the adverse effects of unwanted EM radiation, EM shielding compounds are indispensable. Extensive research has focused on the advancement of polymeric materials that provide inherent flexibility, ease of processing, resistance to chemicals, scalability, and lightweight characteristics. Additionally, shielding materials based on polymers offer benefits as compared to metal-based alternatives, as they can primarily shield electromagnetic waves through absorption. This renders them especially appealing for uses like incorporating camouflage and advanced stealth technology. Intrinsically conducting polymers, such as polyaniline and PPy, are the preferred choice for achieving effective Electromagnetic interference (EMI) shielding. Their good conductivity can be significantly enhanced through doping. This can be achieved through the appropriate addition of magnetic and electrically conducting fillers into the matrix. Notably, carbonaceous materials like CNTs, graphene, carbon black, and graphite fibers, as well as metallic structures like metal nanowires and nanoparticles, have been the primary choices for conducting fillers. These nanomaterials and their composites have demonstrated excellent EMI shielding and also found applications in solar cells, lithium-ion batteries, supercapacitors, electrochemistry, catalysis, sensors, improved mechanical properties, and numerous other potential uses [103].

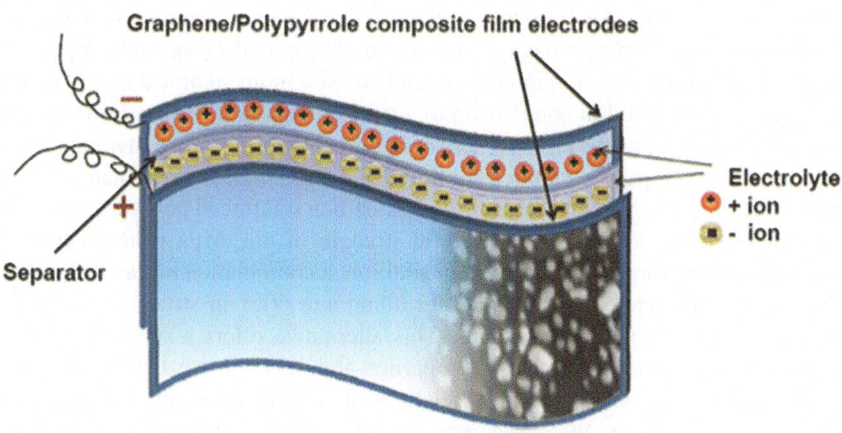

Fig. 9 A diagrammatic representation illustrating the structure of flexible supercapacitors based on Graphene/PPy composite. Reprinted with permission from ref. [102]

3.4 Marine Applications

In recent decades, the marine industry has witnessed widespread adoption of advanced composites owing to their exceptional engineering properties. The key factors driving this surge in usage across various marine sectors include reductions in weight and costs, as well as a commitment to environmental sustainability [5]. The integration of advanced composites marked a significant milestone in the boat and vessel manufacturing industry. This technology has empowered manufacturers to enhance product quality, resulting in rigid yet lightweight structures that contribute to improved sailing performance and longevity. The reduced weight brings advantages such as increased cargo capacity, fuel efficiency, reduced inertia, enhanced ship stability, and buoyancy [104]. The initial use of FRP composite materials in the marine sector was observed in the construction of boats shortly after World War II. A significant portion of maritime vessels is constructed using composites that include glass-reinforced polyester and advanced FRP materials, often incorporating carbon. They are commonly used for high-performance structural applications [105]. Recreational boats, military vessels, and even helicopter landing platforms situated in the middle of the ocean utilize these materials [106]. In the offshore sector, composites are used for various purposes, including repairs, such as submerged structures on spar platforms and pipes as well as in the floating platform components, construction of gas pipelines, and supporting engineering framework. Their suitability in offshore applications stems from their low weight, impressive specific mechanical properties, and resistance to the harsh seawater environment, making composite materials a viable substitute for metals [104]. Mouritz et al. [107] conducted a series of studies aimed at comparing the weight, structural performance, and cost of large patrol boats constructed from different materials, including steel, aluminum, and sandwich composites. Their findings indicated that when utilizing glass-reinforced plastic sandwich composite material, the structural weight of a boat could be reduced by 10% compared to an aluminum boat, and an even more significant reduction of 36% could be achieved in comparison to a steel boat of similar size. Furthermore, the research highlights the effectiveness of hybrid composites in maintaining essential mechanical properties required for prolonged and efficient performance within the marine industry. Notably, the hybrid glass-carbon reinforced polymer composite (GCG2C) exhibits an impressive flexural strength of 462 MPa while displaying minimal water absorption tendencies. In addition, a combination of flax and carbon fiber composites, used as a substitute for aluminum 6061 in structural materials, has shown remarkable improvements. This alternative offers a substantial 141% enhancement in vibration-damping properties and an impressive 252% boost in tensile strength, all while reducing weight by 49%. Moreover, the integration of reinforcements such as jute and carbon fibers in hybridized composites not only contributes to improved vibration-damping properties but also aligns with economic and environmental sustainability goals [5, 108]. An example in Fig. 10 is a Carbon-FRP propeller for application as a podded propulsion unit, which is manufactured using a hand lay-up process [109].

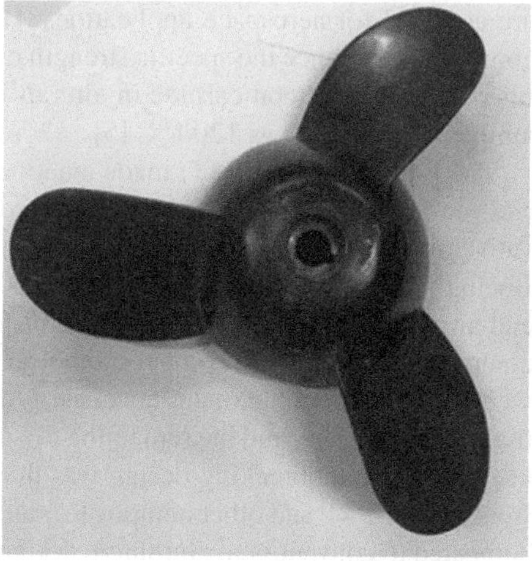

Fig. 10 Carbon fiber propeller fabricated through a hand lay-up process. Reprinted with permission from ref. [109]

Aerospace

Polymer composite materials have become increasingly prevalent in the aerospace industry, with up to 40% of modern airframes being constructed from these materials. The use of composites is expected to continue to grow due to ongoing advancements in technology [110]. Reducing weight, cutting costs, and ensuring radiation protection are key priorities in this sector. The reduction in weight is especially important as it impacts various factors, including fuel efficiency, velocity, number of assembled components, and others. Additionally, the implementation of nanofiller-reinforced polymer matrices has been investigated to enhance radiation shielding capabilities when compared to traditional metal alternatives. The capability of polymer composites to provide effective shielding is attributed to their insulating features and the opportunity to include non-toxic, high-Z fillers, thereby enhancing protection against X-rays. These polymer composites are applied in various components, including aircraft wing boxes, aircraft brakes, blades, fittings, window frames, rotors, brackets, bulkheads, fuselages, airframes, vertical fins, and tail assemblies [5]. In the aerospace and aeronautical fields, where safety and reliability take precedence over cost, the adoption of FRP composites has rapidly gained prominence. When seeking a unique combination of safety, reliability, lightweight characteristics, strength, and efficiency, polymer-reinforced composites emerge as the unparalleled choice [110]. As early as 1983, airbus became the first commercial aircraft manufacturer to incorporate composite materials into the design and construction of the rudders for its A300 and A310 aircraft. Carbon fiber is generally the predominant polymer composite material used in aerospace applications. Polymer composites are extensively utilized in the Eurofighter jet for constructing its wings, fuselage, flaperons, and rudder, with approximately 40% of the structural components of the Eurofighter being composed of carbon-FRP material [111]. Hybrid composites have also gained prominence in recent times, as research has revealed their improved mechanical properties, which are essential for aerospace applications. Hybrid composites, like kenaf/glass FRP composites, enhance the specific strength of aircraft. Additionally, the use of carbon fiber-reinforced silicon carbide in aircraft brake systems allows them to withstand temperatures as high as 1200 °C [5].

The Boeing 787 (Fig. 11) made aviation history by becoming the first commercial jet aircraft predominantly constructed from composite materials. This innovative approach is particularly evident in the main structure and fuselage of the Boeing 787, surpassing the use of composites in any previous Boeing commercial aircraft. The Boeing 787 is distinguished by its remarkable composition, with composites accounting for a substantial 80% of its volume. When considering the weight distribution, the materials consist of 50% composites, 20% aluminum, 15% titanium, 10% steel, and the remaining 5% comprised of various other materials. The key to this groundbreaking design was that nearly half of the fuselage was crafted from carbon-FRP and other composite materials. This revolutionary approach, when compared to conventional aluminum designs, results in an average weight reduction of approximately 20% [112, 113].

Biomedical

The substantial rise in demand for medical products utilizing biomaterials and tissue engineering has spurred significant growth in biomedical research over the last two decades [114–116]. These biomaterials play a crucial role in the development of various biomedical devices, ranging from hydrogel contact lenses to polymer stents, knee and ligament implants, surgical adhesives, polymer vascular grafts, steel joint and hip replacements, artificial heart valves, ceramic dental implants, polymer barrier films, polymer sutures, and porous dialysis membranes, among others. As per the comprehensive global industry analysis, the biomaterials market surpassed USD 94 billion in 2018, with projections indicating it is set to surpass USD 256 billion by 2025. An intriguing category of biomaterials, polymer-based composites, is currently

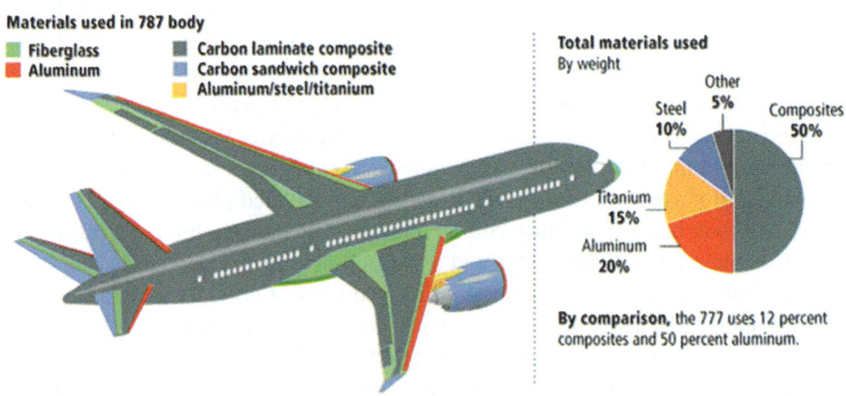

Fig. 11 The comprehensive allocation of composite materials employed in Boeing 787 aircraft. Reprinted with permission from ref. [113]

extensively employed in biomedical applications (Fig. 12) due to their exceptional mechanical and physical characteristics [117]. Furthermore, polymer-based nanocarriers receive the most attention among organic-based photothermal carriers for cancer therapy [118]. It is worth noting that polymer-based nanofibrous materials find application in numerous biomedical fields, including tissue engineering for oral tissues, blood vessels, bones, and wound dressing [119]. Their applications span various fields, encompassing wound dressing, dental practices, medical devices, blood vessel treatments, tissue engineering, oral tissue treatments, antimicrobial materials protein immobilization, surgical implants, drug delivery systems, regenerative medicine, and bone therapy. Within the biomedical realm, natural polymers such as chitosan, collagen, pectin, psyllium, agar, guar gum, starch, and cellulose, as well as synthetic counterparts like polyglycolic acid, polyester amides, PCL, polyamide, PLA, and poly lactic-co-glycolic acid, are widely used predominantly employing fibers as their primary reinforcement mode [5]. In the realm of bio applications, polymer composites offer several benefits, including cost-effectiveness, utilization of readily available natural and synthetic matrices, and tunable fabrication techniques [120]. Various fillers possessing bioactive or non-bioactive properties present opportunities for crafting appropriate scaffolds or implants in the field of regenerative medicine and tissue engineering applications. Modern techniques involving polymer composites typically require refinement to fine-tune the structure needed for specific applications. Polymer composite scaffolds have demonstrated high cell adhesion, biodegradability, and biocompatibility with exceptional bioactivities in terms of tissue formation, stimulation, survival, function, and antimicrobial properties in both in vitro and in vivo experiments [117].

Polymer Composites serve as integral components in a diverse array of medical devices, including MRI and CT scanners, X-ray couches, surgical tables, mammography plates, target tools for surgeries, wheelchairs, and prosthetics. Examples of these applications are outlined below.

Polymer nanocomposites, incorporating CNTs or TiO_2 nanotubes, contribute to the accelerated healing of fractures by serving as a "scaffold" that directs the growth of new bone [122]. The exploration of nanocomposites in diagnostics and therapy is underway, with one example being the integration of magnetic nanoparticles and fluorescent nanoparticles in composite particles. These particles exhibit both magnetic and fluorescent properties, enhancing tumor visibility in pre-surgery MRI tests and potentially aiding surgeons in better-visualizing tumors during surgical procedures. Polymer composites find extensive use in biomedical applications, serving in both soft tissues like skin and hard tissues such as bone. The utilization of both synthetic and naturally degradable polymer composites is prevalent in creating scaffolds for bone repair, leveraging their exceptional biological and mechanical attributes [123].

Researchers are employing biocompatible polymer composites in the realm of skin regeneration these days. Polymeric materials play a crucial role in drug delivery systems due to their high suitability and adaptability. Specifically, polymer-based hydrogels serve as carriers for various drug molecules, including antifungal, antibiotic, and anticancer drugs. These hydrogels not only act as protective barriers at wound sites, expediting the healing process, but also contribute to tissue engineering

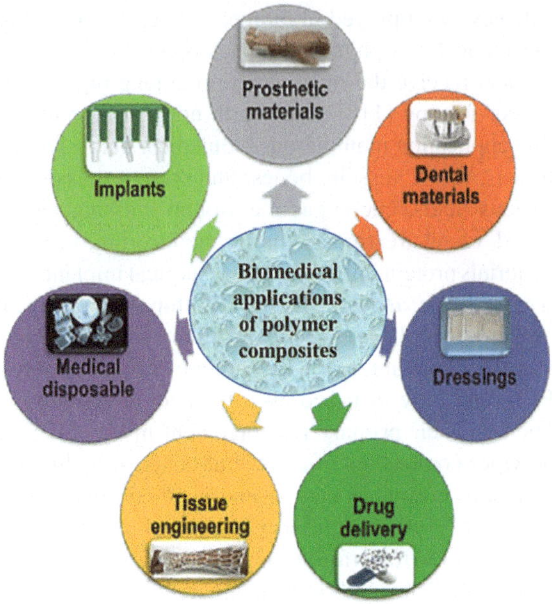

Fig. 12 Areas of applications of polymeric materials in biomedical. Reprinted with permission from ref. [121]

by facilitating the regeneration of damaged or lost tissue through the promotion of new cell growth [124].

4 Theoretical Aspects of Polymers and Polymeric Composites

4.1 Background and Overview

Polymers and polymer composites are a complex class of material that present special challenges to computational scientists because they exhibit fascinating and significant phenomena over a wide range of length scales and time scales. Owing to advancements in hardware and software, an increasing number of polymer simulation studies offer insightful information about both novel and well-established macromolecular materials. Through forecasts, these studies motivate polymer chemists to identify synthetic pathways for the creation of highly promising new materials. In order to facilitate the performance of polymer simulations, this section has been structured so that the order of the sub-sections corresponds to the methodology of computational investigation on polymers and polymer composites. Before considering modeling and simulation as a possible tool to account for a scientific or technical polymer-related problem, it is crucial to know what modeling and simulation study is expected to meet so as to determine which model should be compatible with the simulation. The "optimal model" is contingent upon the system under investigation and/or the topic or questions that need to be addressed. In general, the majority of models for classical molecular simulations of polymers and polymer composites can be classified as either coarse-grained or atomistic [125].

4.2 Models for Polymeric Systems

Atomistic models are generally the best option when the challenge requires a knowledge of the local, monomer-level (re)arrangements, fluctuations, or interactions across a disorganized or organized polymeric structure. Throughout a polymeric nanocomposite, atomic models have demonstrated how monomers and polymeric fragments align themselves close to a particular nanoparticle interface [126–140]. Nevertheless, atomic models are often restricted to length ranges of 1–100 Å and time periods of 1 fs–100 ns because of their substantial computing cost. Therefore, events dealing with polymeric chain reorganization that might happen over longer lengths and time scales are unable to be investigated by these models. However, the force fields as well as the starting configuration that is fed into the simulation and the simulation procedure that is selected all have a significant impact on how accurate these atomistic models are. For the system of interest, the force field along with the right initial configurations are often unavailable or optimized, and using one or both of them incorrectly can produce unintended outcomes [125].

When studying systems are with known chain configuration at a specific condition previously from other computations or tests, it is preferable to utilize atomistic models. Atomistic model simulations are used to characterize certain atomic interactions in a particular morphology, intermolecular associations across the chain and among the chain and solvent/additive/ion that stabilize a secondary structure or crystalline state of the polymer, and interactions between two polymers that are held collectively along with additional interactions. Since the motions of hydrogen atoms are high-frequency and require small time steps in the integrated computation, it may not always be necessary to distinctly represent them in circumstances when atomistic details are required, hence limiting computational efficiency. Then, one can choose to employ unified atomic models having blended hydrogen atoms with the heavier atom they are linked to, rather than being immediately depicted [126–132]. It is crucial to remember that, despite the potential benefits of faster computation, united atom representations need to be modified in order to account for structural and thermodynamic features that may be influenced by hydrogen bonding [127]. Moreover, atomistic dynamics can be deliberately accelerated in the absence of hydrogens, and conclusions involving united atoms may differ from reality because of variations in partial charge representation. Finding the ideal force field among the wide range of options pertinent to polymers is typically the very first step if an atomistic model is the best option for the given system [8]. For organic molecules in general and for particular macromolecules that display secondary structures (like proteins and nucleic acids [133–141] or crystalline structures (like polyethylene [142, 143] and poly(3-hexylthiophene) [144, 145]), computational chemists streamline and define force field parameters. Under particular conditions (such as temperature, pressure, concentration, and composition), these force fields were subjected to evaluation to make sure they accurately replicate pertinent physical properties of polymers that have been observed in experiments, such as melting temperature, density, glass transition temperature, cohesive energy, etc. Force fields have been optimized to operate within a particular temperature and pressure range and shouldn't be applied in environments outside of those parameters without additional validation or adjustments. It could also be worthwhile to test a few different relevant force fields when the "optimum" force field for the exact polymer chemistry and conditions of interest is not very apparent. This will help assure that the important findings and insights are not affected by the force field selection [125, 146–151].

The optimum approach is to begin with coarse-grained (CG) models when the objective is to anticipate structure/morphology for polymeric materials in a wide range of conditions or variables, out of which some may possibly be lacking established atomistic force fields/initial structures. When certain atoms in a monomer, a full monomer, a collection of monomers, or Kuhn segments are grouped to a single CG bead, the coarse graining eliminates some of the degrees of freedom in the system [152–156]. The large separation of length and time that polymers inherently possess makes this technique feasible. Therefore, in order to increase computing efficiency and lower barriers to free energy, CG techniques eliminate part of this atomistic detail. Depending on the system under study, the model's (i.e., resolution's) level of fineness or coarseness must be adjusted. By virtue of CG simulations, one can take greater time steps by lowering the degrees of freedom and eliminating the macromolecule's rapid motions. Hence, large polymers can be equilibrated more quickly. In order to find potentially intriguing or beneficial factors or conditions to accomplish a desired morphology, CG models remain advantageous in investigating small-sized systems with a wide range of polymeric parameters [125].

Polymeric chain CG models consist of "beads," where each bead, which is often spherical, represents a segment of the polymeric chains. One of the most important decisions made when selecting or creating a suitable CG model is the length scale of the mapping between complete atomistic detail and a CG bead, which might differ greatly. More coarse-grained models are typically generalized and offer typical characteristics (e.g., scaling laws relating Rg to molecular weight) for a wide variety of polymers; the more fine-grained models are typically designed for describing distinct polymeric chemistries. Phenomenological CG models, which imitate the length and time scale phenomena far bigger than the intermediate-resolution CG models, have occasionally been constructed by researchers. These simulations resemble intermediate-resolution CG models of particular chemistries. In order to replicate important structural or thermodynamic aspects of interest, interactions among monomer CG beads should be determined after an appropriate spatial mapping has been selected [125].

4.3 Simulation for Polymeric Systems

The next stage after choosing a good model is to choose a simulation approach; just like with model building, this largely relies on the system and research issues of interest. In a nutshell, trial configurations are made in Monte Carlo (MC) simulations by randomly moving atoms or beads. These trial moves are then evaluated according to certain criteria, which are typically a function of the potential energies of the final and initial configurations. Following that, averages over the approved configurations are used to determine thermodynamic and structural parameters. It is important to employ MC moves carefully so that they provide for an adequate sampling of the pertinent configurations for that state and have a decent chance of being accepted [132, 157–160].

Whichever approach is selected, a typical (molecular dynamics) MD or MC simulation can be broadly divided into three phases such as system setup and initialization, equilibration, and sampling. The user then has to decide which kind of thermodynamic ensemble to be assigned to the sample in the simulation after deciding between MC and MD. In its most basic version, MD, for example, samples the microcanonical or NVE ensemble, where the simulation box's volume, total energy, and particle count are kept constant. Gibbs ensemble MC simulations use bead insertion/deletion moves to perform simulations with a fluctuating system size [158, 161–166], enabling the investigation of phase coexistence if the phase behavior of polymer blends and solutions is to be studied. Because they allow to study the phase coexistence without requiring the resources to model the interface between phases, both of these approaches are frequently used to research phase behavior in polymeric systems [158].

The next consideration is the system size (i.e., number of atoms or CG beads and/or starting simulation box size), which is made once the method and ensemble have been chosen. Avoiding system size effects is crucial, as it is related to each computational technique. Fewer chain simulations are suitable in the case of features of diluted solutions, such as chain configurations and scaling laws are to be studied [157]. Inaccuracies resulting from "finite size effects" must be taken into consideration while opting for a system with smaller sizes for achieving simulation results more rapidly. While performing simulations within periodic boundary conditions, finite size effects can arise due to self-interacting periodic boundaries, from the model system's size constraint, or from the size of the box. Selecting smaller system sizes may result in significant variations in stress and pressure [167]. Choosing a suitable system size to maximize computational resources while producing accurate findings is a difficult undertaking that requires expertise and experience. The simplest way to ensure that the size of the system has no effect on the results is to run an identical simulation with varying system sizes; as the system size increases, the results should become insensitive (within error) to the system size beyond a certain size (the optimal system size). When an isotropic system is under study, an arbitrarily initial placement of the system's components might be enough for creating an initial configuration. To avoid needless equilibration time and/or eliminate bias, careful creation of the starting simulation configuration remains crucial. In cases where nonisotropic or phase-separated systems hold significance and the corresponding structures are identified, it might be imperative to build the simulation's initial condition in a configuration that approximates the intended structure. This will mitigate the need to wait an excessively long period during the spontaneous assembling of the required structure to the simulation cage [125].

To eliminate structural artefacts from the system initialization, equilibration must be carried out following the formation of the simulation domain and its components [154]. Since there is always a chance that the system will become stuck far from the true global free energy minimum, it can be pointed out that simulations cannot guarantee the attainment of a "true" equilibrium state. Because relaxation in chain configuration is relatively sluggish as compared to easy-to-compute quantities like pressure or total potential energy in most polymer systems, thus chain-level measurements are sometimes necessary to determine whether the system has equilibrated yet [168].

In order to confirm that for data analysis the collected configurations are noncorrelated, it is imperative to ensure the adequate separation between snapshots that are used for data collection when sampling the equilibrated state. The relaxation times of polymers can be several orders of magnitude longer than those of simple fluids; hence, this is particularly crucial for polymer simulations. Calculating the time autocorrelation function of a property of interest is a simple way to affirm uncorrelated data points. Finally, the sampling step must be conducted long enough so that desired quantities can be estimated correctly, which may be strongly dependent on the system and questions of interest. Systems having significant barriers to rearrangement (e.g., block copolymer micelles, glassy polymers, etc.) are unlikely to experience major changes in structure during a single simulation. As a result, just running a single simulation for a longer time may not enhance the sample statistics. Instead of distinct configurations derived from the same experiment, one may prefer to perform several independent replicates of the simulation protocol and average over independent trials. When performing free energy calculations, it becomes necessary to determine the least free energy path between two states, each of which is specifically described in terms of the collective variables (CVs) and another state. Finite temperature or zero temperature string approaches could be employed for such tasks [169–173].

4.4 Analysis of Simulation for Polymeric Systems

Following the completion of the simulation and the generation of an ensemble of configurations, the challenge shifts to analyzing the molecular "trajectories" in order to offer valuable data. Any simulation attempt must include appropriate quantification, even if the observation under discussion was initially recognized visually. For instance, it is simple to visually categorize the morphologies in studies of amphiphilic copolymer assembly, such as spherical micelles, cylindrical micelles, and bilayers/

vesicles [174–176]. The next step in the quantification process is statistical analysis, which yields the ensemble average value and standard error for the duration of the simulation [177]. The reporting of errors and statistical analysis are also essential components of quantification. All calculations should have error bars provided so that readers can evaluate the variation both inside and between runs. Using the block averaging approach [178], which divides a simulation into (~10 moderate number – 30) of chunks, is a straightforward but successful tactic [125].

4.5 Theory for Polymeric Systems

Some of the main drawbacks of simulations can be avoided by combining theory and simulations. Theoretical techniques can bypass some of these restrictions (because of the peculiarities of their derivation/formalism), with the benefit of being exceptionally fast/efficient computation. Thus, a combination of simulation and theoretical approach should be the ideal practice, where simulations feed information and the theory help in filling in large gaps in the parameter. Generally, theoretical tools employed in polymer investigation are classical density functional theory (DFT) [179–182], self-consistent field theory (SCFT) [183–187], polymer reference interactions site model (PRISM) [188–191], statistical associating fluid theory (SAFT) [192–196], and others [197].

In the classical DFT approach, the free energy of the system is stated as a function of the local density of the components, and minimizing the free energy functional offers insight into the equilibrium density distributions of the system. The wide application of classical DFT is due to its capability to predict structure and thermodynamics in bulk polymer systems and polymeric systems [180]. Yet another frequently utilized theoretical technique for polymer studies is self-consistent field theory (SCFT). SCFT lowers a complex system of several polymeric chains to a simplified problem of a single chain in a field constructed from the other component chains. SCFT has proven especially useful in predicting the order–disorder transition and various ordered morphologies. DFT and SCFT have been coupled in some circumstances to explore phase diagrams of morphologies inside polymer nanocomposites [198].

In addition, the dynamic density functional theory (DDFT) approach has also been used in numerous investigations in the area of polymers. When DDFT is brought to equilibrium, it converges to the solutions of the SCFT equations. Combined "particle-based" and "field-based" simulations, as in "theoretically informed coarse-grained" (TICG) and "single chain in mean-field" simulations, are also used in copolymer studies. TICG is a particle-based MC method that calculates the interchain interaction energy from the local densities [199–202]. The integral equation polymer reference interaction site model (PRISM) theory is yet another theoretical approach that excludes mean-field approximations but takes into account the density and concentration fluctuations. The chemical and physical properties of the system are the inputs to PRISM theory [174, 175, 188, 203–205].

5 Conclusions and Future Trends

Polymer composites have become a predominant choice across a wide array of industries and sectors due to their exceptional properties that significantly enhance material performance and efficiency. Also, these composites are anticipated to stay in focus for further examination due to their lightweight nature, having a lower density compared to many conventional composites, and their environmentally friendly characteristic, particularly their ease of recycling through straightforward melt reprocessing [206]. These versatile materials find application in virtually every facet of human activities, including transportation, civil construction, biomedical applications, military uses, sports and leisure equipment, food packaging, and the electrical and electronics sectors. Considering the ongoing and emerging challenges, it's crucial to explore how polymers can address future requirements and compete effectively with other materials. One promising avenue is nanotechnology, which has the potential to merge the strengths of nanoscience with the benefits of polymeric materials to create materials that are more advanced and dynamic [207]. In light of current developments, it is anticipated that the future trends in the use of PMCs will be closely tied to the integration of nanomaterials as reinforcements. This strategic combination is anticipated to fulfill the ever-evolving and dynamic demands across various sectors. This assumption is grounded in the idea that nanofillers play a pivotal role in promoting strong interactions between the polymer matrix and additive, resulting in exceptional properties. By exclusively incorporating nanotechnology into the production of these advanced composites, materials will exhibit even further enhanced properties, which are essential for meeting the ever-increasing demands of different industries [5]. Numerous sectors, including energy, communication, packaging, household goods, electronics, sports, and leisure, are actively embracing the application of polymer nanocomposites to meet their diverse needs and requirements. There's an urgent need to categorize all composite material grades according to their distinct properties, manufacturing methods, and production expenses. It's crucial to develop cost-effective and portable non-destructive kits for accurately measuring defects in hybrid composites [208].

Acknowledgements S. V. P. thanks DST-Inspire for the research fellowship. The authors also gratefully acknowledge the constant encouragement extended by the Director, CSIR-Indian Institute of Petroleum, Dehradun, India, and the Director, CSIR-Institute of Minerals and Materials Technology, Bhubaneswar, India.

Author Contributions Nishtha Arora, Sachin Dua, and Smruti Vardhan Purohit shared equal contributions. Bibek Dash, Bikash Kumar Jena, and T. Senthilkumar-Supervision.

Conflict of Interest The research, authorship, and publication of this article hold no probable conflicts of interest, as stated by the author(s).

References

1. Afzal A, Nawab Y (2021) Polymer composites. In: Composite solutions for ballistics, 1st edn. Woodhead publishing series in composites science and engineering, pp 139–152. https://doi.org/10.1016/B978-0-12-821984-3.00003-6
2. Jose JP K, Malhotra SK, Thomas S et al (2012) Advances in polymer composites: macro-and micro composites–state of the art, new challenges, and opportunities. In: Part one introduction to polymer composites, 1st edn. Wiley-VCH, pp 1–16
3. Advani SG, Hsiao KT (2012) Introduction to composites and manufacturing processes. In: Manufacturing techniques for polymer matrix composites (PMCs): a volume in Woodhead Publishing series in composites science and engineering
4. Chohan JS, Boparai KS, Singh R, Hashmi MSJ (2022) Manufacturing techniques and applications of polymer matrix composites: a brief review. Adv Mater Process Technol 8:1–11. https://doi.org/10.1080/2374068X.2020.1835012
5. Oladele IO, Omotosho TF, Adediran AA (2020) Polymer-based composites: an indispensable material for present and future applications. Int J Polym Sci 8834518. https://doi.org/10.1155/2020/8834518
6. Nagavally RR (2016) Composite materials-history, types, fabrication techniques, advantages, and applications. Int J Mech Prod Eng 5:82–87
7. Keya KN, Kona NA, Koly FA et al (2019) Natural fiber reinforced polymer composites: history, types, advantages, and applications. Mater Eng Res 1:69–87. https://doi.org/10.25082/MER.2019.02.006
8. Ghosh AK, Dwivedi M (2019) Processability of polymeric composites, 1st edn. Springer
9. Sivasubramanian P, Mayandi K, Arumugaprabu V et al (2021) History of composites and polymers. In: Polymer-based composites, pp 1–22. https://doi.org/10.1201/9781003126300-1
10. https://www.snsinsider.com/reports/composite-packaging-market-1427
11. Guo Y, Ruan K, Wang G, Gu J (2023) Advances and mechanisms in polymer composites toward thermal conduction and electromagnetic wave absorption. Sci Bullet 68:1195–1212. https://doi.org/10.1016/j.scib.2023.04.036
12. Barari B, Pillai KM (2017) Green composites made from cellulose nanofibers and bio-based epoxy: processing, performance, and applications. In: Natural fiber-reinforced biodegradable and bioresorbable polymer composites. Woodhead publishing, pp 31–49. https://doi.org/10.1016/B978-0-08-100656-6.00003-0
13. Patel VK, Kant R, Chauhan PS, Bhattacharya S (2022) Introduction to applications of polymers and polymer composites. In: Trends in applications of polymers and polymer composites. AIP Publishing, pp 1–6. https://doi.org/10.1063/9780735424555_001
14. Wang RM, Zheng SR, Zheng YP (2011) Polymer matrix composites and technology, 1st edn. Woodhead publishing series in composites science and engineering
15. Hamidi YK, Altan CM (2018) Process-induced defects in resin transfer molded composites. In: Comprehensive composite materials, vol 2, pp 95–106. https://doi.org/10.1016/B978-0-12-803581-8.09902-1
16. Islam MS, Shahruzzaman M, Khan MN et al (2022) Composite materials: concept, recent advancements, and applications. In: Renewable polymers and polymer-metal oxide composites: synthesis, properties, and applications. Elsevier, pp 1–43. https://doi.org/10.1016/B978-0-323-85155-8.00011-X
17. Freile-Pelegrín Y, Madera-Santana TJ (2017) Characterization techniques for algae-based materials. In: Algae based polymers, blends, and composites: chemistry, biotechnology and materials science, pp 649–670. https://doi.org/10.1016/B978-0-12-812360-7.00018-5
18. Sivasubramanian P, Mayandi K, Arumugaprabu V et al (2021) History of composites and polymers. In: Polymer-based composites. CRC Press, pp 1–21. https://doi.org/10.1201/9781003126300-1
19. Zhao X, Lu S, Li W et al (2022) Epoxy as filler or matrix for polymer composites. In: Epoxy-based composites. IntechOpen. https://doi.org/10.5772/intechopen.102448

20. Varma IK, Gupta VB, Sini NK (2017) Thermosetting resin-properties. In: Comprehensive composite materials. Pergamon, London, pp 1–56. https://doi.org/10.1016/B0-08-042993-9/00177-7
21. Campbell FC (2003) Commercial composite processes: these commercial processes produce far more parts than the high-performance processes. In: Manufacturing processes for advanced composites, 1st edn. Elsevier Science, pp 399–438. https://doi.org/10.1016/B978-185617415-2/50012-5
22. Hodgkin J (2001) Thermosets: epoxies and polyesters. In: Encyclopedia of materials: science and technology, 2nd edn. Elsevier, pp 9215–9221. https://doi.org/10.1016/B0-08-043152-6/01660-0
23. Mullins MJ, Liu D, Sue HJ (2018) Mechanical properties of thermosets. In: Thermosets: structure, properties, and applications, 2nd edn. Elsevier. https://doi.org/10.1016/C2015-0-06205-0
24. Huang J, Fu S, Gan L (2019) Lignin-modified thermosetting materials. In: Lignin chemistry and applications, 1st edn. Elsevier, pp 163–180. https://doi.org/10.1016/B978-0-12-813941-7.00006-0
25. Frollini E, Silva CG, Ramires EC (2013) Phenolic resins as a matrix material in advanced fiber-reinforced polymer (FRP) composites. In: Advanced fibre-reinforced polymer (FRP) composites for structural applications. Woodhead publishing series in civil and structural engineering, pp 7–43. https://doi.org/10.1533/9780857098641.1.7
26. Hsissou R, Seghiri R, Benzekri Z et al (2021) Polymer composite materials: a comprehensive review. Compos Struct 262:113640. https://doi.org/10.1016/j.compstruct.2021.113640
27. Mallick PK (2010) Thermoplastics and thermoplastic–matrix composites for lightweight automotive structures. In: Materials, design and manufacturing for lightweight vehicles, 2nd edn. Woodhead publishing series in composites science and engineering, pp 174–207. https://doi.org/10.1533/9781845697822.1.174
28. Adeniyi A, Agboola O, Sadiku ER et al (2016) Thermoplastic-thermoset nanostructured polymer blends. In: Design and applications of nanostructured polymer blends and nanocomposite systems. William Andrew Publishing, pp 15–38. https://doi.org/10.1016/B978-0-323-39408-6.00002-9
29. Reis JP, de Moura M, Samborski S (2020) Thermoplastic composites and their promising applications in joining and repair composites structures: a review. Materials 13:5832. https://doi.org/10.3390/ma13245832
30. Gopanna A, Rajan KP, Thomas SP, Chavali M (2019) Polyethylene and polypropylene matrix composites for biomedical applications. In: Materials for biomedical engineering: thermoset and thermoplastic polymers. Elsevier, pp 175–216. https://doi.org/10.1016/B978-0-12-816874-5.00006-2
31. Pleşa I, Noţingher PV, Stancu C et al (2019) Polyethylene nanocomposites for power cable insulations. Polym (Basel) 11:24. https://doi.org/10.3390/polym11010024
32. Nomura M, Shanmuga Ramanan SM, Arun S (2017) Automobile bumpers. In: Comprehensive composite materials, 2nd edn. Elsevier, pp 460–468. https://doi.org/10.1016/B978-0-12-803581-8.03962-X
33. Koncar V (2019) Composites and hybrid structures. In: Smart textiles for in situ monitoring of composites, 1st edn. Elsevier, pp 1–151. https://doi.org/10.1016/B978-0-08-102308-2.00001-2
34. Rulkens R, Koning C (2012) 5.18-Chemistry and technology of polyamides. In: Polymer science: a comprehensive reference. Elsevier, pp 431–467. https://doi.org/10.1016/B978-0-444-53349-4.00147-3
35. Baek YM, Shin PS, Kim JH et al (2020) Thermal transfer, interfacial, and mechanical properties of carbon fiber/polycarbonate-CNT composites using infrared thermography. Polym Test 81:106247. https://doi.org/10.1016/j.polymertesting.2019.106247
36. McKeen LW (2010) Polyesters. In: Fatigue and tribological properties of plastics and elastomers, 2nd edn. William Andrew Publishing. https://doi.org/10.1016/C2009-0-20351-5

37. Borges C, Chícharo A, Araújo A et al (2023) Designing of carbon fiber-reinforced polymer (CFRP) composites for a second-life in the aeronautic industry: strategies towards a more sustainable future. Front Mater 10. https://doi.org/10.3389/fmats.2023.1179270
38. Talikoti RS, Kandekar SB (2019) Strength and durability study of concrete structures using aramid-fiber-reinforced polymer. Fibers 7:11. https://doi.org/10.3390/FIB7020011
39. Dharmavarapu P, Sreekara SR (2022) Aramid fibre as potential reinforcement for polymer matrix composites: a review. Emergent Mater 5. https://doi.org/10.1007/s42247-021-00246-x
40. Sathishkumar TP, Satheeshkumar S, Naveen J (2014) Glass fiber-reinforced polymer composites-a review. J Reinf Plast Compos 33:1258–1275. https://doi.org/10.1177/0731684414530790
41. Sandor RB (1990) PBI (Polybenzimidazole): synthesis, properties and applications. High Perform Polym 2. https://doi.org/10.1177/152483999000200103
42. Anidha S, Latha N, Muthukkumar M (2019) Reinforcement of Aramid fiber with bagasse epoxy bio-degradable composite: investigations on mechanical properties and surface morphology. J Mater Res Technol 8. https://doi.org/10.1016/j.jmrt.2019.05.008
43. Kandekar SB, Talikoti RS (2018) Study of torsional behavior of reinforced concrete beams strengthened with aramid fiber strips. Int J Adv Struct Eng 10:465–474. https://doi.org/10.1007/s40091-018-0208-y
44. Xu G, Jin Y, Song J (2022) Improvement of the mechanical properties by surface modification of $ZnCl_2$ and polydopamine in aramid fiber composites. Appl Sci (Switz) 12:3119. https://doi.org/10.3390/app12063119
45. Kichloo AF, Raina A, Haq MIU, Wani MS (2022) Impact of carbon fiber reinforcement on mechanical and tribological behavior of 3D-printed polyethylene terephthalate glycol polymer composites—an experimental investigation. J Mater Eng Perform 31. https://doi.org/10.1007/s11665-021-06262-6
46. Davidson JR, Quinn JA, Rothmann C et al (2022) Mechanical characterisation of pneumatically-spliced carbon fibre yarns as reinforcements for polymer composites. Mater Des 213. https://doi.org/10.1016/j.matdes.2021.110305
47. Ahmad Fauzi AA, Osman AF, Alrashdi AA et al (2022) On the use of dolomite as a mineral filler and co-filler in the field of polymer composites: a review. Polym (Basel) 14:2843. https://doi.org/10.3390/polym14142843
48. Ramesh M, Rajeshkumar LN, Srinivasan N et al (2022) Influence of filler material on properties of fiber-reinforced polymer composites: a review. E-Polymers 22:898–916. https://doi.org/10.1515/epoly-2022-0080
49. DeArmitt C (2016) Functional fillers for plastics. In: Applied plastics engineering handbook: processing, materials, and applications, 2nd edn. Wiley-VCH
50. Los P, Lukomska A, Jeziorska R (2016) Metal-polymer composites for electromagnetic interference shielding applications. Polim/Polym 61:663–669. https://doi.org/10.14314/polimery.2016.663
51. Xu F, Yan C, Shyng YT et al (2014) Ultra-toughened nylon 12 nanocomposites reinforced with IF-WS2. Nanotechnology 25:325701. https://doi.org/10.1088/0957-4484/25/32/325701
52. Sabirneeza AAF, Subhashini S, Rajalakshmi R (2013) Water soluble conducting polymer composite of polyvinyl alcohol and leucine: an effective acid corrosion inhibitor for mild steel. Mater Corros 64:74–82. https://doi.org/10.1002/maco.201106096
53. Sahu M, Narashimhan L, Prakash O, Raichur AM (2017) Noncovalently functionalized tungsten disulfide nanosheets for enhanced mechanical and thermal properties of epoxy nanocomposites. ACS Appl Mater Interfaces 9:14347–14357. https://doi.org/10.1021/acsami.7b01608.
54. Badawi A (2021) Enhancement of the optical properties of PVP using Zn1-xSnxS for UV-region optical applications. Appl Phys A Mater Sci Process 127. https://doi.org/10.1007/s00339-020-04157-2
55. Ren J, Li Q, Yan L et al (2020) Enhanced thermal conductivity of epoxy composites by introducing graphene@boron nitride nanosheets hybrid nanoparticles. Mater Des 191:108663. https://doi.org/10.1016/j.matdes.2020.108663

56. Luo S, Yu J, Yu S et al (2019) Significantly enhanced electrostatic energy storage performance of flexible polymer composites by introducing highly insulating-ferroelectric microhybrids as fillers. Adv Energy Mater 9:1803204. https://doi.org/10.1002/aenm.201803204
57. Zhao YH, Zhang YF, Wu ZK, Bai SL (2016) Synergic enhancement of thermal properties of polymer composites by graphene foam and carbon black. Compos B Eng 84:52–58. https://doi.org/10.1016/j.compositesb.2015.08.074
58. Zhang Z, Du Y, Zhu C et al (2021) Unprecedented enhancement of wear resistance for epoxy-resin graphene composites. Nanoscale 13:2855–2867. https://doi.org/10.1039/d0nr08600b
59. Jia YW, Zhao X, Fu T et al (2020) Synergy effect between quaternary phosphonium ionic liquid and ammonium polyphosphate toward flame retardant PLA with improved toughness. Compos B Eng 197:108192. https://doi.org/10.1016/j.compositesb.2020.108192
60. Adak B, Joshi M, Butola BS (2019) Polyurethane/functionalized-graphene nanocomposite films with enhanced weather resistance and gas barrier properties. Compos B Eng 176. https://doi.org/10.1016/j.compositesb.2019.107303
61. Fawaz J, Mittal V (2015) Synthesis of polymer nanocomposites: review of various techniques In: Synthesis techniques for polymer nanocomposites, 1st edn. Wiley-VCH. https://doi.org/10.1002/9783527670307.ch1
62. Bhagyaraj S, Oluwafemi OS, Kalarikkal N, Thomas S (2018) Synthesis of inorganic nanomaterials: advances and key technologies, 1st edn. Elsevier
63. Verma D, Goh KL (2019) Functionalized graphene-based nanocomposites for energy applications. In: Functionalized graphene nanocomposites and their derivatives: synthesis, processing and applications. https://doi.org/10.1016/B978-0-12-814548-7.00011-8
64. Maiti S, Suin S, Shrivastava NK, Khatua BB (2014) Low percolation threshold and high electrical conductivity in melt-blended polycarbonate/multiwall carbon nanotube nanocomposites in the presence of poly(ε-caprolactone). Polym Eng Sci 54. https://doi.org/10.1002/pen.23600
65. Lee SK, Seong DG, Youn JR (2005) Degradation and rheological properties of biodegradable nanocomposites prepared by melt intercalation method. Fibers Polym 6. https://doi.org/10.1007/BF02875664
66. Kumar DS (2016) Synthesis and characterization of polymer nanocomposites for biomedical applications-current perspectives and challenges. IJRES 4:1–6
67. Mtibe A, Mokhothu TH, John MJ et al (2018) Fabrication and characterization of various engineered nanomaterials. In: Handbook of nanomaterials for industrial applications. Elsevier, pp 151–171. https://doi.org/10.1016/B978-0-12-813351-4.00009-2
68. Tang LC, Zhao L, Qiang F et al (2019) Mechanical properties of rubber nanocomposites containing carbon nanofillers. In: Carbon-based nanofillers and their rubber nanocomposites: fundamentals and applications. https://doi.org/10.1016/B978-0-12-817342-8.00012-3
69. Akpan EI, Shen X, Wetzel B, Friedrich K (2018) Design and synthesis of polymer nanocomposites. In: Polymer composites with functionalized nanoparticles: synthesis, properties, and applications, 1st edn. Elsevier, pp 47–75. https://doi.org/10.1016/B978-0-12-814064-2.00002-0
70. Vasudeo Rane A, Kanny K, Abitha VK et al (2017) Clay-polymer composites: design of clay polymer nanocomposite by mixing. In: Clay-polymer nanocomposites, 1st edn. Elsevier, pp 113–142. https://doi.org/10.1016/B978-0-323-46153-5.00004-5
71. Dantas de Oliveira A, Augusto Gonçalves Beatrice C (2019) Polymer nanocomposites with different types of nanofiller. In: Nanocomposites-recent evolutions, 1st edn. IntechOpen. https://doi.org/10.5772/intechopen.81329
72. Paszkiewicz S, Szymczyk A (2019) Graphene-based nanomaterials and their polymer nanocomposites. In: Nanomaterials and polymer nanocomposites: raw materials to applications. Elsevier, pp 177–216. https://doi.org/10.1016/B978-0-12-814615-6.00006-0
73. Grothe J, Kaskel S, Leuteritz A (2012) Nanocomposites and hybrid materials. In: Polymer science: a comprehensive reference, vol 10, pp 177–209. https://doi.org/10.1016/B978-0-444-53349-4.00206-5

74. Avella M, Errico ME, Martuscelli E (2001) Novel PMMA/$CaCO_3$ nanocomposites abrasion resistant prepared by an in situ polymerization process. Nano Lett 1:213–217. https://doi.org/10.1021/nl015518v
75. Allen R, Pan L, Fuller GG, Bao Z (2014) Using in-situ polymerization of conductive polymers to enhance the electrical properties of solution-processed carbon nanotube films and fibers. ACS Appl Mater Interfaces 6:9966–9974. https://doi.org/10.1021/am5019995
76. Khan W, Sharma R, Saini P (2016) Carbon nanotube-based polymer composites: synthesis, properties and applications. In: Carbon nanotubes-current progress of their polymer composites. InTech. https://doi.org/10.5772/62497
77. Friedrich K (2018) Polymer composites for tribological applications. Adv Ind Eng Polym Res 1:3–39. https://doi.org/10.1016/j.aiepr.2018.05.001
78. Shields RJ, Bhattacharyya D, Fakirov S (2008) Fibrillar polymer-polymer composites: morphology, properties and applications. J Mater Sci 43:6758–6770. https://doi.org/10.1007/s10853-008-2693-z
79. Sajan S, Philip Selvaraj D (2021) A review on polymer matrix composite materials and their applications. Mater Today: Proc 47:5493–5498. https://doi.org/10.1016/j.matpr.2021.08.034
80. Alhazmi H, Shah SAR, Anwar MK et al (2021) Utilization of polymer concrete composites for a circular economy: a comparative review for assessment of recycling and waste utilization. Polym (Basel) 13:2135. https://doi.org/10.3390/polym13132135
81. Tomar P, Lakhani R, Chhibber VK, Kumar R (2018) Macro-defect free cements: a future oriented polymer composite materials for construction industries. Compos Interfaces 25:1–21. https://doi.org/10.1080/09276440.2018.1439637
82. Zaman A, Gutub SA, Wafa MA (2013) A review on FRP composites applications and durability concerns in the construction sector. J Reinf Plast Compos 32:1966–1988. https://doi.org/10.1177/0731684413492868
83. Lewis JA, Boyer MA (1995) Effects of an organotitanate cross-linking additive on the processing and properties of macro-defect-free cement. Adv Cem Based Mater 2:2–7. https://doi.org/10.1016/1065-7355(95)90033-0
84. Qiao Y (2005) On the fracture toughness of calcium aluminate cement-phenol resin composites. Cem Concr Res 35:220–225. https://doi.org/10.1016/j.cemconres.2004.05.039
85. Humphreys MF (2003) The use of polymer composites in construction. In: International conference on smart and sustainable built environment, pp 1–9
86. Wan B (2014) Using fiber-reinforced polymer (FRP) composites in bridge construction and monitoring their performance: an overview. In: Advanced composites in bridge construction and repair. Woodhead Publishing, pp 3–29. https://doi.org/10.1533/9780857097019.1.3
87. Hollaway LC, Head PR (2001) Advanced polymer composites and polymers in the civil infrastructure, Ist edn. Elsevier. https://doi.org/10.1016/B978-0-08-043661-6.X5000-1
88. Siwowski T, Kaleta D, Rajchel M (2018) Structural behaviour of an all-composite road bridge. Compos Struct 192:555–567. https://doi.org/10.1016/j.compstruct.2018.03.042
89. Kossakowski PG, Wciślik W (2022) Fiber-reinforced polymer composites in the construction of bridges: opportunities, problems and challenges. Fibers 10:37. https://doi.org/10.3390/fib10040037
90. Holbery J, Houston D (2006) Natural-fiber-reinforced polymer composites in automotive applications. JOM 58:80–86. https://doi.org/10.1007/s11837-006-0234-2
91. Muhammad A, Rahman MR, Baini R, Bin Bakri MK (2020) Applications of sustainable polymer composites in automobile and aerospace industry. In: Advances in sustainable polymer composites, Ist edn. Woodhead publishing series in composites science and engineering, pp 185–207. https://doi.org/10.1016/B978-0-12-820338-5.00008-4
92. Veličković S, Stojanović B, Ivanović L et al (2019) Application of nanocomposites in the automotive industry. Mobil Veh Mech 45:323–332. https://doi.org/10.24874/mvm.2019.45.03.05
93. Mohammadi H, Ahmad Z, Mazlan SA et al (2023) Lightweight glass fiber-reinforced polymer composite for automotive bumper applications: a review. Polym (Basel) 15:193. https://doi.org/10.3390/polym15010193

94. Ma R, Tao Y, Chen Y et al (2021) Achieving 16.68% efficiency ternary as-cast organic solar cells. Sci China Chem 64:581–589. https://doi.org/10.1007/s11426-020-9912-0
95. Song R, Wu Y, Lin C et al (2022) High-speed shift register with dual-gated thin-film transistors for a 31-Inch 4K AMOLED display. Micromachines (Basel) 13:1696. https://doi.org/10.3390/mi13101696
96. Cao X, Xiong Y, Sun J et al (2021) Piezoelectric nanogenerators derived self-powered sensors for multifunctional applications and artificial intelligence. Adv Funct Mater 31:2102983. https://doi.org/10.1002/adfm.202102983
97. Sim K, Rao Z, Kim HJ et al (2019) Fully rubbery integrated electronics from high effective mobility intrinsically stretchable semiconductors. Sci Adv 5:1–10. https://doi.org/10.1126/sciadv.aav5749
98. De Rossi F, Renno G, Taheri B et al (2021) Modified P3HT materials as hole transport layers for flexible perovskite solar cells. J Power Sources 494:229735. https://doi.org/10.1016/j.jpowsour.2021.229735
99. Lee H, Kim Y, Kim J et al (2022) Consecutive ink writing of conducting polymer and graphene composite electrodes for foldable electronics-related applications. Polym (Basel) 14:5294. https://doi.org/10.3390/polym14235294
100. Liu H, Li Q, Zhang S et al (2018) Electrically conductive polymer composites for smart flexible strain sensors: a critical review. J Mater Chem C Mater 6:12121–12141. https://doi.org/10.1039/C8TC04079F
101. Kong W, Yang Y, Wang Y et al (2022) An ultra-low hysteresis, self-healing and stretchable conductor based on dynamic disulfide covalent adaptable networks. J Mater Chem A Mater 10:2012–2020. https://doi.org/10.1039/d1ta08737a
102. Shown I, Ganguly A, Chen LC, Chen KH (2015) Conducting polymer-based flexible supercapacitor. Energy Sci Eng 3:2–26. https://doi.org/10.1002/ese3.50
103. Kumar P, Narayan Maiti U, Sikdar A et al (2019) Recent advances in polymer and polymer composites for electromagnetic interference shielding: review and future prospects. Polym Rev 59:1–52. https://doi.org/10.1080/15583724.2019.1625058
104. Rubino F, Nisticò A, Tucci F, Carlone P (2020) Marine application of fiber reinforced composites: a review. J Mar Sci Eng 8:26. https://doi.org/10.3390/jmse8010026
105. Selvaraju S, Ilaiyavel S (2011) Applications of composites in marine industry. J Eng Res Stud 2:89–91
106. Okuma SO, Obaseki M, Okiemute Ofuyekpone D, Endurance Ashibudike O (2023) A review assessment of fiber-reinforced polymers for maritime applications. J Adv Ind Technol Appl 4:17–28. https://doi.org/10.30880/jaita.2023.04.01.003
107. Mouritz AP, Gellert E, Burchill P, Challis K (2001) Review of advanced composite structures for naval ships and submarines. Compos Struct 53:21–42. https://doi.org/10.1016/S0263-8223(00)00175-6
108. Rajak DK, Pagar DD, Menezes PL, Linul E (2019) Fiber-reinforced polymer composites: manufacturing, properties, and applications. Polym (Basel) 11:1667. https://doi.org/10.3390/polym11101667
109. Kumar A, Lal Krishna G, Anantha Subramanian V (2019) Design and analysis of a carbon composite propeller for podded propulsion. In: Lecture notes in civil engineering, vol 22. Springer, pp 203–215. https://doi.org/10.1007/978-981-13-3119-0_13
110. Kesarwani S (2017) Polymer composites in aviation sector a brief review article. IJERT 6:518–525. https://doi.org/10.17577/IJERTV6IS060291
111. Muhammad A, Rahman MR, Baini R, Bin Bakri MK (2020) Applications of sustainable polymer composites in automobile and aerospace industry. In: Advances in sustainable polymer composites, 1st edn. Woodhead publishing series in composites science and engineering, pp 185–207. https://doi.org/10.1016/B978-0-12-820338-5.00008-4
112. Fahrenholtz WG, Hilmas GE, Talmy IG, Zaykoski JA (2007) Refractory diborides of zirconium and hafnium. J Am Ceram Soc 90:1347–1364. https://doi.org/10.1111/j.1551-2916.2007.01583.x

113. Parveez B, Kittur MI, Badruddin IA et al (2022) Scientific advancements in composite materials for aircraft applications: a review. Polym (Basel) 14:5007. https://doi.org/10.3390/polym14225007
114. Hench LL (1998) Biomaterials: a forecast for the future. Biomaterials 19:1419–1423. https://doi.org/10.1016/s0142-9612(98)00133-1
115. Langer R, Tirrell DA (2004) Designing materials for biology and medicine. Nature 428:487-492. https://doi.org/10.1038/nature02388
116. Bose S, Ke D, Sahasrabudhe H, Bandyopadhyay A (2018) Additive manufacturing of biomaterials. Prog Mater Sci 93:45–111. https://doi.org/10.1016/j.pmatsci.2017.08.003
117. Guo Z, Poot AA, Grijpma DW (2021) Advanced polymer-based composites and structures for biomedical applications. Eur Polym J 149:110388. https://doi.org/10.1016/j.eurpolymj.2021.110388
118. Hussein EA, Zagho MM, Nasrallah GK, Elzatahry AA (2018) Recent advances in functional nanostructures as cancer photothermal therapy. Int J Nanomedicine 13:2897–2906. https://doi.org/10.2147/IJN.S161031
119. Al-Enizi AM, Zagho MM, Elzatahry AA (2018) Polymer-based electrospun nanofibers for biomedical applications. Nanomaterials 8:259. https://doi.org/10.3390/nano8040259
120. Salernitano E, Migliaresi C (2003) Composite materials for biomedical applications: a review. J Appl Biomater Biomech 1:3–18. https://doi.org/10.1177/228080000300100102
121. Ibrahim ID, Jamiru T, Sadiku ER, Hamam Y (2019) Development and utilization of polymers in biomedical applications. In: 2019 open innovations conference (OI), pp 165–170. https://doi.org/10.1109/OI.2019.8908248
122. Abbas M, Alqahtani MS, Alhifzi R (2023) Recent developments in polymer nanocomposites for bone regeneration. Int J Mol Sci 24:3312. https://doi.org/10.3390/ijms24043312
123. Ramakrishna S, Mayer J, Wintermantel E, Leong KW (2001) Biomedical applications of polymer-composite materials: a review. Compos Sci Technol 61:1189–1224. https://doi.org/10.1016/S0266-3538(00)00241-4
124. Thang NH, Chien TB, Cuong DX (2023) Polymer-based hydrogels applied in drug delivery: an overview. Gels 9:523. https://doi.org/10.3390/gels9070523
125. Gartner TE, Jayaraman A (2019) Modeling and simulations of polymers: a roadmap. Macromolecules 52:755–786. https://doi.org/10.1021/acs.macromol.8b01836
126. Ndoro TVM, Voyiatzis E, Ghanbari A, Theodorou DN, Böhm MC, Müller-Plathe F (2011) Interface of grafted and ungrafted silica nanoparticles with a polystyrene matrix: atomistic molecular dynamics simulations. Macromolecules 44:2316–2327. https://doi.org/10.1021/ma102833u
127. Pandey YN, Doxastakis M (2012) Detailed atomistic Monte Carlo simulations of a polymer melt on a solid surface and around a nanoparticle. J Chem Phys 136. https://doi.org/10.1063/1.3689316
128. Ndoro TVM, Böhm MC, Müller-Plathe F (2012) Interface and interphase dynamics of polystyrene chains near grafted and ungrafted silica nanoparticles. Macromolecules 45:171–179. https://doi.org/10.1021/ma2020613
129. Lane JMD, Ismail AE, Chandross M, Lorenz CD, Grest GS (2009) Forces between functionalized silica nanoparticles in solution. Phys Rev E 79:050501. https://doi.org/10.1103/PhysRevE.79.050501
130. Kyrychenko A, Pasko DA, Kalugin ON (2017) Poly(vinyl alcohol) as a water protecting agent for silver nanoparticles: the role of polymer size and structure. Phys Chem Chem Phys 19:8742–8756. https://doi.org/10.1039/C6CP05562A
131. Kyrychenko A, Korsun OM, Gubin II, Kovalenko SM, Kalugin ON (2015) Atomistic simulations of coating of silver nanoparticles with poly(vinylpyrrolidone) oligomers: effect of oligomer chain length. J Phys Chem C 119:7888–7899. https://doi.org/10.1021/jp510369a
132. Ganesan V, Jayaraman A (2014) Theory and simulation studies of effective interactions, phase behavior and morphology in polymer nanocomposites. Soft Matter 10:13–38. https://doi.org/10.1039/C3SM51864G

133. Eslami H, Rahimi M, Müller-Plathe F (2013) Molecular dynamics simulation of a silica nanoparticle in oligomeric poly(methyl methacrylate): a model system for studying the interphase thickness in a polymer-nanocomposite via different properties. Macromolecules 46:8680–8692. https://doi.org/10.1021/ma401443v
134. Brown D, Mélé P, Marceau S, Albérola ND (2003) A molecular dynamics study of a model nanoparticle embedded in a polymer matrix. Macromolecules 36:1395–1406. https://doi.org/10.1021/ma020951s
135. Karatrantos A, Clarke N, Kröger M (2016) Modeling of polymer structure and conformations in polymer nanocomposites from atomistic to mesoscale: a review. Polym Rev 56:385–428. https://doi.org/10.1080/15583724.2015.1090450
136. Vogiatzis GG, Theodorou DN (2014) Local segmental dynamics and stresses in polystyrene–C 60 mixtures. Macromolecules 47:387–404. https://doi.org/10.1021/ma402214r
137. Smith JS, Borodin O, Smith GD, Kober EM (2007) A molecular dynamics simulation and quantum chemistry study of poly(dimethylsiloxane)–silica nanoparticle interactions. J Polym Sci Part B Polym Phys 45:1599–1615. https://doi.org/10.1002/polb.21119
138. Peters BL, Lane JMD, Ismail AE, Grest GS (2012) Fully atomistic simulations of the response of silica nanoparticle coatings to alkane solvents. Langmuir 28:17443–17449. https://doi.org/10.1021/la3023166
139. Pavlov AS, Khalatur PG (2016) Filler reinforcement in cross-linked elastomer nanocomposites: insights from fully atomistic molecular dynamics simulation. Soft Matter 12:5402–5419. https://doi.org/10.1039/C6SM00543H
140. Pandey YN, Papakonstantopoulos GJ, Doxastakis M (2013) Polymer/nanoparticle interactions: bridging the gap. Macromolecules 46:5097–5106. https://doi.org/10.1021/ma400444w
141. Kukol A (2009) Lipid models for united-atom molecular dynamics simulations of proteins. J Chem Theory Comput 5:615–626. https://doi.org/10.1021/ct8003468
142. Stubbs JM, Potoff JJ, Siepmann JI (2004) Transferable potentials for phase equilibria. 6. United-atom description for ethers, glycols, ketones, and aldehydes. J Phys Chem B 108:17596–17605. https://doi.org/10.1021/jp049459w
143. Chen B, Potoff JJ, Siepmann JI (2001) Monte Carlo calculations for alcohols and their mixtures with alkanes. Transferable potentials for phase equilibria. 5. United-atom description of primary, secondary, and tertiary alcohols. J Phys Chem B 105:3093–3104. https://doi.org/10.1021/jp003882x
144. Tieleman DP, MacCallum JL, Ash WL, Kandt C, Xu Z, Monticelli L (2006) Membrane protein simulations with a united-atom lipid and all-atom protein model: lipid–protein interactions, side chain transfer free energies and model proteins. J Phys Condens Matter 18:S1221–S1234. https://doi.org/10.1088/0953-8984/18/28/S07
145. Yang L, Tan C, Hsieh M-J, Wang J, Duan Y, Cieplak P, Caldwell J, Kollman PA, Luo R (2006) New-generation amber united-atom force field. J Phys Chem B 110:13166–13176. https://doi.org/10.1021/jp060163v
146. Wick CD, Martin MG, Siepmann JI (2000) Transferable potentials for phase equilibria. 4. United-atom description of linear and branched alkenes and alkylbenzenes. J Phys Chem B 104:8008–8016. https://doi.org/10.1021/jp001044x
147. Martin MG, Siepmann JI (1999) Novel configurational-bias Monte Carlo method for branched molecules. Transferable potentials for phase equilibria. 2. United-atom description of branched alkanes. J Phys Chem B 103:4508–4517. https://doi.org/10.1021/jp984742e
148. González MA (2011) Force fields and molecular dynamics simulations. École Thématique La Société Française La Neutron 12:169–200. https://doi.org/10.1051/sfn/201112009
149. Lindorff-Larsen K, Piana S, Palmo K, Maragakis P, Klepeis JL, Dror RO, Shaw DE (2010) Improved side-chain torsion potentials for the Amber ff99SB protein force field. Proteins Struct Funct Bioinform 78:1950–1958. https://doi.org/10.1002/prot.22711
150. Lopes PEM, Guvench O, MacKerell AD (2015) Current status of protein force fields for molecular dynamics simulations, pp 47–71. https://doi.org/10.1007/978-1-4939-1465-4_3
151. Mackerell AD (2004) Empirical force fields for biological macromolecules: overview and issues. J Comput Chem 25:1584–1604. https://doi.org/10.1002/jcc.20082

152. Hornak V, Abel R, Okur A, Strockbine B, Roitberg A, Simmerling C (2006) Comparison of multiple Amber force fields and development of improved protein backbone parameters. Proteins Struct Funct Bioinform 65:712–725. https://doi.org/10.1002/prot.21123
153. Pérez A, Marchán I, Svozil D, Sponer J, Cheatham TE, Laughton CA, Orozco M (2007) Refinement of the AMBER force field for nucleic acids: improving the description of α/γ conformers. Biophys J 92:3817–3829. https://doi.org/10.1529/biophysj.106.097782
154. MacKerell AD, Bashford D, Bellott M, Dunbrack RL, Evanseck JD, Field MJ, Fischer S, Gao J, Guo H, Ha S, Joseph-McCarthy D, Kuchnir L, Kuczera K, Lau FTK, Mattos C, Michnick S, Ngo T, Nguyen DT, Prodhom B, Reiher WE, Roux B, Schlenkrich M, Smith JC, Stote R, Straub J, Watanabe M, Wiórkiewicz-Kuczera J, Yin D, Karplus M (1998) All-atom empirical potential for molecular modeling and dynamics studies of proteins. J Phys Chem B 102:3586–3616. https://doi.org/10.1021/jp973084f
155. Vanommeslaeghe K, Hatcher E, Acharya C, Kundu S, Zhong S, Shim J, Darian E, Guvench O, Lopes P, Vorobyov I, Mackerell AD (2010) CHARMM general force field: a force field for drug-like molecules compatible with the CHARMM all-atom additive biological force fields. J Comput Chem 31:671–690. https://doi.org/10.1002/jcc.21367
156. Klauda JB, Venable RM, Freites JA, O'Connor JW, Tobias DJ, Mondragon-Ramirez C, Vorobyov I, MacKerell AD, Pastor RW (2010) Update of the CHARMM all-atom additive force field for lipids: validation on six lipid types. J Phys Chem B 114:7830–7843. https://doi.org/10.1021/jp101759q
157. Karasawa N, Dasgupta S, Goddard WA (1991) Mechanical properties and force field parameters for polyethylene crystal. J Phys Chem 95:2260–2272. https://doi.org/10.1021/j100159a031
158. Lacks DJ, Rutledge GC (1994) Simulation of the temperature dependence of mechanical properties of polyethylene. J Phys Chem 98:1222–1231. https://doi.org/10.1021/j100055a030
159. Marcon V, Raos G (2004) Molecular modeling of crystalline oligothiophenes: testing and development of improved force fields. J Phys Chem B 108:18053–18064. https://doi.org/10.1021/jp047128d
160. DuBay KH, Hall ML, Hughes TF, Wu C, Reichman DR, Friesner RA (2012) Accurate force field development for modeling conjugated polymers. J Chem Theory Comput 8:4556–4569. https://doi.org/10.1021/ct300175w
161. Halgren TA (1999) MMFF VII. Characterization of MMFF94, MMFF94s, and other widely available force fields for conformational energies and for intermolecular-interaction energies and geometries. J Comput Chem 20:730–748. https://doi.org/10.1002/(SICI)1096-987X(199905)20:7<730::AID-JCC8>3.0.CO;2-T
162. Mark P, Nilsson L (2001) Structure and dynamics of the TIP3P, SPC, and SPC/E water models at 298 K. J Phys Chem A 105:9954–9960. https://doi.org/10.1021/jp003020w
163. Jorgensen WL, Chandrasekhar J, Madura JD, Impey RW, Klein ML (1983) Comparison of simple potential functions for simulating liquid water. J Chem Phys 79:926–935. https://doi.org/10.1063/1.445869
164. Mu Y, Kosov DS, Stock G (2003) Conformational dynamics of trialanine in water. 2. Comparison of AMBER, CHARMM, GROMOS, and OPLS force fields to NMR and infrared experiments. J Phys Chem B 107:5064–5073. https://doi.org/10.1021/jp022445a
165. Underwood TR, Greenwell HC (2018) The water-alkane interface at various NaCl salt concentrations: a molecular dynamics study of the readily available force fields. Sci Rep 8:352. https://doi.org/10.1038/s41598-017-18633-y
166. Mayo SL, Olafson BD, Goddard WA (1990) DREIDING: a generic force field for molecular simulations. J Phys Chem 94:8897–8909. https://doi.org/10.1021/j100389a010
167. Müller-Plathe F (2002) Coarse-graining in polymer simulation: from the atomistic to the mesoscopic scale and back. ChemPhysChem 3:754–769. https://doi.org/10.1002/1439-7641(20020916)3:9%3c754::AID-CPHC754%3e3.0.CO;2-U
168. Nielsen SO, Lopez CF, Srinivas G, Klein ML (2004) Coarse grain models and the computer simulation of soft materials. J Phys Condens Matter 16:R481–R512. https://doi.org/10.1088/0953-8984/16/15/R03

169. Theodorou DN (n.d.) Equilibration and coarse-graining methods for polymers. In: Computer simulations in condensed matter systems: from materials to chemical biology volume 2, pp. 419–448. Springer, Berlin. https://doi.org/10.1007/3-540-35284-8_16
170. Marrink SJ, de Vries AH, Mark AE (2004) Coarse grained model for semiquantitative lipid simulations. J Phys Chem B 108:750–760. https://doi.org/10.1021/jp036508g
171. Brini E, Algaer EA, Ganguly P, Li C, Rodríguez-Ropero F, van der Vegt NFA (2013) Systematic coarse-graining methods for soft matter simulations–a review. Soft Matter 9:2108–2119. https://doi.org/10.1039/C2SM27201F
172. Hsu H-P, Grassberger P (2011) A review of Monte Carlo simulations of polymers with PERM. J Stat Phys 144:597–637. https://doi.org/10.1007/s10955-011-0268-x
173. Panagiotopoulos AZ (2000) Monte Carlo methods for phase equilibria of fluids. J Phys Condens Matter 12:R25–R52. https://doi.org/10.1088/0953-8984/12/3/201
174. Fried H, Binder K (1991) The microphase separation transition in symmetric diblock copolymer melts: a Monte Carlo study. J Chem Phys 94:8349–8366. https://doi.org/10.1063/1.460067
175. Binder K, Müller M (2000) Monte Carlo simulation of block copolymers. Curr Opin Colloid Interface Sci 5:314–322. https://doi.org/10.1016/S1359-0294(00)00074-1
176. Panagiotopoulos AZ (1992) Direct determination of fluid phase equilibria by simulation in the Gibbs ensemble: a review. Mol Simul 9:1–23. https://doi.org/10.1080/08927029208048258
177. Panagiotopoulos AZ (1987) Direct determination of phase coexistence properties of fluids by Monte Carlo simulation in a new ensemble. Mol Phys 61:813–826. https://doi.org/10.1080/00268978700101491
178. Panagiotopoulos AZ (1995) Gibbs ensemble techniques. In: Observation, prediction and simulation of phase transitions in complex fluids, pp 463–501. Springer Netherlands, Dordrecht. https://doi.org/10.1007/978-94-011-0065-6_11
179. Panagiotopoulos AZ, Quirke N, Stapleton M, Tildesley DJ (1988) Phase equilibria by simulation in the Gibbs ensemble. Mol Phys 63:527–545. https://doi.org/10.1080/00268978800100361
180. Escobedo FA, de Pablo JJ (1996) Expanded grand canonical and Gibbs ensemble Monte Carlo simulation of polymers. J Chem Phys 105:4391–4394. https://doi.org/10.1063/1.472257
181. de Pablo JJ, Yan Q, Escobedo FA (1999) Simulation of phase transitions in fluids. Annu Rev Phys Chem 50:377–411. https://doi.org/10.1146/annurev.physchem.50.1.377
182. Statistical Physics (1980) Elsevier. https://doi.org/10.1016/C2009-0-24487-4
183. Nedelcu S, Werner M, Lang M, Sommer J-U (2012) GPU implementations of the bond fluctuation model. J Comput Phys 231:2811–2824. https://doi.org/10.1016/j.jcp.2011.12.021
184. Rohrdanz MA, Zheng W, Clementi C (2013) Discovering mountain passes via torchlight: methods for the definition of reaction coordinates and pathways in complex macromolecular reactions. Annu Rev Phys Chem 64:295–316. https://doi.org/10.1146/annurev-physchem-040412-110006
185. Weinan E, Ren W, Vanden-Eijnden E (2002) String method for the study of rare events. Phys Rev B 66:052301. https://doi.org/10.1103/PhysRevB.66.052301
186. Weinan E, Ren W, Vanden-Eijnden E (2005) Finite temperature string method for the study of rare events. J Phys Chem B 109:6688–6693. https://doi.org/10.1021/jp0455430
187. Vanden-Eijnden E, Venturoli M (2009) Revisiting the finite temperature string method for the calculation of reaction tubes and free energies. J Chem Phys 130. https://doi.org/10.1063/1.3130083
188. Ting CL, Wang Z-G (2012) Minimum free energy paths for a nanoparticle crossing the lipid membrane. Soft Matter 8:12066. https://doi.org/10.1039/c2sm26377g
189. Lyubimov I, Beltran-Villegas DJ, Jayaraman A (2017) PRISM theory study of amphiphilic block copolymer solutions with varying copolymer sequence and composition. Macromolecules 50:7419–7431. https://doi.org/10.1021/acs.macromol.7b01419
190. Beltran-Villegas DJ, Lyubimov I, Jayaraman A (2018) Molecular dynamics simulations and PRISM theory study of solutions of nanoparticles and triblock copolymers with solvophobic end blocks. Mol Syst Des Eng 3:453–472. https://doi.org/10.1039/C7ME00128B

191. Beltran-Villegas DJ, Jayaraman A (2018) Assembly of amphiphilic block copolymers and nanoparticles in solution: coarse-grained molecular simulation study. J Chem Eng Data 63:2351–2367. https://doi.org/10.1021/acs.jced.7b00925
192. Grossfield A, Zuckerman DM (2009) Quantifying uncertainty and sampling quality in biomolecular simulations, pp 23–48. https://doi.org/10.1016/S1574-1400(09)00502-7. (Chapter 2)
193. Understanding Molecular Simulation (2002) Elsevier. https://doi.org/10.1016/B978-0-12-267351-1.X5000-7
194. Melenkevitz J, Muthukumar M (1991) Density functional theory of lamellar ordering in diblock copolymers. Macromolecules 24:4199–4205. https://doi.org/10.1021/ma00014a038
195. Wu J, Li Z (2007) Density-functional theory for complex fluids. Annu Rev Phys Chem 58:85–112. https://doi.org/10.1146/annurev.physchem.58.032806.104650
196. McMullen WE, Freed KF (1990) A density functional theory of polymer phase transitions and interfaces. J Chem Phys 92:1413–1426. https://doi.org/10.1063/1.458153
197. Woodward CE, Yethiraj A (1994) Density functional theory for inhomogeneous polymer solutions. J Chem Phys 100:3181–3186. https://doi.org/10.1063/1.466409
198. Nakamura I, Wang Z-G (2012) Salt-doped block copolymers: ion distribution, domain spacing and effective χ parameter. Soft Matter 8:9356. https://doi.org/10.1039/c2sm25606a
199. Martin JI, Wang Z-G (1995) Polymer brushes: scaling, compression forces, interbrush penetration, and solvent size effects. J Phys Chem 99:2833–2844. https://doi.org/10.1021/j100009a045
200. Schmid F (1998) Self-consistent-field theories for complex fluids. J Phys Condens Matter 10:8105–8138. https://doi.org/10.1088/0953-8984/10/37/002
201. Müller M, Schmid F (n.d.) Incorporating fluctuations and dynamics in self-consistent field theories for polymer blends. In: Advanced computer simulation approaches for soft matter sciences II, pp 1–58. Springer-Verlag, Berlin. https://doi.org/10.1007/b136794
202. Jiang Y, Li S, Chen JZY (2016) Perspective: parameters in a self-consistent field theory of multicomponent wormlike-copolymer melts. Eur Phys J E 39:91
203. Martin TB, Gartner TE, Jones RL, Snyder CR, Jayaraman A (2018) PyPRISM: a computational tool for liquid-state theory calculations of macromolecular materials. Macromolecules 51:2906–2922. https://doi.org/10.1021/acs.macromol.8b00011
204. Schweizer KS, Curro JG (n.d.) PRISM theory of the structure, thermodynamics, and phase transitions of polymer liquids and alloys. In: Atomistic modeling of physical properties, pp 319–377. Springer, Berlin. https://doi.org/10.1007/BFb0080203
205. Jayaraman A, Nair N (2012) Integrating PRISM theory and Monte Carlo simulation to study polymer-functionalised particles and polymer nanocomposites. Mol Simul 38:751–761. https://doi.org/10.1080/08927022.2012.683528
206. Karger-Kocsis J, Bárány T (2014) Single-polymer composites (SPCs): status and future trends. Compos Sci Technol 92:77–94. https://doi.org/10.1016/j.compscitech.2013.12.006
207. Lakshmi V, Karun AS, Rajan TPD (2022) Polymeric nanocomposites. In: Nanotechnology in the automotive industry. Elsevier, pp 37–63. https://doi.org/10.1016/B978-0-323-90524-4.00003-7
208. Dhiman B, Guleria V, Sharma P (2020) Applications and future trends of carbon fiber reinforced polymer composites: a review. Int Res J Eng Technol 7:1883–1889

Chapter 2
Different Curing Methods Used for Polymer Composites

Pankaj Kumar, Pakanati Siva Prasad, and Juan David Matallana Guerrero

1 Introduction

Composite materials are made up of two or more chemically and physically distinct phases that are separated from one another by a discrete interface [1–5]. Due to its benefits including low weight, excellent fatigue strength, corrosion resistance, and quick assembly, composites—the wonder materials—are increasingly important in modern materials [6–17]. They are widely utilized as building materials for everything from spacecraft to homes, as well as for everything from electronic packaging to medical equipment. Interestingly, the most significant polymeric composites—also referred to as natural composites—can be found in nature [18, 19]. Mammal connective tissues are among the most advanced polymer composites ever created, with collagen, a fibrous protein, serving as reinforcement. It performs both soft and rigid connective tissue functions [20, 21].

Polymer composite materials (PMCs) have proven to have distinct advantages over more traditional metallic materials in a variety of demanding automotive, infrastructural, and aerospace industries. PMCs have been also used in automotive, mechanical, aerospace, and other industries [22, 23]. To produce low-cost, high-efficiency production of polymer composites, which continues to be a difficult task, a wide range of curing techniques are being researched [24, 25]. For the manufacture of polymer composites, researchers and industries have long sought to discover and create low-cost, high-efficiency curing technologies and have looked into various radiation and

P. Kumar (✉)
School of Nano Science and Technology, Indian Institute of Technology Kharagpur, Kharagpur 721302, India
e-mail: pankajbharti434@gmail.com

P. S. Prasad · J. D. M. Guerrero
Department of Metallurgical and Materials Engineering, Indian Institute of Technology Kharagpur, Kharagpur 721302, India

S. K. Sethi et al. (eds.), *Polymer Composites: From Computational to Experimental Aspects*, Materials Horizons: From Nature to Nanomaterials,
https://doi.org/10.1007/978-981-97-0888-8_2

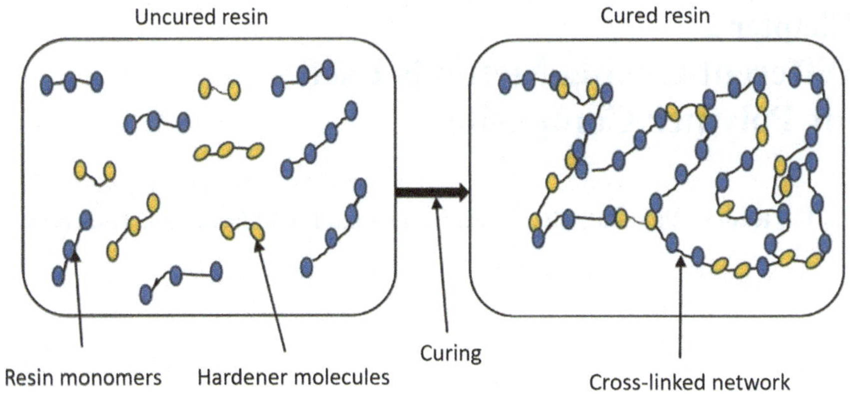

Fig. 1 Curing of resin (Reprinted with permission from [27])

thermal curing alternatives. Other techniques like the autoclave curing process that are used for curing high-performance polymer composites are very costly as well as labor and capital-intensive [26].

In polymer chemistry and process engineering, the phrase "curing" refers to the process of toughening or hardening a polymer material by cross-linking polymer chains, which can be accomplished using heat, electron beams, or chemical additives. The ultimate properties of polymer materials and constructions are significantly influenced by the curing process. The performance of materials, which is essential to the molding process of PMCs, as well as the preparation cycle and material cost are all impacted by the curing process. PMCs are still primarily thermally cured using a variety of thermal heating techniques, with the oven or autoclave being frequently utilized in this process. Figure 1 shows the curing of resin [27].

In this chapter, the current development status of the radiation and thermal curing methods has been discussed. The main advantages and disadvantages of these methods have been discussed comparatively. New challenges and opportunities of these techniques have been covered briefly (Fig. 2).

2 Radiation Curing

Radiation curing is widely used in several industries like biomedical, textile, nuclear plants, coating, aerospace, and food packaging [28, 29]. Not only in the field of plastics but also in the case of rubber and rubber composites, radiation curing is widely used by electron beam, gamma rays, X-rays, and ultraviolet rays. The physical, optical, mechanical, electrical, chemical, and structural properties of the polymer composites are altered and improved upon during radiation curing. Radiation curing also enhances thermal stability [30]. Induced by photons (UV curing) or electrons

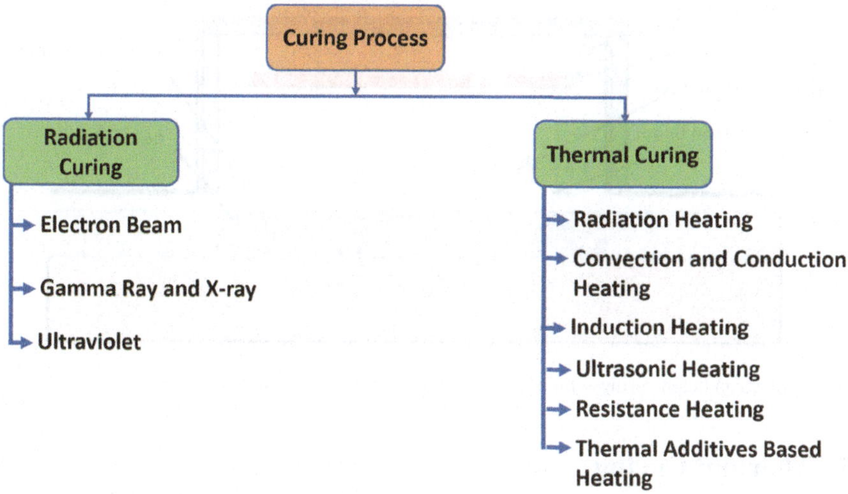

Fig. 2 Curing process of polymer composites

(EB curing), radiation curing entails the polymerization and cross-linking of functional monomers and oligomers (often liquid) into a cross-linked polymer network (typically a solid film). Free radical or cationic polymerization can be used to cure materials, while the former method is more commonly used in industry.

An electron beam is a stream of electrons produced by an electron gun and traveling in a particular direction at a constant speed. Because of their tiny size and great energy, electrons can easily enter materials and perform curing. As seen in Fig. 3, the primary components of the electron beam curing device are an emitting cathode and an anode that have an aperture for the passage of some of the created electrons that travel in a straight line at a constant speed. A tungsten wire or filament serving as the cathodic element is heated to a high temperature of roughly 2400 K in an acceleration chamber. The disadvantages of electron beam curing are it requires trained operators, has high installation costs, and is limited to 2D objects.

UV curing is one of the curing processes under radiation curing. This process is an energy-efficient, environmentally friendly, and reasonably priced curing method that is mostly applied to thin film-coated materials [32, 33]. Compared to other heat and traditional curing procedures, UV curing speeds up the material's curing process. Although UV radiation can come from a variety of sources, mercury arc lamps are often the source of UV radiation. The cross-linking density provides an easy way to forecast the properties of the UV-cured polymer [34]. UV light can only cure composite of a certain thickness because of the material's restricted ability to penetrate it. UV curing works through two mechanisms: cations and free radicals. By being exposed to light, the radicals produce new radicals, which causes polymerization and the cross-linking of monomers. In thin film, UV curing occurs very quickly or instantaneously, polymerizing resin from a liquid or gel state into solid material.

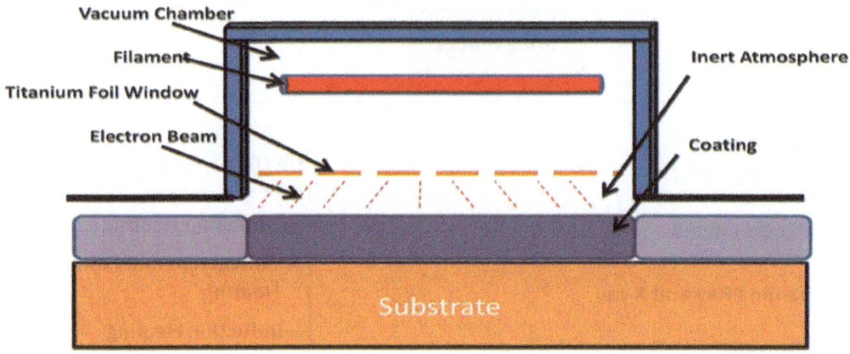

Fig. 3 Electron beam curing setup (Reprinted with permission from [31])

3 Thermal Curing

In developed thermal curing material systems and curing procedures, thermal curing is by far the most common curing method for polymer composites. Today, a wide range of thermal heating technologies, including infrared, laser, microwave, flame, oven, induction, ultrasonic, resistance heating, and others, are used for thermal curing. Radiation heating (infrared, laser, and microwave), convection and conduction heating (hot gas, flame, oven, and hot shoe), induction heating, ultrasonic heating, resistance heating, and thermal additive-based heating are the several categories that fall under [35].

When using thermal curing, heat is used to polymerize the resin. Despite the ease with which resin can be polymerized at room temperature, epoxy and polyester require 24–48 h to cure. Thermal energy is therefore employed to accelerate the rate of polymerization. The automotive and aerospace industries use thermal curing procedures more frequently, although these processes take time. Large volumetric aircraft parts cannot be cured using standard curing techniques.

3.1 Microwave Heating

When compared to conventional heating, microwave processing has often been offered as a way to quickly heat/cure resins or polymer composites in a more homogeneous volumetric manner [36, 37]. As a result, it seems to be highly advised for use in the curing of resins during manufacturing, either alone or in conjunction with reinforced composites, or as a way to cure specific regions of polymer bonds, and consequently for bonded joints. However, exposure to microwaves can cause resins' mechanical performance to decrease [38, 39], and the presence of carbon fibers that impede penetration suggests that microwave technology is not yet a dependable way to produce high-performance composite structures. Microwave curing must be completely controlled due to problems with the quality of composites produced using microwaves and variations in curing parameters (such as temperature, time, and radiation).

By sending microwaves through the laminate's thickness and imparting momentum to the material's molecules, microwave curing is accomplished. Heat is produced as a result of the molecules colliding with the microwave, which transfers energy. This heat aids in the matrix's polymerization, both in the beginning and the end. Heat is absorbed by the composite material during microwave curing, which fully polymerizes the matrix material. The material's dielectric characteristics determine its microwave curing effectiveness.

Altering the microwave's input parameters (frequency and amplitude) can reduce the cure time. For the mass manufacture of polymer composite components, a shorter curing time for composites can be advantageous. About 2.5 GHz is the typical frequency used for microwave curing. This method is widely accessible and reasonably priced. During curing, waves move in two distinct directions: transverse magnetically and transversely electrically. Whereas transverse electric only has electric intensity in the transverse direction, transverse magnetic has both longitudinal and transverse components.

3.2 *Electrical Resistance Heating*

Targeted heating of particular regions is made possible by the Electrical Resistance Heating Curing (ERHC) approach, which may lead to uniform curing and reduce overall energy use. Because of this, the ERHC process has significant application value in the production of high-quality, low-energy PMCs. The fundamental method of heating used in ERHC is to apply electric current to a resistance material. This produces an effect known as "Joule heating," also known as "ohmic heating" or "resistance heating," in which free electrons move in a specific direction when exposed to an external electric field. Conduction electrons can transmit energy when they pass through the collision; this tiny-scale heat generation and energy transfer cause the temperature to rise to produce the heating effect [40–42]. By passing an electric current through conductive fibers, it is possible to employ them as a heating element due to their electric conductive nature. The polymerization of the resin around the surface of conductive fibers is started by the generation of heat around them. The amount of current and resistance provided by the reinforcement or conductive fibers determine how long the curing process takes.

Radiation curing and thermal curing are two different processes used in various industries, including coatings, adhesives, and printing. Each method has its own set of advantages and disadvantages shown in Table 1. The choice between radiation curing and thermal curing depends on various factors, including the specific application, substrate materials, production speed requirements, and cost considerations. Each method has its strengths and weaknesses, and the optimal choice will vary based on the particular needs of the industry and the product being manufactured.

Table 1 Advantages and disadvantages of radiation and thermal curing

Curing	Advantages	Disadvantages
Radiation curing	**Speed**: Radiation curing, particularly UV (Ultraviolet) and EB (Electron Beam) curing, is extremely fast. It can cure fast, leading to high production efficiency **Energy Efficiency**: It typically requires less energy compared to thermal curing methods, as there is no need to heat an entire curing oven **Low VOC Emissions**: Radiation curing produces fewer volatile organic compounds (VOCs) because it does not require the use of solvents that are often needed in thermal curing **Precise Control**: It allows for precise control over the curing process, including the depth of cure and the level of cross-linking, which can result in improved performance and quality **Reduced Heat-Related Damage**: Ideal for heat-sensitive substrates as it doesn't subject materials to high temperatures	**Initial Cost**: The equipment required for radiation curing, especially for electron beam (EB) curing, can be expensive to install **Limited Substrate Compatibility**: Not all materials are suitable for radiation curing, and the substrate must be UV or EB transparent **Safety Concerns**: UV radiation can pose safety hazards to operators, and electron beam equipment may have radiation shielding requirements
Thermal curing	**Wider Substrate Compatibility**: Thermal curing can be used on a wider range of substrates, including heat-sensitive materials **Lower Initial Investment**: The equipment for thermal curing is often less expensive than radiation curing equipment **Simple Formulations**: Thermal curing is often easier to formulate, and fewer chemical additives may be needed	**Slower Curing Speed**: Thermal curing processes are generally slower than radiation curing, which can impact production efficiency **Higher Energy Consumption**: It requires higher energy consumption, as it involves heating ovens and the surrounding environment **Emission of VOCs**: Traditional thermal curing often involves the use of solvents, which can lead to higher emissions of volatile organic compounds **Reduced Control Over Cross-linking**: The degree of cross-linking and the depth of cure may be more challenging to control in thermal curing processes

4 New Challenges and Opportunities

Curing methods used for polymer composites play a crucial role in determining the properties, performance, and quality of the final composite materials. As technology and materials science continue to advance, new challenges and opportunities in curing methods for polymer composites emerge. Here are some of the key challenges and opportunities in this field.

4.1 Challenges

Energy Efficiency: The curing process for polymer composites often requires elevated temperatures, which can be energy-intensive. The challenge is to develop more energy-efficient curing methods to reduce costs and environmental impact.

Uniformity and Consistency: Achieving uniform curing and consistency throughout the polymer composite is a challenge, especially in large or complex-shaped components. Non-uniform curing can result in structural defects and reduced mechanical properties.

Curing Time: Long curing times can slow down production processes. Reducing curing time without sacrificing the quality of the cured polymer composite is a significant challenge.

Environmental Impact: Traditional curing methods may involve the use of hazardous chemicals or produce harmful emissions. There is a growing demand for more environmentally friendly curing processes.

Heat-Sensitive Materials: Some polymer composites, like those incorporating bio-based or thermally sensitive materials, may degrade or experience thermal damage during curing. Finding methods that can cure such materials without compromising their properties is a challenge.

4.2 Opportunities

Advanced Curing Technologies: New curing technologies, such as microwave curing, induction heating, and laser curing, offer opportunities for more precise and efficient curing of polymer composites. These methods can potentially reduce energy consumption and improve control over the curing process.

Nanomaterials and Additives: The incorporation of nanomaterials and additives can enhance the curing process. Nanoparticles, for example, can improve heat transfer and increase the thermal conductivity of composites, leading to faster and more efficient curing.

Sustainability: Opportunities exist for developing curing methods that use fewer or no hazardous chemicals, emit fewer volatile organic compounds (VOCs), and utilize renewable energy sources, contributing to a more sustainable and eco-friendly approach.

In-Situ Curing: In-situ curing methods, where curing takes place during the manufacturing process or at the point of use, can reduce transportation costs and lead to more customized and just-in-time production.

Automation and Process Control: Automation and advanced process control technologies, such as real-time monitoring and feedback systems, can improve the consistency and quality of curing, making it easier to achieve the desired properties in polymer composites.

Adaptive Curing: Developing curing methods that can adapt to the specific requirements of different composite materials and geometries offers opportunities for more tailored and efficient curing processes.

Hybrid Curing Approaches: Combining different curing methods, such as a combination of thermal and radiation curing, can provide enhanced control and flexibility in curing polymer composites.

Curing methods for polymer composites are evolving to address various challenges while taking advantage of emerging opportunities. Innovations in materials, processes, and technologies are continuously shaping the industry, enabling the production of high-performance, sustainable, and cost-effective composite materials. As technology and research progress, researchers and industry professionals need to stay updated and adapt to the changing landscape of curing methods for polymer composites.

5 Conclusion

Curing methods for polymer composites are a critical aspect of composite material production, impacting their mechanical properties, durability, and overall performance. In the ever-evolving landscape of curing methods for polymer composites, researchers, engineers, and manufacturers need to stay informed about the latest developments and adapt to emerging technologies and best practices. Ultimately, the choice of curing method should align with the unique requirements of the composite material, taking into consideration factors such as material compatibility, production efficiency, and sustainability goals. With ongoing research and innovation, the curing process for polymer composites is poised to become more efficient, environmentally friendly, and tailored to meet the demands of diverse industries and applications.

Declarations

Conflicts of Interest The authors confirm that there are no conflicts of interest associated with this work.

Ethical Compliance The authors affirm that this study adheres to all ethical guidelines.

Competing Interests No competing interests are declared by the authors.

Data and Material Availability Data and materials are not applicable to this study.

References

1. Egbo MK (2021) A fundamental review on composite materials and some of their applications in biomedical engineering. J King Saud Univ-Eng Sci 33:557–568. https://doi.org/10.1016/j.jksues.2020.07.007
2. Kumar P (2023) Recent progress in computational techniques in various coating materials. Mater Horiz: Nat Nanomater Part F1086:217–228. https://doi.org/10.1007/978-981-99-3549-9_10/COVER
3. Upadhaya D, Talinungsang P, Kumar DD (2019) Purkayastha, tuning the wettability and photocatalytic efficiency of heterostructure ZnO-SnO_2 composite films with annealing temperature. Mater Sci Semicond Process 95:28–34. https://doi.org/10.1016/J.MSSP.2019.02.009
4. Talinungsang D, Upadhaya P, Kumar DD (2019) Purkayastha, superhydrophilicity of photocatalytic ZnO/SnO_2 heterostructure for self-cleaning applications. J Solgel Sci Technol 92:575–584. https://doi.org/10.1007/S10971-019-05127-8/FIGURES/14
5. Upadhaya D, Kumar P, Dhar Purkayastha D (2019) Superhydrophobic ZnO/TiO_2 heterostructure with significantly enhanced photocatalytic activity. J Mater Sci Mater Electron 30:10399–10407. https://doi.org/10.1007/S10854-019-01381-2/FIGURES/10
6. Sethi SK, Manik G (2021) A combined theoretical and experimental investigation on the wettability of MWCNT filled PVAc-g-PDMS easy-clean coating. Prog Org Coat 151:106092. https://doi.org/10.1016/j.porgcoat.2020.106092
7. Gogoi R, Sethi SK, Manik G (2021) Surface functionalization and CNT coating induced improved interfacial interactions of carbon fiber with polypropylene matrix: a molecular dynamics study. Appl Surf Sci 539. https://doi.org/10.1016/j.apsusc.2020.148162
8. Gogoi R, Sethi SK, Manik G (2021) Surface functionalization and CNT coating induced improved interfacial interactions of carbon fiber with polypropylene matrix: a molecular dynamics study. Appl Surf Sci 539:148162. https://doi.org/10.1016/J.APSUSC.2020.148162
9. Sethi SK, Soni L, Manik G (2018) Component compatibility study of poly(dimethyl siloxane) with poly(vinyl acetate) of varying hydrolysis content: an atomistic and mesoscale simulation approach. J Mol Liq 272:73–83. https://doi.org/10.1016/J.MOLLIQ.2018.09.048
10. Sethi SK, Kadian S, Gogoi R, Manik G (2023) Layer-by-layer fabrication of self-cleaning superhydrophobic surface made from carboxymethylcellulose and ZnO quantum dots: a combined experimental and computational study. Surf Interfaces 37:102752. https://doi.org/10.1016/J.SURFIN.2023.102752
11. Sethi SK, Gogoi R, Verma A, Manik G (2022) How can the geometry of a rough surface affect its wettability?-A coarse-grained simulation analysis. Prog Org Coat 172:107062. https://doi.org/10.1016/J.PORGCOAT.2022.107062
12. Maurya AK, Gogoi R, Sethi SK, Manik G (2021) A combined theoretical and experimental investigation of the valorization of mechanical and thermal properties of the fly ash-reinforced polypropylene hybrid composites. J Mater Sci 56:16976–16998. https://doi.org/10.1007/S10853-021-06383-2/FIGURES/10
13. Dhiman A, Gupta A, Sethi SK, Manik G, Agrawal G (2022) Encapsulation of wax in complete silica microcapsules. J Mater Res 2022:1–14. https://doi.org/10.1557/S43578-022-00865-Y
14. Chaturvedi S, Verma A, Sethi SK, Mavinkere Rangappa S, Siengchin S (2022) Stalk fibers (rice, wheat, barley, etc.) composites and applications. In: Plant fibers, their composites, and applications, pp 347–362. https://doi.org/10.1016/B978-0-12-824528-6.00009-6
15. Khandelwal V, Sahoo SK, Kumar A, Sethi SK, Manik G (2019) Bio-sourced electrically conductive epoxidized linseed oil based composites filled with polyaniline and carbon nanotubes. Compos B Eng 172:76–82. https://doi.org/10.1016/j.compositesb.2019.05.050
16. Agrawal G, Samal SK, Sethi SK, Manik G, Agrawal R (2019) Microgel/silica hybrid colloids: bioinspired synthesis and controlled release application. Polym (Guildf) 178:121599. https://www.sciencedirect.com/science/article/pii/S003238611930583X. Last Accessed 24 June 2019
17. Shankar U, Sethi SK, Singh BP, Kumar A, Manik G, Bandyopadhyay A (2021) Optically transparent and lightweight nanocomposite substrate of poly(methyl methacrylate-co-acrylonitrile)/

MWCNT for optoelectronic applications: an experimental and theoretical insight. J Mater Sci 56:17040–17061. https://doi.org/10.1007/S10853-021-06390-3/FIGURES/17
18. Dixit S, Goel R, Dubey A, Shivhare PR, Bhalavi T (2017) Natural fibre reinforced polymer composite materials-a review. Polym Renew Resour 8:71–78. https://doi.org/10.1177/204124791700800203
19. Lee DW, Yoo BR (2016) Advanced silica/polymer composites: materials and applications. J Ind Eng Chem 38:1–12. https://doi.org/10.1016/j.jiec.2016.04.016
20. Hsissou R, Seghiri R, Benzekri Z, Hilali M, Rafik M, Elharfi A (2021) Polymer composite materials: a comprehensive review. Compos Struct 262:113640. https://doi.org/10.1016/j.compstruct.2021.113640
21. Kangishwar S, Radhika N, Sheik AA, Chavali A, Hariharan S (2023) A comprehensive review on polymer matrix composites: material selection, fabrication, and application. Springer, Berlin. https://doi.org/10.1007/s00289-022-04087-4
22. Oladele IO, Omotosho TF, Adediran AA (2020) Polymer-based composites: an indispensable material for present and future applications. Int J Polym Sci 2020. https://doi.org/10.1155/2020/8834518
23. Kumar VV, Balaganesan G, Lee JKY, Neisiany RE, Surendran S, Ramakrishna S (2019) A review of recent advances in nanoengineered polymer composites. Polym (Basel) 11. https://doi.org/10.3390/polym11040644
24. Neisiany RE, Khorasani SN, Kong Yoong Lee J, Ramakrishna S (2016) Encapsulation of epoxy and amine curing agent in PAN nanofibers by coaxial electrospinning for self-healing purposes. RSC Adv 6:70056–70063. https://doi.org/10.1039/c6ra06434e
25. Neisiany RE, Lee JKY, Khorasani SN, Ramakrishna S (2017) Towards the development of self-healing carbon/epoxy composites with improved potential provided by efficient encapsulation of healing agents in core-shell nanofibers. Polym Test 62:79–87. https://doi.org/10.1016/j.polymertesting.2017.06.016
26. Wang Y, Tao F, Zuo Y, Zhang M, Qi Q (2023) Digital twin enhanced fault diagnosis reasoning for autoclave. J Intell Manuf. https://doi.org/10.1007/s10845-023-02174-5
27. Manral A, Ahmad F, Sharma B (2019) Advances in curing methods of reinforced polymer composites. In: Reinforced polymer composites, pp 77–104. https://doi.org/10.1002/9783527820979.ch5
28. Smirnov YN, Allayarov SR, Lesnichaya VA, Ol'khov YA, Belov GP, Dixon DA (2009) The effect of gamma-radiation on polymer composites based on thermoplastic matrices. High Energy Chem 43:449–455. https://doi.org/10.1134/S001814390906006X
29. Ghosh SK, Das TK, Ghosh S, Ganguly S, Nath K, Das NC (2021) Physico-mechanical, rheological and gas barrier properties of organoclay and inorganic phyllosilicate reinforced thermoplastic films. J Appl Polym Sci 138:1–16. https://doi.org/10.1002/app.49735
30. Siddhartha S, Aarya K, Dev SK, Raghuvanshi JBM, Krishna MA (2012) Wahab, Effect of gamma radiation on the structural and optical properties of polyethyleneterephthalate (PET) polymer. Radiat Phys Chem 81:458–462. https://doi.org/10.1016/j.radphyschem.2011.12.023
31. Sheel Wali A, Kumar S, Khan D (2023) A review on recent development and application of radiation curing. Mater Today Proc 82:68–74. https://doi.org/10.1016/j.matpr.2022.11.342
32. Dadkhah S, Gharieh A, Khosravi M (2024) Eco-friendly UV-curable graphene oxide/fluorinated polyurethane acrylate nanocomposite coating with outstanding anticorrosive performance. Prog Org Coat 186:108020. https://doi.org/10.1016/j.porgcoat.2023.108020
33. Wang Q, Yang Z, Feng X, Liu X (2023) Modification of nanocellulose via atom transfer radical polymerization and its reinforcing effect in waterborne UV-curable resin. Int J Biol Macromol 253:126743. https://doi.org/10.1016/j.ijbiomac.2023.126743
34. Xiao T, Geng L, Dai Y, Zhao J, Liu C (2023) UV-cured polymer aided phase change thermal energy storage: preparation, mechanism and prospects. J Energy Storage 64:107066. https://doi.org/10.1016/j.est.2023.107066
35. Jiang B, Shi X, Zhang T, Huang Y (2022) Recent advances in UV/thermal curing silicone polymers. Chem Eng J 435:134843. https://doi.org/10.1016/j.cej.2022.134843

36. Mijović J, Wijaya J (1990) Review of cure of polymers and composites by microwave energy. Polym Compos 11:184–191. https://doi.org/10.1002/pc.750110307
37. Kwak M, Robinson P, Bismarck A, Wise R (2015) Microwave curing of carbon–epoxy composites: penetration depth and material characterization. Compos Part A Appl Sci Manuf 75:18–27. https://doi.org/10.1016/j.compositesa.2015.04.007
38. Boey FYC, Yap BH (2001) Microwave curing of an epoxy–amine system: effect of curing agent on the glass-transition temperature. Polym Test 20:837–845. https://doi.org/10.1016/S0142-9418(00)00070-2
39. Barkoula NM, Alcock B, Cabrera NO, Peijs T (2008) Flame-retardancy properties of intumescent ammonium poly(phosphate) and mineral filler magnesium hydroxide in combination with graphene. Polym Polym Compos 16:101–113. https://doi.org/10.3390/polym6112875
40. Kim M, Sung DH, Kong K, Kim N, Kim B-J, Park HW, Park Y-B, Jung M, Lee SH, Kim SG (2016) Characterization of resistive heating and thermoelectric behavior of discontinuous carbon fiber-epoxy composites. Compos B Eng 90:37–44. https://doi.org/10.1016/j.compositesb.2015.11.037
41. Noh SH, Eom W, Lee WJ, Park H, Ambade SB, Kim SO, Han TH (2019) Joule heating-induced sp2-restoration in graphene fibers. Carbon N Y 142:230–237. https://doi.org/10.1016/j.carbon.2018.10.041
42. Silva VLM, Santos LMNBF, Silva AMS (2017) Ohmic heating: an emerging concept in organic synthesis. Chem-A Eur J 23:7853–7865. https://doi.org/10.1002/chem.201700307

Chapter 3
Multiscale Modelling of Polymer Composites

Dheeraj Gunwant and Neeraj Bisht

1 Introduction

Polymers have been used widely in industries worldwide. As the technology advanced, new materials with specific desired properties were required, and scientists started developing composites with polymers as matrix material. Composites are materials with a matrix and reinforcements in the form of fibres. These composites were observed to have superior properties to conventional polymers [1–4]. The types and nature of fibres used with polymers have been phenomenal. Fibres have been used in the nanoscale to microscale, ranging from carbon fibres [5, 6] and graphene [7, 8] to natural fibres [9, 10].

The composites developed have proved their efficacy in numerous applications. These composites have been used extensively in the aerospace and automotive industries, film and packaging industries, and electronics industries. Any application, however, will depend upon the mechanical characteristics of the composites developed. Experimental investigation is one of the methods to determine these properties and has been heavily relied upon in the past decade. But experiments are expensive and require a lot of specialised machines. Sometimes, they can be time-consuming, like in weathering or biodegradation studies. Developing powerful computers and computation techniques has led to extensive use of computational methods in the mechanical characterisation of these composites.

D. Gunwant (✉)
Department of Mechanical Engineering, Apex Institute of Technology, Rampur, UP 244901, India
e-mail: mailto.dheerajgunwant@gmail.com

Department of Mechanical Engineering, Invertis University, Bareilly, UP 243001, India

N. Bisht
Department of Mechanical Engineering, Govind Ballabh Pant University of Agriculture and Technology, Pantnagar, Uttarakhand 243145, India

S. K. Sethi et al. (eds.), *Polymer Composites: From Computational to Experimental Aspects*, Materials Horizons: From Nature to Nanomaterials,
https://doi.org/10.1007/978-981-97-0888-8_3

Further, it saves a lot of effort and money by predicting the properties of any novel materials in their ideation stage and only after desired properties are achieved in the actual development of the composites. The computational method is a broad topic covering methodologies at various lengths and time scales. Different computational methods have been used so far for the systems of polymer composites which can be roughly divided into three types: **molecular scale methods** (e.g. molecular dynamics (MD), Monte Carlo (MC)), **microscale methods** (e.g. Brownian dynamics (BD), dissipative particle dynamics (DPD), lattice Boltzmann (LB), time-dependent Ginzburg–Landau method, dynamic density functional theory (DFT) method) and **mesoscale and macroscale methods** (e.g. micromechanics, equivalent-continuum and self-similar approaches, finite element method (FEM)). The next chapter focusses on various researches undertaken in this field.

2 Molecular Scale Methods

The basic units considered at the molecular level are atoms, molecules or their clusters. The reliance is on molecular mechanics, MD, and coarse-grained simulation at the atomistic or molecular level. Molecular models encompassing thousands and perhaps millions of atoms can be solved by these methods and used to predict fundamental, molecular-level material behaviour. The methods are both static and dynamic. For example, molecular mechanics can establish the minimum-energy structure statically, and MD can resolve the nanosecond-scale evolution of a molecule or molecular assembly. These approaches can model bonded and non-bonded forces (e.g. van der Waals and electrostatic) but are not parameterised for bond cleavage.

3 Simulation Methods

The computational methods employed to analyse composite materials are categorised into particle-based and field-based techniques. In the particle-based approach, the polymer, viz. atoms, molecules and monomers or even the polymer chain, is represented as the particles. The interaction of these particles forms a force field. These particles are forced to move within a particular thermodynamic system, simulating a desired process. Two of the most popular particle-based approaches are MD and DPD simulations. Alternatively, the field-based approach mainly comprises the description of the system in the form of electric potential, density fields and dynamic variables. The polymer reference interaction site model is developed for assessing the polymer structure in terms of the density correlation function.

4 Atomic/Molecular-Level Techniques

4.1 Monte Carlo Methods

Monte Carlo methods are indispensable in the comprehensive analysis of polymer composites, offering valuable insights into their thermodynamic behaviour, structural characteristics, and molecular interactions. These simulations play a pivotal role in the prediction of essential material properties, encompassing thermodynamics, mechanical strength [11], electrical conductivity [12], and thermal performance [13]. This probabilistic simulation technique deals with comprehensive fibre distribution within the matrix, performing several calculations to predict material behaviour [14]. By stochastically sampling polymer chains and filler particle configurations, researchers can estimate thermodynamic properties like heat capacity and free energy and predict composite behaviour across temperature and pressure. Monte Carlo simulations also illuminate composite spatial arrangements, filler dispersion, interfacial interactions, and morphology. These methods reveal molecular interactions, assessing adhesion forces and constituent compatibility. Monte Carlo methods allow researchers to optimise composite compositions and processing parameters, enabling them to tailor polymer composites to specific application requirements while minimising cost and energy consumption.

The progressive damage accumulation in fibre-reinforced composites affects their bulk mechanical properties. The understanding of the mechanical behaviour of these composites is incomplete without knowledge about multi-damage evolution throughout the loading duration. Moreover, the involvement of many interconnected phenomena makes the study of the progressive damage of these composites a complex procedure [15]. Modelling of failure behaviour of fibre-reinforced composites requires accurate information regarding the load transfer. The FEM provides a natural choice for determining complex stresses and strains in such materials. Unfortunately, applying FEM to handle the demands of 3D models of fibre-reinforced composites requires high computational efficiency. Micromechanics-based 2D theoretical models for investigating the tensile response of fibre/polymer composites have been reported by Wang et al. [11]. The model considered statistical variations of the matrix shear failure and fibre strength. These researchers proposed an MC method-based simulation scheme to simulate the progressive failure behaviour of the composites. The MC simulation revealed increased stress with the applied strain up to a certain point, followed by reduction due to the dominant microdamage mechanisms. The normalised tensile strength predicted from 200 simulations of varying sized specimens is depicted in the figure—the tensile strength of composites reduced with increased fibres. The strength of a 180-fibre composite specimen was 4% lower than the 10-fibre specimens. The MC simulation revealed a fracture of the weakest fibre at a fracture strain of 0.9%.

A limited number of studies have been reported regarding the application of MC methods in the composite's simulation. A 3D simulation of the 'hard' and 'soft' rectangular platelet-reinforced nanocomposites has been reported Wang and Jayatissa

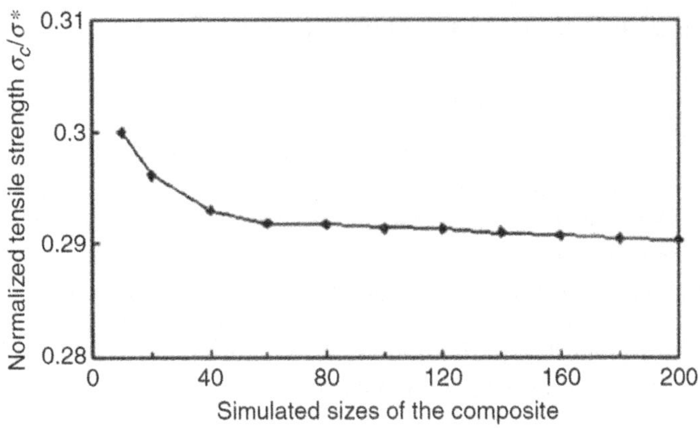

Fig. 1 Effect of simulated specimen size on the normalised tensile strength of the composites [11]

[16]. Similarly, a 3D model was developed to analyse the composites with 2D filler nano-discs [17]. The effect of graphene orientation on the nanocomposite conductivity has also been reported Zabihi and Araghi [18]. Electrical conductivity enhancement is one of the most extraordinary characteristics of graphene/polymer composites (Fig. 1).

However, determining this property is a complex process requiring accurate analysis of the underlying conduction phenomenon. A Monte Carlo simulation was conducted to determine the electrical conductivity of graphene-based composites having a randomly distributed graphene nanoplatelet network [19]. The contact conductance of the network was calculated through the tunnelling effect, followed by the calculation of the nanoplatelets to polymer current flow. The graphene nanoplatelet network was modelled as a short cylindrical element with a short graphene sheet stack. The MC simulation revealed the tunnelling effect between the adjacent nanoplatelets, significantly improving the composite conductivity. The figure depicts the graphene nanoplatelet arrangement within the RVE and the position and orientation of a single nanoplatelet.

One of the problems associated with the composites is the mechanical deterioration of the polymer upon increasing the filler content. Therefore, extensive experimentation may not be sufficient to determine their electrical conductivity. The MC method is a statistical technique for generating random position and shape parameters for polymer-based composites. Parametric analysis of the existing modelling approaches for determining the electrical conductivity of polymeric composites has been reported [20].

The MC simulations investigated the effect of coating layer, fibre-aspect ratio, orientation angle, and solidity on the electrical conductivity. The authors highlighted the suitability of MC methods in predicting electrical conductivity accurately. The thermal conductivities and effective stress–strain curves for CNT/polymer composites were obtained using the FE and MC methods [21] (Fig. 2).

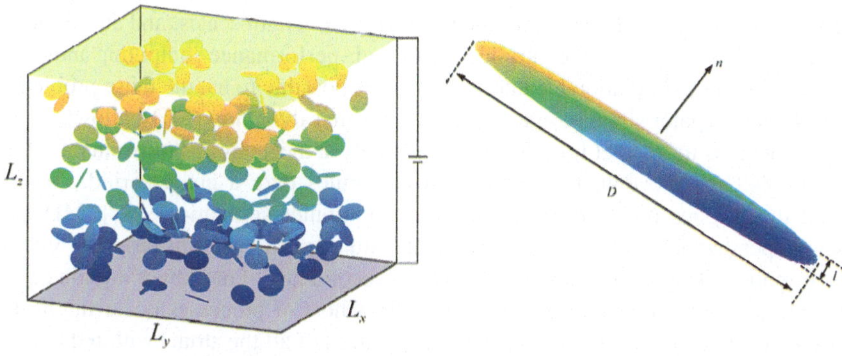

(a) (b)

Fig. 2 Schematic showing the graphene nanoplatelet arrangement within the RVE and **b** orientation and position of the nanoplatelet [19]

Factors such as CNT geometry, alignment, and waviness were considered to improve the prediction accuracy. The CNT interconnectivity was determined using the spanning network algorithm in the RVE. MC simulations of 500 randomly generated microstructures were conducted to record the stochastic fibre generation process, simultaneously obtaining accurate results. The simulation-obtained results were compared to the experimental results from the literature, and an accuracy of 5% was observed. The accuracy decreased for CNT volume fractions >3% due to the complex mechanisms active within the composites at high concentrations. These complex mechanisms were able to affect the thermal conduction at the CNT/polymer interface negatively. The authors used MC simulation to reduce statistical variation in composite thermal, electrical, and mechanical properties.

MC methods are widely employed for modelling polymer composites, serving as versatile tools to simulate and analyse the behaviour of these materials. These methods involve generating random configurations of polymer and filler particles and then calculating the corresponding macroscopic properties through statistical averaging. Using stochastic sampling in MC simulations, researchers can predict mechanical strength, thermal conductivity, and electrical behaviour in complex polymer composite systems. MC simulations facilitate the exploration of broad parameter space, offering valuable guidance for designing and optimising polymer composites with tailored properties.

5 Molecular Dynamics Simulations

Experimental capturing of the molecular and atomic motion in a material is impossible. Moreover, using analytical techniques for determining these motions is also impossible due to the several particles in any molecular system. Researchers struggle understanding material behaviour due to short time, length scales, and atomic-level dynamics. However, with the introduction of high-performance computing and associated computer simulation techniques, it is now possible to handle these problems. One such computer simulation technique, molecular dynamics (MD), allows atoms and molecules to interact for a fixed time to analyse their physical movements. MD can predict the movement of atoms and molecules over a given period, considering the physics involved via atomic-scale computer simulations [22]. MD is a time-dependent technique for the numerical simulation of materials based on atomic coordination. The method employs Newton's second law of motion or motion equations to simulate molecular movements [23]. A host of processes occurring at the atomic level can be captured, revealing the position of all the atoms within the resolution of femtoseconds [24]. MD simulations, combined with X-ray crystallography, nuclear magnetic, and electron paramagnetic resonance, can capture the position of every atom at each time point.

Polymeric materials are subjected to high and low temperatures, abrasive wear, and corrosion, rendering them vulnerable to catastrophic failure [25, 26]. It becomes imperative for the designer to explore different failure mechanisms in these polymer-based materials and the methods to mitigate them. Accurate information about these failure mechanisms can simultaneously improve the mechanical properties of such materials. Experimental and analytical tools and techniques for analysing polymer-based materials have enabled researchers to decipher their behaviour [27]. But, applying these techniques is cumbersome, labour-intensive, time-consuming and inefficient. Besides the experimental approach, continuum mechanics-backed theoretical analysis plays a vital role in analysing various atomic-level phenomena in these materials. The exploration of atomic-scale interaction is formidable for advancing the knowledge about polymeric materials.

Due to their exceptionally high mechanical properties, graphite/epoxy composites are the materials of choice in the aerospace industry. It is well known that the fibre/matrix adhesion plays an essential role in determining the overall mechanical properties of the composite materials. A stronger interfacial implies stiffer and stronger material with lower toughness due to the absence of a crack deflection mechanism. Contrary to this, a weak interface means lower strength and stiffness but higher toughness properties [28]. Experiments provide limited insight into the fibre/matrix physicochemical interaction, necessitating the application of molecular models for tailoring composites with desirable properties. Several researchers have attempted to model the thermosetting systems using MD simulations with Bandyopadhyay et al. [29] and without nanotubes [27, 30], nanoparticles [31], and nanofibres [32, 33]. A course-grained model for simulating the interfacial fracture between a solid body and a highly cross-linked thermoset plastic has been reported [34]. MD simulations identified a 10-Å-thick interfacial region different from the bulk polymer at the interface [35]. The effect of cross-link density on the molecular structure of fibre/matrix in graphite-epoxy composites has been studied by Hadden et al. [36]. The MD simulations are determined the residual stresses, molecular potential energy, and the interfacial region's mass density. These authors also identified a surface region of thickness 10 Å near fibre with a bulk density different from the mass density

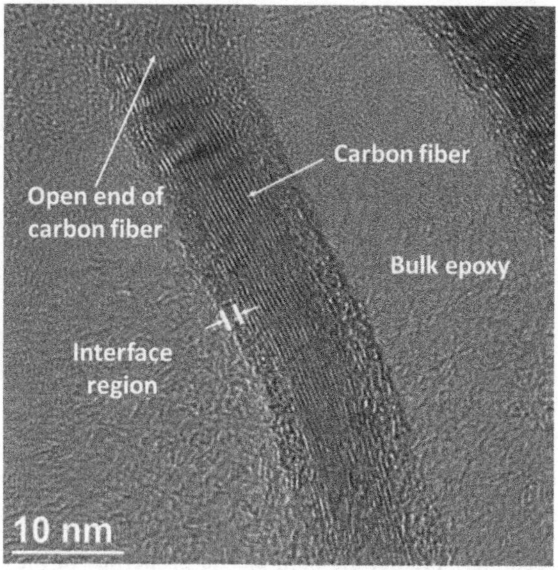

Fig. 3 TEM image of the carbon fibre embedded in the epoxy matrix [36]

of the bulk material. This finding was validated with the TEM images showing the interfacial area of thickness approximately 10 Å near the graphite fibre (see figure). The residual stresses and a higher molecular energy level were higher in this region than in the bulk material. Moreover, no effect of cross-link density was noticed on these stresses. Interestingly, the magnitude of this energy was observed to decrease on increasing the polymer cross-linking density (Fig. 3).

Carbon nanotubes serve as ideal reinforcements for nanocomposite fabrication. MD simulations were performed on nanotubes with different ethylene-di-amine molecule densities to determine how amine functionalisation affects composite Young's, bulk and shear moduli and stresses [30]. MD simulations revealed a reduction in Young's modulus in the carboxylic-functionalised CNTs for different E-NH_2 numbers. The negative effect of amine functionalisation on the elastic modulus is attributed to the introduction of sp^3 carbon in the sp^2 network, leaving undulations on the surface. Contrarily, the tensile strength improved with the number of functionalised groups due to the restriction in the lateral deformation in MWCNTs under tensile loading. CNTs tend towards aggregation and entanglement due to intrinsic van der Waals forces, higher surface energy, and fine size. This drawback can be overcome using functionalised nanotubes providing multiple bonding sites to the epoxy matrix for efficient load transfer.

A continuum-based micromechanical model was created for silica nanoparticle–polyimide composites with different interfacial treatments using molecular modelling with coarse-grained and reverse mapping [37]. The figure shows the molecular model of the silica nanoparticle/polyimide composite. Various molecular structures, viz. interfacial zone, polyimide, and nanoparticle, were incorporated into the model to enhance intuitiveness and accuracy. The presented model could

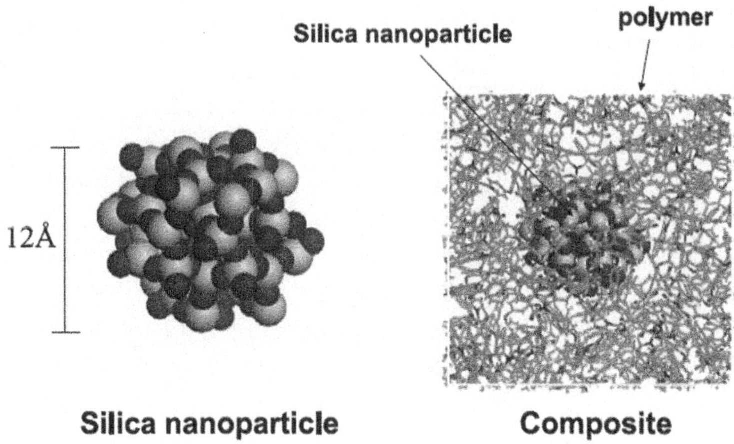

Fig. 4 Molecular model of silica nanoparticle/polyimide composite [37]

accurately predict the bulk elastic properties of the composites with nanoparticle radii ranging from 10 to 1000 Å. For radii >1000 Å, the predicted properties were equal to those predicted by the micromechanical model presented by Mori–Tanaka. Young's and shear moduli of composites increased on increasing silica particle size asymptotically approaching the moduli indicated using the micromechanics-based approach of Mori–Tanaka (Fig. 4).

Moreover, a reduction in the composite's stiffness loss was observed using the MD studies due to the chemical functionalisation between silica and polyimide molecules. This reduction implies enhancement in the silica/polyimide bonding following the functionalisation.

MD simulations have been successfully employed to estimate Young's modulus, solubility parameter, and the glass-transition temperature of SWCNTs/epoxy composites [38]. The build polymer tool available in the software was employed to generate the isolated epoxy chains containing 11 monomers. The amorphous cell module was used to determine the mechanical properties of composites based on SWCNT diameter and composition. The cross-linking procedure was simulated by manually linking the appropriate bonds between the curing agent (ethylene tetra-amine) and epoxy resin chains. The packing stage was performed at a minimal density of 0.01 g/cc for each model due to the presence of the aromatic elements. The simulations revealed the positive impact of SWCNT reinforcement (diameter 10–14 Å) on the composites' glass-transition temperature and mechanical integrity. MD simulations were conducted for graphene/polymer composites to investigate their tribological and mechanical properties [27]. Figure shows the molecular conformations of the various models generated for the MD simulations. An MD model was developed using the condensed-phase-optimised molecular potential for atomistic simulation studies (COMPASS) to predict the composites' tribo-mechanical properties (Fig. 5).

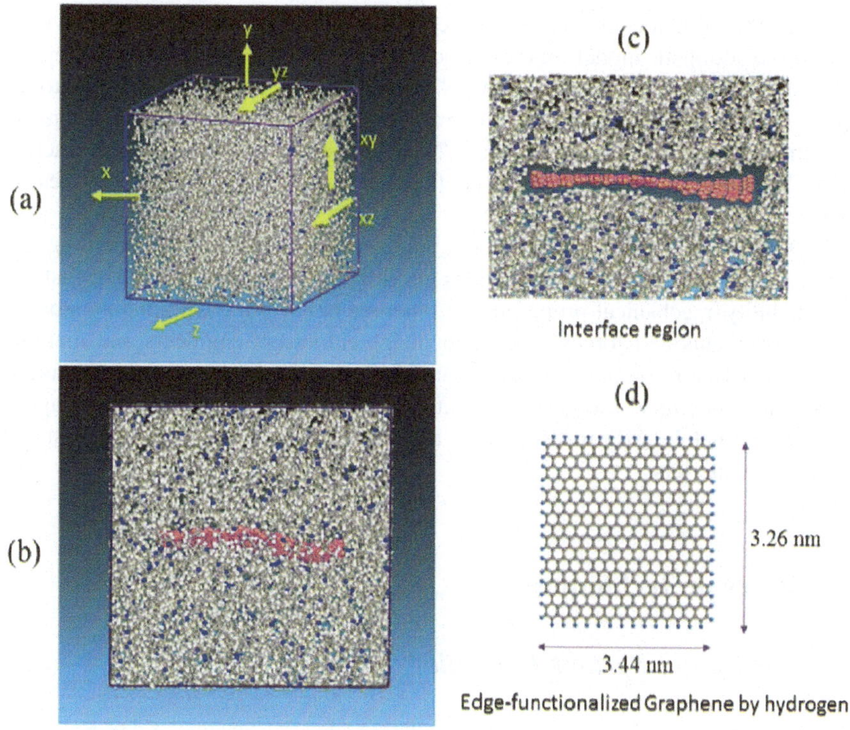

Fig. 5 Molecular conformations of the **a** unreinforced polymer, **b** graphene-reinforced polymer, **c** interface region and **d** edge-functionalised graphene by hydrogen [27]

The periodic boundary conditions were applied in all the molecular models to calculate the properties. The non-bonded interactions were incorporated using the hydrogen bonds, long-range electrostatic interactions, and the van der Waals forces. The strain constant method and layer model were applied to determine the mechanical and tribological properties of the composites, respectively. The packing was accomplished using the amorphous cell packing task in the Accelrys Materials Studio software. The MD simulations showed graphene reinforcement improved composite Young's modulus, shear modulus, and hardness by 150%, 26.7%, and 35%, respectively. Contrarily, 48% and 35% reductions were observed in the abrasion and average friction constant of the polymer composites due to the graphene reinforcement. CNT-reinforced nanocomposites find extensive utilisation in high-temperature ranges due to their thermal resistance. Therefore, the determination of the thermal performance of these materials is imperative for the designers. T_g and thermal expansion coefficient are the primary indicators of polymeric materials' thermal resistance and usable temperature range. Techniques such as FEM can be used to investigate these properties, but such methods cannot be used directly for simulating interatomic/intermolecular forces. In this regard, an MD simulation was performed to determine the coefficient of thermal expansion and T_g of a thermoset polymeric/CNT nanocomposite [23]. The COMPASS27 force field was employed to develop the atomistic models of the diglycidyl ether of bisphenol A (DGEBA) epoxy and diethylenetriamine (DETA). The CNTs were modelled to be embedded into the epoxy matrix. The model consisted of various embedded in the epoxy. The model was utilised to investigate the effect of CNT diameter, chirality and volume fraction on the T_g and the thermal expansion coefficient. The CNTs showed effectiveness in reducing the CTE vis-à-vis neat epoxy.

MD studies are pivotal in advancing polymeric composites' understanding and performance optimisation. MD provides insights into polymeric composites' structural stability, mechanical properties, and thermal behaviour by simulating atomic-scale movements and forces. Such knowledge empowers engineers and material scientists to design composites with enhanced strength, durability, and thermal resistance, which is critical for aerospace, automotive, and structural engineering applications. Ultimately, MD investigations are crucial in seeking superior polymeric composite materials.

6 Mesoscale Techniques

6.1 Dissipative Particle Dynamics

Dissipative Particle Dynamics (DPD) is an off-lattice mesoscopic simulation technique involving particles moving in continuous space and discrete time proposed by Hoogerbrugge and Koelman [39]. It is a powerful tool for simulating and understanding the behaviour of composite materials by representing their components as particles, modelling the interactions between these components, and studying their mechanical and thermodynamic properties. The DPD technique is a particle-based method employed to simulate the hydrodynamic phenomena. DPD applied to composite materials involves using these simulations to represent different components or phases within the composite as particles and studying their interactions and movements. It provides insights into composite materials' mechanical behaviour, thermodynamics, and properties, aiding in designing and optimising composite structures for various applications, particularly in industries like aerospace and automotive.

Elliot and Windle presented a model for representing the geometrical packing pattern of polydisperse materials for both non-spherical and spherical filler particles [40]. These authors applied the DPD technique for modelling soft spheres fused rigidly to generate particulates of different shapes and sizes. They focussed on developing a generalised methodology for investigating the geometrical packing of filler particles having irregular shapes within composites. The DPD technique has also been used for studying the effect of polyethene and poly (l-lactide) on the properties of the PE/PLLA/CNTs in a system [41]. The compressibility and mixing energy parameters were obtained using the MD simulations and transferred by the coarse-grained mapping techniques. The overall modelling methodology consists of MD simulation followed by coarse-grained mapping and DPD simulation.

These authors reported better dispersion of CNTs into polymer matrix at lower CNT concentrations. Contrarily, the polymer-CNT repulsive parameter values increased at higher CNT concentrations, leading to their aggregation. In another study, the dynamics and morphology of CNTs dispersed in a polycarbonate matrix in varying ratios were investigated using DPD simulations. The atomistic model was employed for determining polycarbonate-CNT interaction parameters for performing the mesoscale DPD simulations [42]. These authors attempted to investigate the CNT diffusivity within the matrix for different temporal simulation regimes. The atomistic simulations revealed CNTs to be roughly four times the monomer. Due to this ratio, the matrix monomer was coarse-grained as one and four DPD beads, respectively. The figure represents the coarse-graining scheme used for the simulation. The CNT dispersion slowed when the bundling effect was set in all three concentrations. The figure shows that the small aggregates dispersed discretely into the matrix for the 2% equilibrated mixture (Fig. 6).

DPD was employed to predict the mechanical properties of SWCNT/polyisoprene composites [43]. The computational efficiency was optimised using coarse-grained DPD-based simulations on the large-scale Atomic/Molecular Massively Parallel Simulator. A single bead consisting of 40 DPD beads represented the isoprene monomer unit. The graphite carbon ring was coarse-grained as linear 20 DPD

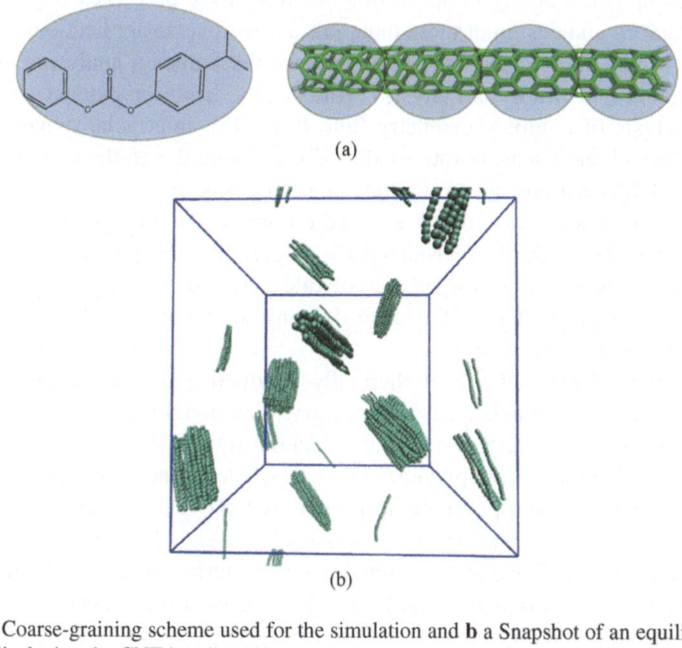

Fig. 6 **a** Coarse-graining scheme used for the simulation and **b** a Snapshot of an equilibrated 2% mixture displaying the CNT bundles [42]

beads. These authors observed restrictions on the motion of polymer movement manifested by decreased end-to-end distance and mean square displacement due to CNT addition. This restricted polymer motion due to SWCNT entanglement was the leading cause of improved mechanical properties. The sulphur cross-linking between polymer chains was represented via a single-site bifunctional particle interlinking two polymer chains.

6.2 Lattice Boltzmann

The lattice Boltzmann method (LBM) is a computational approach to simulate fluid flow and transport phenomena, often in polymer composites. It is a mesoscale computational fluid dynamics technique based on the statistical distribution of the lattice [44]. The method enables the modelling of fluid behaviour within and around composite materials with complex geometries and porous structures. LBM finds application in analysing polymer composites by providing insights into various aspects, such as the transport of fluids like polymer matrices or solvents within the composite, assessing permeability and diffusion properties, predicting interactions between fluids and composite materials, understanding manufacturing processes, including resin flow during composite fabrication, and even analysing the thermal characteristics and microstructure of these materials. Through LBM, researchers and engineers can gain valuable insights into how fluids interact with and move through polymer composites, aiding in optimising manufacturing processes and enhancing the performance and design of these materials across diverse applications.

The LBM has emerged as an attractive numerical scheme for analysing fluid flow simulation and modelling fluid physics. This method has been extensively utilised for the analysis of complex geometry fluid flow [45], interfacial dynamics [46], and flow through multiscale porous media [47]. Compared with the traditional CFD method, the LBM demonstrates a clear advantage in cases involving several interfaces and complex geometric features. It is based on the micro- and mesoscopic kinetic equations in which the fluid is considered a set of particles. The particles interact with the lattice according to a specific set of basic rules. Despite the simple representation of the reality in LBM, it is sufficient in deriving the most crucial features of the complicated fluid flow problem.

Due to the widespread use of thermally conductive particle-filled polymeric composites in the thermos-electronic industry, conductivity prediction is crucial for the scientific community. A multiple-relaxation-time LBM coupled with a half-lattice division scheme was employed to predict the heterogeneous materials' thermal conductivity with anisotropic material properties [48]. The traditional LBM was extended to deal with the anisotropic component interfaces. This technique obtained the local temperature distribution, maintaining the interfacial heat flux and temperature continuity. The simulation results agreed well with the experimental results, validating the applicability of the LBM approach. The internal braiding angle and

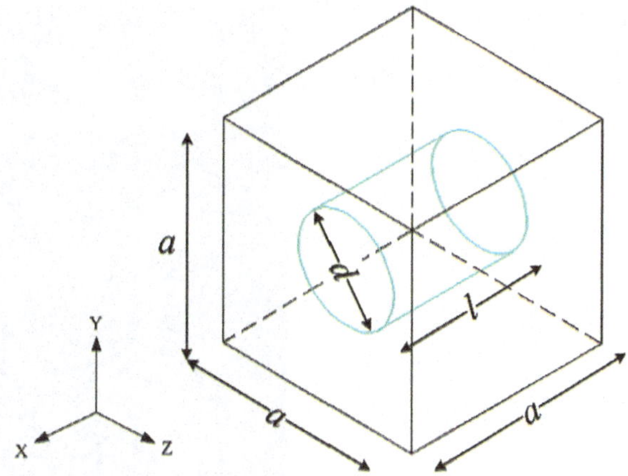

Fig. 7 Cell geometry reinforced with a single short fibre [48]

fibre volume fraction negatively and positively affect the effective conductivity. The figure shows the cell geometry employed in the simulation (Fig. 7).

Many interfaces and nanoscale particles increase the contact thermal resistance, challenging accurate thermal predictions. An LBM-based modelling approach was proposed for predicting carbon black/polymer composites' thermal conductivity [44]. The article introduced a method for reconstructing numerical material with similar filler distribution characteristics to actual material. The LBM technique solved the heat equation in a two-dimensional rectangular discrete structure. The conductivity of each phase and the temperature distribution within each rectangle were employed to determine the overall conductivity of the composite material. The LBM-based predictions aligned well with both theoretical and experimental data. In another work, the researchers attempted to determine the interfacial thermal contact resistance through the LBM approach. For this, a real contact resistance was introduced into the model by re-deriving the heat distribution function. The scheme was used to determine the effective thermal conductivity of the composites at different filler concentrations, particle size and thermal resistance.

The composites' random particle size and location distribution were generated via the quartet generation set algorithm by Xie et al. [49]. Every particle generated by this technique was assigned a unique number to differentiate it from others. This numbering was done to make the identification of particles easy. The LBM-based simulations were performed in a 200 × 200 grid 2D domain of 0.2 × 0.2 mm^2 dimensions. The boundary temperatures were considered to be 10 °C at the top and 0 °C at the bottom, with the left and right sides being adiabatic. These authors reported a decrease in the thermal conductivity in the case of low contact resistance and higher particle size. The results are congruent with the bulk properties of the composite materials. The figure shows the microstructure of the composite generated using the discussed scheme (Fig. 8).

The use of Lattice LBM in the analysis of polymer composites has emerged as a powerful and versatile tool in materials science. LBM is uniquely suited for

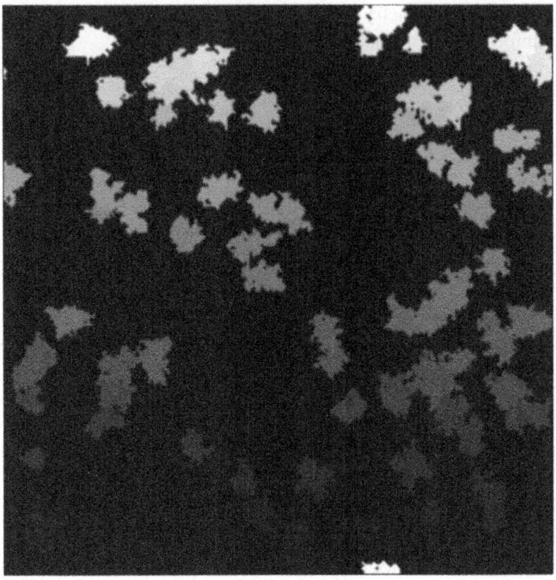

Fig. 8 Composite microstructure generated with particle volume fraction 0.2 [49]

modelling the complex fluid dynamics and transport phenomena often occurring in polymer composite systems. By employing a mesoscopic approach, LBM can simulate the behaviour of polymers at the molecular level, capturing the intricate interactions between the polymer matrix and reinforcing materials such as nanoparticles, fibres, or fillers. This level of detail allows researchers to gain insights into these composites' dispersion, distribution, and mechanical properties, ultimately leading to designing materials with enhanced performance and tailored properties. Furthermore, LBM's computational efficiency and scalability make it an attractive choice for studying large-scale systems. It provides predictions for real-world applications, from lightweight automotive components to advanced aerospace materials.

6.3 Brownian Dynamics

Brownian dynamics (BD) is a mathematical approach for describing the dynamics of molecular systems in the diffusive regime. It is a technique for computer simulations of physical systems driven primarily by thermal fluctuations. The BD simulations play a crucial role in the study and development of polymer composites. Simulating nanoparticle movement and interaction in the composite helps researchers understand dispersion, mechanical properties, thermal and electrical conductivity, barrier properties, and chemical interactions. They also aid in predicting how these composites behave under stress or in specific environments, enabling the design of materials tailored to particular applications. Simulations help optimise polymer composites for aerospace, automotive, electronics, and other industries, saving money and speeding up product development.

Several composite manufacturing techniques involve the consolidation of pre-impregnated preforms. These preforms are available in thin sheets or tapes; the final composite material is formed by putting one such tape over another. However, the optimal performance of manufactured components is ensured via intimate contact between the different layers of these preforms [50]. This intimate contact further provides adequate performance of the overall composite material. If the temperature and pressure are sufficient, the asperities flow, forming a close contact between the two surfaces [51]. Ideal bulk properties are achieved at the contact surface when the desired molecular mobility is obtained, ensuring the molecular reptation at the interface.

The dynamic behaviour of nanographene sheets embedded in a polymer matrix was investigated via the united-atom MD simulations [52]. The nanographene sheet Brownian motion, translational anisotropy, and oriental diffusion were also studied. The study's main aim was to gain insight into the atomic-scale anisotropic dynamic behaviour of the nanographene. Various polymeric matrices and pristine and functionalised graphene sheets were examined. The dynamics depend upon different factors such as the shape and size of the sheet, edge functional groups, and polymer/graphene interaction strength. These authors revealed faster diffusion of the nanographene in the polyethylene matrix vis-à-vis polyethylene oxide matrix owing to the better graphene/polymer interaction.

Similarly, the carboxylated graphene sheets displayed slower movement than the hydrogenated and pristine sheets due to the more significant effect of the carboxyl group. This finding indicates the relatively greater effect of the H-bond and electrostatic interactions between functionalised graphene and the polyethylene oxide matrix. The figure depicts the characteristic configuration of nanographene/polymer nanocomposite (Fig. 9).

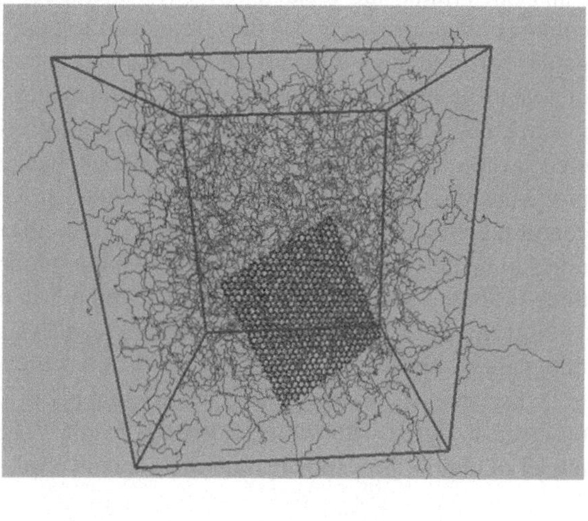

Fig. 9 Schematic of the characteristic configuration of nanographene/polymer nanocomposite [52]

7 Macroscale Techniques

7.1 Finite Element Method

Finite Element Method (FEM) has proven to be an indispensable tool for modelling composite materials, offering a powerful approach to understanding and predicting the mechanical behaviour of these advanced materials. Composite materials, which typically consist of two or more distinct constituents with different properties, exhibit complex and anisotropic behaviour that traditional analytical methods often struggle to capture. FEM can discretise complex geometries and apply material properties at a microstructural level. This ability allows for a detailed and accurate representation of the composite's mechanical response. This method enables engineers and researchers to simulate the stress, strain, and deformation characteristics of composite structures under various loading conditions, aiding in the optimisation of material layup, design, and performance prediction for a wide range of applications, from aerospace and automotive industries to civil engineering and sporting equipment. FEM's versatility and computational power make it an essential tool in the ongoing advancement of composite materials and their engineering applications.

The FEM allows comprehensive analysis of the matrix, fibre, and interfacial damages caused in the composites under service conditions [53, 54]. The simulation of composite materials using FEM can be broadly categorised into micromechanics-based, equivalent homogeneous material-based (EHM) and combinational approaches. The micromechanical approach is highly time-consuming due to the explicit modelling of the fibre/matrix interaction, local defects, and complicated deformation mechanisms in the composites. The meshing requirements are computationally inefficient vis-à-vis the EHM-based approach, demanding high-performance computing machines [53, 55].

On the other hand, although computationally efficient, the EHM-based approach cannot determine the local fibre/matrix interaction and damage mechanics. The combinational approach was thus developed to take advantage of both these techniques for predicting sub-surface damage and the shear force. The FE modelling of composite material behaviour can be further categorised into micro-, meso-, and macroscales. In the microscale, the interaction of constituent materials is considered by the heterogenous material behaviour. In the mesoscale, the interlaminar properties, laminate layup, and ply behaviour are considered for the analysis. In contrast, homogeneous material behaviour allows the effect of constituents on properties to be determined by composite behaviour alone [56]. Figure summarises the three FE-based approaches for modelling composite materials (Fig. 10).

Nurhaniza et al. presented an FE model for analysing the mechanical response of fibreglass composites for aerospace applications under thermo-mechanical loading [57]. The analysis was conducted on different ply orientations, and the results were observed to agree with the experimental results. These authors employed the FE model to obtain the laminates' ultimate failure load, tensile strength, stresses, and

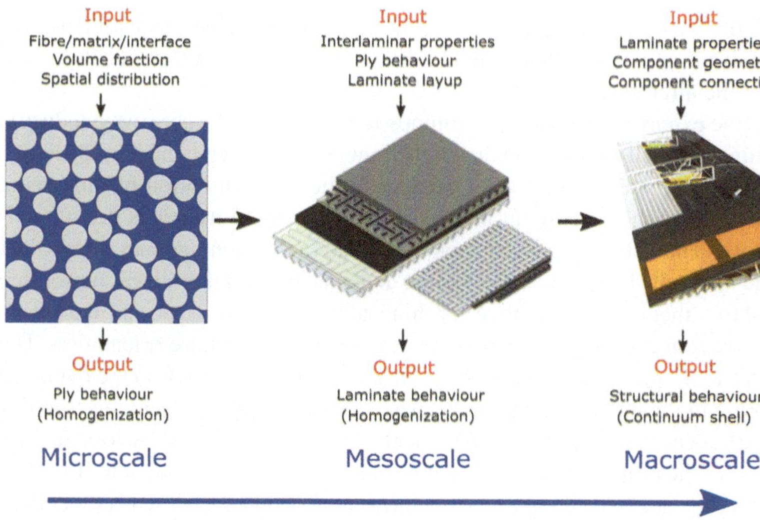

Fig. 10 The three FE-based modelling approaches for composite materials [56]

strains. Figure (a) and (b) shows the loading and boundary conditions used by these authors in the FE model and the strain contours in the specimen (Fig. 11).

The dimensional quality and productivity of aerospace manufacturing processes can be reduced drastically with the help of the FE modelling approach. A novel method for aircraft fuselage shape control based on FE modelling has been proposed to allow precise shape control before assembly [58]. An FE model consisting of the detailed ply design, material properties, fixture structures, and actuator installation was first developed. Via this, these authors imitated the actual fuselage fabrication process for analysing the advanced failure tests conducted on real fuselages.

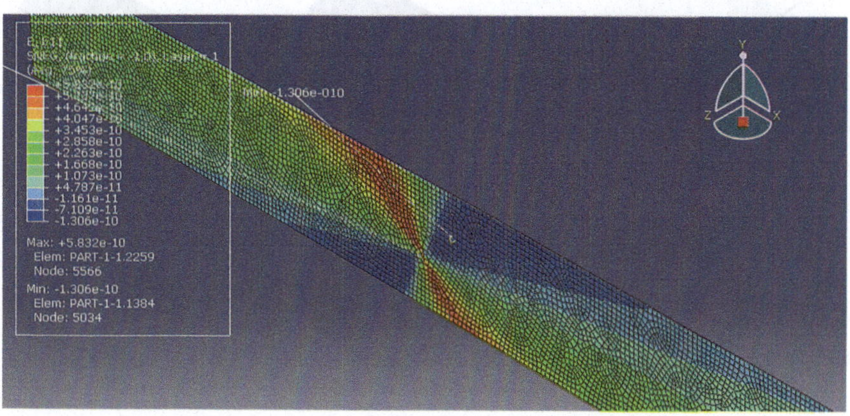

Fig. 11 **a** Strain distribution contour plot determined by the model [57]

The feasibility was then performed, including single- and double-plane dimensional control analysis and failure test. Figure shows the actuator setup and the stress contours in the aircraft fuselage.

Despite the extensive utilisation of composite materials in the aerospace industry, their machining is still challenging due to the inherent anisotropy and heterogeneity. The FE modelling was employed for predicting the chip formation phenomenon, cutting forces, and the machining-induced damage in these materials during end-milling [59]. The CFRP material was modelled as a homogenous equivalent orthotropic material. The plane-stress failure analysis of this material was performed using Hashin's theory. The adaptive meshing technique was used, and the workpiece–tool friction coefficient was assumed to depend on the fibre orientation. The damage onset was predicted by degrading the material stiffness to 0 in the damaged region from 1 in the undamaged region. These authors could represent the complex end-milling process via a simplified 2D model with negligible accuracy loss through this approach. The proposed model is simple and computationally efficient (Fig. 12).

Regarding the analysis of the impact behaviour of composites, the FE modelling can be bifurcated into two approaches: 2D shell modelling and 3D solid element modelling. The 2D modelling approach is computationally efficient and better for thin plates but unsuitable for low-velocity transverse impact loading on composite structures. In such cases, the through-thickness lamina stresses under the impact event are required, demanding a computationally costlier 3D solid modelling approach. Ahmad et al. presented two different 3D-modelling-based FE models for laminated composites under transverse low-velocity impact loading [60]. The LS-DYNA explicit FE method was used for identifying and comparing the low-velocity damage patterns in the laminated composite plates. The model used an impactor, composite plate, steel fixtures, and rubber-stripped clamps to represent the actual experimental conditions

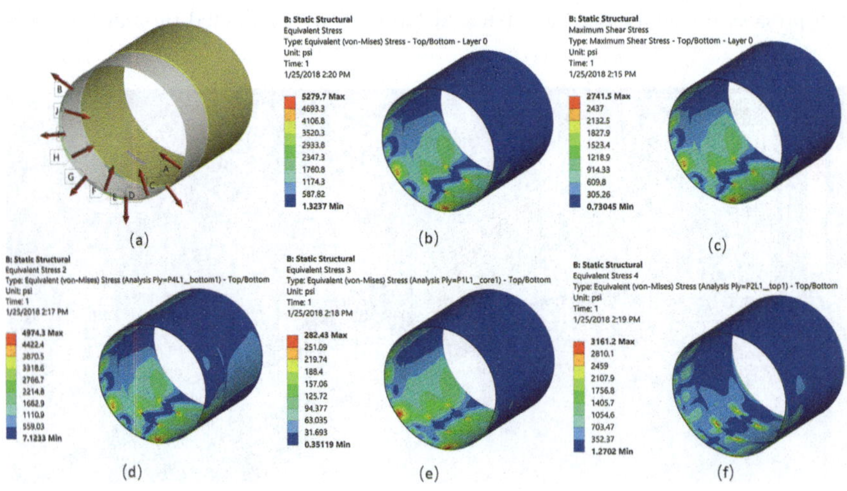

Fig. 12 **a** Actuator setup and the equivalent stress contours in the aircraft fuselage [58]

accurately. The eight-node constant stress solid elements with one-point integration were used to model the composite plate to predict laminate damage.

Wind turbine blades are the main components of a wind turbine. The blades' shape must be optimised to derive maximum efficiency from the turbine [61]. Moreover, the turbine blades are manufactured from composite materials to harness the inherent benefits, such as a high strength-to-weight ratio and excellent fatigue resistance. Wind turbine blades have highly complex structures comprising several composite material plies. However, the blade performance can be altered by altering the fibre orientation, stacking sequence and material to suit the performance criterion. The experimental evaluation of the structural performance of these materials is very challenging for the designers. In this regard, a parametric structural optimisation model based on FEM and a genetic algorithm for optimising the wind turbine blade performance were proposed by Wang et al. [62]. The proposed FE model primarily aims to minimise turbine blade mass under constraints such as stress, deformation, vibration, buckling, and manufacturing continuity. The authors showed that the proposed modelling strategy can reduce turbine blade mass by 17.4% compared to the initial design. Figure shows the meshed model of the turbine blade (Fig. 13).

The fibre/matrix interphase plays a crucial role in the performance of composite materials. Accurately determining stresses and strains in this interphase can yield significant advancement in tailoring the properties of the composite material. Heydari-Meybodi et al. presented an interesting approach based on Python programming coding implemented in the ABAQUS FE package to develop a representative volume element (RVE) [63]. This RVE consisting of the matrix, interphase, and nanoclay as constituents was utilised for the 3D FE modelling of the composite. This model was then used for determining the material Young's modulus under various loading conditions. The nanoclay was modelled as thin cuboid geometry to consider the effect of the third dimension. The effect of loading direction, contact area, thickness and the nanoclay alignment angle on the mechanical properties was

Fig. 13 Meshed model of the turbine blade [62]

also determined. These authors highlighted the use of FE modelling in predicting the mechanical behaviour of composites.

In summary, FEM in multiscale composites modelling is a transformative approach that empowers researchers and engineers to unlock the full potential of composite materials. With its ability to bridge the gap between micro- and macroscales, FEM provides invaluable insights into the behaviour and performance of composites. As computational capabilities and material characterisation techniques advance, this methodology will remain a cornerstone in developing more robust, lighter, and efficient composite structures across various industries.

7.2 Finite Volume Method

The finite volume method (FVM) plays a pivotal role in the realm of multiscale modelling of composites, an essential discipline in materials science and engineering. Composites, consisting of heterogeneous materials, offer superior mechanical, thermal, and electrical properties compared to their constituents. FVM is particularly valuable in this context as it facilitates the accurate simulation of material behaviour and flow phenomena across various length scales with its ability to handle complex geometries. The following section will delve into the significance of FVM in the multiscale modelling of composites, emphasising its role in advancing the design and optimisation of composite materials for a wide range of applications, from aerospace and automotive industries to infrastructure and beyond.

The finite difference method and FEM constitute an important class of numerical techniques for solving partial differential equations representing the real-world problem. However, these techniques suffer from severe limitations when modelling curved domains. One of the most important limitations is the approximation error encountered when representing these domains through rectangular mesh. Another major limitation of these techniques is the lack of an efficient approach to handling internal and natural boundary conditions. Yet another major drawback of these schemes is a loss in accuracy if the developed difference equations represent fewer nodal points. The classical FEM Galerkin's method can handle equations of mechanics and heat transfer but is unsuccessful when one applies it to the fluid dynamics regime.

Due to these limitations, an accurate technique with smaller errors and simple boundary conditions handling, known as the generalised differences method, was developed [64]. The accuracy of GDM is higher than the FDM and close to FEM. These advantages led to the advancement of GDM into fluid dynamics. The FVM is also known as the finite control volume method, providing a discrete approximation of the control equation as an integral. In this way, the GDM is equivalent to the FVM, implying applying the classical difference scheme to the irregular grids [65]. FVM can be used for gaining insight into the resin transfer moulding process. One such attempt has been reported to predict the second-order permeability tensor of 3D-braided preforms using FVM and Darcy's law [66]. Figure shows the pressure

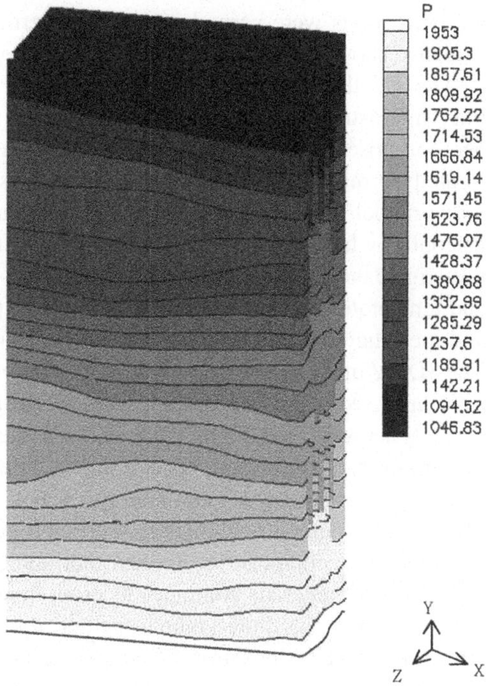

Fig. 14 FVM predicted pressure contours in the unit cell [66]

contours in the unit cell subjected to a pressure gradient in the y-direction. The numerically obtained pressure and velocity distributions were employed to predict the permeability tensor of the preform. The unit cell's 3D flow analysis is performed by considering the yarns as impermeable. Using this assumption, these authors could reduce the simulation's memory requirement and calculation time. However, the complexity of nesting, shifting, and unit cell deformation may lead to discrepancies between experimental and numerical results. This discrepancy may lead to the over-prediction of the permeability (Fig. 14).

The micromechanical modelling of heterogeneous media may play a substantial role in developing novel composite materials [67]. This approach is the identification of exciting material combinations with tailored thermo-mechanical properties. The approach also facilitates the design optimisation of multiscale composite material properties by manipulating the constituent properties and arrangement. The finite volume theory has been demonstrated as a dependable tool for the homogenisation and analysis of heterogeneous solids. Recently, a 3D micromechanical model was proposed to determine the effective thermal conductivity of the composite material possessing periodic microstructure [68]. The proposed tool was tested for fidelity via the solutions of different examples of two- and three-phase composite materials with aligned short and hollow long fibres. The proposed finite volume-based model was able to predict the results accurately with a coarser mesh vis-à-vis FEM. Moreover, the predictions were very close to the experimental results, providing compelling evidence about the high efficiency of the proposed model.

The FV theory can be extended to study the mechanical behaviour of arbitrary microstructures-based composites using 3D domains. One such attempt has been proposed using a triply periodic FV-based direct averaging micromechanics to analyse multiphase random internal microstructure composites [69]. A constitutive model was developed using the state-variable Bodner-Partom to model the viscoplastic behaviour of the composites. The proposed model was then utilised to investigate the composites' uniaxial stress and strain response at various strain rates and temperatures. These authors were able to predict the rate-dependent nonlinear and inelastic tensile behaviour of the SiC/Al composites with reasonable accuracy. The 3D FV model allows any multiphase composite with random-shaped inclusions to be analysed. Such an advantage is unavailable with the 2D model.

In conclusion, the FVM emerges as a vital tool in the multiscale modelling of composites, offering a versatile and accurate approach to capture the intricate behaviour of materials at various length scales. Its ability to simulate physical quantity transport, complex geometries, and heterogeneity makes it vital in understanding and predicting composite materials' thermomechanical properties. As the demand for advanced materials continues to grow in aerospace, automotive, and construction industries, the finite volume method remains an invaluable asset, facilitating the design and optimisation of composite structures that meet the ever-evolving performance requirements of modern engineering applications.

8 Conclusions

Polymer composites can be modeled at various levels from nano to atomistic scale. However, the choice of the simulation technique depends upon the type of polymer composite. Moreover, the accuracy of results is another factor to be considered while selecting the correct simulation technique. After choosing the simulation technique, various factors should be controlled for accurate analysis. It, therefore, becomes imperative to have a solid foundational knowledge of the multiscale modeling of polymer composites. MC methods are essential for analyzing polymer composites and predicting properties like thermodynamics, mechanical strength, electrical conductivity, and thermal performance. They facilitate the exploration of fiber-reinforced composite behavior, aiding in studying progressive damage and mechanical properties. MC simulations determine electrical conductivity in nanocomposites by modeling the nanoplatelet network. These methods also play a crucial role in parametric analysis for predicting electrical conductivity accurately in polymeric composites. MC methods are versatile tools for modeling and optimizing polymer composites, providing valuable insights for tailored material properties. High-performance computing and MD simulations have transformed the understanding and optimization of polymeric composites. MD allows for predicting atomic and molecular movements, providing insights into material behavior, especially in fiber-reinforced composites like graphite/epoxy and carbon nanotube-reinforced materials. These simulations extend to silica nanoparticle-polyimide and SWCNT/epoxy composites, offering bulk elastic properties and glass-transition temperature predictions. MD proves crucial for understanding thermal performance reducing thermal expansion coefficients in CNT-reinforced nanocomposites. This approach is pivotal for designing superior polymeric composites with enhanced strength, durability, and thermal resistance in aerospace, automotive, and structural engineering applications. DPD methods play a crucial role in the multiscale modeling of polymer composites, offering a powerful tool for simulating the complex interactions and dynamics within these materials. DPD is particularly valuable in capturing the mesoscale behavior, where interactions between polymer chains, fillers, and solvents are critical. This method bridges the gap between atomistic and macroscopic scales, enabling a more comprehensive understanding of the composite's behavior. By accounting for the dissipative and random forces at the particle level, DPD allows researchers to explore phenomena such as polymer blending, phase separation, and self-assembly, providing valuable insights for designing and optimizing advanced polymer composites with tailored properties. Integrating Finite Element (FE) and Finite Volume (FV) methods holds significant importance in the multiscale modeling of polymer composites. These numerical techniques excel in capturing the mechanical behavior, thermal performance, and structural responses of composite materials across various length and time scales. FEM are particularly adept at simulating the intricate interactions between individual constituents at the microscale, enabling a detailed examination of stress distributions, failure mechanisms, and overall structural integrity. On the other hand, FVMs are valuable for modeling heat transfer and fluid flow within the composite, providing insights into the thermal performance. Together, these methods facilitate a comprehensive understanding of the complex behavior exhibited by polymer composites, guiding the design and optimization of materials with tailored properties for diverse applications. The LB and BD methods hold crucial significance in the multiscale modeling of polymer composites. The LB method is highly effective in simulating fluid dynamics and flow phenomena within composite materials, providing insights into processes such as permeability and transport properties. Concurrently, the BD method is instrumental in capturing the Brownian motion and particle interactions at the nanoscale, offering a detailed understanding of the behavior of polymer chains and nanoparticles. Together, these two methods contribute to a comprehensive multiscale model, allowing researchers to study the complex interplay between fluid dynamics and particle behavior in polymer composites. This holistic approach aids in predicting material properties guiding the design of composites with enhanced performance tailored to specific applications.

Declarations

Conflicts of Interest The authors confirm that there are no conflicts of interest associated with this work.

Ethical Compliance The authors affirm that this study adheres to all ethical guidelines.

Competing Interests No competing interests are declared by the authors.

Data and Material Availability Data and materials are not applicable to this study.

References

1. Yang HS, Kim HJ, Son J, Park HJ, Lee BJ, Hwang TS (2004) Rice husk flour filled polypropylene composites, mechanical and morphological studies. Compos Struct 63:305–312
2. Premlal HGB, Ismail H, Baharin A (2002) Comparison of the mechanical properties of the rice husk powder filled polypropylene composites with talc filled polypropylene composites. Polym Test 21:833–839
3. Rosa SML, Santos EF, Ferreira CA, Nachtigall SMB (2009) Studies on the properties of rice husk filled PP composites-effect of Maleated PP. Mater Res 12(3):333–338
4. Bisht N, Gope PC (2018) Effect of alkali treatment on mechanical properties of rice husk flour reinforced epoxy bio-composite. Mater Today: Proc 5(11, Part 3):24330–24338
5. Zhang HY, Li JY, Pan Y et al (2022) Flexible carbon fiber-based composites for electromagnetic interference shielding. Rare Met 41:3612–3629. https://doi.org/10.1007/s12598-022-02057-3
6. Das TK, Ghosh P, Das NC (2019) Preparation, development, outcomes, and application versatility of carbon fiber-based polymer composites: a review. Adv Compos Hybrid Mater 2:214–233. https://doi.org/10.1007/s42114-018-0072-z
7. Tarhini A, Tehrani-Bagha AR (2023) Advances in preparation methods and conductivity properties of graphene-based polymer composites. Appl Compos Mater. https://doi.org/10.1007/s10443-023-10145-5
8. Kumar A, Sharma K, Dixit AR (2020) A review on the mechanical and thermal properties of graphene and graphene-based polymer nanocomposites: understanding of modelling and MD simulation. Mol Simul 46(2):136–154. https://doi.org/10.1080/08927022.2019.1680844
9. Sari NH, Fajrin J, Fudholi A (2020) Characterisation of swellability and compressive and impact strength properties of corn husk fibre composites. Compos Commun 18:49–54. https://www.sciencedirect.com/science/article/pii/S2452213920300188. https://doi.org/10.1016/j.coco.2020.01.009. ISSN 2452-2139
10. Ibrahim ID, Jamiru T, Sadiku ER, Kupolati WK, Agwuncha SC, Ekundayo G (2016) Mechanical properties of sisal fibre-reinforced polymer composites: a review. Compos Interfaces 23(1):15–36. https://doi.org/10.1080/09276440.2016.1087247
11. Wang F, Chen ZQ, Wei YQ, Zeng XG (2010) Numerical modeling of tensile behavior of fiber-reinforced polymer composites. J Compos Mater 44(19):2325–2340
12. Lee DC, Kwon G, Kim H, Lee HJ, Sung BJ (2012) Three-dimensional Monte Carlo simulation of the electrical conductivity of carbon nanotube/polymer composites. Appl Phys Express 5(4):045101
13. Yan D, Wen J, Xu G (2016) A Monte Carlo simulation and effective thermal conductivity calculation for unidirectional fiber reinforced CMC. Appl Therm Eng 94:827–835
14. Okabe T, Nishikawa M (2009) GLS strength prediction of glass-fiber-reinforced polypropylene. J Mater Sci 44:331–334
15. Bouaziz A, Zaïri F, Naït-Abdelaziz M, Gloaguen JM, Lefebvre JM (2007) Micromechanical modelling and experimental investigation of random discontinuous glass fiber polymer–matrix composites. Compos Sci Technol 67(15–16):3278–3285
16. Wang W, Jayatissa AH (2015) Computational and experimental study of electrical conductivity of graphene/poly (methyl methacrylate) nanocomposite using Monte Carlo method and percolation theory. Synth Met 204:141–147
17. Oskouyi AB, Mertiny P (2011) Monte Carlo model for the study of percolation thresholds in composites filled with circular conductive nano-disks. Procedia Eng 10:403–408

18. Zabihi Z, Araghi H (2016) Monte Carlo simulations of effective electrical conductivity of graphene/poly (methyl methacrylate) nanocomposite: Landauer-Buttiker approach. Synth Met 217:87–93
19. Fang C, Zhang J, Chen X, Weng GJ (2020) Calculating the electrical conductivity of graphene nanoplatelet polymer composites by a Monte Carlo method. Nanomaterials 10(6):1129
20. Folorunso O, Hamam Y, Sadiku R, Ray SS, Joseph AG (2019) Parametric analysis of electrical conductivity of polymer-composites. Polymers 11(8):1250
21. Spanos P, Elsbernd P, Ward B, Koenck T (2013) Estimation of the physical properties of nanocomposites by finite-element discretization and Monte Carlo simulation. Philos Trans R Soc A Math Phys Eng Sci 371(1993):20120494
22. Karplus M, McCammon JA (2002) Molecular dynamics simulations of biomolecules. Nat Struct Biol 9(9):646–652
23. Hadipeykani M, Aghadavoudi F, Toghraie D (2020) A molecular dynamics simulation of the glass transition temperature and volumetric thermal expansion coefficient of thermoset polymer based epoxy nanocomposite reinforced by CNT: a statistical study. Physica A 546:123995
24. Hollingsworth SA, Dror RO (2018) Molecular dynamics simulation for all. Neuron 99(6):1129–1143
25. Hutchings I, Shipway P (2017) Tribology: friction and wear of engineering materials. Butterworth-Heinemann
26. Hung PY, Lau KT, Cheng LK, Leng J, Hui D (2018) Impact response of hybrid carbon/glass fibre reinforced polymer composites designed for engineering applications. Compos B Eng 133:86–90
27. Li Y, Wang Q, Wang S (2019) A review on enhancement of mechanical and tribological properties of polymer composites reinforced by carbon nanotubes and graphene sheet: molecular dynamics simulations. Compos B Eng 160:348–361
28. Wagner HD, Vaia RA (2004) Nanocomposites: issues at the interface. Mater Today 7(11):38–42
29. Bandyopadhyay A, Valavala PK, Clancy TC, Wise KE, Odegard GM (2011) Molecular modeling of crosslinked epoxy polymers: the effect of crosslink density on thermomechanical properties. Polymer 52(11):2445–2452
30. Singh PK, Sharma K, Kumar A, Shukla M (2017) Effects of functionalization on the mechanical properties of multiwalled carbon nanotubes: a molecular dynamics approach. J Compos Mater 51(5):671–680
31. Wei Q, Zhang Y, Wang Y, Yang M (2017) A molecular dynamic simulation method to elucidate the interaction mechanism of nano-SiO_2 in polymer blends. J Mater Sci 52(21):12889–12901
32. Wang Y, Yang G, Wang W, Zhu S, Guo L, Zhang Z, Li P (2019) Effects of different functional groups in graphene nanofiber on the mechanical property of polyvinyl alcohol composites by the molecular dynamic simulations. J Mol Liq 277:261–268
33. Jiao W, Zheng T, Liu W, Jiao W, Wang R (2019) Molecular dynamics simulations of the effect of sizing agent on the interface property in carbon fiber reinforced vinyl ester resin composite. Appl Surf Sci 479:1192–1199
34. Stevens MJ (2001) Interfacial fracture between highly cross-linked polymer networks and a solid surface: effect of interfacial bond density. Macromolecules 34(8):2710–2718
35. Mansfield KF, Theodorou DN (1991) Atomistic simulation of a glassy polymer/graphite interface. Macromolecules 24(15):4295–4309
36. Hadden CM, Jensen BD, Bandyopadhyay A, Odegard GM, Koo A, Liang R (2013) Molecular modeling of EPON-862/graphite composites: interfacial characteristics for multiple crosslink densities. Compos Sci Technol 76:92–99
37. Odegard GM, Clancy TC, Gates TS (2005) Modeling of the mechanical properties of nanoparticle/polymer composites. Polymer 46(2):553–562
38. Ionita M (2012) Multiscale molecular modeling of SWCNTs/epoxy resin composites mechanical behaviour. Compos B Eng 43(8):3491–3496
39. Hoogerbrugge PJ, Koelman JMVA (1992) Simulating microscopic hydrodynamic phenomena with dissipative particle dynamics. Europhys Lett 19(3):155

40. Elliott JA, Windle AH (2000) A dissipative particle dynamics method for modeling the geometrical packing of filler particles in polymer composites. J Chem Phys 113(22):10367–10376
41. Wang YC, Ju SP, Huang TJ, Wang HH (2011) Modeling of polyethylene, poly (l-lactide), and CNT composites: a dissipative particle dynamics study. Nanoscale Res Lett 6:1–8
42. Chakraborty S, Choudhury CK, Roy S (2013) Morphology and dynamics of carbon nanotube in polycarbonate carbon nanotube composite from dissipative particle dynamics simulation. Macromolecules 46(9):3631–3638
43. Ketkaew R, Tantirungrotechai Y (2018) Dissipative particle dynamics study of SWCNT reinforced natural rubber composite system: an important role of self-avoiding model on mechanical properties. Macromol Theory Simul 27(3):1700093
44. Kim YJ, Tan YF, Kim S (2018) Two-dimensional lattice Boltzmann modeling for effective thermal conductivity in carbon black filled composites. J Compos Mater 52(15):2047–2053
45. Martys NS, Chen H (1996) Simulation of multicomponent fluids in complex three-dimensional geometries by the lattice Boltzmann method. Phys Rev E 53(1):743
46. Kandhai D, Vidal DE, Hoekstra AG, Hoefsloot H, Iedema P, Sloot PMA (1998) A comparison between lattice-Boltzmann and finite-element simulations of fluid flow in static mixer reactors. Int J Mod Phys C 9(08):1123–1128
47. Spaid MA, Phelan FR Jr (1997) Lattice Boltzmann methods for modeling microscale flow in fibrous porous media. Phys Fluids 9(9):2468–2474
48. Fang WZ, Chen L, Gou JJ, Tao WQ (2016) Predictions of effective thermal conductivities for three-dimensional four-directional braided composites using the lattice Boltzmann method. Int J Heat Mass Transf 92:120–130
49. Xie C, Wang J, Wang D, Pan N, Wang M (2015) Lattice Boltzmann modeling of thermal conduction in composites with thermal contact resistance. Commun Comput Phys 17(4):1037–1055
50. Runacher A, Kazemzadeh-Parsi MJ, Di Lorenzo D, Champaney V, Hascoet N, Ammar A, Chinesta F (2023) Describing and modeling rough composites surfaces by using topological data analysis and fractional Brownian motion. Polymers 15(6):1449
51. Chinesta F, Leygue A, Bognet B, Ghnatios C, Poulhaon F, Bordeu F, Barasinski A, Poitou A, Chatel S, Maison-Le-Poec S (2014) First steps towards an advanced simulation of composites manufacturing by automated tape placement. Int J Mater Form 7:81–92
52. Rissanou AN, Bačová P, Harmandaris V (2019) Investigation of the properties of nanographene in polymer nanocomposites through molecular simulations: dynamics and anisotropic Brownian motion. Phys Chem Chem Phys 21(43):23843–23854
53. Barbero Ever J (2013) Finite element analysis of composite materials using ANSYS®
54. Tenek LT, Argyris J (1997) Finite element analysis for composite structures, vol 59. Springer Science & Business Media
55. Dandekar CR, Shin YC (2012) Modeling of machining of composite materials: a review. Int J Mach Tools Manuf 57:102–121
56. Tan W, Naya F, Yang L, Chang T, Falzon BG, Zhan L, Molina-Aldareguía JM, González C, Llorca J (2018) The role of interfacial properties on the intralaminar and interlaminar damage behaviour of unidirectional composite laminates: Experimental characterization and multiscale modelling. Compos B Eng 138:206–221
57. Nurhaniza M, Ariffin MKA, Ali A, Mustapha F, Noraini AW (2010) Finite element analysis of composites materials for aerospace applications. In: IOP conference series: materials science and engineering, vol 11, no 1, p 012010. IOP Publishing
58. Wen Y, Yue X, Hunt JH, Shi J (2018) Feasibility analysis of composite fuselage shape control via finite element analysis. J Manuf Syst 46:272–281
59. Ghafarizadeh S, Chatelain JF, Lebrun G (2016) Finite element analysis of surface milling of carbon fiber-reinforced composites. Int J Adv Manuf Technol 87:399–409
60. Ahmad F, Abbassi F, Park MK, Jung JW, Hong JW (2018) Finite element analysis for the evaluation of the low-velocity impact response of a composite plate. Adv Compos Mater
61. Liu X, Wang L, Tang X (2013) Optimized linearization of chord and twist angle profiles for fixed-pitch fixed-speed wind turbine blades. Renew Energy 57:111–119

62. Wang L, Kolios A, Nishino T, Delafin PL, Bird T (2016) Structural optimisation of vertical-axis wind turbine composite blades based on finite element analysis and genetic algorithm. Compos Struct 153:123–138
63. Heydari-Meybodi M, Saber-Samandari S, Sadighi M (2016) 3D multiscale modeling to predict the elastic modulus of polymer/nanoclay composites considering realistic interphase property. Compos Interfaces 23(7):641–661
64. Li R, Chen Z, Wu W (2000) Generalized difference methods for differential equations: numerical analysis of finite volume methods. CRC Press
65. Rappaz M, Bellet M, Deville MO, Snyder R (2003) Numerical modeling in materials science and engineering. Springer-Verlag, Berlin, pp 448–475
66. Song YS, Chung K, Kang TJ, Youn JR (2004) Prediction of permeability tensor for three dimensional circular braided preform by applying a finite volume method to a unit cell. Compos Sci Technol 64(10–11):1629–1636
67. Bansal Y, Pindera MJ (2006) Finite-volume direct averaging micromechanics of heterogeneous materials with elastic–plastic phases. Int J Plast 22(5):775–825
68. de Sousa Vieira C, Marques SPC (2019) A new three-dimensional finite-volume model for evaluation of thermal conductivity of periodic multiphase composites. Int J Heat Mass Trans 139:412–424. https://doi.org/10.1016/j.ijheatmasstransfer.2019.05.031
69. Qiang, Chen Xuefeng, Chen Zhi, Zhai Zhibo, Yang (2016) A new and general formulation of three-dimensional finite-volume micromechanics for particulate reinforced composites with viscoplastic phases Composites Part B: Engineering 85216–232. https://doi.org/10.1016/j.compositesb.2015.09.014

Chapter 4
Atomistic Modelling of Bio-Nanocomposites for Bone Tissue Engineering Applications

Jashveer Singh and Rajesh Kumar

1 Introduction

Composites comprising biologically derived components and particles with a size distribution between 1 and 100 nm are called bio-nanocomposites [1]. Researchers in the field of materials science are devoting an enormous amount of time to investigate nanohybrids due to their remarkable properties as structural or functional materials. These materials have many potential uses, including as components of heterogeneous catalysts and in various mechanical, electronic, magnetic, and electrochemical devices. Materials with a polymer matrix, whether manufactured or occurring naturally, are known as polymer nanocomposites. Inorganic (like silica clays) or organic (like nanocellulose) nanoparticles may fill the polymer matrix at the nanoscale. The filler also improves performance characteristics over the individual polymers and causes the organic matrix to create extensive interfacial interactions. Preparing polymer nanocomposites often involves using polymers that aren't biodegradable, making them less eco-friendly. Researchers in the field of materials science are putting a lot of energy into creating bio-nanocomposites, which are nanocomposites made of biopolymers. These materials will exhibit identical excellent mechanical properties, thermal stability, and gas barrier properties as nanocomposites made of synthetic polymers [2, 3]. Beyond these characteristics, bio-nanocomposites demonstrate biodegradability, biocompatibility, and, sometimes, functional qualities imparted by inorganic or biological moieties. Bio-nanocomposites, made from

J. Singh
Department of Mechanical Engineering, National Institute of Technology, Hamirpur, Himachal Pradesh 177005, India

R. Kumar (✉)
Department of Mechanical Engineering, National Institute of Technology, Kurukshetra, Haryana 136118, India
e-mail: rpawar@nitkkr.ac.in

S. K. Sethi et al. (eds.), *Polymer Composites: From Computational to Experimental Aspects*, Materials Horizons: From Nature to Nanomaterials,
https://doi.org/10.1007/978-981-97-0888-8_4

a variety of biopolymers, have recently emerged as an eco-friendly material option. Starch, cellulose, poly(hydroxy alkanoates), pectin, chitosan, and other naturally occurring polymers as well as synthetic poly (lactic-co-glycolic acids) (PLGA) and poly(ε-caprolactone) are being employed more and more in bio-nanocomposite technology as alternatives to nonbiodegradable polymers. In the past few decades, researchers have created more synthetic biodegradable polymers, with an emphasis on those made of physiological and nontoxic building components including a-amino acids, fatty diols, carbonic acid, and dicarboxylic acids. Some of the most promising matrices for the construction of bio-nanocomposites are the copolymers of derived poly (ester amide)s, poly (ester urethane)s, poly(ester urea)s, and other related compounds [4–9]. Results indicate that the mechanical, barrier, and thermal characteristics are noticeably better than those of pure polymers. Incorporating an inorganic nanofiller into a polymer alters its biodegradation rate, which may be improved or reduced while keeping its biodegradation capacity.

In addition to their excellent properties, biopolymers have intrinsic qualities like biocompatibility and biodegradability that provide new opportunities for bio-nanocomposite materials in pharmaceuticals and eco-friendly products. The use of bio-nanocomposite materials in the field of regenerative medicine has great promise. These materials have a wide range of potential uses for public goods and more. Bio-nanocomposites are a new multidisciplinary subject that brings together research on biodegradable polymers, biomineralization, bioinspired materials, biomimetic systems, and their synthesis, investigation, and applications. Characterizing the mechanical characteristics of bio-nanocomposites may be challenging using traditional experimental approaches, despite their intrinsic value, because of the complexity and dynamic nature of these materials [10–19]. One effective method for filling this knowledge gap is the use of atomic simulations, especially molecular dynamics simulations, which may reveal intricate details about the mechanical behaviour of biopolymers on an atomic scale [20–22]. To better understand the mechanisms controlling these processes, researchers are using computational methods to describe the interactions among different atoms and to replicate their dynamic movements.

2 Atomistic Modelling Techniques

Researchers have employed an extensive number of approaches to predict the mechanical properties of nanofillers. When it comes to characterizing nanofillers, experimental procedures continue to be employed even though they are more expensive and take more time. Most experimental methods rely on imaging techniques including atomic force microscopy (AFM) and transmission electron microscopy (TEM) [23, 24]. Alternative computer approaches based on density functional theory (DFT), molecular dynamics (MD), and the finite element method (FEM) are emerging as a viable alternative to older methods due to their excessive cost and low efficiency. Atomic truss frame models based on the finite element method

(FEM) are often used by researchers to study carbon–carbon bonding in carbon nanotubes (CNTs) and graphene [25, 26]. Methods for predicting hBN nanofillers using molecular mechanics (MM) have also been developed by researchers [27]. The general opinion is that these numerical approaches are sufficient for modelling global features. Higher fidelity simulations, such as density functional theory (DFT) [28, 29], are often used by researchers to assess the atomic-level properties of nanofillers, alongside the previously mentioned numerical methods. Using a quantum mechanical framework, DFT-based methods assess the elastic properties of nanofillers [30]. Due to the high computing cost, density functional theory can only be applied to a limited number of atoms simultaneously.

Although there are more accurate numerical simulations and experimental methods, molecular dynamics (MD)-based simulations have a lot of promise. Improving our understanding of interatomic potentials has been crucial in predicting how nanofillers would behave [31–33]. One of the many substance characteristics that stands out as a reliable and well-established parameter is Young's modulus. Strain energy (U) is a frequent element in MD-based simulations that may be used to predict the stress–strain behaviour and elastic modulus of the nanofiller under consideration. The ensuing equations that determine Young's modulus (Y, measured in N/m^2) and 2D Young's modulus (YS, measured in N/m) provide insight into the relationship between strain energy (U) and strain (ε).

$$\mathrm{Y} = \left(\partial^2 U/\partial \varepsilon^2\right)/V \tag{1}$$

$$\mathrm{Y} = \left(\partial^2 U/\partial \varepsilon^2\right)/S \tag{2}$$

To this atomic system, V denotes its volume and S is its surface area.

2.1 Interatomic Potentials

Interatomic potentials that have been either found empirically or created with the help of experiments or higher fidelity numerical simulations are often used in MD-based simulations. Interatomic potentials are mathematical expressions that depict the potential energy (E) of an atomic system. Many bodies, one body, and two bodies are the most common methods to classify potentials. Some well-known interatomic pair potentials include Tersoff [34], LJ [35], the embedded atom method (EAM) [36], REBO [37], AIREBO [38], CHARMM [39], and COMPASS [40]. Tersoff, REBO, and EAM are potentials that are bonded, while van der Waals and LJ are potentials that are electrostatic but not bonded. Graphene, carbon nanotubes (CNTs), and hexagonal boron nitride (hBN) are mechanical and thermal nanofillers that are often modelled using Tersoff-type interatomic potentials [31–33]. For potentials of the Tersoff type, the potential energy (E) may be mathematically represented as

$$E = \sum_i E_i = \left(\sum_{i \neq j} V_{ij} \right) /2, \tag{3}$$

$$V_{ij} = f_c(r_{ij})\left[a_{ij} f_R(r_{ij}) + b_{ij} f_A(r_{ij})\right]. \tag{4}$$

The site energy (E_i) and the bond energy (V_{ij}) are carefully separated from the overall energy (E) of the system here. While r_{ij} measures the distance between particles, i and j encompass all particles in the system.

The functions f_R and f_A, which stand for the repulsive and attractive potentials, respectively, are generated when the wave functions of two atoms intersect. They are formally stated as

$$f_R(r) = A \exp(-\lambda_1 r), \tag{5}$$

$$f_A(r) = -B \exp(-\lambda_2 r). \tag{6}$$

The computational cost can be maintained low by limiting these atomic interactions to a certain cutoff distance, denoted by a cutoff function f_c.

$$f_c(r) = \begin{cases} 1, & r < R - D \\ (1/2) - \sin\left(\dfrac{\pi(r - R)}{2D}\right)/2, & R - D < r < R + D \\ 0, & r > R + D \end{cases} \tag{7}$$

It is common practice to choose the cutoff radius (R) so that just the first neighbouring shell is considered. A measure of the bond order is given by the function b_{ij}. Furthermore, b_{ij} contains expressions that restrict the interaction limit to the first neighbour shell. These range-limiting expressions are the only part of the function a_{ij}. Within the CHARMM27 framework, the potential energy equation for nucleic acids is given by

$$\begin{aligned} V(r) = & \sum_{bonds} K_b(b - b_0)^2 + \sum_{angles} K_\theta(\theta - \theta_0)^2 \\ & + \sum_{dihedrals} K_\chi(1 + \cos(n\chi - \delta)) + \sum_{impropers} K_\phi(\phi - \phi_0)^2 \\ & + \sum_{Urey-Bradley} K_{UB}(S - S_0)^2 \\ & + \sum_{nonbondedatompairs} \left(\varepsilon_{ij} \left[\left(\frac{R_{min,ij}}{r_{ij}} \right)^{12} - \left(\frac{R_{min,ij}}{r_{ij}} \right)^{6} \right] + \frac{q_i q_j}{\varepsilon_D r_{ij}} \right) \end{aligned} \tag{8}$$

The variables b, θ, χ, ϕ, and S represent the bond length, bond angle, dihedral torsion angle, improper dihedral angle, and Urey-Bradley 1,3 distance in Eq. (8), while K_b, K_θ, K_χ, K_ϕ, and K_{UB} stand for the bond, valance angles, dihedral angle, force constant, and improper dihedral angle, respectively. A particular term's value at equilibrium is shown by the subscript value. Also, the dihedral terms have a phase (δ) and a multiplicity (n) that indicate the minimum and maximum values.

The following is the mathematical expression of the total energy of the COMPASS force field potential, as it pertains to atomic interactions in organic compound materials.

$$E_{total} = E_b + E_\theta + E_\phi + E_\chi + E_{bb'} + E_{b\theta} + E_{b\phi} + E_{\theta\theta'} + E_{\theta\theta'\phi} + E_{elec} + E_{LJ} \tag{9}$$

There are two types of potential terms: those involving diagonal and off-diagonal cross-coupling, and those about nonbonded interactions. The valence terms for the bond, angle, torsion, and out-of-plane angle coordinates are E_b, E_θ, E_ϕ, and E_χ. The vibrational frequency and structural differences related to conformational variations may be predicted with the use of cross-coupling terms. The cross-coupling terms occurring between internal coordinates are denoted as $E_{bb\prime}$, $E_{b\theta}$, $E_{b\phi}$, $E_{\theta\theta\prime}$, and $E_{\theta\theta\prime\phi}$. Conversely, E_{elec} and E_{LJ} are used in the case of nonbonded interactions.

Predicting the properties of these new materials by MD-based simulations is an emerging potential. Enhancements to these potentials are also being developed by researchers to better improve these calculations. The initial Tersoff potential indicated in the research article [31] has been improved upon by Brenner [34]. Many researchers, such as Brenner [41], Matsunaga et al. [42, 43], Albe et al. [44, 45], Nord et al. [46], and Sevik et al. [47], have proposed optimized Tersoff potential parameters to improve the accuracy of these atomistic studies.

3 Atomistic Studies of Mechanical Properties of Biopolymers

The mechanical features of bio-nanocomposites play a crucial role in various kinds of areas, including biomedicine, biosensors, and biotechnology. To investigate the mechanical behaviour of polyhydroxyalkanoates (PHAs), Bejagam et al. [22] used molecular dynamics (MD) simulations. They found that PHAs decrease mechanical characteristics as the number of carbon atoms in the core and side chains increases. The mechanical properties of chitosan, a biodegradable biopolymer reinforced with hydroxyapatite nanoparticles, were characterized by Rath et al. [48] using MD simulations. Hydrogen bonding and electrostatic interactions between hydroxyapatite and chitosan were responsible for the improved mechanical characteristics observed in the final nanocomposite. Nevertheless, the wear characteristics of this nanocomposite were inadequate. Kulasinski et al. [49] used MD simulations to investigate the physical characteristics of amorphous cellulose and hemicellulose, two hydrophilic biopolymers. Scientists investigated how nanocrystalline biopolymers' mechanical properties were affected by water adsorption. The interactions at the interface determine the mechanical characteristics of these biopolymers when reinforced with nanocrystalline features. To investigate the interactions of protein–mineral nanocomposites at their interfaces, Lai et al. [50] used MD simulations. Their focus was on studying hydroxyapatite (HA) reinforcement of osteopontin (OPN), specifically looking at surface defects on the HA. The result showed that the pulling force and energy dissipation were both decreased in HA layers that had more nano-sized grooves. Patel et al. [51] used the CHARMM27 force field to examine the mechanical characteristics of Bombyx mori silk fibroin biopolymer in a separate investigation. Their tensile strength (340 MPa) and elastic modulus (7.4 GPa) were higher than those reported in the literature. Researchers added nanoparticles to these biopolymers to improve their mechanical properties even further. Using MD simulations employing COMPASS force fields, Pandele et al. [52] examined graphene composites containing chitosan and found that the elastic modulus increased from 8.55 GPa (for chitosan alone) to 11.55 GPa (for the composite with graphene). By using bond-stretch, bond-bending, and AIREBO potentials, Ebrahimi et al. [17] performed MD simulations to examine the interfacial interactions of graphene–collagen nanocomposites. The results showed that the nanocomposite's pull-out forces were exceeded when the graphene was wrinkled instead of flat. Using molecular dynamics simulations with force fields associated with polymer composites, Rahman et al. [53] improved the mechanical characteristics of cellulose biopolymer by adding graphene nanosheet reinforcements to it. In comparison with pure cellulose biopolymer, Young's modulus was improved also. For biophysical characterization, Bayoumy et al. [54] conducted DFT simulations after adding functionalized graphene quantum dots to a chitosan biopolymer. According to their research, the most stable structure was created by armchair hexagonal graphene quantum dots reinforced to chitosan. The removal of heavy metals from wastewater was shown using this nanocomposite. Additionally, Menazea et al. [55] investigated, using DFT simulations, graphene-reinforced chitosan nanocomposites. Using density functional theory (DFT) simulations, Masoumi et al. [56] studied the chitosan biopolymer–carbon nanotube (CNT) interaction. They found that chitosan tends to chemisorb onto CNT surfaces, which might lead to its use in bone tissue engineering and medication delivery. Omar et al. [57] used chitosan biopolymers to strengthen functionalized graphene oxide, which finds application in biosensors. The computational models that used DFT simulations brought attention to the structural stability of this dopamine sensor. Graphene reinforcement resulted in an exponential rise in Young's modulus, as shown in an MD simulation study by Kossovich et al. [58] that examined chitosan-based nanocomposites reinforced with different nanofillers. Biopolymers such as chitin, cyclodextrins, and cellulose that have been reinforced with graphene nanoparticles have the potential to be more effective, more biocompatible, more biodegradable, and less poisonous, as reported by Shende et al. [59]. Jayakumar et al. [60] showed that adding hexagonal boron nitride (hBN) nanosheets to cellulose improved its characteristics even more, suggesting that this material has great promise as a sensing technology. Using DFT simulations, Anota et al. [61] investigated how point defects affected the adsorption characteristics of chitosan reinforced with hBN nanosheets. They found that hBN surfaces with divacancies had an enhanced chemical adsorption capacity.

4 Conclusion and Future Perspective

Biopolymers are polymers that occur naturally and are derived from living organisms. They provide a crucial function as fundamental components in humans and other living organisms. Their utilization in biological applications, including medication delivery and bone tissue engineering, is becoming more widespread. On one hand, nanoparticles are used for drug delivery at the nanoscale, while, on the other hand, they serve as supportive structures in bone tissue engineering. This chapter examines the mechanical properties of the material, which were determined by atomistic simulations [62–66]. Atomistic simulations are a valuable tool for studying nanoscale phenomena in our living organisms that are difficult to observe using the experimental approaches [67–70]. The increasing use of atomistic methods in investigating the mechanical properties of biopolymers and bio-nanocomposites will facilitate their application in biomedical fields [71–74], including the development of biocompatible cardiac stents and scaffolds for tissue engineering.

5 Conflicts of Interest

“The authors have no conflicts of interest to declare”.

Acknowledgements We express our gratitude to NIT Hamirpur for their invaluable academic assistance.

References

1. Shchipunov Y (2012) Bionanocomposites: green sustainable materials for the near future. Pure Appl Chem 84(12):2579–2607
2. Mittal V (2011) Nanocomposites with biodegradable polymers: synthesis, properties, and future perspectives, vol. 68. Oxford University Press
3. Darder M, Aranda P, Ruiz-Hitzky E (2007) Bionanocomposites: a new concept of ecological, bioinspired, and functional hybrid materials. Adv Mater 19(10):1309–1319
4. Díaz A, Katsarava R, Puiggalí J (2014) Synthesis, properties and applications of biodegradable polymers derived from diols and dicarboxylic acids: from polyesters to poly (ester amide) s. Int J Mol Sci 15(5):7064–7123

5. Laurencin C, Deng M (2014) Natural and synthetic biomedical polymers. Newnes
6. Fonseca AC, Gil MH, Simoes PN (2014) Biodegradable poly (ester amide) s–a remarkable opportunity for the biomedical area: review on the synthesis, characterization and applications. Prog Polym Sci 39(7):1291–1311
7. Lendlein A, Sisson A (2011) Handbook of biodegradable polymers: isolation, synthesis, characterization and applications. Wiley & Sons
8. Rodriguez-Galan A, Franco L, Puiggali J (2010) Degradable poly (ester amide) s for biomedical applications. Polym (Basel) 3(1):65–99
9. Felton GP (2011) Biodegradable polymers: processing, degradation, and applications. Nova Science Publishers
10. Gao C, Peng S, Feng P, Shuai C (2017) Bone biomaterials and interactions with stem cells. Bone Res 5(May):1–33. https://doi.org/10.1038/boneres.2017.59
11. Müller FA, Müller L, Hofmann I, Greil P, Wenzel MM, Staudenmaier R (2006) Cellulose-based scaffold materials for cartilage tissue engineering. Biomaterials 27(21):3955–3963. https://doi.org/10.1016/j.biomaterials.2006.02.031
12. Naseri N, Poirier JM, Girandon L, Fröhlich M, Oksman K, Mathew AP (2016) 3-Dimensional porous nanocomposite scaffolds based on cellulose nanofibers for cartilage tissue engineering: tailoring of porosity and mechanical performance. RSC Adv 6(8):5999–6007. https://doi.org/10.1039/c5ra27246g
13. Ferreira AM, Gentile P, Chiono V, Ciardelli G (2012) Collagen for bone tissue regeneration. Acta Biomater 8(9):3191–3200. https://doi.org/10.1016/j.actbio.2012.06.014
14. Saravanan S, Leena RS, Selvamurugan N (2016) Chitosan based biocomposite scaffolds for bone tissue engineering. Int J Biol Macromol 93:1354–1365. https://doi.org/10.1016/j.ijbiomac.2016.01.112
15. Newman P, Minett A, Ellis-Behnke R, Zreiqat H (2013) Carbon nanotubes: their potential and pitfalls for bone tissue regeneration and engineering. Nanomedicine Nanotechnol Biol Med 9(8):1139–1158. https://doi.org/10.1016/j.nano.2013.06.001
16. Gao C, Feng P, Peng S, Shuai C (2017) Carbon nanotube, graphene and boron nitride nanotube reinforced bioactive ceramics for bone repair. Acta Biomater 61:1–20. https://doi.org/10.1016/j.actbio.2017.05.020
17. Ebrahimi S, Montazeri A, Rafii-Tabar H (2013) Molecular dynamics study of the interfacial mechanical properties of the graphene–collagen biological nanocomposite. Comput Mater Sci 69:29–39. https://doi.org/10.1016/j.commatsci.2012.11.030
18. Sharma A, Molla MS, Katti KS, Katti DR (2017) Multiscale models of degradation and healing of bone tissue engineering nanocomposite scaffolds. J Nanomechanics Micromechanics 7(4):1–14. https://doi.org/10.1061/(asce)nm.2153-5477.0000133
19. Scocchi G, Posocco P, Fermeglia M, Pricl S (2007) Polymer-clay nanocomposites: a multiscale molecular modeling approach. J Phys Chem B 111(9):2143–2151. https://doi.org/10.1021/jp067649w
20. Heller H, Grubmüller H, Schulten K (1990) Molecular dynamics simulation on a parallel computer. Mol Simul 5(3–4):133–165
21. Zhang C, Coasne B, Guyer R, Derome D, Carmeliet J (2020) Moisture-induced crossover in the thermodynamic and mechanical response of hydrophilic biopolymer. Cellulose 27:89–99
22. Bejagam KK, Gupta NS, Lee K-S, Iverson CN, Marrone BL, Pilania G (2022) Predicting the mechanical response of polyhydroxyalkanoate biopolymers using molecular dynamics simulations. Polym (Basel) 14(2):345
23. Golberg D et al (2007) Direct force measurements and kinking under elastic deformation of individual multiwalled boron nitride nanotubes. Nano Lett 7(7):2146–2151. https://doi.org/10.1021/nl070863r
24. Arenal R, Wang M-S, Xu Z, Loiseau A, Golberg D (2011) Young modulus, mechanical and electrical properties of isolated individual and bundled single-walled boron nitride nanotubes. Nanotechnology 22(26):265704. https://doi.org/10.1088/0957-4484/22/26/265704
25. Parashar A, Mertiny P (2012) Study of mode i fracture of graphene sheets using atomistic based finite element modeling and virtual crack closure technique. Int J Fract 176(1):119–126. https://doi.org/10.1007/s10704-012-9718-y

26. Li C, Chou T-W (2003) A structural mechanics approach for the analysis of carbon nanotubes. Int J Solids Struct 40(10):2487–2499. https://doi.org/10.1016/S0020-7683(03)00056-8
27. Boldrin L, Scarpa F, Chowdhury R, Adhikari S (2011) Effective mechanical properties of hexagonal boron nitride nanosheets. Nanotechnology 22(50):505702. https://doi.org/10.1088/0957-4484/22/50/505702
28. Kudin KN, Scuseria GE, Yakobson BI (2001) C_2F, BN, and C nanoshell elasticity from *ab initio* computations. Phys Rev B 64(23):235406. https://doi.org/10.1103/PhysRevB.64.235406
29. Andrew RC, Mapasha RE, Ukpong AM, Chetty N (2012) Mechanical properties of graphene and boronitrene. Phys Rev B 85(12):125428. https://doi.org/10.1103/PhysRevB.85.125428
30. Peng Q, Ji W, De S (2012) Mechanical properties of the hexagonal boron nitride monolayer: *ab initio* study. Comput Mater Sci 56:11–17. https://doi.org/10.1016/j.commatsci.2011.12.029
31. Tersoff J (1986) New empirical model for the structural properties of silicon. Phys Rev Lett 56(6):632–635. https://doi.org/10.1103/PhysRevLett.56.632
32. Tersoff J (1988) New empirical approach for the structure and energy of covalent systems. Phys Rev B 37(12):6991–7000. https://doi.org/10.1103/PhysRevB.37.6991
33. Tersoff J (1988) Empirical interatomic potential for carbon, with applications to amorphous carbon. Phys Rev Lett 61(25):2879–2882. https://doi.org/10.1103/PhysRevLett.61.2879
34. Tersoff J (1989) Modeling solid-state chemistry: interatomic potentials for multicomponent systems. Phys Rev B 39(8):5566–5568. https://doi.org/10.1103/PhysRevB.39.5566
35. Jones JE (1924) On the determination of molecular fields.—II. From the equation of state of a gas. Proc R Soc Lond Ser A, Contain Pap Math Phys Character 106(738):463–477
36. Daw MS, Baskes MI (1984) Embedded-atom method: derivation and application to impurities, surfaces, and other defects in metals. Phys Rev B 29(12):6443–6453. https://doi.org/10.1103/PhysRevB.29.6443
37. Jeong B-W, Lim J-K, Sinnott SB (2007) Tensile mechanical behavior of hollow and filled carbon nanotubes under tension or combined tension-torsion. Appl Phys Lett 90(2)
38. Stuart SJ, Tutein AB, Harrison JA (2000) A reactive potential for hydrocarbons with intermolecular interactions. J Chem Phys 112(14):6472–6486. https://doi.org/10.1063/1.481208
39. MacKerell AD Jr, Banavali N, Foloppe N (2000) Development and current status of the CHARMM force field for nucleic acids. Biopolym Orig Res Biomol 56(4):257–265
40. Sun H, Ren P, Fried JR (1998) The COMPASS force field: parameterization and validation for phosphazenes. Comput Theor Polym Sci 8(1):229–246. https://doi.org/10.1016/S1089-3156(98)00042-7
41. Brenner DW (1990) Empirical potential for hydrocarbons for use in simulating the chemical vapor deposition of diamond films. Phys Rev B 42(15):9458–9471. https://doi.org/10.1103/PhysRevB.42.9458
42. Matsunaga K, Iwamoto Y (2001) Molecular dynamics study of atomic structure and diffusion behavior in amorphous silicon nitride containing boron. J Am Ceram Soc 84(10):2213–2219. https://doi.org/10.1111/j.1151-2916.2001.tb00990.x
43. Matsunaga K, Fisher C, Matsubara H (2000) Tersoff potential parameters for simulating cubic boron carbonitrides. Jpn J Appl Phys 39(Part 2, No. 1A/B):L48–L51. https://doi.org/10.1143/jjap.39.l48
44. Albe K, Möller W (1998) Modelling of boron nitride: atomic scale simulations on thin film growth. Comput Mater Sci 10(1):111–115. https://doi.org/10.1016/S0927-0256(97)00172-9
45. Albe K, Möller W, Heinig K-H (1997) Computer simulation and boron nitride. Radiat Eff Defects Solids 141(1–4):85–97. https://doi.org/10.1080/10420159708211560
46. Nord J, Albe K, Erhart P, Nordlund K (2003) Modelling of compound semiconductors: analytical bond-order potential for gallium, nitrogen and gallium nitride. J Phys Condens Matter 15(32):5649–5662. https://doi.org/10.1088/0953-8984/15/32/324
47. Sevik C, Kinaci A, Haskins JB, Çağın T (2011) Characterization of thermal transport in low-dimensional boron nitride nanostructures. Phys Rev B 84(8):85409. https://doi.org/10.1103/PhysRevB.84.085409
48. Rath A, Mathesan S, Ghosh P (2016) Nanomechanical characterization and molecular mechanism study of nanoparticle reinforced and cross-linked chitosan biopolymer. J Mech Behav Biomed Mater 55:42–52. https://doi.org/10.1016/j.jmbbm.2015.10.005

49. Kulasinski K, Guyer R, Keten S, Derome D, Carmeliet J (2015) Impact of moisture adsorption on structure and physical properties of amorphous biopolymers. Macromolecules 48(8):2793–2800. https://doi.org/10.1021/acs.macromol.5b00248
50. Lai ZB, Bai R, Lei Z, Yan C (2018) Interfacial mechanical behaviour of protein–mineral nanocomposites: a molecular dynamics investigation. J Biomech 73:161–167. https://doi.org/10.1016/j.jbiomech.2018.03.044
51. Patel M, Dubey DK, Singh SP (2020) Phenomenological models of Bombyx mori silk fibroin and their mechanical behavior using molecular dynamics simulations. Mater Sci Eng C 108:110414. https://doi.org/10.1016/j.msec.2019.110414
52. Pandele AM, IONIŢĂ M, Iovu H (2014) Molecular modeling of mechanical properties of the chitosan based graphene composites. UPB Sci Bull Ser B Chem Mater Sci 76:107–112
53. Rahman R, Foster JT, Haque A (2013) Molecular dynamics simulation and characterization of graphene-cellulose nanocomposites. J Phys Chem A 117(25):5344–5353. https://doi.org/10.1021/jp402814t
54. Bayoumy AM et al (2019) Functionalization of graphene quantum dots (GQDs) with chitosan biopolymer for biophysical applications. Opt Quantum Electron 52(1):16. https://doi.org/10.1007/s11082-019-2134-z
55. Menazea AA et al (2020) Chitosan/graphene oxide composite as an effective removal of Ni, Cu, As, Cd and Pb from wastewater. Comput Theor Chem 1189:112980. https://doi.org/10.1016/j.comptc.2020.112980
56. Masoumi M, Jahanshahi M, Ahangari MG, Darzi GN (2020) Density functional theory study on the interaction of chitosan monomer with TiO_2, SiO_2 and carbon nanotubes. Mater Chem Phys 255:123576. https://doi.org/10.1016/j.matchemphys.2020.123576
57. Omar A, Bayoumy AM, Aly AA (2022) Functionalized graphene oxide with chitosan for dopamine biosensing. J Funct Biomater 13(2). https://doi.org/10.3390/jfb13020048
58. Kossovich EL, Safonov RA (2016) Predictive analysis of chitosan-based nanocomposite biopolymers elastic properties at nano- and microscale. J Mol Model 22(4):75. https://doi.org/10.1007/s00894-016-2942-z
59. Shende P, Pathan N (2021) Potential of carbohydrate-conjugated graphene assemblies in biomedical applications. Carbohydr Polym 255:117385. https://doi.org/10.1016/j.carbpol.2020.117385
60. Jayakumar A et al (2023) Energy harvesting using high-strength and flexible 3D-printed cellulose/hexagonal boron nitride nanosheet composites. ACS Appl Nano Mater 6(15):14278–14288. https://doi.org/10.1021/acsanm.3c02233
61. Anota EC, Rodríguez LD, Cocoletzi GH (2013) Influence of point defects on the adsorption of chitosan on graphene-like BN nanosheets. Graphene 1(2):124–130
62. Verma A, Parashar A (2017) The effect of STW defects on the mechanical properties and fracture toughness of pristine and hydrogenated graphene. Phys Chem Chem Phys 19(24):16023–16037
63. Verma A, Parashar A (2018) Molecular dynamics based simulations to study the fracture strength of monolayer graphene oxide. Nanotechnology 29(11):115706
64. Verma A, Parashar A, Packirisamy M (2018) Tailoring the failure morphology of 2D bicrystalline graphene oxide. J Appl Phys 124(1)
65. Verma A, Parashar A (2018) Molecular dynamics based simulations to study failure morphology of hydroxyl and epoxide functionalised graphene. Comput Mater Sci 143:15–26
66. Verma A, Parashar A (2018) Reactive force field based atomistic simulations to study fracture toughness of bicrystalline graphene functionalised with oxide groups. Diam Relat Mater 88:193–203
67. Verma A, Parashar A, Packirisamy M (2018) Atomistic modeling of graphene/hexagonal boron nitride polymer nanocomposites: a review. Wiley Interdiscip Rev Comput Mol Sci 8(3):e1346
68. Verma A, Zhang W, Van Duin AC (2021) ReaxFF reactive molecular dynamics simulations to study the interfacial dynamics between defective h-BN nanosheets and water nanodroplets. Phys Chem Chem Phys 23(18):10822–10834

69. Singla V, Verma A, Parashar A (2018) A molecular dynamics based study to estimate the point defects formation energies in graphene containing STW defects. Mater Res Express 6(1):015606
70. Kataria A, Verma A, Sethi SK, Ogata S (2022) Introduction to interatomic potentials/forcefields. In: Forcefields for atomistic-scale simulations: materials and applications, pp 21–49. Springer Nature Singapore, Singapore. https://doi.org/10.1007/978-981-19-3092-8_2
71. Verma A, Kumar R, Parashar A (2019) Enhanced thermal transport across a bi-crystalline graphene–polymer interface: an atomistic approach. Phys Chem Chem Phys 21(11):6229–6237
72. Verma A, Parashar A, Packirisamy M (2019) Effect of grain boundaries on the interfacial behaviour of graphene-polyethylene nanocomposite. Appl Surf Sci 470:1085–1092
73. Chaurasia A, Verma A, Parashar A, Mulik RS (2019) Experimental and computational studies to analyze the effect of h-BN nanosheets on mechanical behavior of h-BN/polyethylene nanocomposites. J Phys Chem C 123(32):20059–20070
74. Shankar U, Sethi SK, Verma A (2022) Forcefields and modeling of polymer coatings and nanocomposites. In: Forcefields for atomistic-scale simulations: materials and applications, pp 81–98. Springer Nature Singapore, Singapore. https://doi.org/10.1007/978-981-19-3092-8_4

Chapter 5
Modelling and Simulation of Composite Structure Using Ansys Composite PrePost (ACP) Tool

Subhankar Chakraborty, Bhakti S. Malve, and Rajesh Govindan

1 Introduction

Effective material composition with precision analysis is needed to attain the desired strength and thermodynamic characteristics within the given environment. In ancient times, the Egyptians left visual records by documenting the earliest materials which were made from straw and mud to create composite for constructing buildings. A composite material also known as a (Composition Material) is formed by combining two or more materials which together produce desirable properties that cannot be achieved with a single constituent alone. In the mid-1920s, researchers embarked to discover materials possessing unique characteristics such as lightweight properties, high resistance to melting, and corrosion resistance. Initially, in the fabrication of aerospace structures materials like aluminum and its alloys have been used. After the mid-1950s, composition of materials like Vanadium and Titanium alloys started replacing aerospace sector, this evolution played a crucial role in enabling the successful execution of the Apollo mission in 1969. Composite material has two phases, matrix and the reinforcing phase. The reinforcement fibers and matrices maintain the unique properties in composites. So therefore, these materials have found in many applications like spanning from kitchen utensils to complex structures such as aircraft bodies. In the realm of composite materials, glass, carbon, and aramid fibers serve as the primary reinforcement components. These fibers are combined with matrices, which are responsible for binding the fibers together and are made up of thermoplastics and thermosetting materials [1–4].

In this chapter, the procedure for analysis of Simple and Curved Plate by using Ansys Composite PrePost (ACP) module for the static analysis is presented. Ansys

S. Chakraborty · B. S. Malve · R. Govindan (✉)
Department of Aerospace Engineering, Chandigarh University, Punjab, India
e-mail: raajgovindan@gmail.com

S. K. Sethi et al. (eds.), *Polymer Composites: From Computational to Experimental Aspects*, Materials Horizons: From Nature to Nanomaterials,
https://doi.org/10.1007/978-981-97-0888-8_5

ACP is a software tool within Ansys that is specifically designed for composite analysis. It helps engineers and designers to predict behavior of the composite material under various conditions. It is beneficial in design optimization, fatigue analysis, buckling and stability analysis, material modeling and failure analysis. ACP Pre is an add-on module dedicated to the modeling of layered composite structures. It is included with the Mechanical Enterprise licenses. ACP Pre allows users to define materials, import models, and set up composite fabrics, laminate assemblies, and element orientations. It also enables users to define the ply sequence for groups of elements. The module is integrated into the Workbench platform, allowing for the proper definition of stacking sequences to cover the entire structure. Whereas, ACP Post is the phase where the results of the model are processed. After the model is set up in ACP Pre and computed by the standard ANSYS solver, the results are post-processed in ACP Post. This includes deformations, thicknesses, and failure criteria. ACP Post offers efficient post-processing of composite-specific results quantities. ACP Post offers several advantages like it can accurately compute stress and deformation distributions over structural elements, helping engineers make informed design decisions. It allows for the incorporation of different materials and designs, helping engineers give flexibility in their analysis and optimization processes. It also provides comprehensive support for composite material analysis, which allows engineers to accurately simulate and predict the behavior of composite structures and offers advanced post-processing capabilities, allowing users to visualize and analyze simulation results in a clear and insightful manner [5, 6].

The chapter provides in-depth knowledge on ACP Pre/Post software tools and analysis of stress and deformation in a structure under a static load using composite material. The ACP (PrePost) Section is shown in Fig. 1. It also delves into Preprocessing in ANSYS ACP which includes Geometry Import, Material selection, Layups, Meshing, Boundary Conditions, Linear and Nonlinear Analysis. Post-processing involves individual behavior of plies within the composite layup which is been discussed with proper linkage of overall analysis in this chapter. The process flow chart of composite analysis in ACP is shown in Fig. 2.

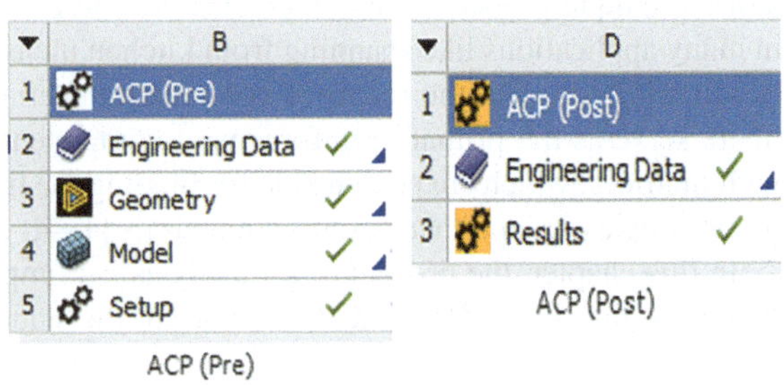

Fig. 1 ACP (PrePost) section

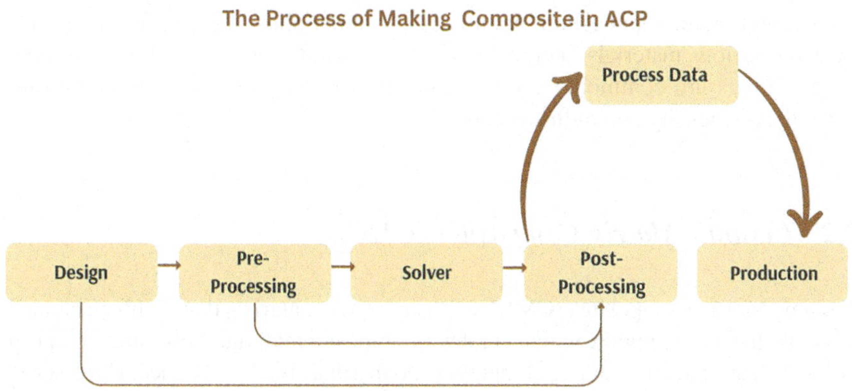

Fig. 2 Process flow chart of making composite in ACP

2 Fiber-Reinforced Composites

Fiber-reinforced composites can be categorized into four distinct groups according to their matrices: metal matrix composites (MMCs), ceramic matrix composites (CMCs), and polymer matrix composites (PMCs) or polymeric composites [7] as shown in Fig. 3.

2.1 Metal Matrix Composites (MMC)

In material science, the Metal Matrix Composites is a composite material that consists of a metal matrix reinforced with fibers to improve the material property such as strength, stiffness and thermal conductivity. The matrix in MMCs is typically a metal

Fig. 3 Types of composites

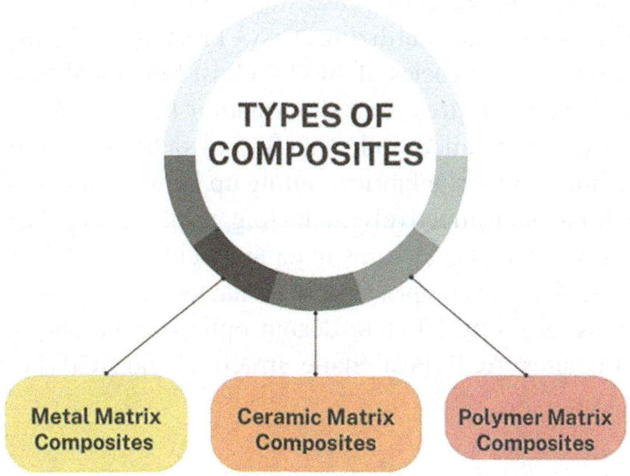

such as aluminum, magnesium, titanium, or a similar alloy. The reinforcing is done by using various materials, including ceramics, carbides or even carbon nanotubes. Carbon fibers are commonly used in aluminum matrix to synthesize composites showing low density and high strength.

2.2 Ceramic Matrix Composites (CMC)

Ceramic Matrix Composites (CMC) are the advance materials that combine ceramic fibers with ceramic matrix which creates a composite material with improved properties such as crack resistance, elongation and thermal shock resistance. CMC serves to strengthen the fracture toughness of the combined material system. It typically, consist of high-strength ceramic fibers like silicon carbide (SiC), alumina (Al_2O_3) and mullite ($Al_2O_3{-}SiO_2$). CMC has application where weight reduction is essential, such as in aircraft components as they are often lighter than metals and some other ceramics.

2.3 Polymer Matrix Composites (PMC)

Polymer Matrix Composites (PMC) also known as polymeric composites are the type of composite materials that consist of a polymer matrix reinforced with high-strength fibers. The polymer matrix is typically a thermosetting or thermoplastic polymer, such as epoxy, polyester, polypropylene, or nylon.

3 Case Study

The section presents a case study to perform static analysis of Simple and Curved Composite Plates utilizing Ansys PrePost (ACP) module. The focus lies on elucidating the intricacies of ACP PrePost functionalities to accurately model, simulate, and interpret stress and deformation behavior in composite structure. It aims to provide a detailed understanding of defining materials, importing models, establishing composite fabrics, setting up laminate assemblies, specifying element orientations, and effectively analyzing results using ACP Post. By delving into these aspects, the chapter aims to equip engineers and designers with the necessary skills to predict and comprehend the behavior of composite materials under varying conditions, enabling efficient design optimization, and failure prediction in composite structures. ACP (Pre), Static Structural and ACP (Post) Linkage is shown in Fig. 4

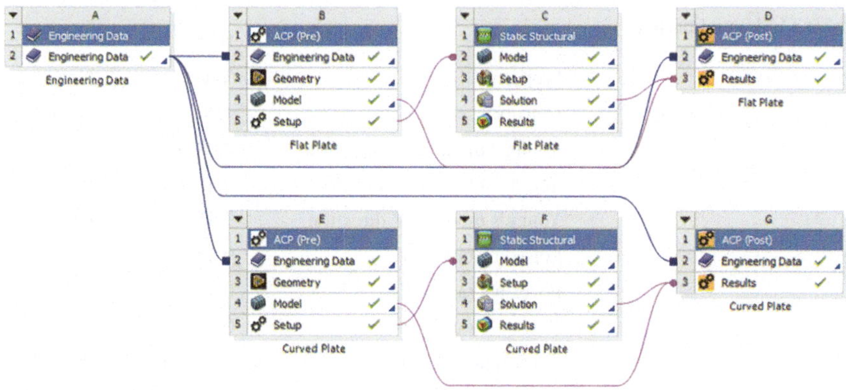

Fig. 4 ACP (Pre), static structural and ACP (Post) linkage

3.1 *ACP Pre*

The ACP Pre module is organized into distinct sections, containing Engineering Data, Geometry, Model, and Setup. After choosing the composite material from the Engineering Data the geometry is designed and is imported in the model section for mesh generation. The meshed model is then imported into setup. The fabric is assigned with thickness 0.125 mm. The stack-up angle for both the plates is defined as $[0, 90]_S$. The Rosettes properties are specified for the global coordinate system. Oriented Selection Sets are created by selecting all elements. The ply orientation and stacking sequence of composite laminates is defined using Modelling Groups.

3.1.1 Engineering Data

ACP has the PRE and POST processing mode. In the PRE-processing mode, the initial process commences with the selection of the material to be applied on the structure from Engineering Data library. The material chosen for the analysis of Simple and Curved Composite Plate is Epoxy Carbon unidirectional (UD) Prepreg (Table 1). Epoxy Carbon UD Prepreg is a composite material that consists of unidirectional carbon fiber reinforcements impregnated with an epoxy resin matrix which acts as a binding agent, enhancing the overall structural integrity of the composite. It a lightweight material known for its exceptional strength, stiffness, resistance to fatigue. The average density of the material is 1540 g/cc and the volume fraction range is 60–70%.

Carbon fibers are known for their exceptional strength-to-weight ratio. Epoxy resin provides protection against corrosion and environmental degradation. This property makes it a popular choice in industries where high strength and low weight are crucial [4].

Table 1 Material property

A	B	C
Property	Value	Unit
Density	1540	Kg m^{-3}
Young's modulus X direction	2.09×10^{11}	Pa
Young's modulus Y direction	9.45×10^{9}	Pa
Young's modulus Z direction	9.45×10^{9}	Pa
Poisson's ratio XY	0.27	
Poisson's ratio YZ	0.4	
Poisson's ratio XZ	0.27	
Shear modulus XY	5.5×10^{9}	Pa
Shear modulus YZ	3.9×10^{9}	Pa
Shear modulus XZ	5.5×10^{9}	Pa
Tsai—Wu Constants		
Coupling coefficient XY, YZ, XZ	−1	
Puck constants		
Material classification	Carbon	
Compressive inclination XZ	0.3	
Compressive inclination YZ	0.25	
Tensile inclination XZ	0.35	
Tensile inclination YZ	0.25	

3.1.2 Geometry

After the completion of material selection, the Simple and Curved Plate is designed in geometry section in Ansys SpaceClaim as shown in Fig. 5. The flat and curved surface has been generated for further composite analysis. The overall dimension for Simple Plate and Curved Plate is given in Table 2.

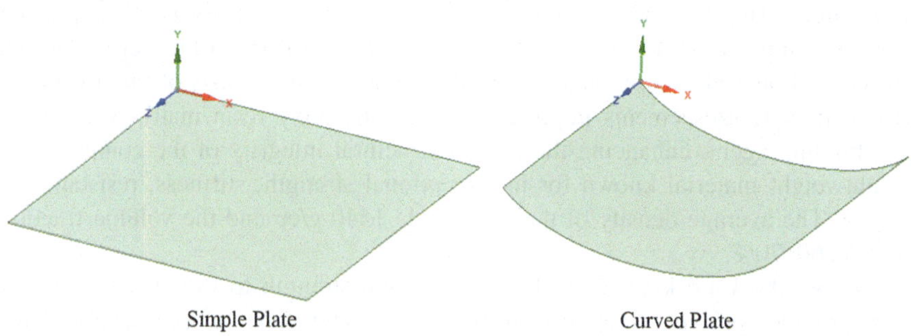

Fig. 5 Geometry of simple and curved plate

Table 2 Dimensions of simple and curved plate

Dimension	Simple plate (mm)	Curved plate (mm)
Length X	50	50
Length Y	0	6.0047
Length Z	50	50

Fig. 6 Mesh models

3.1.3 Model

The above geometry is then imported into model section for mesh generation. A thickness of 2.5 mm is given to the structure. The co-ordinate system chosen for the analysis is the global co-ordinate system. In order to achieve a high-quality mesh with fine details, an element size of 1 mm is set for mesh generation in sizing under the 'details of mesh' panel. High smoothing has been assigned to the mesh under 'Quality' with average skewness value that comes out to be 8.7412×10^{-5} for Simple Plate whereas, for Curved Plate it is 1.4983×10^{-2}. The mesh models are shown in Fig. 6. In the meshing process, the statistical data shows that 2601 nodes and 2500 elements have been generated in Simple Plate and in case of Curved Plate 2630 nodes and 2531 are generated.

3.1.4 Setup

The model is imported into the setup for composite analysis. Initially, the fabric is been assigned with thickness of 0.125 mm. With stack-up we can predefine a layup pattern and apply it to laminate referring to the arrangement of individual composite material layers in a composite structure. The stack-up angle given to both the plates are $[0, 90]_s$ as shown in Table 3. Rosettes are coordinate systems that are used to set the Reference Direction of Oriented Selection Sets. By specifying the material orientation using rosettes, one can accurately predict how the material will behave under various loads. It is used to define in-plane 0-degree reference direction. The Rosette's properties for the global coordinate system are given in the below Table 4.

Table 3 Stack-up angle

Fabric	Angle
Fabric. 1	0.0
Fabric. 2	90.0
Fabric. 3	90.0
Fabric. 4	0.0

Table 4 Rosette's properties

Direction	Co-ordinates
Origin (x, y, z)	(0.0000, 0.0000, 0.0000)
1 Direction (x, y, z)	(1.0000, 0.0000, 0.0000)
2 Direction (x, y, z)	(0.0000, 1.0000, 0.0000)

Oriented Selection Sets helps in creating a group of elements with a specific orientation. The oriented selection is set by selecting all elements as an element set. Oriented Selection properties for Simple and Curved Plates are shown in Fig. 7. The modelling groups are then used to define the ply orientation and stacking sequence of composite laminates. Modelling groups are created by selecting the oriented selection sets. Modelling ply properties are shown in Fig. 8. The fiber orientation of simple and curved plates are shown in Fig. 9.

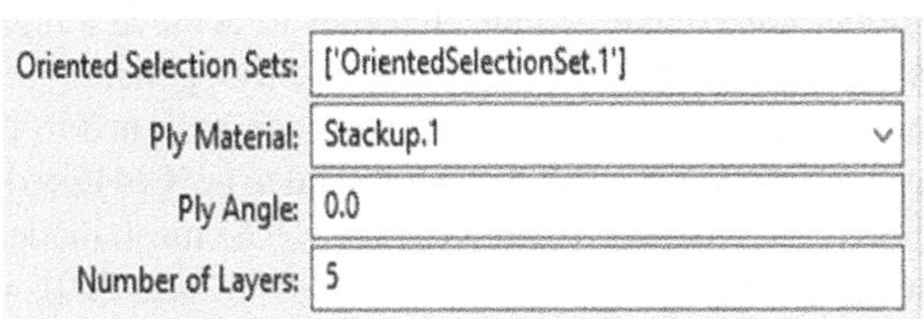

Fig. 7 Oriented selection sets properties

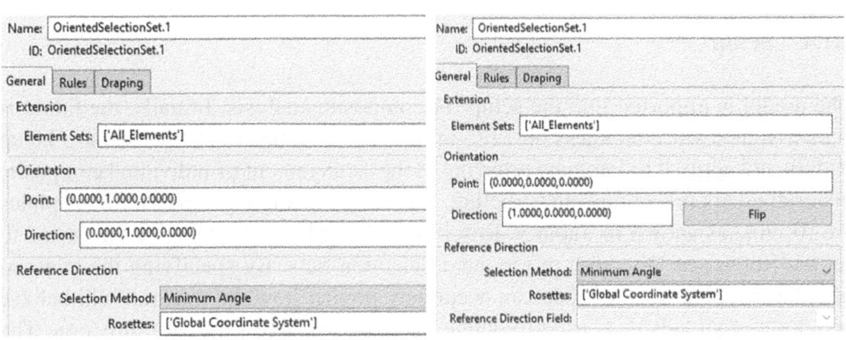

Fig. 8 Modeling ply property

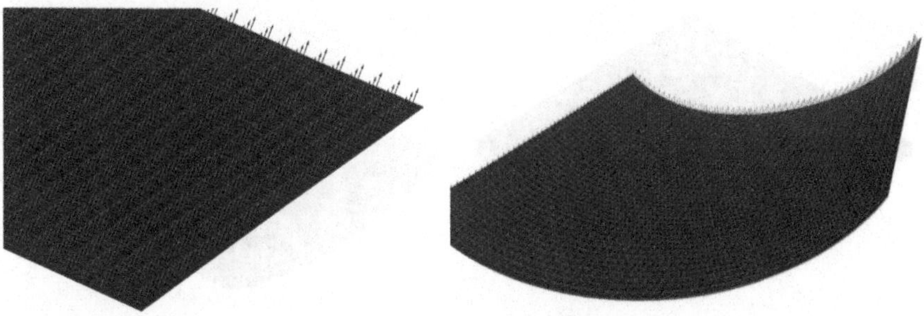

Simple Plate Fiber Orientation   Curved Plate Fiber Orientation

Fig. 9 Orientation of fiber

3.2 Analysis

The ACP PRE setup has been transferred into Ansys Static Structural. Following the application of the plies, the resulting geometric properties containing the mass and volume has been increased in both the plates. Table 5 shows the change in geometrical properties, nodes and elements of the structure. The discretized FE model for Simple and Curved Plates with coarse mesh are illustrated in Fig. 10. The plates have fixed support boundary conditions on each of its four faces as shown in Fig. 11. Both simple and curved plates are subjected to the load of magnitude 5000 N being applied on the top surface of the plate.

Table 5 Change in geometric properties of Simple and Curved Plate

Property	Simple plate	Curved plate
Mass (Kg)	1.125×10^{-2}	1.1392×10^{-2}
Volume (mm^3)	6250	6328.8
Elements	50000	50620
Nodes	54621	55230

Simple Plate   Curved Plate

Fig. 10 Mesh Models after application of plies

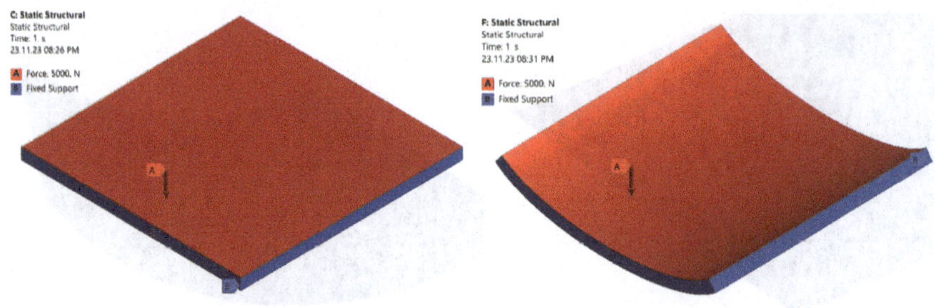

Fig. 11 Boundary conditions applied on the simple and curved plates

3.2.1 Solution

Figure 12 demonstrates the total deformation in Simple and Curved shape plate. The maximum deformation experience by the Simple Plate is 0.17886 mm while for the Curved plate it is 5.0476×10^{-2} mm. Figure 13 illustrates the total equivalent von-Mises stress for Simple and Curved shape plates. The maximum equivalent von-Mises stress for simple Plate is 555.42 MPa while for curved shape plate it is 180.99 MPa. The red zone indicates the area of maximum deflection and the blue zone indicates the area of minimum deflection.

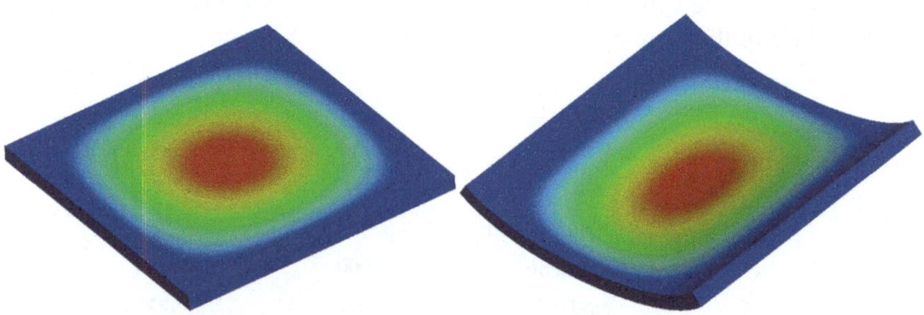

Fig. 12 Total deformation in the simple (Left) and curved (Right) plates

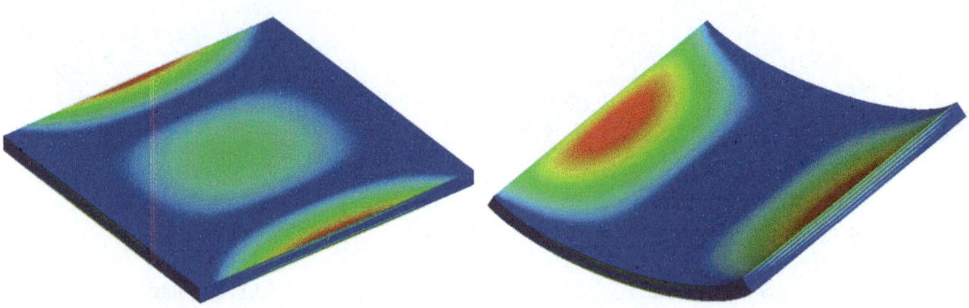

Fig. 13 Equivalent von-Mises stress distribution in the simple (Left) and curved (Right) plates

3.3 ACP Post

ACP (Post) facilitates to study the laminate failure and stress acting on each laminate for better understanding of the structural strength. Figure 4 shown above represents the linkage of ACP (pre) and Static Structural module with the ACP (Post) module to obtain proper results.

One of the most important things to evaluate in the ANSYS Composite Prep-Post (ACP) study is the stress distribution in each laminate. The stress distribution at the bottom layer is shown in Fig. 14, which sheds light on the composite structure's probable failure sites and load-bearing capacity. Similar to this, Fig. 15 shows the stress at the top layer and provides a thorough understanding of how the material responds to applied stresses. Furthermore, an important factor in assessing the material's safety margin is the Inverse Reverse Factor (IRF). The ratio of stress to material strength is used to compute the IRF. Potential failure is indicated by a value greater than 1. For evaluating the effectiveness and dependability of composite laminates, this critical measure is essential [1, 8, 9].

ACP (Post) can also export the results to other formats, such as Excel, CSV, or text files, for further processing or reporting. ACP (Post) is a powerful and user-friendly

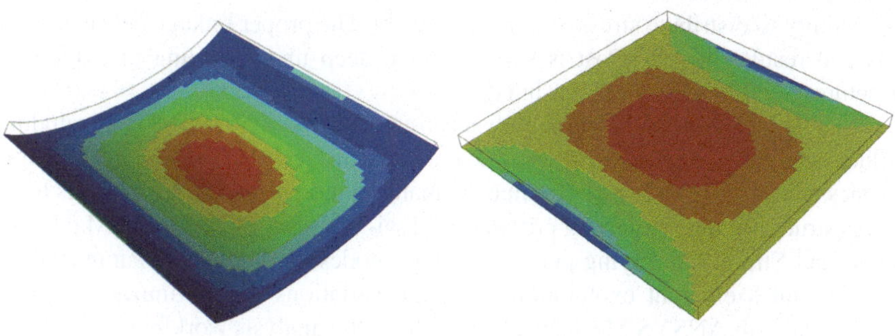

Fig. 14 Stress distribution at bottom layer

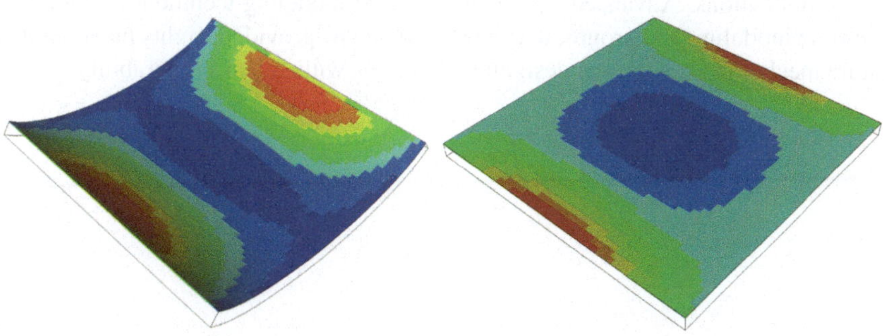

Fig. 15 Stress distribution at top layer

tool that can enhance the understanding and optimization of composite laminate designs.

In addition to the IRF, ACP Post provides a range of other results that can be used to analyze the performance of composite materials. For instance, ACP Post allows for efficient post-processing of composite-specific results quantities. This includes the ability to plot the worst-case failure criteria over all layers, offering insights into areas under maximum stress and prone to failure.

4 Conclusions

In conclusion, this chapter provides a detailed overview of the Modeling and Analysis process of Composite Structures using Ansys ACP Pre/Post Software. It emphasizes the significance of factors such as Stackup Angle, Rosettes Orientation, Oriented Selection properties, and Modelling Groups in accurately defining ply orientation and stacking sequence. These elements play a crucial role in determining the structural behavior of composite structures under various loading conditions.

The analyses compare the performance of simple and curved plates under static loading, highlighting the enhanced structural performance of curved plates due to their ability to distribute stresses more effectively. The proper linkage between analysis and results in ACP Post is stressed for a deep understanding of composite structures, enabling informed design decisions.

The chapter also underscores the advanced capabilities of ACP Pre/Post, including failure analysis, parameterization, integration with ANSYS Mechanical, multi-physics analysis, and advanced composite material modeling. Failure analysis tools assess structural integrity using criteria like Tsai-Wu, Tsai-Hill, Hashin, Maximum Stress and Strain, identifying potential failure modes and locations. Parameterization facilitates efficient exploration of design variations and optimization, while integration with ANSYS Mechanical streamlines the analysis workflow.

Multi-physics analysis supports the examination of composite structures under coupled mechanical, thermal, and electromagnetic loading scenarios, crucial for real-world applications. Advanced composite material modeling techniques, like cohesive zone modeling and progressive damage analysis, provide insights into complex failure mechanisms, aiding in designing structures with enhanced durability.

In summary, Ansys ACP Pre/Post Software emerges as an invaluable tool for engineers working with composite structures, offering a comprehensive suite of features for modeling, analysis, and optimization. Its ability to handle the complexities of composite materials and provide deep insights into their behavior makes it an essential tool for designing and analyzing composite structures in modern engineering challenges.

Declarations

Conflicts of Interest The authors confirm that there are no conflicts of interest associated with this work.

Ethical Compliance The authors affirm that this study adheres to all ethical guidelines.

Competing Interests No competing interests are declared by the authors.

Data and Material Availability Data and materials are not applicable to this study.

References

1. Kanani AY (2020) Ansys tutorial for ACP (Full composite tutorial in ANSYS) adhesive on multi material joints view project. (January 2017) (Online)
2. Ho K (2017) Fiber reinforced composites. C. E. Materials, U. Kingdom. https://doi.org/10.1016/B978-0-08-101871-2.00003-5
3. Balasubramanian K, Sultan MTH, Rajeswari N (2018) Manufacturing techniques of composites for aerospace applications. Elsevier Ltd. https://doi.org/10.1016/B978-0-08-102131-6.00004-9
4. Sapuan S, Ap CH (2017) Chapter 3-composite materials, vol 72. Elsevier Inc. https://doi.org/10.1016/B978-0-12-802507-9/00003-9
5. Nurhaniza M, Ariffin MKA, Ali A, Mustapha F, Noraini AW (2010) Finite element analysis of composites materials for aerospace applications. IOP Conf Ser Mater Sci Eng 11:012010. https://doi.org/10.1088/1757-899x/11/1/012010
6. Yancey RN (2016) Challenges, opportunities, and perspectives on lightweight composite structures: aerospace versus automotive. Elsevier Ltd. https://doi.org/10.1016/B978-1-78242-325-6.00002-5
7. Sharma AK, Bhandari R, Aherwar A (2019) Matrix materials used in composites: a comprehensive study. Mater Today Proc (xxxx):0–3. https://doi.org/10.1016/j.matpr.2019.11.086
8. de Sousa Burgani T (2021) Modeling flexural failure in composites using ANSYS ACP. MSE, The University of Texas at Austin
9. Venkatesan K, Ramanathan K, Vijayanandh R, Selvaraj S, Raj Kumar G, Senthil Kumar M (2019) Comparative structural analysis of advanced multi-layer composite materials. Mater Today Proc 27(xxxx):2673–2687. https://doi.org/10.1016/j.matpr.2019.11.247

Chapter 6
An Overview on the Joining Processes of Polymer Composites

Jayant Kumar, Pawan Kumar Rakesh, and Inderdeep Singh

1 Introduction

The development of composite materials has been a major emphasis in the past several decades in the fields of materials science and engineering. Composites are a preferred material over other ones because they provide a better balance of mechanical and physical qualities [1–3]. The traditional materials are now under fierce competition from composite materials. Industries including aeroplanes, automobiles, sports, and construction regularly employ composite materials to replace several components made of traditional materials. Their demand has increased in various structural and non-structural applications because of their high strength to weight ratio [4–6]. Materials that combine two or more elements are called composites. These elements are mixed and physically bonded to one another. At the bulk size, the components don't interact chemically. As a result, each component's physical importance is unaltered. Composite material components may be divided primarily into two groups: matrix and reinforcement. The reinforcing element is implanted in the matrix, which is a continuous media. Typically, it just supports the reinforcement; it does not contribute to load carrying. As matrix materials, metal, ceramic, and polymers are often employed. In composite materials, reinforcements are a sporadic phase that

J. Kumar
Department of Mechanical Engineering, Katihar Engineering College, Katihar-854109, Bihar, India

P. K. Rakesh (✉)
Department of Mechanical Engineering, National Institute of Technology Uttarakhand, Srinagar-246174, Uttarakhand, India
e-mail: pawankumarrakesh@gmail.com

I. Singh
Department of Mechanical and Industrial Engineering, Indian Institute of Technology Roorkee, Roorkee-247667, Uttarakhand, India

S. K. Sethi et al. (eds.), *Polymer Composites: From Computational to Experimental Aspects*, Materials Horizons: From Nature to Nanomaterials,
https://doi.org/10.1007/978-981-97-0888-8_6

bears the load. It could take the shape of sheets, particles, or threads. The high strength to weight ratio of composite materials is its most important quality. Many different types of matrixes and reinforcing materials are now accessible. By choosing various combinations of reinforcing and matrix material, one may readily build composite materials based on the needs or applications of a product. Many composite materials may be moulded into intricate forms, giving designers and manufacturers more creative freedom [7–9].

Man-made fibres are known as synthetic fibres (Fig. 1). These fibres (like carbon and glass fibres) are derived from petroleum-based resources. Synthetic fibre reinforced composites are widely employed in a variety of engineering applications; however, these materials confront challenges due to the rapid depletion of petroleum-based resources. The inability of synthetic fibres to decompose biologically is another issue [10–12]. As a result, scientists are investigating the use of natural fibres in fibre reinforced composites. Natural fibres are biodegradable by nature, and unlike synthetic fibres, their resources are not scarce. Renewable resources are used to produce natural fibres. These are widely accessible and readily available. Natural fibres are mostly derived from plants, animals, and minerals. The majority of natural fibres come from plants. Natural fibres are abundant in plants. Frequently utilised natural fibres for processing NFRCs include jute, sisal, bagasse, banana, bamboo, hemp, kenaf, and bagasse [13–15]. These fibres are taken from a variety of plant elements, including the wood, seed, and leaves. Lignin, the outermost component of plant fibre, encloses every other component [16–18]. Natural plant fibre is hydrophilic because of the presence of hydroxyl groups. The moisture in the air easily reacts with natural fibres. This poses a significant problem for natural fibres. In comparison to the other components of plant fibre, cellulose makes up a larger portion. Its percentage ranges from 40 to 80% [19–21]. The strength of plant fibre rises along with the proportion of cellulose. It has also been seen with NFRCs, larger mechanical strength in terms of tensile, flexural, and impact is seen when fibre with a larger proportion of cellulose is employed to build the composites [22–24]. However, the hydrophilicity of plant fibres poses the most problem. Researchers have experimented with chemically treating fibres to increase their hydrophilicity. The most popular kind of therapy for this is alkaline. The hydrophilicity of the fibres may be decreased with the use of these treatments. Natural fibres are more compatible with polymers after the treatment. Natural fibres may sometimes exhibit poor adherence to polymers when left untreated. Natural fibres are said to have a lower tensile strength than synthetic fibres. Natural fibres are less dense than E-glass fibres in terms of density. As a result, the specific tensile strength of natural fibres may be higher than that of E-glass fibres [25–27]. In many non-structural applications, natural fibres offer a significant potential to replace synthetic fibres. Both thermosets and thermoplastics may be used as the matrix materials for NFRC development. Although both thermosets and thermoplastics are polymers, they vary in terms of their physical and chemical characteristics. Polymeric chains may be branched or linear in thermoplastics. They tend to soften after heating and revert to their previous condition after chilling, for this reason. Thermosets, on the other hand, are made of crosslinked and network-type molecular arrangements. Heat is needed to cure them. Thermosets cannot return to their original condition after curing. As a result, the basic processing methods for the two kinds of polymers vary [28–30]. Selected main processing methods for polymer matrix composites are shown in Fig. 2. Hand lay-up is the processing method for thermoset-based NFRCs. Any sort of fibre, including short and long fibres as well as particulate, woven, unidirectional, and bidirectional fibres, may be laid up by hand. The procedure of hand lay-up is simple, affordable, and flexible. Although fundamental processes produce products that are almost net-shaped, the intricacy of the product sometimes requires that the end product be produced in segments. The components produced by primary processing or manufacturing are subject to secondary processing, which includes machining, drilling, and connecting [31–33].

Although joining polymer matrix composites seems straightforward, it is a really difficult procedure. Improper joint design causes stress to concentrate in the joint

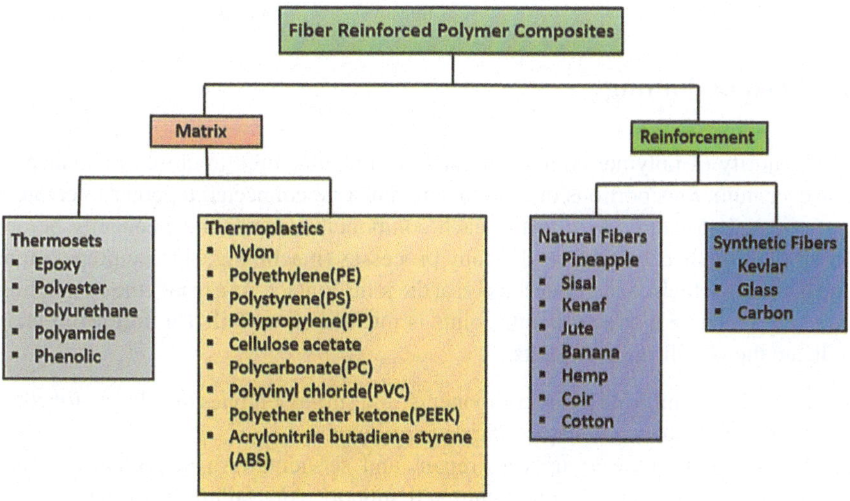

Fig. 1 Classification of fibre reinforced polymer composites

Fig. 2 Primary processing techniques for PMCs

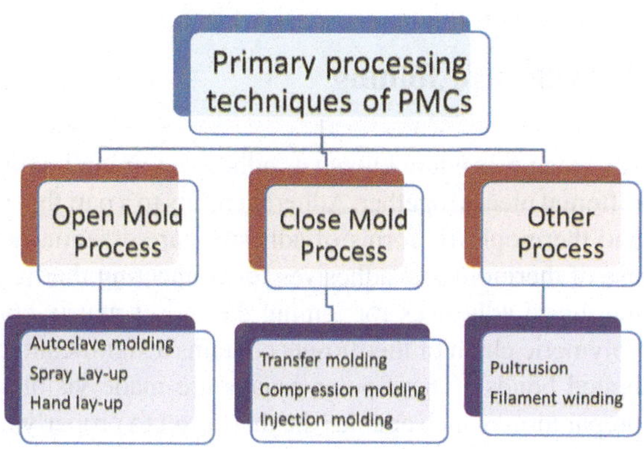

area and might result in failure over the product's service life. The joining behaviour of polymer matrix composites must thus be studied. Polymer matrix composites may be joined using one of three fundamental techniques: adhesive joining, mechanical fastening, or fusion bonding. The final product's intended usage and application will determine the particular procedure that is used. Due to the lower stress operating on the joint section, adhesive and fusion joining is preferred for non-structural applications. Mechanical fastening is used for structural applications because to its great load bearing capability. The joining behaviour of synthetic fibre reinforced polymer composites utilising all three joining procedures has been the subject of several research studies. Researchers have also looked at the behaviour of hybrid composites when they are joined to other materials, such as metal and plastic. For synthetic fibre reinforced composites, advanced joining methods such as induction joining, and microwave joining have been documented in the literature [34–36].

2 Need of Joining

The majority of polymer composite parts are manufactured in close to net shapes, however numerous portions or components must be connected together to create a complicated structure. As a result, each component is made independently before being put together. Joining needs many processes (machining, surface preparation, and heating), which concentrate tension at the joint contact. As a result, the connecting interface is where a structure with joints is most likely to fail. The following facts indicate the significance of joints.

1. Individual composite parts, components, and sub-assemblies may be put together to create complex and large composite products.
2. The product's ease of upkeep, repair, and servicing. An individual modular product is challenging to maintain and support. Any failure or damage necessitates the replacement of the whole product. Any defective component in a manufactured product may be quickly and simply removed and fixed.

3 Adhesive Joining

A joining procedure known as adhesive joining involves glue to affix two similar or different pieces together. Adherent refers to a part that will be attached. The thermoset and thermoplastic forms of adhesives are essentially the two types employed. The use of thermoplastic adhesives for connecting thermoplastic matrix composites and thermoset adhesives for joining thermoset matrix composites are compatible. The polymeric chain of thermosets contains a significant number of polar groups (unsaturated bonds). These polar groups are made visible to the surface during surface preparation, which enables the thermoset to properly bind with adhesives [37]. Thermoplastics, on the other hand, feature lengthy, saturated polymeric chain structures.

In comparison to thermosets, these polymeric chains contain fewer polar groups than thermosets. The connection between thermoplastics and glue becomes weak as a result. Advanced surface preparation procedures, such as plasma treatment, corona discharge treatment, and ion treatment are needed to improve the surface property of thermoplastics [38]. The procedure boosts the thermoplastic's ability to connect with glue by raising the number of free radicals on its surface. The use of solvent at the joining surface, which chemically interacts with the surface and softens the thermoplastic, and applying pressure between the joining surfaces are two additional techniques to improve the bonding of thermoplastics with the adhesive. Between adherends and adhesive, a solid connection is formed as a result. Following are the procedures for connecting polymer matrix composites using adhesives:

a. Making composite elements that will be connected.
b. Choosing an adhesive based on the items being adhered to.
c. Joint design is chosen in accordance with the needs of the product.
d. Surface preparation for adherends in accordance with procedure.
e. Apply the appropriate quantity of glue to the chosen design arrangement.
f. Using pressure and temperature to consolidate properly.
g. Adhesive curing at the joint contact.

Advantages of adhesive joining

Adhesive joining is a straightforward joining method with low startup costs and there is no need for large machines or equipment to attach the parts or components. It is a method of permanent connecting. It may be used to combine adherends that are thin and thick. Since it is challenging to attach components with big surfaces using other joining methods, adhesive joining is a simple way to do it. In the case of mechanical joining for lap joint design, stress concentration is greater at the edges as well as close to the hole where the fastener is fastened. In the case of adhesive joining for lap joint configuration, load is evenly distributed at the joint section except at the edge of the joint. Heat Affected Zone (HAZ) is absent in adherend-like samples that are bonded using the fusion bonding approach in adhesive joining. These are a few benefits of adhesive joining over alternative joining methods [39–41].

Limitations of adhesive bonding

Surface preparation of adherends is necessary for adhesive joining, which is a laborious job. A joint failure at an early stage may result from inappropriate adherend surface preparation. Due to its permanent nature, adhesive joints are difficult to inspect, and using these joining methods makes it impossible to repair a damaged component. Long curing times for glue are required for adhesive joining, which reduces productivity. Adhesives may sometimes be sensitive to their working conditions and chemical reactions. Working temperature is also a factor in adhesive stability. At higher temperatures, it can deteriorate. As a result, the working temperature determines where adhesive joining may be used.

4 Mechanical Joining

The first technique used to link metallic components is mechanical joining. It may also be used to link composite materials. To stop unwanted movement and inadvertent disassembly, mating surfaces of connecting components must interlock or interfere with one another. Mechanical fastening is the process of adding an additional supplement (mechanical fastener) when it is necessary for the interference force and subsequent interlocking. Integral mechanical attachment occurs when geometrical characteristics are offered to the mating surfaces for interference forces and interlocking. On the basis of the kinds of fasteners employed, mechanical fastening is further classified into two categories: (a) threaded fasteners; and (b) unthreaded fasteners. Unthreaded fasteners produce interference by bearing action, but threaded fasteners need preloading to provide interference between mated components. To apply a mechanical fastener, a hole must be drilled in the specimen that is being joined. Given that composites are anisotropic by nature, drilling a hole in a composite specimen is a difficult process. The production of holes is influenced by the kinds of fibre, matrices, laminate stacking order, and processing methods. As a result, choosing specific drilling parameters without first optimising them is challenging [42–44].

Drilling may occasionally be seen with the naked eye and leads to a number of flaws in fibre reinforced composites, including edge chipping, delamination, spalling, and fuzzing. Under a microscope, imperfections like fracture development and damage to the peripheral zone are seen. Types of fibre, tool geometry, drilling settings, and drilling environment all affect drilling-induced flaws. Because carbon and glass fibres, which are often employed in composites as reinforcement, are abrasive by nature, the majority of flaws result from a change in tool geometry during drilling operations. When a drilling tool comes into touch with fibres, erosion occurs and the instrument's dimensions alter. When drilling a hole, an eroded tool produces larger drilling forces. When using a worn-out tool to drill, compressive force rather than shear force is produced, leaving a hole that is damaged and faulty [45–47].

Advantages of mechanical joining

The benefits of mechanically fastened joints are that these can be disassembled, repaired, and easily inspected. Mechanical fasteners in mechanical joining provide these advantages which make them popular among other joining processes. There is no need of surface preparation. It requires machining operation for fixing the fastener on the joining component. In mechanical joining, thermal stress and HAZ is also absent near the joint interface.

Limitations of mechanical joining

The major limitation of mechanical joining in composite materials is drilling of holes in composites. Drilling holes in composite materials is still under investigation. Drilling in composite materials causes several types of defects, like, spalling, fuzzing and delamination. Apart from that, if drilling is done on NFRCs, then natural fibres are exposed to environment. After that, natural fibres easily react with moisture and chemicals present in environment. Another limitation is that the weight of the mechanical fasteners used during mechanical joining increases the overall weight of the components or parts. In mechanical joining, the strength of the material is degraded due to material removal during machining. These are some important limitations of mechanical joining [48–50].

5 Fusion Bonding

Fusion bonding is a method of connecting that is mostly used with thermoplastics. Since thermosets cannot be remelted or remoulded and cure irrevocably, it does not apply to them. Thermosets cannot be reversed after they have hardened, but thermoplastics have a reversible nature. So, using fusion bonding methods, thermoplastic-based composites may be linked with ease.

Advantages of fusion welding

Materials may be connected quickly using the fusion welding process. Surface preparation is not necessary in this method for connecting specimens. Microwave welding, induction welding, resistance welding, ultrasonic welding, and friction welding are some of the fusion welding methods that are accessible. For both polymers and metals, each of these procedures is well-established. Metals and polymers are joined using these methods on an industrial basis. Fusion welding processes may be mechanised and employed in large-scale manufacturing [51–53].

Limitations of fusion welding

The method of fusion welding is only appropriate for usage with thermoplastics matrix composites. The attaching of polymers to ceramics or metals are two examples of materials that cannot be joined using this method. Due to their dissimilar chemical makeup, distinct kinds of polymers (thermoplastics) are challenging to combine even using this technique. In this method, a difficult-to-control HAZ is produced on the parent material during the welding process. A susceptor material may sometimes be needed during the fusion welding process to generate heat at the joining interface.

6 Joining of Natural Fiber Reinforced Composites

6.1 Adhesive Joining

The most common method for combining various sorts of materials during the past several decades is adhesive joining. A variety of adhesives that are used for connecting have been produced in the past. These adhesives are used to link polymer composites (with thermosets as the matrix), alloys, and metals. The literature that is now available on adhesive joining demonstrates that this method is also workable for attaching NFRCs. Modern materials known as NFRCs have gained popularity recently due to their qualities of being lightweight and having a high specific strength. Understanding the joining behaviour of NFRCs using adhesives has been the subject of research studies. The most important choice is choosing an appropriate adhesive for a certain NFRC. It relies on the price of the glue, matrix, and reinforcement used in the composite production process, as well as the applications and service conditions. The majority of adhesives used for polymer-based composites are structural adhesives [54–56].

Bonding mechanism in adhesive joining

The adhesive joining method essentially involves four bonding processes. These include chemical bonds, physical bonds, diffusion, and mechanical interlocking. Figure 3. depicts the various mechanisms schematically. Surface energy, reactive energy, and surface texture all play a role in the bonding procedures that take place when adherend surfaces are joined. Each component has a significant impact. High surface energy makes it simple to glue the connecting surfaces together. These bonding processes might be chemical or physical. The rough roughness of the connecting surfaces facilitates mechanical interlocking during the joining process [57, 58].

Mechanical interlocking

On the adherend surface, cavities and imperfections may be seen at the microscopic level. The terms "crest" and "trough" refer to these voids and irregularities. Adhesive is applied to the adherend surface during adhesive joining, filling any gaps (troughs) or imperfections that may be present. Following the application of pressure to release the trapped air, glue is used to adequately seal the void cavities. Adhesive produces interlocking between adherend surfaces in this manner, forming a junction. Therefore, it is clear that the presence of voids and other abnormalities on adherend surfaces directly affects joint strength. If not, surface preparation techniques must be used to generate them. The adherend's surface becomes rougher as a result of surface preparation techniques. It involves techniques including sanding,

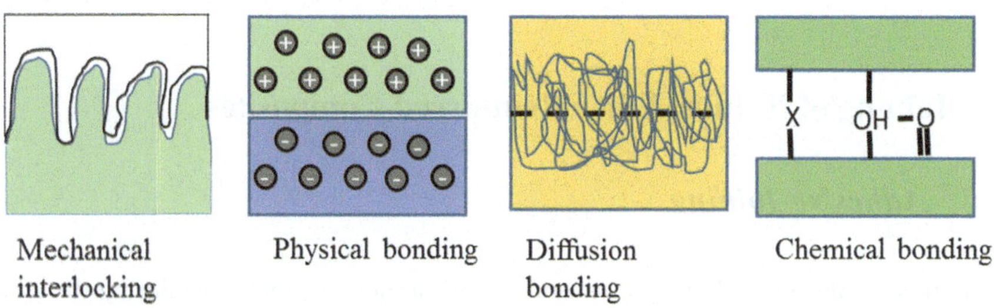

Fig. 3 Bonding mechanisms during adhesive joining

grit blasting, and mechanical abrasion. Different grades of emery paper are used in mechanical abrasion and sanding to create a rough surface, and abrasives are used in grit blasting to create a rough surface [59, 60].

Physical bonding

Physical bonds between the adherend and adhesive are created by low energy molecule interactions such as hydrogen bonds, van der Waal forces, and dipolar interactions. The electronegative differential between the adherend and the adhesives is crucial to this kind of bonding action.

Diffusion bonding

Thermoplastics are the primary materials for diffusion bonding. In this bonding method, the adherend's surface is treated with a solvent or chemical to make it partly or completely soluble. One adherend diffuses into another adherend at the connecting contact once pressure is applied. When identical kinds of adhesives and adherend are employed, diffusion bonding effectiveness improves.

Chemical bonding

High surface energy of the adherend improves wettability and bonding strength. Due to the presence of polar groups in thermosets, the adherend surface has a high surface energy. Due to their high reactivity, these polar groups form chemical bonds with adhesives very quickly. In the case of thermoplastics, these polar groups are smaller. The polymeric chain of thermosets has an unsaturated bond and a network structure. Network or polymeric chains break during surface preparation, providing polar groups at the surface. The only polymeric chains found in thermoplastics are linear and branched chains, which after surface preparation give fewer polar groups at the surface. Thus, corona discharge, ion, and plasma treatments all increase the surface energy of thermoplastics.

6.1.1 Types of Adhesive Joint

Adhesives are used to create many sorts of joins. In particulate-filled composites reinforced with jute and silk fibres, Jeevan et al. [61] investigated the joining behaviour of single and double lap joints. Particulates from eggs and coconuts were employed as filler in the composite. The strength of the joints was enhanced by the filler. Composites with egg shell filling have been reported to have greater lap shear strength. At the adhesive adherend joint contact, the adhesive joint failure may be seen in Fig. 4. Inadequate surface preparation, the inappropriate adhesive choice, and excessive peel tensions all contribute to adhesive failure. The mechanical interlock between the adhesive and adherend is reduced by poor surface preparation, which may lead to adhesive failure. A second reduction in adhesion between the adhesive and adherend due to improper adhesive selection results in an adhesive mode of failure. It is because the adhesive and adherend have a strong relationship rather than because each of them

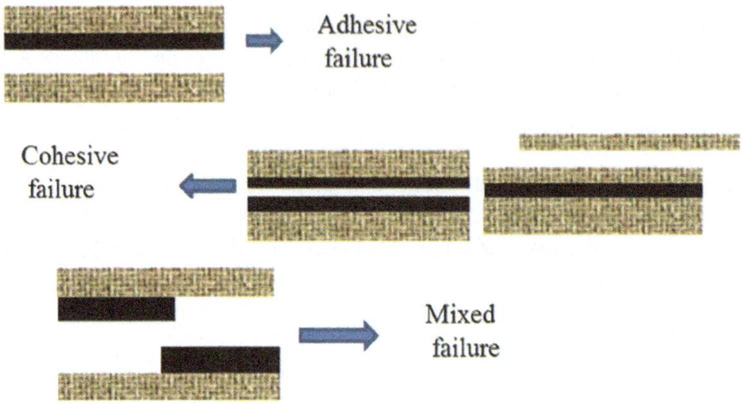

Fig. 4 Common failure mechanisms in adhesive joints

is particularly strong. Cohesive and adhesive failure occur concurrently in mixed types of failure [62, 63].

6.2 *Mechanical Joining*

One of the earliest techniques for combining various kinds of materials is mechanical joining. It is widely used to link composite materials as well. The mechanical connection of thermoset and thermoplastic matrix composites is possible. The most popular method of mechanical joining is mechanical fastening. It makes it simple to put together and take apart assembly components. As a result, in the case of mechanical fastening, joint maintenance is quite simple. Drilling holes in components to secure the fasteners is a significant problem during mechanical fastening. The presence of a hole intensifies the localised stress at the joining contact, which further contributes to joint failure. Drilling a hole through polymeric composites is more difficult than doing so through metal. Composites are not homogeneous, while metals are uniformly homogeneous. Throughout their thickness or between layers, their qualities change. The qualities depend on the kinds of fibre, matrices, laminate stacking order, and processing methods. As a result, choosing the best process parameters for drilling composites is challenging. The best process parameters for drilling holes in composites and minimising drilling-related flaws in composites made of both synthetic and natural fibres have been the subject of several research studies. According to reports, synthetic fibres are more abrasive than natural fibres and generate a high rate of tool wear. Poor quality holes are also the result of tool wear. Another significant flaw associated with drilling laminated composites is delamination.

6.3 Drilling Induced Damage

Drilling leads to a number of flaws in fibre-reinforced composites, including edge chipping, fuzzing, delamination, oval hole shapes, and matrix burning. These are even visible to the unaided eye. Under a microscope, defects such as fracture growth and damage on the hole's periphery are visible. Because carbon and glass fibres, which are often employed in composites as reinforcement, are abrasive by nature, the majority of flaws result from changes in tool geometry during drilling operations. Tool wear occurs during drilling when it comes into touch with fibres, changing the tool's size. Delamination in composites causes layers or lamina to separate from composites, degrading the material's properties. It lessens the composites' ability to support loads. Composites are susceptible to delamination zone failure under loading or service circumstances. It might do serious harm. Two kinds of delamination flaws that have been documented for composites: peel up and push down. Delamination defects of the push-down and peel-up types may both be seen near the hole's entrance and exit, respectively. Surface layers begin to separate when the drilling tool contacts the composite material at the entry because there is a peeling force present between drill tool and composite. Additionally, it causes delamination close to the entering side. Following that, the drilling instrument keeps cutting as it moves into the deepest strata. When a drilling instrument penetrates the lowest strata, the thrust force it produces may result in the separation of those layers. Drilling tools provide compressive force to composite laminate's uncut layers in the presence of thrust force rather than shearing force, which causes uncut layers to bend. When bending stress created by compressive force on uncut layers reaches a threshold thickness and is larger than the interlaminar bond strength, internal fractures are generated. Furthermore, the separation or breaking of the bottom layers is caused by these internal fissures. As a result, the push down kind of delamination is the name given to this form of delamination fault. Using Image J1.42 software and an image analysis approach, delamination may be identified. It is measured by the delamination factor. Numerous studies have been conducted to reduce the delamination that occurs when polymer composites are being drilled for holes. Ismail et al. [64] investigated surface roughness and delamination in drill-cut samples of composites reinforced with hemp and carbon fibres. To drill holes in the specimen, an HSS drill was employed. Glass/epoxy and sisal/epoxy composites were examined for delamination behaviour, and Durao et al. [65] found that delamination rises with feed rate. In jute fibre reinforced composites, delamination behaviour after drilling holes was investigated by Aravindh et al. [66]. Using the Taguchi method's L9 orthogonal array, the ideal process parameters were determined. Delamination behaviour in coir fibre reinforced polyester composites was investigated by Jayabal et al. [67]. As a result, treated coir fibre reinforced composite may have less delamination than untreated coir fibre reinforced composite. Babu et al. [68] investigated the behaviour of delamination during drilling operations in composites reinforced with glass, hemp, banana, and jute fibres, and found that delamination in these composites was very low. Delamination in polyester composites made of rice husk was investigated by

Azuan et al. [69]. Three distinct kinds of drilling equipment, including twist drills, brad drills, and end mill drills, were used to drill specimens. In comparison to twist and end drill, brad drill reported higher delamination. Glass/hemp fibre composites were damaged as they were being drilled, according to research by Naveen et al. [70]. According to reports, the major reason for damage in composites was a high feed rate. For drilling composites, high speed and low feed rate were advised. Banana fibre reinforced epoxy composites' delamination during drilling was investigated by Venkateshwaran et al. [71]. It has been noted that when speed and feed increase, delamination in composite material also rises. According to reports, feed is the main element influencing drilling-related damage. Abilash et al. [72] investigated the delamination behaviour of polyester composites made from bamboo, glass, and royal fibre. Three drilling factors, including speed, feed, and cutting tool diameter, were examined in order to analyse delamination. The best drilling conditions for drilling bamboo composites were found to be the lowest delamination. The basic elements of standard tool geometry are clearance angle, rake angle, chisel angle, point angle, and helix angle. Twist drills are often used to create holes in any form of the material. It is made up of a 118° point angle. According to reports, high point angles for composite materials made of polymers result in high thrust and torque. Therefore, it is advised to drill polymer-based composites at low point angle values. Helix angle and clearance angle, two further drill geometry components, are advised to be in the ranges of 10°–36° and 6°–16°, respectively. Twist drills, however, are unable to create a hole in composite materials without causing any harm. For drilling damage-free holes in polymer matrix composites, researchers have concentrated on designing and creating various drill tip shapes. Several tests have been conducted to lessen drilling-induced damage to polymer composites made of synthetic and natural fibres.

7 Fusion Bonding

Typically, fusion bonding procedures are used to link thermoplastic-based composites. For thermoset-based composites, it is often not relevant since thermosets cure permanently. These cannot return to their original condition after cure. However, following solidification, thermoplastics may alter their state. Thermoplastics begin to soften when heated and begin to solidify as soon as the heat is removed. The fusion bonding method makes advantage of this phenomena. There are several different fusion bonding methods. Friction welding, heat welding, and electromagnetic welding are a few of them. Ultrasonic welding, vibration welding, and stir welding are all types of fiction welding. Polymeric chains that are long and straight are seen in thermoplastics. When polymeric chains are heated, their weak secondary connections dissolve, allowing the chains to move about. Chain mobility is influenced by the rate of heating. At the connecting contact, bonds are formed by these polymeric chains. Secondary connections between polymeric chains are broken by heating, allowing the chains to move freely. Polymeric chains traverse the joint interface during the joining processes and form bonds with polymeric chains of other adherends by entanglement, diffusion, or chemical reactions. Both crystalline and amorphous zones may be found in semi-crystalline thermoplastics. In crystalline areas, polymeric chains are organised in a certain way. More heat must be applied during heating in order to dissolve the bonds between polymeric chains in the crystalline zone. Chain mobility is reached very close to the thermoplastic's melting point. Diffusion is influenced by the viscosity, crystallinity, and molecular weight of thermoplastics as well as the molecular weight of the thermoplastics themselves. Entanglement of polymeric chains is also influenced by these factors. Due to the creation of bonds between polymeric chains at the joining contact, chemical bonding occurs [73, 74].

7.1 Microwave Joining

An electromagnetic heating approach for fusion bonding is microwave joining. Because microwaves are electromagnetic waves, they come under the umbrella of electromagnetic heating. By putting the material within the microwave cavity, it gets heated in this procedure after which, the substance is affected by microwaves. This distinguishes microwave heating from other types of heating. In traditional heating, heat is supplied from an outside source to the substance being heated. Normally, heat energy moves from a material's outside perimeter to its inside core, whereas microwave heating heats molecules directly. As a result, heat is transferred from the material's interior to outer edges (Fig. 5.). Insulating chamber and a microwave port are both parts of a microwave cavity. Microwaves enter the cavity via the microwave port [75–77].

The magnetic and electrical characteristics of the material affect the heating process. When exposed to microwave radiation, the four different kinds of materials exhibit distinct behaviours. As a result, several heating techniques exist. The

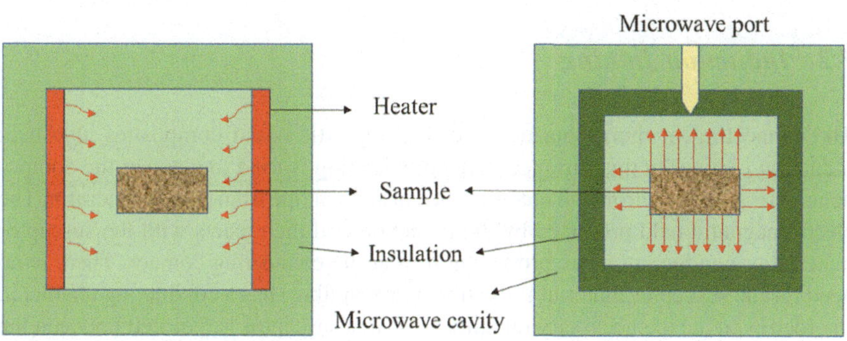

Fig. 5 Conventional versus microwave heating

hysteresis loss and residual loss cause heating in non-conductive and magnetic materials. The disruption caused by the external magnetic field in the orientated magnetic domain is what causes hysteresis loss. At first, the magnetic domains are organised such that the material has no overall magnetic effect. Domains arrange themselves in the magnetic field's direction when it is applied. Domains do not return to their original place after removing the magnetic field. Part of the magnetic energy involved in this misalignment is later transformed into heat. The disruption in the electron spin caused by an external magnetic field or the shift of the domain wall resonance are both causes of residual loss. In domain wall resonance, domains that are parallel to the magnetic field expand, whereas domains that are not parallel shrink. Additionally, reversing the direction of the magnetic field results in the contraction of other domains and the growth of the domain parallel to the magnetic field. As a result, constant contraction and expansion heat up the material. Ferromagnetic materials often exhibit electron spin disruption.

When a substance is both magnetic and conductive, heat is produced in the material as a result of the interaction of the two forces. It involves all of the conduction- and magnetism-based heat loss modes. Only the junction part of thermoplastic composites is exposed to microwaves during microwave joining; the remaining portion is masked. To stop the whole sample from heating up, masking is necessary. Heating just the joint portion is the primary goal. Yarlagadda and Chai, [78] looked at thermoplastic joining utilising focussed microwave energy. The thermoset matrix effectively absorbs microwave energy at ambient temperature, according to Martinelli et al. [79]. However, when the matrix solidifies, the rising viscosity limits adjustments to the orientation of the dipoles in the oscillating electric field. Because of the greater rate of cross linking, this decreases thermosets' microwave absorption. Bajpai et al. [80] examined the joining behaviour of four distinct kinds of natural fibre composites using glue and microwaves, and they used COMSOL Multiphysics software to model the microwave heating process. According to research, the temperature distribution around the joining contact relies on how long the specimen was exposed to the microwave cavity. Through the use of the COMSOL Multiphysics programme, the temperature distribution was anticipated.

7.2 Induction Joining

The connecting of thermoplastics and thermoplastic-based composites also uses induction joining. In this instance, induction heating is used to connect the components. In Fig. 6, induction coil receives current and voltage from power generator. The electromagnetic field produced by the induction coil then reacts with the susceptor material or conducting fibre, producing heat at the connecting contact. There is no need for the susceptor material if the fibre (carbon fibre) has a conducting character. In order to create the necessary quantity of heat, a susceptor is required between the connecting interfaces. The induction coil must be cooled since it heats up during usage and needs a cooling system [81, 82].

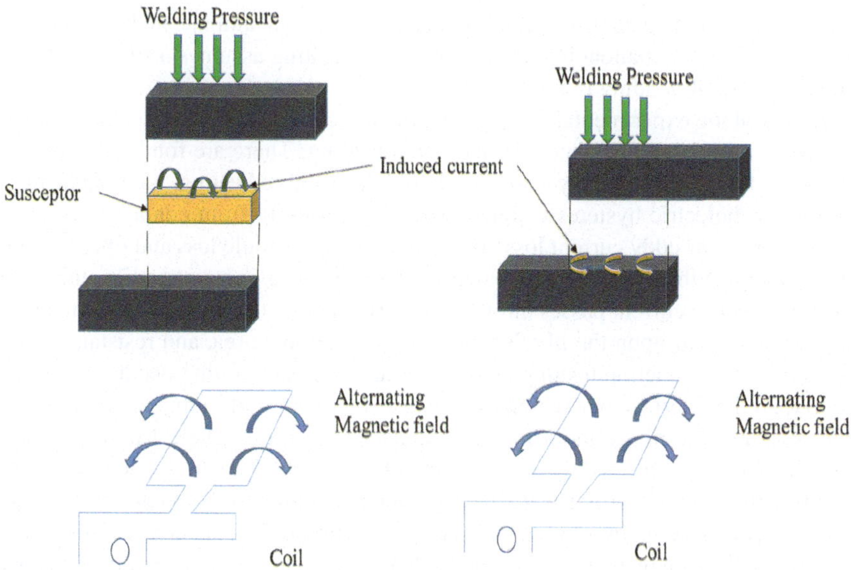

Fig. 6 Induction heating with susceptor and without susceptor

7.2.1 Heating Mechanism During Induction Joining

When a conductive coil is supplied with an alternating voltage, it generates an alternating current. Moreover, the application of alternating current results in the generation of an alternating magnetic field that exhibits a frequency comparable to that of the applied alternating current. Subsequently, the alternating magnetic field engages in interaction with a susceptor that is positioned in close proximity to the coil. This pertains to the topic of non-conductive reinforcement. The heating method for conductive fibre differs. Conducting fibres, mostly composed of carbon, are predominantly used in various experimental and theoretical investigations pertaining to heating mechanisms. Rudolf et al. [83] proposed an induction heating process that relies on the joule losses occurring in the carbon fibres during the passage of electric current in the laminate. The heat-affected zone inside the plane of the workpiece exhibited a symmetrical pattern that mirrored the shape of the coil. The orientation of the coil did not have any discernible impact on the rate of heating. However, as the frequency increased, the time required to heat the composite laminate to the target temperature fell in a quadratic manner. Kim et al. [84] proposed an alternative theoretical framework to explain the heating process of carbon fibre reinforced composites by induction. The researchers reached the conclusion that the heating effects at the junction, namely, dielectric hysteresis or contact resistance, had a greater influence compared to fibre heating. Additionally, it was discovered that the design of conductive fibres plays a pivotal role in selecting the primary mechanism. Woven textiles may exhibit fibre dominance as a result of direct contact between fibres, while laminated prepreg-based systems tend to exhibit junction heating dominance.

The presence of the matrix material between the fibres in a laminate makes achieving fibre–fibre contact challenging. Hence, dielectric heating assumes a significant role in the process of welding [85–87].

Based on the experimental findings and theoretical analysis, four distinct heating processes have been postulated for induction heating. There are four main sources of energy loss in electrical systems: joule loss due to fibre heating, junction heating caused by dielectric hysteresis, junction heating resulting from contact resistance, and hysteresis or eddy current loss. The phenomenon of Joule loss and fibre heating in conducting fibres may be attributed to the resistance encountered by the fibre when an electric current passes through it. Consequently, the quantity of heat generated is contingent upon the fibre's length, cross-sectional area, and resistance. The phenomenon of junction heating and dielectric hysteresis heating occurs inside the matrix. At the interface where two materials meet, there exists a tiny layer of matrix that separates the fibres and serves as a dielectric medium. Dielectric loss occurs when a matrix is subjected to an alternating electric field. The dissipation of energy occurs in the form of thermal heat. The phenomenon of junction heating/contact resistance is seen at the point of contact between two fibres. The occurrence of voltage drop is mostly seen in angle ply composite laminates, namely at the points of contact between the fibres. Hysteresis and eddy current losses occur when a susceptor, which has magnetic and conductive properties, is used at the joining contact.

7.3 *Resistance Welding*

Resistance welding is classified as a kind of fusion bonding technique. It is primarily used in the context of thermoplastic materials and thermoplastic matrix composites. In this scenario, a heating element with electrical resistance is positioned inside the interfaces that are being joined together. The passage of electric current through an implant result in the dissipation of energy in the form of heat, known as joule loss. Moreover, the heat that is produced at the interface of the joining process causes the thermoplastic material to undergo a melting phase. Subsequently, the specimen is subjected to the necessary pressure for appropriate consolidation. The process of connecting occurs between several categories of thermoplastic materials and thermoplastic matrix composites. The configuration for resistance welding is shown in Fig. 7. The primary components of the system are a power supply, clamping device, implant, wires, voltmeter, and ammeter. Thermocouples are often used for the purpose of temperature measurement. The heating element facilitates the flow of electric current, resulting in the production of thermal energy. The process of heat production leads to an elevation in temperature at the interface where two materials are joined. It is necessary for amorphous plastics to be in close proximity to the glass transition temperature, whereas semi-crystalline polymers should be around their melting point. The heat generation should be optimised to minimise the extent of the heat impacted zone (HAZ) in the specimen. Following the deactivation of the current, a defined amount of pressure is then provided. The use of pressure serves to eliminate

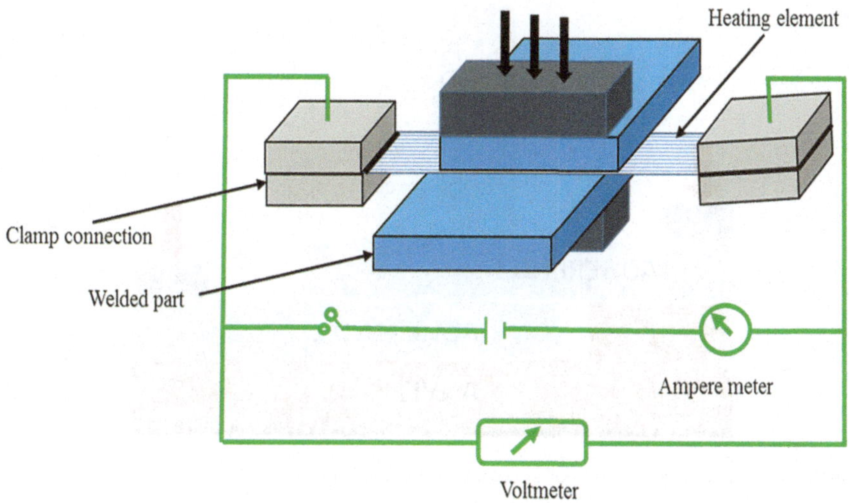

Fig. 7 Resistance heating for thermoplastic composites

gaps, cavities, and air bubbles present at the interface where two materials are joined together. Additionally, it facilitates appropriate contact between the adherends being joined. In order to mitigate heat dissipation from a heating element, it is necessary to implement insulation for the said element. Various materials, such as ceramic, asbestos, silicon rubber, kapton, and teflon, may be used for the purpose of insulation [88].

The heating element plays a crucial role in the process of connecting thermoplastic materials and thermoplastic matrix composites by resistance welding. This term is alternatively referred to as an implant. The heat energy required for the joint contact is provided and then confined inside the joint during the welding process. It is possible to use any material with electrical conductivity as an implant. In the field of implantation, it is often seen that two distinct categories of materials are used. There are two materials under consideration, namely carbon fibre and stainless-steel metal mesh.

7.4 Ultrasonic Joining

Ultrasonic joining may be classified as a fusion bonding method as well. This technique finds primary use in the welding of metallic materials with thermoplastic polymers (Fig. 8). Currently, there is an introduction of this technique in the domain of thermoplastic matrix composites joining. Interfacial heating is created by the use of high frequency ultrasonic vibration during the process of joining. The technology in question is characterised by its rapid joining process, which requires a shorter time commitment and lends itself well to automation. Hence, the use of the ultrasonic

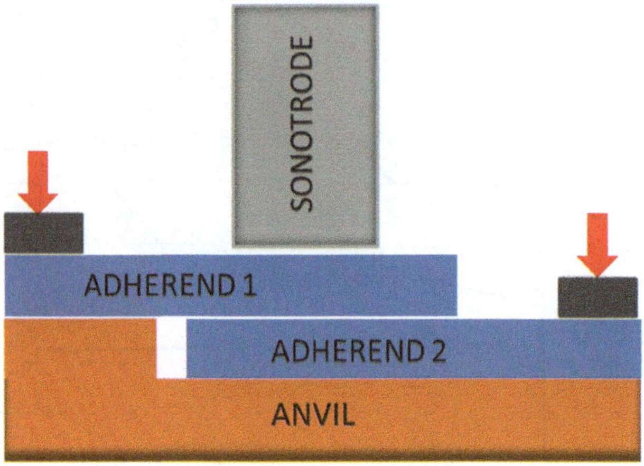

Fig. 8 Schematic representation of ultrasonic welding set up

welding procedure is prevalent within the electronics, computer, and automotive sectors. The researchers, Palardy et al. [89] established a predictive model to assess the impact of the sonotrode's hammering effect on the process of ultrasonic welding of thermoplastic composites. To verify the accuracy of their model, they conducted experiments using an appropriate set-up. The model was constructed via COMSOL Multiphysics software, with the primary objective of investigating the behaviour of a composite material consisting of carbon fibre and polyetherimide. According to the findings, the characteristic that has the greatest significance in achieving ideal weld qualities is welding energy [90–93].

The study revealed that the strength of samples united without the use of electron beam welding (ED) was comparatively lower than the strength of samples welded using ED. The occurrence of incomplete fusion, cracks, and voids has been documented in instances when the welding period is short. The predicted values of heat production and adhesion were compared to the corresponding experimental data, revealing a significant level of similarity between the two sets of values. The joint strength of carbon fibre thermoplastic composites welded using ultrasonic techniques was experimentally examined by Goto et al. [94]. The investigation focused on evaluating the tensile and shear behaviour of these joints. The experiment used carbon fibre reinforced polymer (CFRP) laminates with cross ply and twill weave configurations. The joint strength of cross ply laminates was seen to reach a saturation point at high welding energy levels. Conversely, it was reported that the joint strength of twill laminates exhibited an increase with increasing weld energy values [95–97]. The study conducted by Villegas et al. [98] investigated the impact of sonotrode displacement and wasted power on the strength of welds, as well as the failure mechanisms and morphology of the resulting joints. The study conducted by Villegas et al. [99] examined the bonding characteristics of carbon/epoxy and carbon/PEEK composites by the application of a PEI coupling layer. The study documented the occurrence of very brief heating durations in the process of ultrasonic welding, specifically in the context of connecting different composites. Notably, no thermal damage was seen during this event. In their study, Kumar et al. [100] conducted an optimisation of process parameters, including amplitude, weld pressure, hold time, and weld duration, for the ultrasonic welding of glass fiber/PA6T composites. The essential parameters among all the process parameters were found to be weld pressure.

8 Conclusions and Future Scope

The reduction in effective bond area due to hole creation and drilling-induced damage may have a negative effect. However, the epoxy-filled holes in the overlap region of the adhesive joints had an overall favourable influence. Further study inquiry is needed to conduct an economic evaluation of the efforts involved in producing holes and the greater quantity of epoxy utilised, in relation to the improvement in joint strength.

- The stress distribution inside the overlap area of the lap joint is homogeneous, whereas it reaches its highest value in the neighbouring areas or margins of the joint. Therefore, the failure of ultrasonic welded connections at edges may be ascribed to this cause. Furthermore, it is important to consider the potential occurrence of many failure mechanisms. These possibilities cannot be disregarded and need additional examination in future studies.
- The decreased ductility of composites resulting from the accelerated cooling of un-welded regions is a significant factor contributing to the failure of a majority of ultrasonic welded joints around the interfaces of the joints.
- The presence of sharp edges and uncut fibres, which were seen in specimens that underwent drilling, was not detected in specimens that underwent moulding.
- The production of holes during the moulding process obviates the need for additional secondary operations, such as drilling, or the use of accompanying machinery or equipment.

Declarations

Conflicts of Interest The authors confirm that there are no conflicts of interest associated with this work.

Ethical Compliance The authors affirm that this study adheres to all ethical guidelines.

Competing Interests No competing interests are declared by the authors.

Data and Material Availability Data and materials are not applicable to this study.

References

1. Clark DE, Folz DC, West JK (2000) Processing materials with microwave energy. Mater Sci Eng A 287:153–158. https://doi.org/10.1016/S0921-5093(00)00768-1
2. Ku HS, Siores E, Ball JAR (2000) Relationship between microwave irradiation and constituents of composites during joining process. HKIE Trans. Hong Kong Inst. Eng. 7(3):43–49. https://doi.org/10.1080/1023697X.2000.10667833
3. Ahmed TJ, Stavrov D, Bersee HEN, Beukers A (2006) Induction welding of thermoplastic composites—An overview. Compos Part a-Applied Sci Manuf. 37(10):1638–1651. https://doi.org/10.1016/j.compositesa.2005.10.009
4. Suwanwatana W, Yarlagadda S, Gillespie JW (2006) Hysteresis heating based induction bonding of thermoplastic composites. Compos Sci Technol 66(11–12):1713–1723. https://doi.org/10.1016/j.compscitech.2005.11.009
5. Bayerl T, Schledjewski R, Mitschang P (2012) Induction heating of thermoplastic materials by particulate heating promoters. Polym Polym Compos 20(4):333–342. https://doi.org/10.1177/096739111202000401
6. Rakesh PK, Davim JP (2021) Joining processes for dissimilar and advanced materials, ISBN: 9780323853996. Elsevier
7. Kumar R, Rakesh PK, Sreehari D, Kumar D, Naik TP (2023) Experimental investigations on material properties of alkali retted Pinus Roxburghii Fiber. Biomass Convers Biorefinery 23:1–17
8. Pokhriyal M, Rakesh PK (2023) Processing and characterization of novel Himalayacalamus falconeri fiber reinforced biodegradable composites. Biomass Convers Biorefinery 23:1–16
9. Kumar R, Rakesh PK, Sreehari D (2023) Investigation on Physico-Chemical, mechanical and thermal properties of extracted novel Pinus Roxburghii Fiber. J Nat Fibers 20(1):2157924
10. Pokhriyal M, Rakesh PK, Rangappa SM, Siengchin S (2023) Effect of alkali treatment on novel natural fiber extracted from Himalayacalamus falconeri culms for polymer composite applications. Biomass Convers Biorefinery 23:1–17
11. Lalit R, Gangil B, Rakesh PK, Tej S, Shubham S, Changhe Li RA, Ilyas OM (2022) Effect of surface treatment and fiber loading on the physical, mechanical, sliding wear, and morphological characteristics of tasar silk fiber waste-epoxy composites for multifaceted biomedical and engineering applications: fabrication and characterizations. J Market Res 19(2022):2863–2876
12. Zhou Y, Fan M, Chen L (2016) Interface and bonding mechanisms of plant fibre composites: An overview. Compos Part B Eng 101:31–45. https://doi.org/10.1016/j.compositesb.2016.06.055
13. Pokhriyal M, Rakesh PK (2022) Mechanical and Microstructural behaviour of NaOH treated Himalayacalamus falconeri fibers as biodegradable reinforcing material in polymer-based composites. Mater Today Proc 62(2):1078–1082
14. Ranakoti L, Rakesh PK, Gangil B (2021) Effect of Tasar silk waste on the mechanical properties of Jute/Grewia optiva fibers reinforced epoxy laminates. J Nat Fibers 19(15):10462–10474
15. Adarsh C, Lalit R (2021) Rakesh Pawan Kumar, Mishra Niraj Kumar, (2021) Experimental investigations on mechanical properties of walnut shell and pine needle ash polylactic acid biocomposites. Compos Theory Pract 21(3):114–120
16. Rakesh PK, Singh I, Kumar D (2012) Flexural Behavior of Glass Fiber Reinforced Plastic Laminates with Drilled Hole. Proceedings of the Institution of Mechanical Engineers, Part L: Journal of Materials Design and Applications 226(2):149–158
17. Kumar S, Chauhan SR, Rakesh PK, Singh I, Davim JP (2012) Drilling of Glass Fiber/Vinyl Ester Composites with Fillers. Mater Manuf Processes 27(3):314–319
18. Kishore RA, Tiwari R, Rakesh PK, Singh I, Bhatnagar N (2011) Investigation of Drilling in Fiber Reinforced Plastics using Response Surface Methodology. Proceedings of the Institution of Mechanical Engineers, Part B: Journal of Engineering Manufacture 225(3):453–457

19. Lalit R, Rakesh PK (2020) Physio-mechanical characterisation of tasar silk waste/jute fiber hybrid composites. Compos Commun 22:100526
20. Ranakoti Lalit; Gupta Manoj; Rakesh PK (2019) Analysis of mechanical and tribological behavior of wood flour filled glass fiber reinforced epoxy composite, Mater Res Express 6 085327
21. Sandeep G, Somit G, Hitesh S (2019) Rakesh pawan kumar, impact behavior of pine needle Fiber/Pistachio shell filler based epoxy composite. J Phys: Conf Ser 1240(012096):1–7
22. Rakesh PK, Davim JP (2022) Cellulose Composites: Processing and Characterization" ISBN: 978-3-11-076869-5. De Gruyter, Germany
23. Rakesh PK, Singh Inderdeep (2018) Processing of green composites. Springer, ISBN 978–981–13–6019–0
24. Ranakoti L, Gupta MK, Mangal D, Rakesh PK Recycling of polymers and its application, Cellul Compos: Process Charact. ISBN: 978–3–11–076869–5, De Gruyter, Germany 15, 137–150
25. Kumar R, Rakesh PK, Sreehari D Effect of potassium permagnate treatment on physico-chemical and mechanical properties of extracted pine needle fiber. In: E3S Web of Conference, 391, 01036
26. Jayant K, Kumar V, Singh I, Rakesh PK (2021) Joining behavior of polymeric composites fabricated using agricultural waste as fillers. J Adhes Sci Technol 35(15):1652–1663
27. Ranakoti, L, Rakesh PK, Gangil B (2021) Role of wood flour on physical and mechanical properties in polymer matrix composites-a critical review. Revue des Composites et des Matériaux Avancés-Journal of Composite and Advanced Materials, 31(2) 81–92. https://doi.org/10.18280/rcma.310203
28. Pokhriyal M, Prasad L, Rakesh PK, Raturi HP (2018) Influence of fiber loading on physical and mechanical properties of Himalayan nettle fabric reinforced polyester composite. Materials Today: Proceedings 5(9):16973–16982
29. Agarwal A, Garg S, Rakesh PK, Singh I, Mishra BK (2010) Tensile behavior of glass fiber reinforced plastics subjected to different environmental conditions. Indian J Eng Mater Sci 17:471–476
30. Villegas IF (2014) Strength development versus process data in ultrasonic welding of thermoplastic composites with flat energy directors and its application to the definition of optimum processing parameters. Compos Part A Appl Sci Manuf 65:27–37. https://doi.org/10.1016/j.compositesa.2014.05.019
31. Tao W, Su X, Wang H, Zhang Z, Li H, Chen J (2019) Influence mechanism of welding time and energy director to the thermoplastic composite joints by ultrasonic welding. J Manuf Process 37(2018):196–202. https://doi.org/10.1016/j.jmapro.2018.11.002
32. Levy A, Corre S, Le VI (2014) Fernandez, "Modeling of the heating phenomena in ultrasonic welding of thermoplastic composites with flat energy directors." J Mater Process Technol 214(7):1361–1371. https://doi.org/10.1016/j.jmatprotec.2014.02.009
33. Rakesh PK, Sharma AK, Singh Inderdeep (2020) Advances in engineering design. Springer, ISBN 978–981–334–018–3
34. Rakesh PK, Ranakoti L, Gupta MK Mechanical properties of chemically treated cellulosic fiber-reinforced polymer composites, Cellul Compos: Process Charact. ISBN: 978–3–11–076869–5, De Gruyter, Germany. 151–157
35. Singh Y, Kumar J, Singh I, Rakesh PK (2022) Joining behaviour of natural fiber reinforced polymer composites, Joining Processes for dissimilar and advanced materials, 33–63, Elsevier 9780323860208
36. Rakesh Pawan Kumar (2022) Joining processes of advanced materials, Joining processes for dissimilar and advanced materials, 1–11. Elsevier 9780323860208
37. Chaturvedi Adarsh, Rakesh Pawan Kumar, (2021) Study on the degradation behaviour of natural fillers nased PLA composites, Advances in biobased fiber: moving towards a green society, 499–506. Elsevier 978–0–12–824544–6
38. Ranakoti Lalit, Gangil Brijesh, Rakesh Pawan Kumar, Agrawal Nikita (2021) Synthesis and utilization of biodegradable polymers, biobased composites: processing, Characterization, Properties, and Applications, John Wiley & Sons, Inc. 167–174. 978–1–119–64179–7

39. Ranakoti Lalit, Rakesh PK (2019) Friction and wear analysis of reinforced polymer composites. Reinforced Polymer Composites: Processing Characterization and Post life cycle assessment, Wiley-VCH, Verlag Gmbh, Germany, ISBN: 9783527345991
40. Sharma H, Komal UK, Singh I, Misra JP, Rakesh PK (2018) Introduction to green composites, processing of green composites, 1–13, 2018, Springer, ISBN 978–981–13–6019–0
41. Ranakoti L, Gupta MK, Rakesh PK (2018) Silk and silk-based composites: opportunities and challenges. Process Green Compos, 91–106, Springer, ISBN 978–981–13–6019–0
42. Singh Inderdeep, Rakesh PK, Malik J (2012) Soft computing techniques for predicting forces and damage in drilling of polymer composites. Mechatrons Manuf Eng: Res Dev, Woodhead Publishing, UK. ISBN: 978 0 85709 150 5, 2012. 227–255
43. Baldan A (2004) Adhesively-bonded joints and repairs in metallic alloys, polymers and composite materials: Adhesives, adhesion theories and surface pretreatment. J Mater Sci 39(1):1–49. https://doi.org/10.1023/B:JMSC.0000007726.58758.e4
44. Brewis DMM et al (2009) Testing of adhesives for bonding of polymer composites. Int J Adhes Adhes 26(4):291–295. https://doi.org/10.1016/S0143-7496(02)00006-4
45. Rakesh PK, Singh I, Kumar D (2012) Drilling of composite laminates with solid and hollow drill point geometries. J Compos Mater 46:3173–3180
46. Rakesh PK, Singh I, Kumar D (2010) Failure prediction in glass fiber reinforced plastics laminates with drilled hole under Uni-Axial Loading. Mater Des 31(6):3002–3007
47. Nikoi R, Sheikhi MM, Arab NBM (2015) Experimental analysis of effects of ultrasonic welding on weld strength of polypropylene composite samples. Int J Eng Trans A Basics 28(3):447–453. https://doi.org/10.5829/idosi.ije.2015.28.03c.15
48. Liu SJ, Chang IT, Hung SW (2001) Factors affecting the joint strength of ultrasonically welded polypropylene composites. Polym Compos 22(1):132–141. https://doi.org/10.1002/pc.10525
49. Panneerselvam K, Aravindan S, Noorul Haq A (2012) Study on resistance welding of glass fiber reinforced thermoplastic composites. Mater. Des., 41, 453–459. https://doi.org/10.1016/j.matdes.2012.05.025
50. Yee LS, Ahmad H (2016) Experimental strength of single-lap hybrid joints on woven fabric kenaf fiber composites under quasi static condition, MATEC Web Conf., 47,0–5. https://doi.org/10.1051/matecconf/20164702003
51. Gonzalez-Murillo C, Ansell MP (2010) Co-cured in-line joints for natural fibre composites. Compos Sci Technol 70(3):442–449. https://doi.org/10.1016/j.compscitech.2009.11.017
52. Yallew TB, Kumar P, Singh I (2015) A study about hole making in woven jute fabric-reinforced polymer composites. Proc Inst Mech Eng Part L J Mater Des Appl. 1–11, https://doi.org/10.1177/1464420715587750.
53. Debnath K, Singh I (2017) Low-frequency modulation-assisted drilling of carbon-epoxy composite laminates. J Manuf Process 25:262–273. https://doi.org/10.1016/j.jmapro.2016.12.009
54. Chandramohan D, Marimuthu K (2011) Drilling of natural fiber particle reinforced polymer composite material. Int J Adv Eng Res Stud. I(I), 134–145
55. Bajpai PK, Debnath K, Singh I (2015) Hole making in natural fiber-reinforced polylactic acid laminates: An experimental investigation. J Thermoplast Compos Mater, 1–17, https://doi.org/10.1177/0892705715575094
56. Jayabal S, Natarajan U (2010) Optimization of thrust force, torque, and tool wear in drilling of coir fiber-reinforced composites using Nelder-Mead and genetic algorithm methods. Int J Adv Manuf Technol 51(1–4):371–381. https://doi.org/10.1007/s00170-010-2605-7
57. Nasir AAA, Azmi AI, Khalil ANM (2015) Parametric study on the residual tensile strength of flax natural fibre composites after drilling operation. Procedia Manuf 2:97–101. https://doi.org/10.1016/j.promfg.2015.07.017
58. Vinayagamoorthy R, Rajeswari N, Karuppiah B (2014) Optimization studies on thrust force and torque during drilling of natural fiber reinforced sandwich composites. Jordan J Mech Ind Eng 8(6):385–392
59. Yarlagadda S, Kim HJ, Gillespie JW, Shevchenko NB, Bruce K (2002) Journal of composite materials. J Compos Mater 36(04):401–421. https://doi.org/10.1106/002199802023171

60. Gouin OP, Dubé M, Villegas I (2016) Fernandez, Modeling and experimental investigation of induction welding of thermoplastic composites and comparison with other welding processes. J Compos Mater 50(21):2895–2910. https://doi.org/10.1177/0021998315614991
61. Jeevan BD, Arun KV, Prashanth SJ (2016) Strength of adhesive bonded natural hybrid composites with fillers. Int J Eng Sci Comput. 6(6), 7151–7156, https://doi.org/10.4010/2016.1701
62. Lionetto F, Pappadà S, Buccoliero G, Maffezzoli A (2017) Finite element modeling of continuous induction welding of thermoplastic matrix composites. Mater Des 120:212–221. https://doi.org/10.1016/j.matdes.2017.02.024
63. Ramesh M, Palanikumar K, Reddy KH (2014) Influence of tool materials on thrust force and delamination in drilling Sisal-glass Fiber Reinforced Polymer (S-GFRP) composites. Procedia Mater Sci 5:1915–1921. https://doi.org/10.1016/j.mspro.2014.07.513
64. Ismail SO, Dhakal HN, Popov I, Beaugrand J (2016) Comprehensive study on machinability of sustainable and conventional fibre reinforced polymer composites. Eng Sci Technol an Int J 19(4):2043–2052. https://doi.org/10.1016/j.jestch.2016.07.010
65. Durão LMP et al. (2013) Drilling delamination outcomes on glass and sisal reinforced plastics. Mater Sci Forum 730–732, https://doi.org/10.4028/www.scientific.net/MSF.730-732.301
66. Aravindh S, Umanath K (2015) Delamination in drilling of natural fibre reinforced polymer composites produced by compression moulding. Appl Mech Mater, 766–767. https://doi.org/10.4028/www.scientific.net/AMM.766-767.796
67. Jayabal S, Natarajan U (2011) Drilling analysis of coir-fibre-reinforced polyester composites. Bull Mater Sci 34(7):1563–1567. https://doi.org/10.1007/s12034-011-0359-y
68. Babu J, Sunny T, Paul NA, Mohan KP, Philip J, Davim JP (2016) Assessment of delamination in composite materials: A review. Proc Inst Mech Eng Part B J Eng Manuf. 230(11), 1990–2003, https://doi.org/10.1177/0954405415619343
69. Azuan SAS, Juraidi JM, Muhamad WMW (2012) Evaluation of delamination in drilling rice husk reinforced polyester composites. Appl Mech Mater 232:106–110. https://doi.org/10.4028/www.scientific.net/AMM.232.106
70. Naveen PNE, Yasaswi M, Prasad RV (2012) Experimental investigation of drilling parameters on composite materials. J Mech Civ Eng 2(3):30–37
71. Venkateshwaran N, Perumal A (2012) Elaya, "Hole quality evaluation of natural fiber composite using image analysis technique." J Reinf Plast Compos 32(16):1188–1197. https://doi.org/10.1177/0731684413486847
72. Abilash N, Sivapragash M (2016) Optimizing the delamination failure in bamboo fiber reinforced polyester composite. J King Saud Univ—Eng Sci 28(1):92–102. https://doi.org/10.1016/j.jksues.2013.09.004
73. Sosa ED et al (2016) Thermal and non-thermal interaction of microwave radiation with materials. J Mater Process Technol 30(1–3):5321–5327. https://doi.org/10.1016/S0924-0136(03)00238-3
74. Mishra RR, Sharma AK (2016) Microwave-material interaction phenomena: Heating mechanisms, challenges and opportunities in material processing. Compos Part A Appl Sci Manuf 81:78–97. https://doi.org/10.1016/j.compositesa.2015.10.035
75. Lionetto F, Mele C, Leo P, D'Ostuni S, Balle F, Maffezzoli A (2018) Ultrasonic spot welding of carbon fiber reinforced epoxy composites to aluminum: mechanical and electrochemical characterization. Compos Part B Eng 144:134–142. https://doi.org/10.1016/j.compositesb.2018.02.026
76. Bhattacharya M, Basak T (2016) A review on the susceptor assisted microwave processing of materials. Energy 97:306–338. https://doi.org/10.1016/j.energy.2015.11.034
77. Bae DH et al (2015) Heating behavior of ferromagnetic Fe particle-embedded thermoplastic polyurethane adhesive film by induction heating. J Ind Eng Chem 30:92–97. https://doi.org/10.1016/j.jiec.2015.05.007
78. Yarlagadda PKD, Chai TC (1998) An investigation into welding of engineering thermoplastics using focused microwave energy. J Mater Process Technol 74(1–3):199–212. https://doi.org/10.1016/S0924-0136(97)00269-0

79. Martinelli M, Rolla PA, Tombari E (1985) A method for dielectric loss measurements by a microwave cavity in fixed resonance condition. IEEE Trans Microw Theory Tech 33(9):779–783. https://doi.org/10.1109/TMTT.1985.1133126
80. Bajpai PK, Singh I, Madaan J (2012) Joining of natural fiber reinforced composites using microwave energy: Experimental and finite element study. Mater Des 35:596–602. https://doi.org/10.1016/j.matdes.2011.10.007
81. Bayerl T, Mitschang P (2014) Heating of Polymer-Polymer composites by inductive means. Comp Part A: Applied Sci Manuf. 57, 1–6, https://doi.org/10.1016/j.compositesa.2013.10.024
82. Stavrov D, Bersee HEN (2005) Resistance welding of thermoplastic composites-an overview. Compos Part A Appl Sci Manuf 36(1):39–54. https://doi.org/10.1016/j.compositesa.2004.06.030
83. Rudolf R, Mitschang P, Neitzel M (2000) Induction heating of continuous carbon-fibre-reinforced thermoplastics. Compos Part A Appl Sci Manuf 31(11):1191–1202. https://doi.org/10.1016/S1359-835X(00)00094-4
84. Kim H, Yarlagadda S, Gillespie JW, Shevchenko NB, Fink BK (2002) A study on the induction heating of carbon fiber reinforced thermoplastic composites. Adv Compos Mater 11(2014):71–80. https://doi.org/10.1163/156855102753613309
85. Awaja F, Gilbert M, Kelly G, Fox B, Pigram PJ (2009) Adhesion of polymers. Prog Polym Sci 34(9):948–968. https://doi.org/10.1016/j.progpolymsci.2009.04.007
86. Bayerl T, Duhovic M, Mitschang P, Bhattacharya D (2014) The heating of polymer composites by electromagnetic induction-A review. Compos Part A 57:27–40
87. Custódio J, Broughton J, Cruz H (2009) A review of factors influencing the durability of structural bonded timber joints. Int J Adhes Adhes 29(2):173–185. https://doi.org/10.1016/j.ijadhadh.2008.03.002
88. Taib AA, Boukhili R, Achiou S, Gordon S, Boukehili H (2006) Bonded joints with composite adherends. Part I. Effect of specimen configuration, adhesive thickness, spew fillet and adherend stiffness on fracture. Int J Adhes Adhes 26(4):226–236. https://doi.org/10.1016/j.ijadhadh.2005.03.015
89. Palardy G, Shi H, Villegas IF, Levy A, LeCorre S (2018) Experimental investigation of amplitude transmission in ultrasonic welding of thermoplastic composites. Annu Tech Conf—ANTEC, Conf Proc, 339–349
90. Tsiangou E, Freitas S Teixeira de, Villegas I Fernandez, Benedictus R (2019) Investigation on energy director-less ultrasonic welding of polyetherimide (PEI)- to epoxy-based composites. Compos. Part B Eng, 173, 107014, https://doi.org/10.1016/j.compositesb.2019.107014
91. Ageorges C, Ye L, Hou M (2001) Advances in fusion boding techniques for joining thermoplastics materials composites: a review. 32, 839–857. https://doi.org/10.1016/S1359-835X(00)00166-4
92. Banea MD, Silva LFM da, (2009) Adhesively bonded joints in composite materials: an overview. Proc Inst Mech Eng Part LJ Mater Des Appl. 223, 1–18, https://doi.org/10.1243/14644207JMDA219
93. Yogesh TL, Arunkumar N (2015) Failure mode and analysis of the bonded/bolted joints between a hybrid fibre reinforced polymer and aluminium alloy. J. of Adv. Mater. and Proc. 3(2):49–59
94. Goto K, Imai K, Arai M, Ishikawa T (2019) Shear and tensile joint strengths of carbon fiber-reinforced thermoplastics using ultrasonic welding. Compos Part A Appl Sci Manuf 116(2018):126–137. https://doi.org/10.1016/j.compositesa.2018.10.032
95. Campilho RDSG, Moura DC, Gonçalves DJS (2013) Adhesive joints in natural fibre composites: Estimation of fracture properties. Cienc. e Tecnol. dos Mater. 25(1):31–37. https://doi.org/10.1016/j.ctmat.2013.12.007
96. Yoon HC, Lim JK (2007) Effects of fiber layer on failure strength properties of Adhesive—Joined natural fiber reinforced composites. Key Eng Mater 353–358:1965–1969. https://doi.org/10.4028/www.scientific.net/KEM.353-358.1965
97. Quieroz HD, Banea MD, Silva LFM da (2019) Experimental analysis of adhesively bonded joints in synthetic- and natural fibre-reinforced polymer composites. J Comp MAT. 54, 1245-1255, 10.1177%2F0021998319876979

98. Villegas IF, Moser L, Yousefpour A, Mitschang P, Bersee HEN (2012) Process and performance evaluation of ultrasonic , induction and resistance welding of advanced thermoplastic composites. J Thermoplast Compos Mater. 1007–1024, https://doi.org/10.1177/0892705712456031
99. Villegas IF, van Moorleghem R (2018) Ultrasonic welding of carbon/epoxy and carbon/PEEK composites through a PEI thermoplastic coupling layer. Compos Part A Appl Sci Manuf 109(2017):75–83. https://doi.org/10.1016/j.compositesa.2018.02.022
100. Kumar R Kalyan, Omkumar M (2020) Ultrasonic welding of GF/PA6T composites: Experimental investigation and process optimization. Mater Today Proc, 39, 1180–1186, https://doi.org/10.1016/j.matpr.2020.03.462

Chapter 7
Solid State Additive Manufacturing: An Unconventional Technique to Fabricate Composite

Manish Maurya, Ambrish Maurya, and Sudhir Kumar

Abbreviations

Words	Abbreviation Used
Friction stir additive manufacturing	FSAM
Additive manufacturing	AM
Friction stir processing	FSP
Friction stir deposition	FSD
Friction stir welding	FSW
Additive friction stir deposition	AFSD

1 Introduction

In earlier decades, handcraft and agriculture was the main chauffeurs of economies. This drift was changed through the industrial revolution. Industrial revolution can be categorized into four sessions. First industrial revolution (before 1830) was shifting of manual production to the machinery-based production. Second industry revolution (1840–1940) involves the progression of material and energy production. The third industrial revolution began in 1945 known as digital manufacturing era. The era after 2011, is known as fourth revolution. This era includes the developed techniques like internet of things (IOT), cyber systems, the internet of services (IOS), cloud

M. Maurya · A. Maurya (✉)
National Institute of Technology, Patna, India
e-mail: amaurya.me@nitp.ac.in

S. Kumar
Harlal Institute of Management and Technology, Greater Noida, India

S. K. Sethi et al. (eds.), *Polymer Composites: From Computational to Experimental Aspects*, Materials Horizons: From Nature to Nanomaterials,
https://doi.org/10.1007/978-981-97-0888-8_7

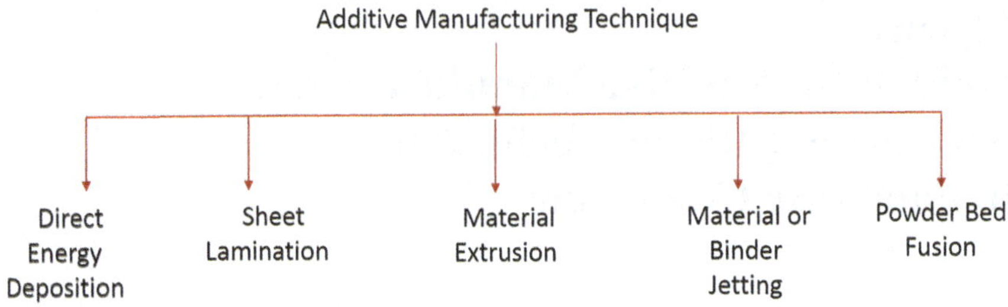

Fig. 1 Classification of additive manufacturing technique

manufacturing, robotics and big data. Presently, we are living in the fourth industrial revolution known as IR4.0.

Traditional AM technique comes under fourth industrial revolution which includes electron beam melting, selective laser melting, direct energy deposition that are based on application of high energy beam (such as electron or laser beam) as a source of heat to melt and solidify the material with various formed layers [1]. Materials fabricated using these techniques possess the static strength analogous to forging. Yet these materials have got the solid structural defects, that tend to fatigue and low plasticity characteristics [2, 3]. Porosity, development of residual stresses was generally found in these fabrication techniques. Moreover, these techniques were not applied to magnesium or aluminum alloys [4]. AM can be categorized into five common methods. The categorization of additive manufacturing technique is given in Fig. 1.

Friction stir-based additive manufacturing techniques have a lot of benefits in comparison to fusion based additive manufacturing techniques. In solid state manufacturing techniques, the attained temperature varies from 60 to 90% of the melting temperature. Defects such as porosity and hot cracking are not found in solid state additive manufacturing technique. FSAM can use the feed material inform of plates/sheets for FSAM. Homogenous recrystallized microstructure is achieved in friction stir-based processes. Limited rate of production, limited geometrical shape due to tool dimensions, more tool wear and formation of residual stress are the main disadvantages of friction stir-based AM techniques.

FSP is a plastic deformation methodology that can instantaneously homogenize, densify and refine the microstructural properties of the material [5]. One solicitation of FSP is AM technique. The additive friction stir-based manufacturing methodology is called as FSAM is a newly developed technique. In a lap joint design and FSW method, by the use of heat and the plastic deformation during the process-two sheets can be joined. Several research articles [6, 7] have been published by the researchers on the preparation of two- or multilayer composites by the application of FSAM technique. This technique had attracted the researchers to produce various improved composites/material of required mechanical properties. In the FSP method, grooves are filled with reinforcement and the stirring action is performed on the above surface [8, 9]. In friction stir additive manufacturing of alloys, plates are stacked with each other and stirring is performed. While for producing composites of two or multi-layer, reinforcement is filled between two layers and stirring action is accomplished [10]. A hole or narrow grooves is made in the top sheet and reinforcement content is filled into it. Reinforcements are covered with another layer of sheets from the top of the surface. Then, the FSP tool is plunged and rotated into the sheets. Due to friction heat between the tool and plates, grain refinement occurs. Owing to refinement of grains, an improvement in microstructural and mechanical properties was observed [11, 12].

Friction stir welding, an unconventional methodology is used for the joining of two similar or dissimilar materials in a continued process. In the FSW process, tools having shoulder with a pin (cylindrical, conical or tapered) is plunged into the joint line and rotated between the area to be joined. FSAM utilizes the principle of FSW, where multiple plates are joined through FSW, keeping one on the top of other plates/sheets. FSW is a set of solid phase processing methodology, which had enabled to join sheets/plates that are problematic to join by conventional means. The various processes used by various authors like Elfishawy et al. [13], Tan et al. [14], Wang et al. [15], Perry et al. [16], etc. have used various types of processes like FSD, FSAM, cold spray process and AFSD etc. to produce the composite, surface composites or laminated composites of improved mechanical properties. The fundamental of all these processes was the sufficient heat produced by the rotating tool and the surface of the material. The important parameters for all the processes were the required substrate height, tool rotational and transverse speed, tilt angle and the depth. Moreover, joining of two sheets by Roodgari et al. [17] through additive manufacturing technique has improved microstructural and mechanical properties. Nowadays, friction stir additive manufacturing technique is an advancement of friction stir processing technique. Friction stir processing was derived from friction stir welding technique. FSAM involves joining of sheets and forming of one layer over other layer (the material of two or more than two layers may be same/different). Through FSAM, the micro structural characteristics are improved due to grain refinement. FSAM may also be used to fabricate the composite. Particles are reinforcement between the sheets. The process of reinforcing the ceramic particle may be different. Particles may be reinforced by making a rectangular groove or by making the holes in the sheet. Rectangular grooves or holes are filled with the ceramic particles. During the stirring action, these particles are mixed with the alloy sheets and come on the surface of the sheet. These hardened particles will improve the surface properties of the fabricated material. FSAM technique has got its industrial application in aerospace and automobile industries. The other variants of solid-state additive manufacturing technique like cold spray FSP, cladding and friction stir deposition are also discussed in this chapter. The various types of tools used with the variation of process parameters on a different material by various researchers and their findings is compiled in tabular form. Table 1 shows the various process parameters, dimensions and the results achieved by authors found in various additive manufacturing processes. The data provided on the table will help the young researchers to decide the parameters of the tool and sheets while conducting the research. The various process parameters, dimensions and the results can be observed in one sight.

Table 1 Process parameters and findings from previous published works

S. No	Authors	Aluminium alloy/ composites	Sheets dimension	Tool rotational speed and/ transverse speed	Prominent findings	Process used
1	Elfishawy et al. [13]	Al Si + Al substrate	20 mm diameter	Spindle Rotation speed = 122 rpm Feed Speed = 3–5 mm/ min	Fine grain structure ranging between 2 and 5 μm was found. Al–Si AMPs was developed by FSD in multiple layers of thickness up to 30 mm	FSD
2	Tan et al. [14]	AA 6061/ Al_2O_3	(200 × 110 × 4) mm	Rotating speed of tool = 1000 rpm Transverse speed = 100 mm/ min	Re-stirring has reduced the accumulation of Al_2O_3 particles Hardness = 87.34 HV	FSAM
3	Wang et al. [15]	Al 6061/ Aluminum powder		Rotating speed of tool = 1500 rpm Transverse speed = 23.5 mm/ min	YS and UTS was enhanced by 38 and 45%	Cold spray FSP
4	Perry et al. [16]	AA 2024 and AA 6061		Rotating speed of tool = 300 rpm Transverse speed = 2 m/sec	Accumulation of material on advancing side was observed	AFSD
5	Roodgari et al. [17]	2-layer laminated steel	(150 × 50) mm, 0.7 and 2 mm thick	Rotating speed of tool = 600 rpm Transverse speed = 70 mm/ min	Hardness = 225 Hv	FSAM

(continued)

Table 1 (continued)

S. No	Authors	Aluminium alloy/ composites	Sheets dimension	Tool rotational speed and/ transverse speed	Prominent findings	Process used
6	Ho et al. [18]	AZ31B alloy/ hydroxyapatite	20 × 127 × 2.3 mm	Rotating speed of tool = 900 rpm Transverse speed = 30 mm/ min	Grain refinement leads to less corrosion	FSP
7	Mukhopadhyay and Saha [19]	Aluminium powder		Rotating speed of tool = 1700 rpm Transverse speed = 22 mm/ min	Hardness = 18 Hv	FSAM
8	Xie et al. [20]	Nano-TiB_2/ AlSi10Mg	(140 × 80 × 4 mm)	Rotating speed of tool = 1500 rpm Transverse speed = 500 mm/ min	Dense structure was found	Cold spray FSP
9	Sharma et al. [21]	Al/TiC		Rotating speed of tool = 1000 rpm Transverse speed = 50 mm/ min	Hardness = 41 Hv	FSP
10	Huang et al. [22]	Brass/Al_2O_3/ SiC			Micro-hardness = 243 Hv	CSAM

2 Additive Manufacturing by FSP Technique

FSAM technique can be categorized as several process designs to fabricate a 3D component. Here, deposition of the material takes place in layer-by-layer, using the FSP method. Based on this arrangement, five routes can be reflected as the main FSAM technique to prepare 3D components. FSAM is by now fairly well-defined process to fabricate components, with more improved mechanical properties.

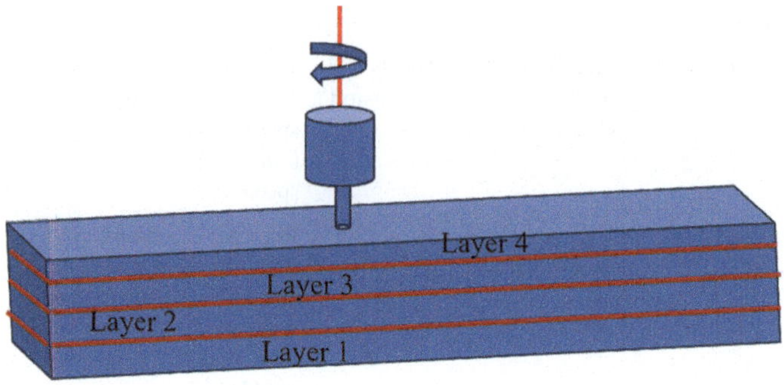

Fig. 2 Formation of various layers in FSAM process

2.1 Solid-Rod Powder Feeding Method

MELD manufacturing company has invented the FSAM technology [23]. FSAM is based on feeding of the material in the form of solid-rod or powder via a rotating hollow tool under pressure for deposition of alloys/metals/ceramics. He et al. [24] observed the microstructural evolution in FSAM of Al-Zn-Mg alloy. 4-mm thick 7N01-T4 Al sheets were used to perform FSAM. After natural aging for six months, the average strength of the build was increased to 352.7 MPa, being about 73% of the BM strength (483.1 MPa). Non-consumable tool promotes the frictional-heating between the material and the shoulder and the plastic deformation at the interface exist [15]. Elfishawy et al. [13] performed the AM process using the FSD method from die-cast Al–Si bar of 20 mm diameter against Al substrate. AM of three cylindrical parts of 25 mm diameter of different feeding rates at different heights were prepared. Ultra-fine grains were found for the additive manufactured parts. The formation of various layers formed is shown in Fig. 2. Initially, at the top of layer 1, the tool is stirred over the surface and the powder is fed in the solid rod form. Due to heat generated between the tool and the work piece, the powder is stuck over the surface of the material. Again, the second layer is placed over the layer 1, and same process is repeated. The tool dimensions are changed accordingly to the thickness of the sheet. Similarly, layer 3 and layer 4 are added over the sheets. This completes the FSAM process.

2.2 Friction Surfacing

Friction surfacing method is used for depositing materials together with hard facings, while tool steels and stainless steels enhance the quality of the formed metallic surface. Friction-surfacing uses a consumable rotating tool to create frictional contact between the interface under axial applied load. The heat produced at the tip of the tool tends to form a plasticized layer. Rotating consumable tool tends to deposit this

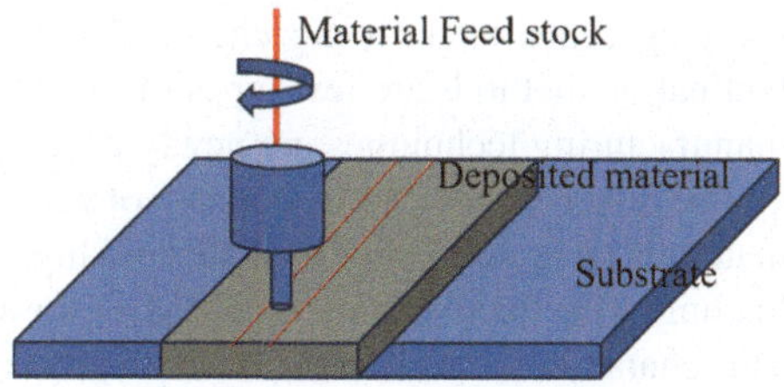

Fig. 3 FSAM by friction surfacing technique

plasticized material onto the substrate. It is based on the high-temperature severe plastic deformation [16]. Friction surfacing can be used in automobile and aerospace sectors.

Due to this, a visco-plastic layer is formed, thus establishing the inter-diffusion mechanisms between the interfaces. The thickness of the deposited material along with quality can be well-ordered in terms of rotational speed, its traveling velocity, axial forging pressure, and consumable tool. Aggarwal et al. [25] deposited Ti-6Al-4 V alloy chips using AFSD. AFSD of the recycled metal offers openings to prepare the sound components exhibiting better tensile properties. Bararpour et al. [26] evaluated the mechanical characteristics of friction surfaced AA5083-15wt%Zn composites. The maximum hardness after solid solution treatment was 63.41 BHN and without treatment was 56.90 BHN occurring at 185 °C, and then the reduction of hardness was decreased. Bararpour et al. [27] observed the influence of Zn powder of Al–Mg alloy friction surfaced on the AA 5052 alloy substrate. With the increase of heat treatment time, the agglomeration of Zn particles was reduced. FSAM by friction surfacing technique is shown in Fig. 3.

Friction surfacing have got no melting of the substrate material. Thus, no dilution of the substrate into the deposit is formed. The deposit's composition is the same as the substrate material. The deposit material is homogenous in nature and has good mechanical properties. Bond quality formed nearbythe deposit is poor. This requires removal of the nearby deposit. Thickness of the deposit is always small. It is a solid phase method that allows the use of metallurgically incompatible materials. Material properties of special combinations can be found by this technique. Friction surfacing process can be utilized to form aluminum/magnesium metal matrix composites by incorporating hard ceramic particles into the consumable material.

2.3 Fabrication of Functionally-Graded Materials

Functionally graded materials consist of a spatial distinction in microstructure/ composition for particular use by the control of variations in functional or structural and thermal characteristics. Nowadays, there are various types of FGMs available. Microstructure gradient functionally graded materials, chemical composition gradient functionally graded materials and porosity gradient functionally graded materials are the types of functionally graded materials. Powder metallurgy, plasma spraying, centrifugal casting, chemical and physical vapor deposition, infiltration and lamination methods are few types of methods to produce FGM. Solid state additive manufacturing techniques are newly developed techniques to fabricate functionally graded materials. FGMs are capable of withstand high strength and toughness. These materials have got ability to withstand high temperature gradient under mechanical loading. The main feature of FGM is the capability to hinder crack propagation. This characteristic had enabled to use this technique makes it useful in armor plates, defense application, bullet-proof vests and energy conversion devices.

Khodabakhshi et al. [10] have reviewed the FGMs prepared through FSAM. FGMs are the advancements of FSP. Different grooves of different dimensions are preplaced with different powders inside the grooves. Tools, having different pin length are stirred in the processed zones. Multiple FSP passes results in intensifying the density with substantial grain refinement and uniform dispersal of the inclusions. This is a type of AM formed with layer-by-layer that changes the grain structure and chemistry to prepare the final component [28]. Thus, it can be a method of FSAM technique to prepare FGMs. Figure 4 shows the encapsulation of reinforcements in the base metal/alloys. This method is capable of producing the composite using FGM technique. Gandra et al. [29] produced AA 5083 based FGM composite reinforced by SiC particles by FSP. 5.15 GPa was the hardness found in the retreating side and 1.05 GPa in the advancing side. 1.45 GPa was the hardness in bead surface center. Bikkina et al. [30] prepared Al/SiC FGM via FSP route. Three passes of FSP were performed on the surface. Micro-hardness was improved by increasing the number of passes by 51.2%. Wear resistance was improved due to SiC content embedded in the Al alloy.

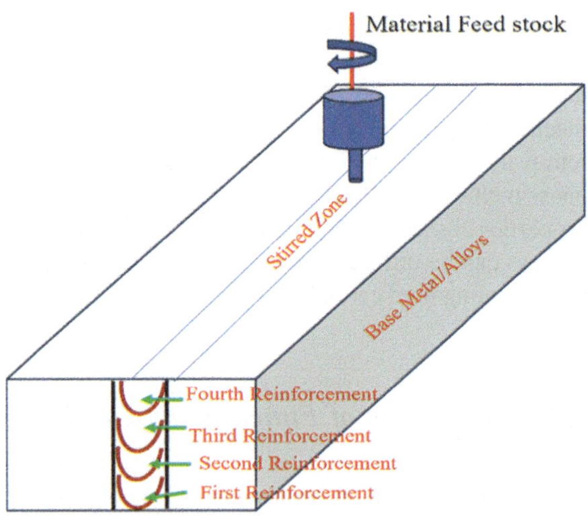

Fig. 4 Functionally graded composite material

2.4 Cladding Strategies

Friction stir welding is a newly developed solid-state joining technique for alloys/metals. No solidification defects are produced, thus enabling the suitable method to join aluminium alloys/metals. Generally, for corrosion protection, alloys are cladded with a thin layer of pure aluminium. The FSW/FSP tool is plunged and inserted into the top surface of overlapped sheets, and rotated along the transverse direction. Due to frictional-heating, the feed material is deposited on the surface of the substrate. Selection of tools dimension is a tedious task in making a 3D component. The method can be performed by multi-passes to cover the whole surface.

Friction stir cladding is an advanced solid-state additive manufacturing method that creates 3D components built on deformation bonding, irrespective of melting and solidification. Friction stir cladding method can be used in automobile industries. Friction stir cladding method depends upon the substrate distortion and cladding quality. AFSD can build high-quality cladding on the substrate of 1.4 mm. No type of buckling is found in the thin substrate. Sometimes anticlastic curvature may be formed on the plate due to improper clamping. The classical lamination theory and curvature measurement can be used to evaluate the residual stress formed by this newly developed technique.

Perry et al. [16] performed AFSD on a single layer of AA 2024 alloy. Experimental results concluded that both the deposited AA 2024 and AA 6061 substrate surfaces illustrated grain refinement produced due to recrystallization. Roodgari et al. [17] performed fabrication of a 2-layer laminated steel composite via FSAM. The maximum hardness of 225 HV was found in the layer at transverse speed and rpm of 70 mm/min and 600 rpm. Hartley et al. [31] performed cladding on thin automotive sheet metals by AFSD. The flat tool produced better surface quality in comparison to the protruding tool. UTS of 250 MPa along with uniform micro-hardness was observed. Results revealed that AFSD has better control than friction surfacing in terms of layer height, surface finish, and flash on the edges of deposition tracks. Mahto et al. [32] performed friction stir cladding of Cu on the aluminium substrate. Cladding of 3 mm Cu on 6 mm thick aluminium sheet was performed. Good intermixing and interlocking across the interface between the substrate and clad layer was formed with the improved hardness. Through the performed experiments by Perry et al. [16], Roodgari et al. [17], Hartley et al. [31] and Mahto et al. [32], AFSD technique was capable to clad the layer of the hard material into other materials. An enhancement in mechanical properties was observed by these performed experiments.

2.5 FSP as an Alteration to Other Methods

Another method for application of FSP is used for variation of properties of deposited materials, coating or those prepared by other methods, such as, cold gas dynamic spray deposition [33], electro-deposition [34], thermal spraying [35], etc. In such cases, plastic deformation of the material is created by the rotating the FSP tool. This tends to alter the modification of the deposited coatings by aiding their densification, structural homogenization, grain refinement, and enhancement in mechanical characteristics. Figure 5 shows the FSP process where cold spray of the reinforcement through a nozzle is done. The tool is rotated over the sprayed reinforcement on the surface of the material. Certain modification in traditional FSP set up is done when cold spray of the ceramic particle is made. To spray the particles in the stirred zone, a nozzle is used. Due to heat produced by the rotation action of tool, sprayed particles are mixed and embedded on the top of the surface. Improvement in mechanical properties can be observed on the processed material.

Tan et al. [14] fabricated the Al/Al_2O_3 composite via FSAM. Experimental results revealed that the re-stirring method in FSAM lead to uniform distribution of Al_2O_3 content. This has improved the hardness to 87.34 HV in comparison to the base metal having 67 HV. Wang et al. [15] prepared bulk Al by cold spray FSP composite additive manufacturing. The grain size was recrystallized to 4.2 μm. The yield strength and ultimate tensile strength was enhanced by 38 and 45%. Ho et al. [18] evaluated the bio-corrosion behavior of AZ31B alloy hydroxyapatite composite fabricated via FSAM. Due to grain refinement, corrosion behavior was improved as compared to non-treated alloy. Mukhopadhyay and Saha [19] performed FSAM on pure aluminum alloy. 7 mm thick 3 D structure of aluminum sheet was fabricated. UTS, YS and % elongation was enhanced by 36, 3 and 54%. Puleo et al. [36] performed FSAM on the Al 7055 alloy. 60% reduction in strength was observed. This may be due to warm holes or cold lapping present in the weld. From the results obtained by Tan et al. [14], Wang et al. [15], Ho et al. [18], Mukhopadhyay and Saha [19], etc., it can be

Fig. 5 FSP process followed by cold spray of the reinforcement through a nozzle

concluded that FSP technique was able to refine the grain size of the surface. This methodology has brought the revolution in the mechanical industries. This technique can be efficiently applied to the aluminium and magnesium alloys.

Palanivel et al. [37] performed FSAM on the Mg-based WE43 alloy. The strength of the alloy was improved to 400 MPa with 17% improvement in ductility. Xie et al. [20] produced the nano TiB_2/AlSi10Mg composite by cold spray additive manufacturing. Experimental results revealed enhancement in UTS (365 ± 35 MPa) and ductility (16.0 ± 1.2%). Sharma et al. [21] fabricated the functionally graded composite by friction stir processing of Al plate reinforced with TiC particles. The maximum hardness of 41 HV was obtained for one pass FSP. Huang et al. [22] prepared brass-based composite using cold spray additive manufacturing technique. Al_2O_3 and SiC particles were used. Performed coating resulted in the improvement of micro-hardness. Santhosh et al. [38] performed the numerical analysis of Al 6061 alloy and SiC in-situ metal matrix fabricated via FSP. The exact linear variation in the thermal property over the entire volume of the material produced was observed for center-to-center distance. Basak et al. [39] fabricated the AA 5086/Gr/SiC surface composite through FSP. Nanoparticles of Gr and Sic were used. Grain refinement of 1.59 μm was found for the surface composite. Compressive strength was improved due to microstructural development. Compressive strength for the fabricated surface composite was enhanced to 193.3 MPa, in comparison to the base material having 147.8 MPa. Due to the integration of Gr and SiC nano-particulate, the material possesses ductile fracture with improved plasticity. Hasan et al. [40] used FSAM technique to develop a 4-layered laminate aluminum alloy Al-7075 sheet. Various profiles of tool (cylindrical threaded, taper threaded and plain cylindrical) at different tool traverse speeds (20, 40 and 60 mm/min) were rotated at 500 and 1200 rpm. Improvement in microhardness was noticed with the increase in traverse speed. Xiao et al. [41] performed numerical and experimental investigation of FSAMed on AA 2024. FEM was used to perform numerical investigation. With an increase in the tool's transverse speed, the maximum value of the thermal cycle was decreased. With a decrease in the tool transverse speed, the high-temperature residence time was increased. Chaudhary et al. [42] performed friction stir powder additive manufacturing of Al 6061 alloy to improve mechanical and microstructure properties. Fine and equiaxed grains of Al 6061 alloy multi-layer deposition was found due to continuous dynamic recrystallization of the alloy. Reduced porosity of 0.19% was observed due to inherent compressive forces in the FSPAM process. This tends to intimate the contact among the powder particles. No agglomeration of the alloying element was found. This was confirmed through Energy dispersive spectroscopy test. Minimal variations in microhardness (8.8%) was found along the build direction due to reduced thermal gradient. Significant number of dimples had indicated the ductile nature of Al 6061 alloy. Elongation of 16.7% was found for the processed material. Jiang et al. [43] used FSAM technique to study the microstructural and mechanical properties of AA 2060 alloy performed at different process parameters. This research work uses 2 mm thick plates with process parameters in the range of (1500–1800) rpm, (300–500) mm/min rotational and translational speed. Optimum results were obtained at 1600 rpm and 300 mm/min of rotational and translational speed. Finer equiaxed grains of size (2–5) μm were found in the nugget zone due to dynamic recrystallization. 135 HV was the maximum microhardness.

Sharma et al. [21], Santhosh et al. [38], Xiao et al. [41] and Jiang et al. [43] fabricated composites using FGM, FSP or an advanced FSP process like the cold spray FSP process. All these processes were suitable and capable to improve the mechanical, thermal and microstructural properties.

3 Implementation of Machine Learning (ML) in Solid State Additive Manufacturing Technique

Nowadays, use of ML in solid state additive manufacturing technique is done for the optimization of various process parameters. Artificial intelligence (AI) in the form of ML is used. A lot of work has been performed in Friction stir processing, friction stir spot welding for parametric optimization of process parameters using Taguchi technique. Kumar et al. [44] have used Taguchi technique to optimize the process parameters. Several ML techniques like Support Vector Machine (SVM), Regression model (RSM), Adaptive Neuro-Fuzzy Inference System (ANFIS) and artificial neural network (ANN) may be used to optimize the process parameters. To achieve a desired build quality and microstructure, ML techniques will enable to select the process parameters. To reduce the material and experimental cost, ML can be used as a newly developed tool in the industrial era. Additive manufacturing technique have several problems before it comes to commercialization. Foremost challenges associated with AM techniques include manufacturing of complicated shapes and geometries. Tool wear may cause inappropriate stirring action. Development of residual stress due to thermal cycling is a major problem. Special clamps and custom-built fixtures are needed to fabricate complex shapes. ML may help the researchers to optimize the process parameters which will reduce material and experimental cost and attain better mechanical and microstructural properties.

4 Conclusion

This reported work sightsees the progress and advancement of the solid-state additive manufacturing technique. This chapter focuses on mechanical and microstructural properties, process parameters and defects produced, and potential applications, along with the present research work carried out by the researchers. The various techniques discussed in this chapter foresight the future prospective of the advanced techniques. Through literature survey, it is revealed that the additive based FSP is a protruding tool for manufacturing purpose. It has enabled to transform the raw material into a material enhanced with superior mechanical and microstructural properties. Other characteristics like negligible emission, non-porous components, non-consumable tool, acceptability and reproducibility, are present for larger components. FSAM has proved to be a better substitute for most of the engineering applications. Moreover, the solid-state manufacturing technique can be utilized in the application of all medical and engineering fields.

FSAM addresses the failings of existing melting-based additive manufacturing processes. FSAM was used to produce defect-free parts with equiaxed refined grain structure, with excellent homogeneous mechanical properties. This technique may be used to fabricate the bulk material with ultrafine grain structure. Parts with a low level of complexity, multi-material and new alloys parts with simple design can be prepared by the FSAM process. Nowadays, FSAM has achieved a lot of consideration in the research area during current years.

5 Future Perspective

Nowadays, Friction additive manufacturing technique, a novel route to fabricate the composite had attracted the researchers to use this technique. Problems oriented in liquid metallurgical route are reduced. With the variation of process parameters of processing and machine, new enhanced mechanical properties may be observed. Meanwhile tool design also played a vital role. With the variation of tool profile, further experiments may be performed to achieve better experimental results.

To eliminate defects by controlling process parameters, and parameters like rotation speed, transverse speed, and tilt angle, is the main task of this game-changing process. Mechanical properties such as tensile strength, hardness and compressive strength are extensively observed for various materials produced through FSAM, failure properties under various types of loadings, still need to be investigated. Generally, aluminium and magnesium are more investigated materials. Ni and Ti material can be further explored using this novel technique.

6 Summary

In this chapter, the aim was on focusing the aspects of FSAM via solid-state additive technology. This chapter reviews the reported work performed by various researchers. FSAM was classified into five routes using FSP to deposit metals/alloys/ceramics as layer by-layer based using the concepts of: (i) surface cladding, (ii) feeding of material through wire/powder, (iii) friction-surfacing, (iv) functionally graded materials, and (v) FSP as variation to the other methods. FSAM methodology has the prospect to refine the grain structure of materials during deposition, owing to the existence of severe plastic deformation created by the elevating temperature during the process or stirring action of the rotating tool. Also, components manufactured by FSAM technique are free from defects like void, cracks, porosity and undesirable phases. Main advantages of this technique are less residual stresses along with distortion in manufactured parts are produced. Complexity and flexibility to fabricate the counterparts are the main disadvantages of this technique in comparison to other fusion-based ones.

Friction stirring played an important role in additive manufacturing technique. Several applications include CSAM, FSAM and FSD etc. Nowadays, FSAM has proved itself the growing manufacturing technique in the field of industrialization. Through vast literature survey, it can be concluded that FSAM had improved the mechanical and microstructural properties of the prepared material in the nugget zone.

To obtain the desirable properties of the material, process parameters like rotational speed, transverse speed, tool tilt angle or tool pin profile played an important role. This technique is a solid-state technique, thus free from several defects like porosity or cavity. Holding fixtures and large axial forces are required for the operation. However, proper selection of parameters will definitely enable to prepare the material of desired shape.

Declarations

Conflicts of Interest The authors confirm that there are no conflicts of interest associated with this work.

Ethical Compliance The authors affirm that this study adheres to all ethical guidelines.

Competing Interests No competing interests are declared by the authors.

Data and Material Availability Data and materials are not applicable to this study.

References

1. Ngo TD, Kashani A, Imbalzano G, Nguyen KTQ, David H (2018) Additive manufacturing (3D printing): a review of materials, methods, applications and challenges. Compos B 143:172–196
2. Leon A, Levy GK, Ron T, Shirizly A, Aghion E (2020) The effect of strain rate on stress corrosion performance of Ti6Al4V alloy produced by additive manufacturing process. J Mater Res Technol. https://doi.org/10.1016/j.jmrt.2020.02.035
3. Pereira LAT, Martinez LG, Mucsi CS, Reis LAM, Rossi JL (2020) On the melting of zirconium alloys from scraps using electron beam and induction furnaces-recycling process viability. J Mater Res Technol. https://doi.org/10.1016/j.jmrt.2020.03.006
4. Rahman NU, Capuano L, Cabeza S, Feinaeugle M, Garcia-Junceda A, Rooij MBD et al (2019) Directed energydeposition and characterization of high-carbon high speedsteels. Addit Manuf 30
5. Maurya M, Kumar S, Maurya NK (2020) Composites prepared via friction stir processing technique: a review. Revue des Composites et des Matériaux Avancés 30
6. García-Vázquez F, Vargas-Arista B, Muñiz R, Ortiz JC, García HH, Acevedo J (2016) Soldagem Inspecao 21:508

7. Azizieh M, Iranparast D, Dezfuli M, Balak Z, Kim H (2017) Trans Nonferrous Met Soc China 27:779
8. Maurya M, Kumar S, Bajpai V (2019) Assessment of the mechanical properties of aluminium metal matrix composite: a review. J Reinf Plast Compos 38(6):267–298
9. Srivastava AK, Maurya NK, Dixit AR, Dwivedi SP, Saxena A, Maurya M (2021) Experimental investigations of A359/Si_3N_4 surface composite produced by multi-pass friction stir processing. Mater Chem Phys 257:123717
10. Khodabakhshi F, Gerlich AP (2018) Potentials and strategies of solid-state additive friction-stir manufacturing technology: a critical review. J Manuf Process 36:77–92
11. Srivastava AK, Maurya NK, Maurya M, Dwivedi SP, Saxena A (2020) Effect of multiple passes on microstructural and mechanical properties of surface composite AI 2024/SiC produced by friction stir processing. Annales de Chimie-Science des Matériaux 44(6):421–426
12. Srivastava AK, Dixit AR, Maurya M, Saxena A, Maurya NK, Dwivedi SP, Bajaj R (2021) 20th Century uninterrupted growth in friction stir processing of lightweight composites and alloys. Mater Chem Phys 266:124572
13. Elfishawy E, Ahmed MMZ, El-Sayed Seleman MM (2020) Additive manufacturing of aluminum using friction stir deposition. In: The Minerals, Metals & Materials Society (eds) TMS 2020 149th annual meeting & exhibition supplemental proceedings. The Minerals, Metals & Materials Series. Springer, Cham. https://doi.org/10.1007/978-3-030-36296-6_21
14. Tan Z, Li J, Zhang Z (2021) Experimental and numerical studies on fabrication of nanoparticle reinforced aluminum matrix composites by friction stir additive manufacturing. J Market Res 12:1898–1912
15. Wang W, Han P, Wang Y, Zhang T, Peng P, Qiao K, Wang Z, Liu Z, Wang K (2020) High-performance bulk pure AI prepared through cold spray-friction stir processing composite additive manufacturing. J Mater Res Technol 9(4):9073–9079
16. Perry MEJ, Griffiths RJ, Garcia D, Sietins JM, Zhu Y, Yu HZ (2020) Morphological and microstructural investigation of the non-planar interface formed in solid-state metal additive manufacturing by additive friction stir deposition. Addit Manuf. https://doi.org/10.1016/j.addma.2020.101293
17. Roodgari MR, Jamaati R, Aval HJ (2020) Fabrication of a 2-layer laminated steel composite by friction stir additive manufacturing. J Manuf Process 51:110–121
18. Ho YH, Joshi SS, Wu TC, Hung CM, Ho NJ, Dahotre NB (2020) In-vitro bio-corrosion behavior of friction stir additively manufactured AZ31B magnesium alloy-hydroxyapatite composites. Mater Sci Eng C 109:110632
19. Mukhopadhyay A, Saha P (2020) Mechanical and microstructural characterization of aluminium powder deposit made by friction stir based additive manufacturing. J Mater Process Technol 281:116648
20. Xie X, Chen C, Chen Z, Wang W, Yin S, Ji G, Liao H (2020) Achieving simultaneously improved tensile strength and ductility of a nano-TiB_2/AlSi10Mg composite produced by cold spray additive manufacturing. Compos B Eng 202:108404
21. Sharmaa A, Vijendra B, Ito K, Kohama K, Ramjia M, Sai BV (2017) A new process for design and manufacture of tailor-made functionally graded composites through friction stir additive manufacturing. J Manuf Process 26:122–130
22. Huang CJ, Wu HJ, Xie YC, Li WY, Verdy C, Planche MP, Montavon G (2019) Advanced brass-based composites via cold-spray additive-manufacturing and its potential in component repairing. Surf Coat Technol 371:211–223
23. MELD brochure, Aeroprobe corporation (2018). http://www.aeroprobe.com/meld/
24. He C, Li Y, Zhang Z, Wei J, Zhao X (2020) Investigation on microstructural evolution and property variation along building direction in friction stir additive manufactured Al–Zn–Mg alloy. Mater Sci Eng A 777:139035
25. Agrawal P, Haridas RS, Yadav S, Thapliyal S, Gaddam S, Verma R, Mishra RS (2021) Processing-structure-property correlation in additive friction stir deposited Ti-6Al-4V alloy from recycled metal chips. Addit Manuf 47:102259

26. Bararpour SM, Aval HJ, Jamaati R (2020) Effect of non-isothermal aging on microstructure and mechanical properties of friction surfaced AA5083–15wt% Zn composites. Surf Coat Technol 384:125307
27. Bararpour SM, Aval HJ, Jamaati R (2020) Effects of Zn powder on alloying during friction surfacing of Al–Mg alloy. J Alloy Compd 818:152823
28. Miranda RM, Santos TG, Gandra J, Lopes N, Silva RJC (2013) Reinforcement strategies for producing functionally graded materials by friction stir processing in aluminium alloys. J Mater Process Technol 213(9):1609–1615
29. Gandra J, Miranda R, Vilaça P, Velhinho A, Teixeira JP (2011) Functionally graded materials produced by friction stir processing. J Mater Process Technol 211(11):1659–1668
30. Bikkina V, Talasila SR, Adepu K (2020) Characterization of aluminum based functionally graded composites developed via friction stir processing. Trans Nonferrous Metals Soc China 30(7):1743–1755
31. Hartley WD, Garcia D, Yoder JK, Poczatek E, Forsmark JH, Luckey SG, Dillard DA, Hang ZY (2021) Solid-state cladding on thin automotive sheet metals enabled by additive friction stir deposition. J Mater Process Technol 291:117045
32. Mahto MK, Kumar A, Raja AR, Vashista M, Yusufzai MZ (2022) Friction stir cladding of copper on aluminium substrate. CIRP J Manuf Sci Technol 36:23–34
33. Khodabakhshi F, Marzbanrad B, Shah L, Jahed H, Gerlich A (2017) Friction-stir processing of a cold sprayed AA7075 coating layer on the AZ31B substrate: structural homogeneity, microstructures and hardness. Surf Coat Technol 331:116–128
34. Ma ZY (2008) Friction stir processing technology: a review. Metall Mater Trans A 39(3):642–658
35. Rahbar-kelishami A, Abdollah-zadeh A, Hadavi MM, Banerji A, Alpas A, Gerlich AP (2015) Effects of friction stir processing on wear properties of WC-12%Co sprayed on 52100 steel. Mater Des 86:98–104
36. Puleo SM (2016) Additive friction stir manufacturing of 7055 aluminum alloy, p 75. Senior Honors theses. https://scholarworks.uno.edu/honors_theses/75
37. Palanivel S, Nelaturu P, Glass B, Mishra RS (2015) Friction stir additive manufacturing for high structural performance through microstructural control in an Mg based WE43 alloy. Mater Des 1980–2015(65):934–952
38. Santhosh V, Agasia Prakash DN, Murugan K, Babu N (2020) Thermo-mechanical analysis of Tailor-made functionally graded materials through Friction stir processing. Mater Today Proc 33:4445–4449
39. Basak AK, Pramanik A, Prakash C, Shankar S, Sehgal SS (2023) Microstructure and micro-mechanical properties of friction stir processed Al 5086-based surface composite. Mater Today Commun 35:105830
40. Hassan A, Awang M, Pedapati SR, Altaf K, Marode RV, Ahmed SW (2023) Experimental investigation on tool pin profile for defect-free multi-layered laminates using friction stir additive manufacturing. Results Eng 20:101516
41. Xiao Y, Li Y, Shi L, Wu C, Liu H, Sun Z (2023) Experimental and numerical analysis of friction stir additive manufacturing of 2024 aluminium alloy. Mater Today Commun 35:105639
42. Chaudhary B, Jain NK, Murugesan J, Patel V (2023) Friction stir powder additive manufacturing of Al 6061 alloy: enhancing microstructure and mechanical properties by reducing thermal gradient. J Mater Res Technol 26:1168–1184
43. Jiang T, Jiao T, Dai G, Shen Z, Guo Y, Sun Z, Li W (2023) Microstructure evolution and mechanical properties of 2060 Al-Li alloy via friction stir additive manufacturing. J Alloy Compd 935:168019
44. Kumar S, Maurya M, Sharma T et al (2023) Optimization and ANFIS-based modeling of two step FSSW process parameters on tensile strength for triple-sheet joining of aluminum alloy. Proc Inst Mech Eng Part C J Mech Eng Sci 0(0). https://doi.org/10.1177/09544062231202143

Chapter 8
Polymer Composites for Environmental Pollution and Remediation

Ankita Saini and Sunil Kumar Saini

1 Introduction

1.1 Overview of Environmental Pollution

Environmental pollution poses enduring global challenges that demand urgent attention and innovative solutions for the well-being of our planet's ecosystems and human populations. From air and water pollution to soil contamination, the detrimental effects of pollution call for transformative approaches that emphasize novel strategies and emerging technologies. Nanotechnology, with its ability to manipulate materials at the atomic and molecular levels, offers exciting possibilities for pollution control. Nano-enabled solutions provide enhanced adsorption capacities, catalytic activities, and selective filtration capabilities. Incorporating different polymers and their composites improves pollutant removal and treatment efficiency. Responsible and safe application of nanomaterials necessitates careful assessment of their potential environmental and health impacts. By embracing the innovative remediation techniques, approaches, and solutions, we can address the complexity of environmental pollution and pave the way for a cleaner, healthier, and more sustainable future for our planet and all its inhabitants.

A. Saini (✉)
Department of Applied Sciences and Humanities, Ganga Institute of Technology and Management, Kablana, Jhajjar, Haryana 124104, India
e-mail: ankitasaini53@gmail.com

S. K. Saini
Department of Zoology, Faculty of Life Sciences, P.D.M. University, Bahadurgarh, Haryana 124507, India

S. K. Sethi et al. (eds.), *Polymer Composites: From Computational to Experimental Aspects*, Materials Horizons: From Nature to Nanomaterials,
https://doi.org/10.1007/978-981-97-0888-8_8

1.2 Importance of Remediation Techniques

To mitigate the adverse effects of pollution and restore environmental quality, remediation techniques play a crucial role. Such techniques play a vital role in mitigating environmental pollution by eliminating or reducing pollutants, restoring ecosystems, safeguarding human health, and promoting sustainable development. By applying these techniques, we can work towards creating a cleaner, healthier, and more sustainable environment for current and future generations. The choice of technique depends on factors such as the type and extent of pollution, site conditions, cost-effectiveness, and regulatory requirements. Remediation techniques for environmental pollution encompass a wide range of approaches tailored to address specific types of pollution. Physical techniques involve the use of physical processes such as soil vapor extraction, capping, and lining to remove or contain pollutants. Chemical techniques focus on chemical reactions, including chemical precipitation, oxidation, and reduction, to transform or treat pollutants. Biological techniques utilize the power of living organisms, employing bioremediation and phytoremediation to break down or stabilize contaminants using microorganisms and plants, respectively. Thermal techniques employ heat to treat contaminated materials, with thermal desorption and incineration being commonly used methods. Electromagnetic techniques utilize electromagnetic radiation or fields, such as UV radiation and electrocoagulation, for the disinfection of water or removal of suspended particles. Nanotechnology-based remediation techniques utilize nanomaterials to enhance adsorption, catalytic activity, and filtration of pollutants in various media [1]. In situ techniques treat pollution directly at the site without excavation, including in situ chemical oxidation and in situ bioremediation, which introduce chemicals or microorganisms to break down contaminants in the soil or groundwater.

Here are some commonly used remediation techniques for different types of pollution:

- Soil Remediation Techniques:

Soil Vapor Extraction: Using a vacuum system to extract volatile contaminants from the soil.

Bioremediation: Utilizing microorganisms to break down organic pollutants into harmless substances.

Phytoremediation: Using plants to absorb, degrade, or stabilize contaminants through processes like phytoextraction, rhizofiltration, and phytostabilization.

Soil Washing: Treating contaminated soil through physical separation, often using water or chemical solutions, to remove or reduce pollutants.

Thermal Desorption: Applying heat to contaminated soil to vaporize and separate contaminants, allowing for further treatment or disposal.

- Water Remediation Techniques:

Activated Carbon Filtration: Adsorbing pollutants onto activated carbon to remove them from water.

Biological Treatment: Utilizing microorganisms for the breakdown of organic contaminants via biodegradation or bioaugmentation.

Chemical Precipitation: Adding chemicals to water to induce the formation of insoluble precipitates that can be removed.

Membrane Filtration: Passing water through membranes with specific pore sizes to separate and remove contaminants.

Electrocoagulation: Applying an electrical current to water to destabilize and aggregate suspended particles, facilitating their removal.

- Air Remediation Techniques:

Air Filtration: Removing pollutants from the air using filters like high-efficiency particulate air (HEPA) filters or activated carbon filters.

Scrubbers: Removing gaseous pollutants through chemical reactions or absorption processes.

Biofiltration: Passing air through a bed of microorganisms that degrade or adsorb pollutants.

Thermal Oxidation: Decomposing pollutants at high temperatures through processes such as incineration or thermal desorption.

Absorption: Absorbing gaseous pollutants into a liquid or solid medium to effectively remove them from the air.

- Groundwater Remediation Techniques:

Pump-and-Treat: Pumping groundwater to the surface, treating it to remove contaminants, and then re-injecting it into the ground.

In situ Chemical Oxidation: Injecting chemicals into groundwater to oxidize and break down contaminants.

Permeable Reactive Barriers: Placing reactive materials in the subsurface to intercept and treat groundwater, removing contaminants as the water flows through them.

Enhanced Bioremediation: Enhancing the natural degradation capacity of microorganisms in groundwater by adding nutrients or electron acceptors.

Electrokinetic Remediation: Applying electric currents to groundwater to induce movement and migration of contaminants toward electrodes for extraction or treatment.

These remediation techniques are selected based on their effectiveness, feasibility, and compatibility with the specific pollution scenario and environmental conditions. Implementing a combination of techniques may be necessary to achieve successful remediation outcomes.

1.3 Role of Polymer Composites in Environmental Remediation

Polymer composites have emerged as critical components in environmental remediation, providing innovative solutions for the treatment and elimination of pollutants. These composites, consisting of a polymer matrix reinforced with filler materials, possess unique properties that make them highly effective in a wide range of remediation applications [12]. Let us delve into the significant contributions of polymer composites in environmental remediation.

1. Enhanced Adsorption Capabilities:

One of the key advantages of polymer composites is their ability to exhibit high surface area-to-volume ratios and tailored surface properties. This enables them to efficiently adsorb and remove pollutants from air, water, and soil. By incorporating specific adsorbent materials like activated carbon or nanomaterials into the polymer matrix, these composites demonstrate enhanced adsorption capacities, making them exceptionally efficient in capturing and immobilizing contaminants.

2. Selective Filtration and Separation:

Polymer composites can be designed with precise pore sizes, allowing for the selective filtration and separation of pollutants based on their size or chemical properties. The matrix can be engineered to possess controlled porosity or incorporate membranes, enabling the filtration of particulate matter, organic compounds, heavy metals, and other contaminants.

3. Chemical Transformation and Degradation:

The incorporation of functionalized polymers or composite materials containing catalysts within polymer composites enables them to facilitate chemical reactions for the transformation and degradation of pollutants. These composites can promote specific chemical reactions such as oxidation or reduction, leading to the breakdown of contaminants into less harmful substances. For instance, in advanced oxidation processes, the polymer matrix acts as a support for photocatalysts, expediting the degradation of organic pollutants.

4. Soil Stabilization and Containment:

Polymer composites find extensive applications in soil remediation by providing stabilization and containment measures. When used as liners or capping materials, polymer composites create impermeable barriers, effectively preventing the migration of contaminants from the soil to water sources or underlying structures. Moreover, they enhance soil stabilization by improving its structural integrity, reducing erosion, and minimizing the leaching of pollutants.

5. Sustainable Remediation Approaches:

Polymer composites promote sustainable remediation approaches by emphasizing resource efficiency and minimizing waste generation. They can be designed for reusability, allowing for repeated use in pollutant removal processes. Additionally, incorporating recycled or biodegradable materials in composite formulations reduces their environmental impact. The longevity and durability of polymer composites contribute to long-term solutions for environmental remediation, thereby reducing the need for frequent replacements.

6. Integration with Advanced Technologies:

Polymer composites seamlessly integrate with advanced technologies, bolstering environmental remediation efforts. They can be combined with nanotechnology, sensors, and monitoring systems to create smart composites capable of real-time monitoring, targeted pollutant detection, and efficient remediation. The compatibility of polymer composites with these technologies facilitates the development of innovative and responsive remediation strategies.

Polymer composites play a vital role in environmental remediation by offering enhanced adsorption capabilities, selective filtration and separation, chemical transformation and degradation, soil stabilization, and containment. Their sustainable nature and integration with advanced technologies make them invaluable tools for addressing environmental pollution challenges.

1.4 Types of Polymer Composites Used in Environmental Remediation

Polymer composites have shown significant attention in the realm of environmental remediation due to their distinctive properties and versatility. These composites blend the advantages of polymers with other reinforcing materials like nanoparticles, fibers, or fillers, to create advanced materials that exhibit enhanced performance in various remediation applications. Below are several types of polymer composites commonly employed in environmental remediation and depicted in Fig. 1.

1. Polymer Nanocomposites:

Polymer nanocomposites are crafted by incorporating nanoparticles, typically ranging from 1 to 100 nm, into a polymer matrix. These nanoparticles, including carbon nanotubes, graphene oxide, or clay nanoparticles [4, 9, 18, 33], bolster the composite's properties such as mechanical strength, thermal stability, and adsorption capacity. Polymer nanocomposites find utility in environmental remediation for tasks such as water purification, soil remediation, and air pollution control [1, 11, 21, 29].

2. Fibrous Polymer Composites:

Fibrous polymer composites involve reinforcing a polymer matrix with fibers like carbon fibers, glass fibers, or natural fibers such as hemp or jute [2, 16, 17, 27]. These

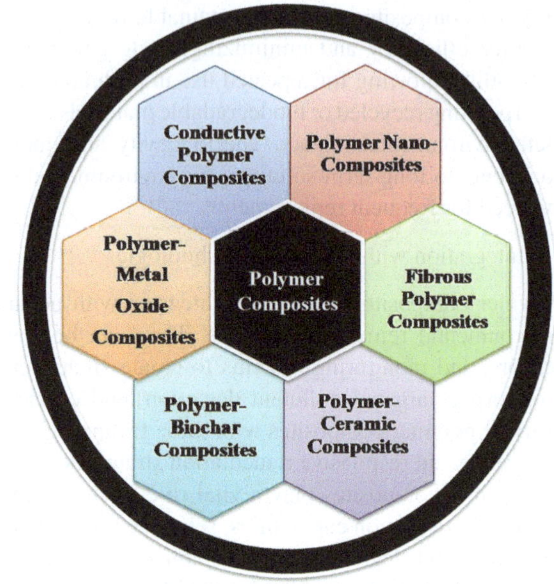

Fig. 1 Types of polymer composites commonly employed in environmental remediation

fibers provide the composite with high tensile strength, stiffness, and durability. Fibrous polymer composites are commonly utilized in environmental remediation for activities like oil spill cleanup, wastewater treatment, and air filtration [32].

3. Polymer–Ceramic Composites:

Polymer–ceramic composites combine the properties of polymers with ceramics to yield robust materials that possess enhanced chemical resistance and mechanical strength. These composites are applied in scenarios that necessitate the removal of heavy metals, organic pollutants, and contaminants from soil, water, or air.

4. Polymer–Biochar Composites:

Biochar, a carbon-rich material derived from biomass pyrolysis, is combined with polymers to create polymer–biochar composites, which demonstrate excellent adsorption capabilities [26, 31]. These composites find widespread use in eliminating organic pollutants, heavy metals, and dyes from wastewater and soil.

5. Polymer–Metal Oxide Composites:

Polymer–metal oxide composites entail integrating metal oxide nanoparticles like titanium dioxide, iron oxide, or zinc oxide into a polymer matrix [5]. These composites possess photocatalytic properties, enabling them to degrade organic pollutants under light irradiation. They are employed in applications such as water purification, air pollution control, and self-cleaning coatings.

6. Conductive Polymer Composites:

Conductive polymer composites combine a polymer matrix with conductive fillers such as carbon nanotubes, graphene, or metallic particles [6, 15, 28]. These composites exhibit electrical conductivity and find utility in electrochemical remediation processes like the electrochemical degradation of pollutants or electrokinetic soil remediation.

The specific selection of a polymer composite depends on factors such as the nature of the contaminants, the remediation process, and the desired properties required for effective environmental remediation. These innovative materials hold promising potential for addressing environmental challenges and contributing to sustainable remediation practices.

2 Adsorption and Filtration Applications of Polymer Composites

Polymer composites encompass materials composed of two or more distinct components, with a polymer matrix fortified by filler substances like particles, fibers, or alternative polymers. These composite materials exhibit distinct characteristics that render them well-suited for adsorption and filtration purposes.

2.1 Adsorption Mechanisms of Polymer Composites

The adsorption mechanisms of polymer composites depend on several factors, including the properties of the polymer matrix, the nature of the adsorbate, and the characteristics of the nanofiller or reinforcement material. Here are some common adsorption mechanisms observed in polymer composites:

1. Physical Adsorption: Physical adsorption, also known as physisorption or van der Waals adsorption, occurs when weak intermolecular forces, such as London dispersion forces, electrostatic interactions, or hydrogen bonding, attract the adsorbate molecules to the surface of the polymer composite. This mechanism is generally reversible and is influenced by factors such as surface area, surface energy, and the nature of the adsorbate. Physical adsorption is typically observed when the nanofillers or reinforcement materials have high surface area and exhibit strong intermolecular forces.
2. Chemical Adsorption: Chemical adsorption, also referred to as chemisorption, involves the formation of covalent or ionic bonds between the adsorbate and the polymer matrix. This mechanism occurs when the adsorbate molecules react with functional groups or reactive sites present on the surface of the composite. Chemical adsorption is often stronger and more selective than physical adsorption, leading to a higher affinity for specific adsorbates. It can involve processes

such as ion exchange, complexation, or surface reactions, depending on the nature of the adsorbate and the functional groups present in the polymer matrix.

3. Ion Exchange: In polymer composites containing ion exchange resins or functionalized nanofillers, ion exchange mechanisms can occur. The functional groups on the polymer matrix or nanofiller surface can interact with ions present in the solution, leading to the displacement of ions and their subsequent adsorption onto the composite. Ion exchange adsorption is commonly used for the removal of heavy metals, ions, and other charged species from water and other solutions.
4. Surface Complexation: Surface complexation involves the formation of complexes or coordination compounds between the adsorbate and the surface functional groups of the polymer composite. This mechanism is particularly relevant when the polymer matrix or nanofiller surface has specific functional groups, such as carboxyl, amine, or hydroxyl groups. Surface complexation can enhance adsorption selectivity and affinity for certain contaminants, especially metal ions or organic compounds that can form coordination complexes.
5. Micropore Filling: In polymer composites with nanoporous structures, adsorption can occur through micropore filling. The adsorbate molecules can penetrate into the nanopores of the composite, and depending on the size and shape of the pores, physical or chemical adsorption mechanisms may dominate. Micropore filling can contribute to increased surface area, enhanced adsorption capacity, and improved access to adsorption sites within the composite.

It is important to note that the adsorption mechanisms in polymer composites can be influenced by factors such as temperature, pH, concentration, and the presence of competing adsorbates. The understanding of these mechanisms is crucial for the design and optimization of polymer composites for various adsorption applications, including water purification, air filtration, and environmental remediation.

2.2 *Removal of Heavy Metals*

Polymer composites have been extensively studied and utilized for the removal of heavy metals from various aqueous environments. The unique properties of polymer composites, such as high surface area, tunable surface chemistry, and the presence of functional groups, make them effective adsorbents for heavy metal ions. Here are some common strategies employed by polymer composites for the removal of heavy metals:

1. Ion Exchange Resins: Ion exchange resins are polymer composites specifically designed for ion exchange processes. These composites contain functional groups, such as sulfonic acid, amino, or carboxyl groups, which have a high affinity for heavy metal ions. The resin beads or particles exchange metal ions in the solution with ions on the resin surface. The adsorbed heavy metal ions can then be desorbed by treating the resin with an appropriate regenerant solution. Ion

exchange resins offer high selectivity and can effectively remove heavy metals from wastewater, industrial effluents, and contaminated groundwater.
2. Chelating Resins: Chelating resins are polymer composites with ligands or chelating agents attached to the polymer matrix. These ligands have a strong affinity for heavy metal ions and can form stable complexes with them. The chelating agents, such as iminodiacetic acid, thiourea, or polyamino-polycarboxylic acid, can effectively sequester heavy metals through chelation, preventing their release back into the environment. Chelating resins are widely used for the removal of heavy metals from various industrial wastewater streams.
3. Modified Natural Polymers: Natural polymers, such as chitosan, alginate, or cellulose, can be modified to enhance their heavy metal adsorption capabilities. Functional groups can be introduced onto the polymer backbone or crosslinked with other compounds to form composite materials with improved adsorption properties. The modified natural polymers offer cost-effective and environmentally friendly alternatives for heavy metal removal, particularly in water treatment applications.
4. Nanocomposites: Polymer nanocomposites incorporating nanoscale fillers, such as activated carbon, graphene oxide, or metal nanoparticles, have shown exceptional heavy metal adsorption capabilities [19, 20]. The high surface area and tailored surface properties of the nanofillers significantly enhance adsorption efficiency and selectivity. Moreover, the combination of nanofillers with a polymer matrix provides stability, easy separation, and regeneration options. Nanocomposites have been applied for the removal of heavy metals from aqueous solutions, including wastewater and contaminated groundwater.
5. Magnetic Polymer Composites: Magnetic polymer composites combine the adsorption properties of polymer matrices with the magnetic properties of nanoparticles, such as magnetite or maghemite. These composites can be easily separated from the solution using an external magnetic field, facilitating the removal and recovery of heavy metal ions. Magnetic polymer composites are effective for the removal of heavy metals from aqueous systems, and their magnetic properties offer additional advantages in terms of separation and reusability.

The choice of polymer composite for heavy metal removal depends on factors such as the target heavy metal, the concentration and pH of the solution, and the specific requirements of the application. The development of efficient and sustainable polymer composites continues to advance the field of heavy metal removal and contributes to environmental remediation efforts.

Polymer composites have found numerous applications in the removal of heavy metals from various environments due to their unique properties and adsorption capabilities [34, 35]. Here are some key applications of polymer composites in heavy metal removal:

1. Water and Wastewater Treatment: Polymer composites are extensively used in water and wastewater treatment processes for the removal of heavy metal contaminants. They can be employed in various treatment systems, such as adsorption

columns, fixed-bed reactors, or membrane filtration units [3]. The high adsorption capacity and selectivity of polymer composites allow for efficient removal of heavy metals, ensuring compliance with regulatory standards for drinking water or wastewater discharge.
2. Industrial Effluent Treatment: Industries, such as mining, metal plating, electronics, and chemical manufacturing, generate effluents containing high concentrations of heavy metals. Polymer composites, including ion exchange resins, chelating resins, or modified natural polymers, are employed in effluent treatment plants to selectively remove heavy metals from the wastewater. These composites help industries meet environmental regulations and minimize the environmental impact of their operations.
3. Groundwater Remediation: Groundwater contaminated with heavy metals poses a significant environmental concern. Polymer composites, particularly those with strong adsorption capabilities, are utilized in groundwater remediation processes. The composites can be injected into the subsurface to adsorb heavy metals, preventing their migration and reducing their concentration in the groundwater. The choice of polymer composite depends on factors such as the type and concentration of heavy metals present and the hydrogeological conditions of the site.
4. Soil Remediation: Polymer composites also play a role in the remediation of heavy metal-contaminated soil. They can be applied to contaminated soil as amendments or incorporated into soil stabilization techniques. The composites help in immobilizing heavy metals, reducing their bioavailability, and preventing their leaching into groundwater. Additionally, polymer composites can facilitate the phytoextraction process, where plants uptake heavy metals from the soil with the assistance of the composite, enhancing the effectiveness of soil remediation.
5. Industrial Processes and Recovery: Polymer composites are employed in various industrial processes to recover and recycle heavy metals. In processes such as electroplating, mining, or metal finishing, polymer composites can be used as selective adsorbents to recover valuable heavy metals from wastewater or process streams. The composites allow for the separation and purification of heavy metals, minimizing waste and enabling their reuse in industrial applications [7, 8].

The versatility and effectiveness of polymer composites in heavy metal removal have made them valuable tools in environmental remediation. Ongoing research and development in this field continue to enhance the adsorption capacity, selectivity, and recyclability of polymer composites, contributing to more efficient and sustainable solutions for heavy metal removal and environmental protection.

Polymer composites can remove a wide range of heavy metals from various aqueous environments [13, 39]. Some common heavy metals that polymer composites can effectively remove include:

1. Lead (Pb): Lead is a toxic heavy metal commonly found in industrial wastewater, contaminated soils, and drinking water sources. Polymer composites, such as ion exchange resins, chelating resins, or modified natural polymers, have shown effective adsorption capabilities for lead ions.

2. Cadmium (Cd): Cadmium is a highly toxic heavy metal often present in industrial wastewater, particularly from electroplating, battery manufacturing, and metal processing industries. Polymer composites, including chelating resins or modified natural polymers, have been successfully utilized for the removal of cadmium from aqueous solutions.
3. Mercury (Hg): Mercury is a hazardous heavy metal that poses significant risks to human health and the environment. Polymer composites have been employed in the removal of mercury from various sources, including contaminated water bodies and industrial effluents. Chelating resins and modified natural polymers are commonly used for the adsorption and sequestration of mercury ions.
4. Copper (Cu): Copper is a common heavy metal found in wastewater from industries such as mining, electroplating, and metalworking. Polymer composites, including ion exchange resins and chelating resins, have demonstrated efficient removal of copper ions from aqueous solutions.
5. Zinc (Zn): Zinc is another heavy metal often present in industrial wastewater, particularly from galvanizing, metal plating, and manufacturing processes. Polymer composites have been utilized for the removal of zinc ions from aqueous solutions, aiding in wastewater treatment and minimizing environmental impact.
6. Nickel (Ni): Nickel is a heavy metal found in various industrial processes, including electroplating, metal finishing, and battery manufacturing. Polymer composites, such as ion exchange resins and chelating resins, can effectively remove nickel ions from wastewater and prevent their release into the environment.

It is important to note that the effectiveness of polymer composites in removing heavy metals may depend on factors such as pH, concentration, presence of other contaminants, and specific properties of the polymer composite. Additionally, the design and formulation of the polymer composite can be tailored to target specific heavy metals or multiple heavy metals simultaneously, offering versatility in addressing heavy metal contamination.

2.3 Removal of Organic Contaminants

A. **Organic Contaminants that Polymer Composites Can Remove by Adsorption**

Polymer composites have demonstrated effectiveness in the adsorption and removal of a wide range of organic contaminants from aqueous solutions [14]. Here are some examples of organic contaminants that polymer composites can remove by adsorption:

1. Dyes: Polymer composites have been successfully used to remove various types of dyes, including synthetic dyes used in textile, dyeing, and printing industries.

The high adsorption capacity and selectivity of polymer composites make them suitable for the removal of dye molecules from wastewater.

2. Pharmaceuticals and Personal Care Products (PPCPs): PPCPs, including pharmaceutical residues and personal care products such as antibiotics, hormones, and fragrances, are increasingly recognized as emerging contaminants in water sources. Polymer composites have been utilized for the adsorption and removal of PPCPs, aiding in water treatment and ensuring the protection of aquatic ecosystems and human health.
3. Volatile Organic Compounds (VOCs): VOCs are organic compounds with low boiling points that easily vaporize into the air. Some common VOCs, such as benzene, toluene, ethylbenzene, and xylene are toxic and can contaminate water sources. Polymer composites have been employed to adsorb and remove VOCs from water, particularly in applications such as groundwater remediation or industrial wastewater treatment.
4. Pesticides and Herbicides: Pesticides and herbicides, including insecticides, fungicides, and weed killers, are organic contaminants frequently found in agricultural runoff and industrial wastewater. Polymer composites have shown efficacy in removing pesticides and herbicides from water, contributing to the reduction of these pollutants' impact on ecosystems and human health.
5. Polycyclic Aromatic Hydrocarbons (PAHs): PAHs are a group of organic compounds that are typically formed during the incomplete combustion of organic materials. They can be found in industrial effluents, contaminated soils, and runoff from urban areas. Polymer composites have been employed for the removal of PAHs from water, helping to mitigate their potential harmful effects on the environment.
6. Organic Solvents: Organic solvents, such as acetone, methanol, or ethyl acetate, are commonly used in industrial processes and can contaminate water sources. Polymer composites have been utilized for the adsorption and removal of organic solvents from industrial wastewater, ensuring compliance with environmental regulations and preventing their release into the environment.

These are just a few examples of the organic contaminants that polymer composites can remove by adsorption. The adsorption capacity, selectivity, and surface chemistry of the polymer composite play a vital role in determining their effectiveness for specific organic pollutants. The design and selection of the polymer composite depend on the nature of the contaminants present and the desired removal efficiency.

B. **Polymer Composites as Adsorbants and Filters for Organic Contaminants**

Polymer composites have proven to be effective adsorbents and filters for the removal of various organic contaminants from aqueous solutions. The adsorption capacity, selectivity, and tunable surface chemistry of polymer composites make them well suited for this application. Here are some examples of polymer composites used for the removal of organic contaminants by adsorption and filtration:

1. Activated Carbon–Polymer Composites: Activated carbon is widely recognized for its excellent adsorption and filtration properties. When incorporated into a

polymer matrix, such as polymeric resins or nanocomposites, activated carbon–polymer composites exhibit enhanced adsorption capacity for a wide range of organic pollutants, including dyes, pharmaceuticals, pesticides, and industrial chemicals [4, 8, 9]. The composite structure provides stability, ease of handling, and improved separation from the treated water.

2. Molecularly Imprinted Polymers (MIPs): MIPs are synthetic polymer composites designed to selectively recognize and bind specific organic molecules. They are created by polymerizing functional monomers in the presence of a target molecule, creating cavities in the polymer matrix that mimic the shape, size, and functional groups of the target compound. MIPs can effectively remove specific organic contaminants, such as environmental pollutants, toxins, or emerging contaminants, from water or other aqueous environments.
3. Zeolite–Polymer Composites: Zeolites are porous materials with high surface area and adsorption capacity. When incorporated into a polymer matrix, zeolite–polymer composites exhibit improved mechanical properties, stability, and selectivity for organic pollutants. These composites have been employed for the removal of organic compounds, including volatile organic compounds (VOCs), chlorinated solvents, and aromatic hydrocarbons, from water and industrial effluents.
4. Graphene–Polymer Composites: Graphene, a two-dimensional sheet of carbon atoms, possesses high surface area and excellent adsorption properties. Graphene–polymer composites have been used for the adsorption and filtration of organic contaminants, such as dyes, hydrocarbons, and emerging pollutants. The large surface area and unique properties of graphene enhance the adsorption capacity and facilitate the efficient removal of organic compounds from water.
5. Nanofiber-Reinforced Composites: Polymer nanocomposites incorporating nanofibers, such as carbon nanofibers or electrospun polymer nanofibers offer increased surface area and porosity which enhances the adsorption capacity for organic contaminants. These composites have shown effectiveness in removing and filtering various organic pollutants, including dyes, pesticides, and pharmaceuticals, from water and wastewater [1, 9].
6. Biopolymer Composites: Biopolymers such as chitosan have inherent adsorption properties for organic compounds [8, 26]. When modified or combined with other polymers, biopolymer composites can exhibit enhanced adsorption capacities and selectivity for specific organic contaminants. Biopolymer composites have been utilized for the removal of dyes, heavy metals complexed with organic ligands, and other organic pollutants from water and industrial effluents [10].

These examples demonstrate the versatility of polymer composites in adsorbing organic contaminants from water and wastewater. Polymer composites continue to be a promising approach for efficient and sustainable removal of organic contaminants, contributing to water purification and environmental remediation efforts.

3 Membrane-Based Applications of Polymer Composites

Membrane-based applications of polymer composites have gained significant attention due to their versatile properties and diverse range of uses. These applications leverage the combination of polymers and reinforcing materials, such as fibers or nanoparticles, to create membranes with enhanced performance characteristics.

Membrane-based applications of polymer composites offer versatile tools for tackling environmental pollution [36]. One notable domain is water treatment, where polymer composite membranes exhibit exceptional separation and filtration capabilities. These membranes effectively eliminate impurities, particles, and even ions from water sources, proving invaluable for processes like desalination, wastewater treatment, and the purification of potable water. They find application in cleaning water sources, purifying air, preventing the spread of contaminants in soil and groundwater, and even addressing plastic pollution. Through their selective permeability and filtration capabilities, these membranes contribute to preserving and restoring the quality of our natural environment.

Another application of polymer composites as membranes is gas separation. These membranes showcase selective permeation properties, enabling the isolation of specific gases from mixtures. This quality holds significance in industries like petrochemicals and natural gas processing.

Membrane-based applications of polymer composites have instigated a transformative impact across various industries, providing solutions for water treatment and environmental pollution remediation.

3.1 Polymer Membrane Development

Polymer membrane development has played a crucial role in addressing environmental pollution and facilitating effective remediation strategies. These membranes, designed with specific properties and structures, offer innovative solutions for tackling a range of pollution-related challenges [1, 30]. Below we will discuss in brief about the different usage of polymer membrane in pollution control and remediation.

Water Purification: Polymer membranes have revolutionized water purification efforts by enabling precise filtration and separation processes. Microfiltration, ultrafiltration, nanofiltration, and reverse osmosis membranes are examples of how polymer composites are tailored to remove contaminants, particles, and pollutants from water sources. These membranes effectively treat wastewater, desalinate seawater, and remove heavy metals and organic pollutants, contributing to improved water quality.

Air Quality Improvement: Polymer membranes are employed in air purification systems to capture airborne pollutants [37]. These membranes are designed to trap particulate matter, volatile organic compounds (VOCs), and gases, enhancing indoor and outdoor air quality [38]. Their high surface area and selective permeability allow for effective pollutant removal, making them valuable components in air filtration technologies.

Soil and Groundwater Remediation: In soil and groundwater remediation, polymer membranes are used to create barriers that prevent the migration of contaminants. By installing these membranes underground, the spread of pollutants is controlled, protecting clean areas from further contamination. This approach is particularly useful for isolating hazardous substances and preventing their infiltration into surrounding ecosystems.

Apart from the above-listed applications, polymer membranes are also being explored for emerging pollution-related challenges. For instance, they are being utilized to address microplastic pollution by selectively capturing microplastics from water bodies. Additionally, advanced materials are being integrated into membranes to enable photocatalytic degradation of pollutants, offering an eco-friendly approach to pollution mitigation. Nanostructured membranes, mixed-matrix membranes with nanoparticles, and smart membranes that respond to specific pollutants are some of the innovative trends in membrane development. These innovations are expected to drive even more effective and efficient pollution control and remediation methods. Thus, polymer membrane development has significantly contributed to addressing environmental pollution and enabling efficient remediation strategies. These membranes have found applications in water purification, air quality improvement, soil and groundwater remediation, and even in emerging challenges like microplastic pollution. With ongoing advancements and innovation, polymer membranes are poised to continue playing a vital role in safeguarding the environment and promoting sustainable practices.

Let's now discuss the two broad applications of polymer membrane in water treatment and desalination, gas separation, and filtration.

3.2 Water Treatment and Desalination

Water treatment and desalination by polymer membrane composites are advancing to address water scarcity and improve water quality [3]. Below we will discuss in brief the research that explores the development, characterization, and application of polymer membrane composites in various water treatment processes, including desalination.

1. **Membrane Fabrication and Characterization**: Research has been done on the fabrication methods of polymer membrane composites using techniques such as phase inversion, interfacial polymerization, electrospinning, and layer-by-layer assembly. Characterization of these polymer membranes, including analyses of porosity, morphology, permeability, mechanical strength, and surface chemistry, was performed.

2. **Membrane Performance in Desalination**: Many researchers assess the performance of polymer membrane composites in desalination processes, such as reverse osmosis (RO) and nanofiltration (NF). They explore factors affecting desalination efficiency, including salt rejection, water flux, fouling resistance, and membrane stability under high-pressure conditions.
3. **Fouling and Scaling Mitigation**: Researchers investigate strategies to mitigate fouling and scaling, common challenges in desalination processes. They explore surface modification techniques, incorporation of nanoparticles, and anti-fouling coatings to enhance the long-term performance and maintenance of polymer membrane composites.
4. **Selective Ion Removal and Contaminant Removal**: Focus on the selective removal of specific ions or contaminants from water sources using tailored polymer membrane composites is explored by many research groups. The studies highlight the development of membranes with ion-specific binding sites, allowing for efficient removal of heavy metals, organic pollutants, and other harmful substances.
5. **Energy-Efficient and Sustainable Desalination**: Work has been emphasized on the importance of developing energy-efficient and environmentally sustainable desalination processes. They explore the use of novel polymer materials, hybrid systems, and advanced polymer membrane configurations to reduce energy consumption and minimize the environmental footprint of desalination operations.

Researchers have also explored novel membrane materials, such as graphene-based composites, polymers of intrinsic microporosity (PIMs), and responsive polymers to push the boundaries of membrane performance and desalination efficiency.

3.3 Gas Separation and Filtration

Recent research on gas separation and filtration by polymer membrane composites has been instrumental in advancing the field of gas separation technology. Polymer membrane composites offer the potential to achieve high selectivity, permeability, and stability, making them attractive candidates for various gas separation and filtration applications. Here we will be discussing some notable aspects of recent research in this area.

1. **Selective Gas Separation**: Recent studies have focused on designing polymer membrane composites with enhanced selectivity for specific gases. These composites are engineered to have tailored structures and functional groups that enable them to preferentially permeate certain gases over others. Researchers are investigating new polymers, nanocomposites, and hybrid materials to achieve superior selectivity in applications like CO_2, CH_4 capture, natural gas purification, and air separation.

2. **High Permeability and Low Resistance**: Efforts are being directed toward developing polymer membrane composites with high gas permeability while minimizing resistance to gas flow. Achieving this balance is crucial for maintaining efficient separation processes and minimizing energy consumption. Researchers are exploring novel approaches to enhance gas transport kinetics within the membrane structure, resulting in improved overall performance.
3. **Nanoparticle Incorporation**: Incorporating nanoparticles into polymer matrices is a novel area of research. Nanoparticles can modify membrane properties by enhancing selectivity, improving mechanical strength, and reducing gas diffusion pathways. Recent studies have investigated various nanoparticle types, including metal–organic frameworks (MOFs), zeolites, and graphene derivatives to impart unique gas separation capabilities to polymer membrane composites.
4. **Mixed-Matrix Membranes (MMMs)**: Mixed-matrix membranes which combine traditional polymers with inorganic fillers, have gained attention for their potential to synergistically combine the strengths of both components. Recent research has focused on optimizing the dispersion of fillers within the polymer matrix to prevent agglomeration, improve compatibility, and enhance gas separation performance.
5. **Advanced Characterization Techniques**: Recent advancements in characterization techniques such as atomic force microscopy (AFM), positron annihilation lifetime spectroscopy (PALS), and neutron scattering have provided researchers with deeper insights into the structure–property relationships of polymer membrane composites. These techniques enable a better understanding of factors influencing gas transport and separation behavior at the nanoscale.
6. **Energy Efficiency and Sustainability**: In line with broader environmental concerns, recent research aims to develop energy-efficient and environmentally sustainable gas separation processes. Researchers are investigating new membrane designs, such as facilitated transport membranes and membrane modules with tailored flow patterns in order to reduce energy consumption and enhance process sustainability.
7. **Application to Industrial Processes**: Recent studies have explored the application of polymer membrane composites in diverse industrial processes, including natural gas processing, greenhouse gas mitigation, and hydrogen purification. These applications hold promise for addressing energy and environmental challenges by providing cost-effective and efficient separation solutions.

Gas separation and filtration by use of polymer membrane composites is focused on designing advanced materials, optimizing performance, and exploring new applications for these membranes in various gas separation processes. As technology continues to advance, these studies contribute to the development of more efficient, selective, and sustainable gas separation technologies with a wide range of industrial applications.

4 Catalytic Applications of Polymer Composites

Catalysis is a fundamental process across various industries. In recent years, there has been growing interest in harnessing the catalytic potential of polymer composites. These composite materials, formed by integrating polymers with catalytically active components, exhibit unique synergistic properties that find applications in diverse catalytic processes. The catalytic applications of polymer composites encompass several key areas:

1. **Environmental Remediation**: Polymer composite catalysts play a vital role in mitigating environmental pollutants. These catalysts can facilitate the degradation of organic pollutants in wastewater and air through advanced oxidation processes. By incorporating catalytic nanoparticles or metal complexes into polymer matrices, these composites efficiently break down harmful compounds, contributing to cleaner ecosystems and improved air quality.
2. **Sustainable Energy Production**: In the realm of energy production, polymer composite catalysts are employed in processes such as fuel cells, water electrolysis, and photovoltaics. These catalysts enhance reaction kinetics, increase energy conversion efficiency, and enable the production of clean energy sources like hydrogen. By integrating catalytic functionalities into polymeric materials, researchers are driving advancements in sustainable energy technologies.
3. **Chemical Synthesis and Transformations**: Polymer composite catalysts find application in chemical synthesis and transformations, offering advantages like enhanced selectivity and improved reaction rates. These catalysts facilitate various transformations, such as hydrogenation, oxidation, and polymerization, across a wide range of industries including pharmaceuticals, fine chemicals, and petrochemicals.
4. **Biocatalysis and Enzyme Immobilization**: In biotechnology and enzymatic processes, polymer composite catalysts enable efficient enzyme immobilization. These composites provide a stable environment for enzymes to retain their catalytic activity, enhancing biocatalytic reactions used in pharmaceutical synthesis, food processing, and biofuel production.
5. **Photocatalysis**: Polymer composite catalysts with photocatalytic properties are utilized in harnessing solar energy for chemical transformations. These composites enable light-driven reactions, such as pollutant degradation and hydrogen generation, by combining light-absorbing materials with catalytic components.
6. **CO_2 Conversion and Green Chemistry**: Addressing the challenge of CO_2 emissions, polymer composite catalysts are being explored for CO_2 conversion into valuable products. These catalysts promote CO_2 reduction reactions, contributing to the development of green and sustainable chemical processes.
7. **Selective Catalysis and Reaction Control**: The tunability of polymer composites allows for the design of catalysts with specific selectivity and reaction control. By modifying the composite's structure, composition, and surface properties, researchers can tailor catalytic performance to suit particular reactions and desired outcomes.

Catalytic applications of polymer composites represent an evolving frontier in materials science and catalysis. These composites offer a versatile platform for developing efficient, selective, and sustainable catalysts across a range of industries. As research continues to explore novel compositions, structures, and functionalities, the catalytic potential of polymer composites holds promise for driving innovation in various applications with positive environmental and economic impacts.

4.1 Polymer-Supported Catalysts

One promising avenue in remediation strategies is essential to mitigate the impact of pollutants with utilization of polymer-supported catalysts. These catalysts offer numerous advantages, including enhanced catalytic activity, reusability, and versatility in tackling various pollutants. Herein, we will explore the application of polymer-supported catalysts for environmental pollution control and remediation.

Polymer-supported catalysts are heterogeneous catalysts where catalytically active species are immobilized on polymer matrices. The polymers provide a stable and porous support structure that facilitates the dispersion and retention of catalytic sites. Various types of polymers can be employed, including organic, inorganic, and hybrid materials.

The catalytic mechanisms of polymer-supported catalysts vary depending on the specific reactions and pollutants targeted. Key mechanisms include adsorption of pollutants onto the polymer matrix followed by catalytic transformation and direct interactions between the catalytic species and pollutants in solution. Common catalytic reactions include oxidation, reduction, hydrolysis, and degradation.

Advantages of Polymer-Supported Catalysts:

a. High Catalytic Activity: Polymer-supported catalysts often exhibit higher catalytic activity than their homogeneous counterparts due to increased surface area and accessibility of active sites.
b. Reusability: The immobilization of catalytic species on polymer supports allows for easy separation and recycling, reducing waste generation and cost.
c. Selectivity: Tailoring the polymer support and catalyst design allows for selectivity in pollutant removal, minimizing the generation of unwanted by-products.
d. Versatility: Polymer-supported catalysts can be designed to target a wide range of pollutants, including organic compounds, heavy metals, and emerging contaminants.

Applications in Environmental Pollution Remediation:

a. Organic Pollutants: Polymer-supported catalysts have been successfully employed in the removal of organic contaminants like pesticides, dyes, pharmaceuticals, and petroleum hydrocarbons.

b. Heavy Metals: These catalysts are used for the immobilization and subsequent removal of heavy metals such as lead, mercury, and cadmium from water and soil.
c. Wastewater Treatment: Polymer-supported catalysts are integral in advanced oxidation processes (AOPs) for wastewater treatment, where they facilitate the degradation of recalcitrant pollutants.
d. Air Pollution Control: These catalysts find applications in catalytic converters for the removal of harmful gases like nitrogen oxides (NOx) and volatile organic compounds (VOCs) from automotive exhaust.

Challenges and Future Directions:

a. Catalyst Stability: Ensuring long-term stability and robustness of polymer-supported catalysts remains a challenge, especially in harsh environmental conditions. b. Catalyst Design: Tailoring polymer support materials and catalytic species for specific pollutants and reactions is an ongoing research area.
b. Scaling up: Transitioning from laboratory-scale studies to large-scale environmental remediation applications requires further optimization.
c. Environmental Impact: Evaluating the environmental footprint of polymer-supported catalysts, including the production and disposal of polymer materials, is essential.

Polymer-supported catalysts hold immense promise in the field of environmental pollution control and remediation. Their unique combination of catalytic activity, reusability, and versatility makes them valuable tools for addressing diverse environmental pollution challenges. Continued research and development in this field are crucial for advancing sustainable solutions to protect our environment.

4.2 Photocatalytic Activity of Polymer Composites in for Environmental Remediation and Pollution

Photocatalytic polymer composites have emerged as promising materials for addressing environmental pollution and remediation challenges [24]. Their unique ability to harness solar or artificial light to initiate chemical reactions makes them valuable tools for mitigating pollution and promoting sustainable environmental practices.

1. **Photocatalytic Materials:**
a. Titanium Dioxide (TiO_2): TiO_2 is one of the most commonly used photocatalytic materials. It absorbs UV light and generates electron–hole pairs, leading to the production of reactive oxygen species (ROS) that can degrade organic compounds.
b. Zinc Oxide (ZnO): ZnO exhibits photocatalytic activity in both UV and visible light regions, making it a versatile choice for polymer composites.

c. Other Semiconductor Nanoparticles: In addition to TiO_2 and ZnO, other semiconductor materials like CdS, WO_3, and g-C_3N_4 are also used as photocatalytic components in composites.

2. **Factors Influencing Photocatalytic Activity:**

a. Photocatalyst Loading: The concentration of photocatalytic nanoparticles in the polymer matrix significantly impacts the photocatalytic activity. Higher loading can enhance activity but may also affect the composite's mechanical properties.
b. Band Gap Energy: The band gap energy of the photocatalytic material determines the range of light it can absorb. Materials with narrower band gaps can utilize visible light, expanding their range of applications.
c. Surface Area and Morphology: A higher surface area and specific surface morphology of photocatalytic nanoparticles promote better light absorption and efficient electron–hole separation.
d. Interfacial Compatibility: Achieving good compatibility between the polymer matrix and photocatalytic material is essential to maintain structural integrity and ensure efficient electron transfer.
e. Light Source: The choice of light source, whether UV or visible light, should align with the photocatalyst's absorption properties.

3. **Polymer Matrix Selection:**

Polymer matrices such as polyethylene, polyvinyl chloride (PVC), and polyvinyl alcohol (PVA) are commonly used in photocatalytic composites due to their flexibility and processability. The polymer should provide good dispersion and stabilization of photocatalytic nanoparticles and offer chemical stability during photocatalysis. Let's now see through into the specific applications and benefits of photocatalytic polymer composites in environmental remediation and pollution control.

1. Water Pollution Remediation:

a. Degradation of Organic Pollutants: Photocatalytic polymer composites, often incorporating materials like TiO_2 or ZnO, are used to treat contaminated water by efficiently degrading organic pollutants, including dyes, pesticides, pharmaceuticals, and industrial chemicals. These composites break down these contaminants into less harmful or non-toxic by-products through oxidative reactions.
b. Removal of Pathogens: Polymer composites embedded with photocatalytic nanoparticles can also be employed for water disinfection, inactivating bacteria, viruses, and other pathogens. This technology has applications in drinking water treatment and wastewater disinfection.

2. Air Pollution Control:

a. VOC Removal: Photocatalytic polymer coatings applied to surfaces like building exteriors, roadways, and industrial equipment can help remove volatile organic compounds (VOCs) and other airborne pollutants from the atmosphere. These coatings break down the pollutants upon exposure to sunlight or artificial UV light.
b. NOx Abatement: Photocatalytic materials incorporated into cement or asphalt surfaces in urban areas can reduce nitrogen oxide (NOx) emissions by converting them into less harmful compounds, contributing to air quality improvement.

3. Self-Cleaning Surfaces:
a. Photocatalytic polymer composites are used to create self-cleaning surfaces for buildings, glass, and textiles. These surfaces effectively degrade organic contaminants, such as dirt, oil, and organic stains, when exposed to light, thus reducing the need for manual cleaning and maintenance.
4. Photocatalytic Water Splitting for Hydrogen Production:

Some photocatalytic polymer composites are designed to split water into hydrogen and oxygen under sunlight. This technology has the potential to provide a clean and renewable source of hydrogen fuel, which can be used in various applications, including transportation and energy storage.

5. Remediation of Contaminated Soils:

Photocatalytic polymer composites can be used in soil remediation efforts to degrade persistent organic pollutants and break down heavy metals. These composites are applied to contaminated soils as coatings or incorporated into geotextiles to enhance the remediation process.

Benefits and Advantages of Photocatalytic Polymer Composites:

a. Sustainable Remediation: Photocatalytic polymer composites provide a sustainable approach to pollution control by utilizing readily available light sources, minimizing the need for chemical additives or energy-intensive processes.
b. Versatility: These composites can be tailored for specific applications and pollutant types, offering versatility in addressing a wide range of environmental challenges.
c. Cost-Effective: Once installed, photocatalytic surfaces and materials operate without the need for frequent replacement or maintenance, making them cost-effective solutions over the long term.
d. Reduced Environmental Impact: The use of photocatalytic polymer composites can contribute to reducing the environmental impact of pollution and contamination, aiding in the preservation of ecosystems and human health.

Challenges and Future Prospects of Photocatalytic Polymer Composites:

a. Composite Durability: Ensuring the long-term stability and durability of photocatalytic polymer composites, especially in harsh environmental conditions, remains a challenge.
b. Scalability: Developing scalable manufacturing methods for these composites is crucial to their widespread adoption.
c. Standardization: The establishment of standardized testing protocols and guidelines for assessing the performance and safety of photocatalytic materials is essential.
d. Maintaining Mechanical Properties: Balancing photocatalytic activity with the mechanical properties of the composite remains a challenge, especially for load-bearing applications.

e. New Photocatalytic Materials: Research is ongoing to discover and develop new photocatalytic materials with improved performance and wider absorption spectra.

In conclusion, photocatalytic polymer composites offer innovative and sustainable solutions for environmental remediation and pollution control. As research and development in this field continue, these materials hold the potential to play a significant role in addressing global environmental challenges and improving the quality of the environment.

5 Nanotechnology and Polymer Composites

Nanotechnology and polymer composites are two fields that have significantly impacted various industries and have the potential to revolutionize material science and engineering. By combining the unique properties of nanomaterials with polymer matrices, researchers have developed a class of advanced materials with enhanced mechanical, thermal, electrical, and optical properties.

Nanotechnology involves the manipulation and control of matter at the nanoscale. At this scale, materials exhibit novel properties that differ from their bulk counterparts. Nanomaterials, such as nanoparticles, nanofibers, nanotubes, and graphene, possess high surface area-to-volume ratios, quantum effects, and improved mechanical strength [10].

Polymer composites, on the other hand, are materials composed of a polymer matrix and reinforcement fillers. These fillers can be in the form of fibers, particles, or sheets and are added to enhance the mechanical properties of the polymer. Traditional fillers include glass fibers, carbon fibers, and natural fibers like jute or hemp.

When nanomaterials are incorporated into polymer matrices, they create nanocomposites, which combine the unique properties of both components. The addition of nanofillers in small quantities can lead to significant improvements in strength, stiffness, toughness, thermal stability, and barrier properties of the resulting composites. Furthermore, nanofillers can enhance electrical and thermal conductivity, provide flame resistance, and offer tailored surface properties.

One example of nanotechnology and polymer composites working together is the use of carbon nanotubes (CNTs) as reinforcement fillers. CNTs are cylindrical structures made up of carbon atoms arranged in a unique hexagonal lattice. Due to their exceptional mechanical strength, CNTs can reinforce the polymer matrix, improving its tensile strength and stiffness [11, 25]. The high aspect ratio of CNTs enables load transfer across the composite and enhances interfacial bonding between the nanofillers and the polymer.

In addition to mechanical improvements, nanotechnology can provide functional enhancements to polymer composites. For instance, the incorporation of nanoparticles with unique optical properties can lead to composites with improved transparency, making them suitable for applications in displays, sensors, and optics.

Magnetic nanoparticles can enable composites with magnetically responsive properties, opening up possibilities in fields such as data storage, robotics, and biomedical engineering.

The development of nanotechnology and polymer composites has resulted in advancements in various industries. Aerospace, automotive, construction, electronics, energy, and healthcare sectors have all benefited from these materials. For example, lightweight polymer composites with nanofillers have been used to manufacture fuel-efficient and structurally robust aircraft components. In the automotive industry, nanocomposites contribute to weight reduction and improved fuel efficiency. Polymer composites also find applications in sports equipment, consumer goods, and protective coatings.

Despite the significant progress, there are still challenges to overcome in the field of nanotechnology and polymer composites. The dispersion of nanofillers within the polymer matrix, achieving a strong interfacial bond, and scaling up production methods are ongoing areas of research. Additionally, the potential environmental and health impacts of nanomaterials require careful consideration and regulation.

5.1 Polymer Composites as Nanoparticles for Environmental Remediation

Polymer composites when designed as nanoparticles can play a significant role in addressing environmental pollution and facilitating remediation processes [22]. These nanocomposites possess unique properties that can be used for various applications related to pollution control and environmental remediation.

One area where polymer nanocomposites are being explored is in the removal of pollutants from water and soil. The high surface area-to-volume ratio and tailored surface properties of nanoparticles allow for efficient adsorption and catalytic degradation of contaminants [1]. Nanocomposites can be designed to have specific functional groups or surface coatings that enhance their adsorption capacity and selectivity for particular pollutants.

For instance, polymer nanocomposites containing activated carbon nanoparticles can effectively remove organic pollutants, heavy metals, and dyes from water [11]. The porous structure of activated carbon provides a large surface area for adsorption, while the polymer matrix provides stability and facilitates easy separation from water. Similarly, nanocomposites incorporating metal nanoparticles, such as zero-valent iron (ZVI) or palladium, exhibit catalytic activity and can degrade organic pollutants through processes like oxidation or reduction reactions [5].

Polymer nanocomposites can also be used for the remediation of contaminated soil and groundwater. In situ applications involve injecting nanocomposites into the subsurface to treat contaminated sites. For example, nanocomposites containing nanoscale zero-valent iron (nZVI) have been used for the remediation of chlorinated solvents, heavy metals, and other organic contaminants. The high reactivity and mobility of nZVI nanoparticles enable the degradation or immobilization of contaminants, reducing their toxicity and mobility in the environment.

Moreover, polymer nanocomposites can contribute to the development of efficient filtration systems for air and water purification. Nanocomposites with nanofiber or nanoporous structures can act as filters or membranes, removing particulate matter, bacteria, and even nanoparticles themselves from the environment. These filters can find applications in industrial settings, indoor air purification, and wastewater treatment, aiding in the reduction of pollutants and improving overall air and water quality.

While polymer nanocomposites offer promising solutions for environmental pollution and remediation, it is crucial to consider their potential environmental impact. The release of nanoparticles into the environment raises concerns about their long-term fate, mobility, and toxicity. Research efforts are directed toward understanding and mitigating these risks, such as through the development of encapsulation techniques or designing nanoparticles with reduced toxicity.

Polymer composites as nanoparticles have the potential to address environmental pollution and aid in remediation efforts. Their unique properties, such as high surface area, reactivity, and tunable surface chemistry, enable effective pollutant removal and degradation. However, it is essential to continue research into the environmental implications of these nanocomposites and ensure their safe and responsible use in environmental applications.

5.2 Polymer-Matrix Nanocomposites

Polymer-matrix nanocomposites have shown promising applications in addressing environmental pollution and facilitating remediation processes. Here are some examples of how these materials can be utilized:

1. Water Filtration and Purification: Polymer-matrix nanocomposites can be designed as advanced filtration membranes for water treatment. Nanocomposite membranes with tailored nanoporous structures or functionalized surfaces can effectively remove contaminants such as heavy metals, organic pollutants, and bacteria from water sources. These membranes offer high selectivity, improved flux rates, and long-term stability, making them valuable for water purification and desalination processes.
2. Adsorption of Contaminants: Polymer-matrix nanocomposites can act as efficient adsorbents for the removal of pollutants from water and wastewater. The incorporation of nanofillers, such as activated carbon, zeolites, or metal nanoparticles, into the polymer matrix significantly enhances the adsorption capacity and selectivity of the composite. These nanocomposites can be employed for the removal of organic pollutants, dyes, heavy metals, and other harmful substances, contributing to the remediation of contaminated water sources [23].

3. Soil Remediation: Polymer-matrix nanocomposites play a role in the remediation of contaminated soil and the stabilization of hazardous substances. Nanocomposites containing nanoscale zero-valent iron (nZVI) or other reactive nanoparticles can be introduced into the soil to facilitate the degradation or immobilization of organic pollutants, heavy metals, and other contaminants. The high reactivity and mobility of nanocomposites enable targeted and effective remediation of contaminated sites.
4. Environmental Sensing: Polymer-matrix nanocomposites can be designed as sensors for the detection and monitoring of environmental pollutants. The incorporation of functional nanoparticles or nanofibers into the polymer matrix enables the development of highly sensitive and selective sensors for gases, heavy metals, volatile organic compounds (VOCs), and other pollutants. These sensors can be utilized for real-time monitoring of air and water quality in industrial settings, urban environments, and other areas prone to pollution.
5. Catalytic Degradation: Polymer-matrix nanocomposites incorporating catalytic nanoparticles, such as palladium or titanium dioxide, can be utilized for the degradation of organic pollutants through advanced oxidation processes. The nanocomposite's catalytic properties, combined with the large surface area and stability of the polymer matrix, enhance the efficiency of pollutant degradation. These nanocomposites are used in the treatment of wastewater, industrial effluents, and air pollution control.
6. Environmental Remediation Films and Coatings: Polymer-matrix nanocomposite films and coatings can be applied to contaminated surfaces or areas to facilitate remediation. These films can serve as protective barriers, preventing the migration of contaminants, and providing controlled release of remediation agents. Additionally, nanocomposite coatings with photocatalytic properties can aid in the degradation of pollutants when exposed to sunlight or UV radiation.

By harnessing the unique properties of nanomaterials within polymer matrices, polymer-matrix nanocomposites offer versatile and efficient solutions for environmental pollution control and remediation. Ongoing research and development in this field continue to expand the applications and effectiveness of these materials in addressing environmental challenges.

6 Trends and Challenges

Polymer composites have gained significant attention in environmental remediation due to their unique properties and versatility. Recent advancements and trends in the utilization of polymer composites for pollution control and environmental remediation encompass a range of innovative approaches. Let's discuss the recent advancements and trends in the use of polymer composites.

1. The integration of nanotechnology, including nanoparticles like CNTs and graphene improved pollutant capture and degradation.

2. The incorporation of biodegradable polymers aligns with sustainability goals by ensuring reduced environmental persistence.
3. Hybrid composites, coupling different polymers or materials, offer synergistic benefits in pollutant removal.
4. In situ remediation strategies employing polymer composites, such as permeable reactive barriers, have gained attention for their potential to mitigate pollution directly at contamination sources.
5. Surface modification and functionalization techniques have been explored to optimize pollutant interaction.
6. Monitoring methods are also being refined to assess the long-term effectiveness and stability of these polymer composite-based solutions.

These exciting trends underscore the ongoing commitment to developing innovative and sustainable methods for pollution control and environmental remediation, all made possible by the versatile and adaptable nature of polymer composites. While polymer composites provide significant advantages such as enhanced strength, weight reduction, and customized properties, they also come with several challenges. These challenges include:

1. The complex manufacturing process requiring specialized equipment
2. The anisotropic nature of their properties complicating design and engineering
3. Potential higher production costs especially with advanced reinforcement materials
4. Concerns about their environmental impact due to non-renewable resources and recycling difficulties
5. Susceptibility to degradation from environmental factors and the critical need for a deep understanding of material properties for appropriate material selection and design.

Balancing these benefits and challenges is crucial for effectively utilizing polymer composites across a wide range of applications. The field of polymer composites for pollution control and environmental remediation is dynamic and continues to evolve.

7 Conclusions

Polymer composites offer a promising and versatile solution to combat environmental pollution and address the need for effective remediation strategies. These advanced materials possess numerous advantages, including enhanced adsorption capacities, increased mechanical strength, and exceptional durability, making them highly effective in tackling various forms of pollution. Extensive research and development in polymer composites have resulted in tailored solutions for specific pollutants, such as heavy metals, organic contaminants, and emerging threats like microplastics. Their adaptability allows customization to fit diverse environmental conditions, making them invaluable tools for global pollution challenges. Additionally, many polymer composites are environmentally friendly, aligning with the goal of sustainable remediation. However, challenges persist, requiring ongoing research to optimize formulations, manufacturing processes, and deployment strategies. As we continue forward, unlocking the full potential of these materials, alongside a commitment to sustainability promises a cleaner, healthier, and more resilient planet for future generations.

8 Conflicts of Interest

The authors confirm that there are no conflicts of interest associated with this work.

9 Ethical Compliance

The authors affirm that this study adheres to all ethical guidelines.

10 Competing Interests

No competing interests are declared by the authors.

11 Data and Material Availability

Data and materials are not applicable to this chapter.

References

1. Agboola O, Fayomi OSI, Ayodeji A, Ayeni AO, Alagbe EE, Sanni SE, Okoro EE, Moropeng L, Sadiku R, Kupolati KW, Oni BA (2021) A review on polymer nanocomposites and their effective applications in membranes and adsorbents for water treatment and gas separation. Membranes 11(2):139. https://doi.org/10.3390/membranes11020139
2. Alonso A, Muñoz-Berbel X, Vigués N, Macanás J, Muñoz M, Mas J, Muraviev DN (2011) Characterization of fibrous polymer silver/cobalt nanocomposite with enhanced bactericide activity. Langmuir 28(1):783–790. https://doi.org/10.1021/la203239d
3. Berber MR (2020) Current advances of polymer composites for water treatment and desalination. J Chem 2020:1–19. https://doi.org/10.1155/2020/7608423
4. Carrado KA, Bergaya F (2007) Clay-based polymer nanocomposites (CPN)

5. Domènech B, Ziegler KK, Carrillo F, Muñoz M, Muraviev DN, Macanás J (2013) Development of novel catalytically active polymer-metal-nanocomposites based on activated foams and textile fibers. Nanoscale Res Lett 8(1). https://doi.org/10.1186/1556-276x-8-238
6. El Rhazi M, Majid S, Elbasri M, Salih FE, Oularbi L, Lafdi K (2018) Recent progress in nanocomposites based on conducting polymer: application as electrochemical sensors. Int Nano Lett 8(2):79–99. https://doi.org/10.1007/s40089-018-0238-2
7. Fu L, Zhou N (2021) The influence of biodegradable polymer materials on heavy metal activity in soil based on analysis of big data. J Phys Conf Ser 1744(3):032154. https://doi.org/10.1088/1742-6596/1744/3/032154
8. Goci MC, Leudjo Taka A, Martin L, Klink MJ (2023) Chitosan-based polymer nanocomposites for environmental remediation of mercury pollution. Polymers 15(3):482. https://doi.org/10.3390/polym15030482
9. Hussain CM, Mishra AK (2018) New polymer nanocomposites for environmental remediation. Elsevier
10. Kalia S, Kaith BS, Kaur I (2011) Cellulose fibers: bio- and nano-polymer composites. Springer Science & Business Media
11. Khan F, Kausar A, Siddiq M (2015) A review on properties and fabrication techniques of polymer/carbon nanotube composites and polymer intercalated buckypapers. Polym-Plast Technol Eng 54(14):1524–1539. https://doi.org/10.1080/03602559.2015.1021486
12. Khodakarami M, Bagheri M (2021) Recent advances in synthesis and application of polymer nanocomposites for water and wastewater treatment. J Clean Prod 296:126404. https://doi.org/10.1016/j.jclepro.2021.126404
13. Kostal J, Prabhukumar G, Lao UL, Chen A, Matsumoto M, Mulchandani A, Chen* W (2005) Customizable biopolymers for heavy metal remediation. J Nanoparticle Res 7(4–5):517–523. https://doi.org/10.1007/s11051-005-5132-y
14. Kyzas GZ, Bikiaris DN, Seredych M, Bandosz TJ, Deliyanni EA (2014) Removal of dorzolamide from biomedical wastewaters with adsorption onto graphite oxide/poly(acrylic acid) grafted chitosan nanocomposite. Biores Technol 152:399–406. https://doi.org/10.1016/j.biortech.2013.11.046
15. Lakard B (2020) Electrochemical biosensors based on conducting polymers: a review. Appl Sci 10(18):6614. https://doi.org/10.3390/app10186614
16. Liu Y, Park M, Ding B, Kim J, El-Newehy M, Al-Deyab SS, Kim HY (2015) Facile electrospun polyacrylonitrile/poly(acrylic acid) nanofibrous membranes for high efficiency particulate air filtration. Fibers Polym 16(3):629–633. https://doi.org/10.1007/s12221-015-0629-1
17. Lv S, Zhao X, Shi L, Zhang G, Wang S, Kang W, Zhuang X (2018) Preparation and properties of *sc*-PLA/PMMA transparent nanofiber air filter. Polymers 10(9):996. https://doi.org/10.3390/polym10090996
18. Ma PC, Siddiqui NA, Marom G, Kim JK (2010) Dispersion and functionalization of carbon nanotubes for polymer-based nanocomposites: a review. Compos A Appl Sci Manuf 41(10):1345–1367. https://doi.org/10.1016/j.compositesa.2010.07.003
19. Mahmoud ME, Nabil GM, Abdel-Aal H, Fekry NA, Osman MM (2018) Imprinting "Nano-SiO_2-Crosslinked Chitosan-Nano-TiO_2" polymeric nanocomposite for selective and instantaneous microwave-assisted sorption of Hg(II) and Cu(II). ACS Sustain Chem Eng 6(4):4564–4573. https://doi.org/10.1021/acssuschemeng.7b03215
20. Mamba G, Mbianda X, Govender P, Mamba B, Krause R (2010) Application of multiwalled carbon nanotube-cyclodextrin polymers in the removal of heavy metals from water. J Appl Sci 10(11):940–949. https://doi.org/10.3923/jas.2010.940.949
21. Matei E, Predescu AM, Râpă M, Țurcanu AA, Mateș I, Constantin N, Predescu C (2022) Natural polymers and their nanocomposites used for environmental applications. Nanomaterials 12(10):1707. https://doi.org/10.3390/nano12101707
22. Mittal V (2011) In-situ synthesis of polymer nanocomposites. Wiley & Sons
23. Nayak SK, Dutta K, Gohil JM (2022) Advancement in polymer-based membranes for water remediation. Elsevier

24. Pandikumar A, Manonmani S, Ramaraj R (2012) TiO_2–Au nanocomposite materials embedded in polymer matrices and their application in the photocatalytic reduction of nitrite to ammonia. Catal Sci Technol 2(2):345–353. https://doi.org/10.1039/c1cy00298h
25. Potts JR, Dreyer DR, Bielawski CW, Ruoff RS (2011) Graphene-based polymer nanocomposites. Polymer 52(1):5–25. https://doi.org/10.1016/j.polymer.2010.11.042
26. Sardar UR, Bhargavi E, Devi I, Bhunia B, Tiwari ON (2018) Advances in exopolysaccharides based bioremediation of heavy metals in soil and water: a critical review. Carbohydr Polym 199:353–364. https://doi.org/10.1016/j.carbpol.2018.07.037
27. Sahoo S, Sahoo P, Sharma A, Satpati A (2020) Interfacial polymerized RGO/$MnFe_2O_4$/polyaniline fibrous nanocomposite supported glassy carbon electrode for selective and ultrasensitive detection of nitrite. Sens Actuators B Chem 309:127763. https://doi.org/10.1016/j.snb.2020.127763
28. Salipira K, Krause R, Mamba B, Malefetse T, Cele L, Durbach S (2008) Cyclodextrin polyurethanes polymerized with multi-walled carbon nanotubes: synthesis and characterization. Mater Chem Phys 111(2–3):218–224. https://doi.org/10.1016/j.matchemphys.2008.03.026
29. Suraj Belgaonkar M, Kandasubramanian B (2021) Hyperbranched polymer-based nanocomposites: synthesis, progress, and applications. Eur Polymer J 147:110301. https://doi.org/10.1016/j.eurpolymj.2021.110301
30. Tajik S, Beitollahi H, Nejad FG, Dourandish Z, Khalilzadeh MA, Jang HW, Venditti RA, Varma RS, Shokouhimehr M (2021) Recent developments in polymer nanocomposite-based electrochemical sensors for detecting environmental pollutants. Ind Eng Chem Res 60(3):1112–1136. https://doi.org/10.1021/acs.iecr.0c04952
31. Tao X, Li A, Yang H (2017) Immobilization of metals in contaminated soils using natural polymer-based stabilizers. Environ Pollut 222:348–355. https://doi.org/10.1016/j.envpol.2016.12.028
32. Thakur VK, Thakur MK (2014) Processing and characterization of natural cellulose fibers/thermoset polymer composites. Carbohyd Polym 109:102–117. https://doi.org/10.1016/j.carbpol.2014.03.039
33. Thomas S, Zaikov G, Valsaraj M (2010) Recent advances in polymer nanocomposites: synthesis and characterisation. CRC Press
34. Xiao M, Hu JC, Liu WL, Nie FM (2014) Application of biopolymer-based adsorbents in removal of heavy metals. Adv Mater Res 1048:373–377. https://doi.org/10.4028/www.scientific.net/amr.1048.373
35. Xing H, Xu J, Zhu X, Duan X, Lu L, Wang W, Zhang Y, Yang T (2016) Highly sensitive simultaneous determination of cadmium (II), lead (II), copper (II), and mercury (II) ions on N-doped graphene modified electrode. J Electroanal Chem 760:52–58. https://doi.org/10.1016/j.jelechem.2015.11.043
36. Yang Y, Zhang S, Zhao X, Yu J, Ding B (2015) Sandwich structured polyamide-6/polyacrylonitrile nanonets/bead-on-string composite membrane for effective air filtration. Sep Purif Technol 152:14–22. https://doi.org/10.1016/j.seppur.2015.08.005
37. Zhang R, Liu C, Hsu PC, Zhang C, Liu N, Zhang J, Lee HR, Lu Y, Qiu Y, Chu S, Cui Y (2016) Nanofiber air filters with high-temperature stability for efficient PM2.5 removal from the pollution sources. Nano Lett 16(6):3642–3649. https://doi.org/10.1021/acs.nanolett.6b00771
38. Zhang S, Liu H, Yin X, Yu J, Ding B (2016) Anti-deformed polyacrylonitrile/polysulfone composite membrane with binary structures for effective air filtration. ACS Appl Mater Interfaces 8(12):8086–8095. https://doi.org/10.1021/acsami.6b00359
39. Zhao L, Yang S, Yilihamu A, Ma Q, Shi M, Ouyang B, Zhang Q, Guan X, Yang ST (2019) Adsorptive decontamination of Cu2+-contaminated water and soil by carboxylated graphene oxide/chitosan/cellulose composite beads. Environ Res 179:108779. https://doi.org/10.1016/j.envres.2019.108779

Chapter 9
Natural and Synthetic Fiber-Filled Polymer Composites Used as Anticorrosive Materials

Pakanati Siva Prasad, Juan David Matallana Guerrero, and Pankaj Kumar

1 Introduction

Corrosion poses a significant threat to the integrity, performance, and longevity of materials in numerous industrial applications. The detrimental effects of corrosion are observed across various sectors, including infrastructure, transportation, Oil and gas, marine, and many others. The need for effective and durable anticorrosive materials has never been more critical.

In recent years, the development and utilization of fiber-filled polymer composites have emerged as a promising solution to combat corrosion-related challenges. These composites, composed of a polymer matrix reinforced with natural or synthetic fibers, offer a unique combination of properties that enhance their resistance to corrosion. By harnessing the inherent advantages of fibers and polymers, these materials provide improved performance, longevity, and cost-effectiveness in harsh corrosive environments.

The objective of this chapter is to provide an in-depth exploration of natural and synthetic fiber-filled polymer composites as efficient and sustainable anticorrosive materials. We will delve into the properties, fabrication techniques, corrosion resistance mechanisms, and real-world applications of these composites. By understanding the potential of fiber-filled polymer composites, researchers, engineers, and industry professionals can make informed decisions to protect against corrosion and extend the service life of materials.

P. S. Prasad (✉) · J. D. M. Guerrero
Department of Metallurgical and Materials Engineering, Indian Institute of Technology Kharagpur, Kharagpur 721302, India
e-mail: psiva432@gmail.com

P. Kumar
School of Nano Science and Technology, Indian Institute of Technology, Kharagpur 721302, India

S. K. Sethi et al. (eds.), *Polymer Composites: From Computational to Experimental Aspects*, Materials Horizons: From Nature to Nanomaterials,
https://doi.org/10.1007/978-981-97-0888-8_9

Throughout this chapter, we will examine the benefits of natural fibers, such as jute, sisal, bamboo, and coir, and synthetic fibers, including carbon fibers, glass fibers, and aramid fibers. We will explore the fabrication methods employed to produce fiber-reinforced polymer composites and the mechanisms through which these composites demonstrate enhanced resistance to corrosion. Furthermore, we will highlight various industries and applications where fiber-filled polymer composites have proven effective in combating corrosion.

By comprehensively assessing the advancements, challenges, and future perspectives of natural and synthetic fiber-filled polymer composites used as anticorrosive materials, we aim to contribute to the growing body of knowledge in this important field. Ultimately, this chapter strives to foster innovation, promote sustainable practices, and inspire further research in the development of effective anticorrosive solutions.

2 Fiber-Reinforced Polymer Composites (FRPCs)

Fiber-reinforced polymer (FRP) composites are composite materials that consist of high-strength fibers embedded in a polymeric matrix. These composites offer numerous advantages over monolithic polymer materials, including high strength, stiffness, long fatigue life, and adaptability to structural requirements [1]. Additionally, FRPCs can be tailored to enhance various properties such as corrosion resistance, wear resistance, appearance, temperature-dependent behavior, environmental stability, thermal insulation, and conductivity [1]. Furthermore, the use of natural fibers in reinforcing composites provides technical and ecological benefits, offering a promising alternative to synthetic FRPCs [1]. This section provides an overview of the use of fiber-reinforced polymer composites as anticorrosive materials, examining the incorporation of synthetic and natural fibers, as well as the role of nanoparticles and polymer matrices in these composites.

Polymer matrices used in FRPCs can be classified into two types: thermoplastics and thermosetting resins. Thermoplastic materials, such as polypropylene (PP), polyethylene, and polyvinyl chloride (PVC), are commonly used as matrices for biofibers. On the other hand, thermosetting matrices like phenolic, epoxy, and polyester resins find extensive use in FRPCs due to their excellent mechanical and thermal properties [2]. The selection of an appropriate polymer matrix plays a crucial role in determining the overall performance and corrosion resistance of the composite material.

Due to the fast-growing demands for eco-friendly, low-cost, and easy production of new materials, many attempts have been made to produce materials with excellent physical, mechanical, and corrosion-resistant properties for on-site applications in which the engineered components are required to be shielded from physical, environmental and chemical effects from its surroundings. Polymer nanocomposite materials, comprising a polymeric matrix embedded with nanoparticles, have gained significant interest due to their unique properties compared to neat polymers [3,

4]. Nanoparticles, typically ranging from 1 to 100 nm in size, provide enhanced mechanical, optical, electrical, and magnetic properties to the composite materials. The incorporation of nanoparticles into polymer matrices can further improve the corrosion resistance of fiber-reinforced polymer composites, making them suitable for anticorrosive applications.

The FRP is broadly divided into natural-reinforced-fiber composites (NRFC) and synthetic-reinforced-fiber composites (SRFC), as shown in Fig. 1 [5, 6]. The FRP has been considered a potential replacement for metallic or ceramic composite components in aerospace, automobile, electrical, and building infrastructure. They offer high specific strength, lightweight, good stiffness, improved corrosion resistance, low material cost, and easy installation. The performance of the FRP depends greatly on the type of fiber employed, as they modify the composite intrinsic properties.

Synthetic FRPCs are fabricated using high-strength fibers, such as glass, aramid, carbon, and other synthetic fibers, which are embedded in a polymer matrix. These composites offer superior mechanical properties, including high strength and stiffness. Moreover, synthetic FRPCs can be modified to enhance their corrosion resistance, wear resistance, and other desired properties, making them suitable for various applications [1]. However, they have certain limitations, such as high density, high cost compared to polymers, and poor recycling and nonbiodegradable properties.

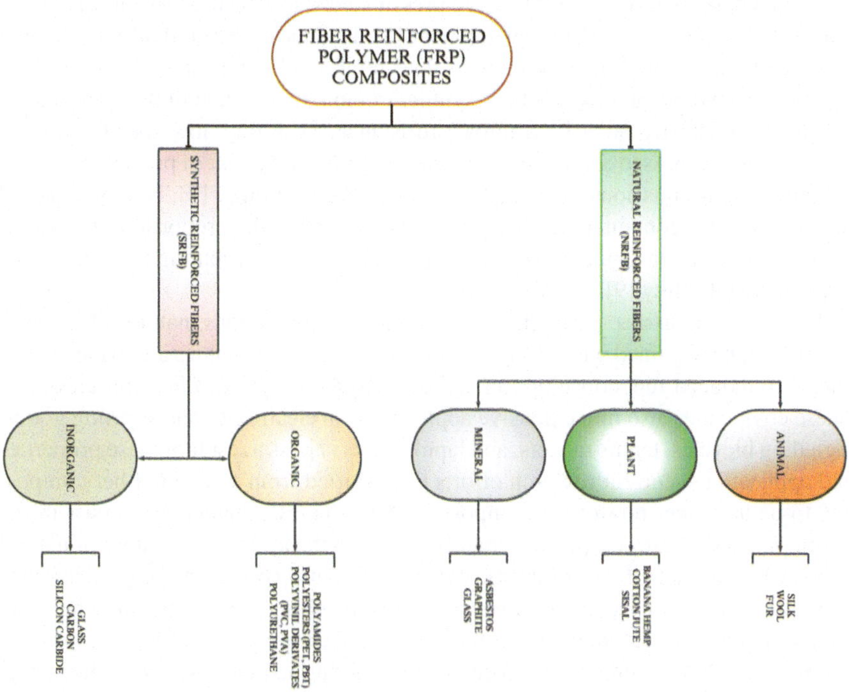

Fig. 1 Classification of the fiber-reinforced polymers (FRP)

Natural fibers, derived from renewable resources, have gained attention as a viable alternative to synthetic fibers in FRPCs. A wide range of natural fibers, including jute, straw, flax, hemp, wood, sugarcane, bamboo, grass, kenaf, sisal, coir, rice husks, and various agricultural by-products, have been investigated for their reinforcement potential in polymer composites [1]. The use of natural fibers not only provides mechanical strength but also offers economic and ecological advantages, such as cost savings and reduced environmental impact. These natural fiber-reinforced composites have shown promising results in terms of mechanical properties and can be considered for anticorrosion applications.

2.1 *Natural Fiber-Reinforced Polymer Composites*

In earlier times, composite materials predominantly featured synthetic fiber reinforcements, primarily glass, and carbon fibers, owing to their cost-effectiveness and excellent mechanical properties [7]. However, the modern era has witnessed a significant shift towards sustainability, becoming a central concern in both research and industry. Industries are now challenged to replace non-sustainable products with environmentally friendly alternatives [7]. As conventional energy resources dwindle and environmental concerns grow, there is a pressing need to transition to renewable raw materials for new component development [8], leading to the gradual replacement of synthetic fibers with natural ones in fiber-reinforced composites. Natural fibers offer numerous advantages over their synthetic counterparts, including lower relative density, cost-effectiveness, high-impact resistance, flexibility, low specific gravity, reduced abrasiveness to equipment, minimal health hazards, ease of processing, lower greenhouse gas emissions, recyclability, and carbon neutrality [7]. Notably, natural fiber composites contain a higher fiber content, reducing the proportion of environmentally harmful base polymers in the composite, thus contributing to their overall environmental safety [9].

In various industries, particularly in automotive applications, natural fiber-based composites have gained widespread use due to their lightweight nature, which translates into reduced fuel consumption and decreased emissions of harmful gases [7]. These composites also find extensive applications in electronics and sporting goods, including bicycles, tennis rackets, and laptop cases [7]. As a result of these properties and applications, current research efforts have shifted from synthetic fiber composites to natural fiber-reinforced composites. Nevertheless, challenges persist in the development of these composites, mainly related to the hydrophilic nature of natural fibers, which reduces their suitability as reinforcements in polymer composites due to poor moisture resistance and limited wettability, resulting in weak bonding between the matrix and reinforcement and inferior mechanical properties [10]. Additionally, natural fibers have a tendency to form aggregates during composite manufacturing, limiting their usability, along with issues related to reduced dimensional stability and thermal stability [7].

The natural fibers derived from plants and animals can be categorized into two main groups: primary and secondary plants, based on whether they are primarily cultivated for their fiber content or yield fibers as by-products. Primary plants, such as jute, hemp, kenaf, and sisal, are cultivated primarily for their fiber content, while secondary plants, like pineapple and coir, yield fibers as by-products [7].

Various types of natural fibers have proven suitable for reinforcing natural fiber-reinforced polymer composites, with some demonstrating exceptional performance as composite reinforcements. Examples include pineapple leaf fiber, bamboo, jowar, banana, jute, hemp, flax, and sisal, each falling into one of six categories: reed fibers (e.g., wheat, corn, rice), leaf fibers (e.g., abaca, sisal, pineapple), bast fibers (e.g., jute, flax, hemp, ramie, kenaf), seed fibers (e.g., coir, cotton, kapok), core fibers (e.g., kenaf, hemp, jute), and other types (e.g., woods and roots). For instance, jowar fiber's low density makes it ideal for manufacturing lightweight materials used in housing, automobile body construction, and packaging [7, 11].

Polymer matrices used in composites can be broadly classified into two types: thermosetting and thermoplastics. Thermoplastics include polymers like polypropylene, polyurethane, and nylon, while thermosetting matrices encompass epoxy and polyester, which are commonly used. Epoxy resin is preferred as the matrix in polymer composites due to its desirable properties, including low shrinkage, low toxicity, strong adhesion characteristics, and wide industrial applicability. However, epoxy's brittleness and low fracture toughness limit its industrial use [7].

One of the primary challenges with natural fiber-reinforced composites lies in the interaction between the matrix and fibers, their durability, and resistance to water. Natural fibers exhibit high hydrophilicity, while the polymeric resins they are combined with tend to be hydrophobic and non-polar. This incongruity leads to poor adhesion and interfacial bonding, resulting in a noticeable decrease in composite properties and hindering their industrial applications [7].

2.1.1 Fabrication Techniques for NFRPCs

NFRPs can be produced using various fabrication techniques, each offering distinct advantages and limitations. Among these methods are the hand lay-up technique, compression molding, resin transfer molding, vacuum bagging, filament winding, hot press method, single-screw extrusion, twin-screw extrusion, and injection molding [5].

The hand lay-up technique stands out as the simplest and most cost-effective approach. Nevertheless, it suffers from drawbacks, including uneven thickness and a high occurrence of voids within the composite structure. On the other hand, the hot press method is employed to enhance the strength of the composites but tends to introduce variations in thickness and results in a less-than-ideal surface finish. Consequently, it is not commonly favored for industrial applications. Compression molding provides a more straightforward processing route and delivers composites with improved surface quality, rendering them suitable for structural purposes. However, it may not be the method of choice for intricate structural designs. In contrast, the vacuum bagging technique, while more expensive than compression molding, produces composites with superior surface finish and minimal void content. It is, therefore, widely employed in the production of large and complex composite components for industrial applications. Filament winding is a specialized method utilized for manufacturing both open and closed structures. It excels in providing excellent reinforcement but is noted for being a time-consuming process. For thermoplastic composite materials, hot press methods, compression molding, extrusion, and injection molding techniques are commonly employed. Twin-screw extrusion and injection molding, in particular, are favored for crafting advanced and intricate composite parts tailored to meet specific industrial requirements [5, 12, 13].

2.1.2 Properties and Advantages of Natural Fibers in Terms of Corrosion Resistance

Natural fibers offer certain properties and advantages related to corrosion resistance, although they may not possess the same level of corrosion resistance as synthetic fibers. Here are some noteworthy properties and advantages of natural fibers concerning corrosion resistance.

Natural fibers exhibit inherently low electrical conductivity, making them suitable for applications where electrical conductivity must be minimized, such as non-conductive reinforcement in concrete structures in the construction industry to reduce the risk of galvanic corrosion between dissimilar metals [14]. Additionally, natural fibers typically do not undergo chemical reactions with metals, which can be advantageous when maintaining contact between a fiber-reinforced composite material and a metal substrate is necessary without inducing galvanic corrosion [15]. Their hygroscopic properties, allowing absorption and release of moisture, are valuable for corrosion protection. For instance, in the timber industry, natural fibers like flax or hemp, when combined with resins, create moisture-absorbing layers in wooden structures, reducing moisture content and minimizing the risk of corrosion in embedded metal components. Furthermore, natural fibers are environmentally sustainable as renewable and biodegradable resources, aligning with eco-friendly practices. Incorporating natural fiber composites in corrosion protection applications reduces reliance on synthetic materials and contributes to lowering the environmental impact associated with corrosion protection strategies. Finally, the cost-effectiveness of natural fibers, being readily available and locally sourced in many regions, makes them an attractive option for corrosion protection applications, particularly when budget constraints are a consideration. It's essential to consider natural fibers' limitations in terms of strength, durability, and resistance to harsh environments compared to synthetic fibers. The suitability of natural fibers for corrosion protection applications depends on specific project requirements, environmental conditions, and expected service life. Proper material selection and design considerations are crucial to ensure optimal performance and durability [14–16].

2.1.3 Role of Natural Fiber Reinforcement in Enhancing the Corrosion Resistance of Composites

The role of natural fiber reinforcement in enhancing the corrosion resistance of composites is multifaceted and can provide certain benefits in specific applications. Here are some ways in which natural fiber reinforcement can contribute to improved corrosion resistance in composites.

Physical barrier: Natural fibers can act as a physical barrier between the corrosive environment and the underlying metal substrate. By embedding natural fibers in a polymer matrix, a protective layer is formed that hinders the direct contact of moisture, chemicals, and other corrosive agents with the metal surface. This barrier effect helps to mitigate the risk of corrosion initiation and progression [17].

Moisture absorption and desorption: Natural fibers possess hygroscopic properties, allowing them to absorb and release moisture from the surrounding environment. This property can help regulate the moisture content within a composite material. In corrosion protection applications, natural fiber composites can absorb excess moisture, reducing the moisture content that could potentially lead to corrosion. By maintaining an optimal moisture level, natural fiber reinforcements can contribute to enhanced corrosion resistance [18].

Galvanic corrosion prevention: Natural fibers, being non-reactive with metals, can help prevent galvanic corrosion. When dissimilar metals come into direct contact in the presence of an electrolyte (e.g., moisture), galvanic corrosion can occur. By incorporating natural fibers as reinforcement in composite materials, the risk of galvanic corrosion between the metal components can be reduced or eliminated. This is particularly beneficial in applications where dissimilar metals are present and contact with each other cannot be avoided [19].

Alkaline resistance: Some natural fibers, such as jute and flax, exhibit good resistance to alkalis. In certain corrosive environments where alkaline substances are prevalent, the use of natural fiber reinforcements can provide added resistance to corrosion caused by these substances. By incorporating alkaline-resistant natural fibers in composites, the corrosion resistance of the overall material can be enhanced [20].

Environmental sustainability: The use of natural fiber reinforcements promotes environmental sustainability compared to synthetic fibers. By utilizing renewable and biodegradable materials, the overall environmental impact is reduced. This can be particularly relevant in corrosion protection applications where sustainability and eco-friendliness are important considerations [21].

It's important to note that natural fiber reinforcements may have limitations in terms of strength, durability, and resistance to harsh environments compared to synthetic fibers. Their effectiveness in enhancing corrosion resistance depends on the specific application, environmental conditions, and the expected service life. Proper material selection, composite design, and protective coatings may be required to optimize the corrosion resistance performance of composites reinforced with natural fibers [22].

2.1.4 Examples of NFRPCs Specifically Developed for Anticorrosive Applications

Natural fiber-filled polymer composites (NFRPCs) have gained attention as environmentally friendly alternatives in various industrial applications, including those requiring anticorrosive properties. The incorporation of natural fibers into polymer matrices offers unique advantages and specific properties that contribute to enhanced corrosion resistance. Here are a few notable examples of natural fiber-filled polymer composites developed for anticorrosive applications.

Flax Fiber-Reinforced Polymer (FRP) Composites: Flax fiber composites have shown promise in providing corrosion protection due to the inherent characteristics of flax fibers. Flax fibers exhibit good mechanical properties [23], including high tensile strength and stiffness, making them suitable for reinforcing polymer matrices. The composite's resistance to corrosion is attributed to the hygroscopic nature of flax fibers, allowing them to absorb and release moisture, thus reducing the moisture content in the composite and mitigating the risk of corrosion[24]. Flax FRP composites find applications in the construction industry for reinforcing concrete structures, such as bridge decks and beams, to enhance their durability and resist corrosion in harsh environments [25, 26].

Hemp Fiber-Reinforced Polymer (FRP) Composites: Hemp fiber composites have gained attention as a sustainable and corrosion-resistant alternative [27]. Hemp fibers offer high strength, low weight, and good resistance to alkalis and moisture. These properties make hemp FRP composites suitable for anticorrosive applications, such as reinforcing pipes, tanks, and storage vessels in the chemical processing industry [28]. Hemp FRP composites provide mechanical strength, reduce the risk of galvanic corrosion, and exhibit resistance to various chemicals, contributing to the long-term protection of metal substrates.

Coir Fiber-Reinforced Polymer (FRP) Composites: Coir fiber composites have demonstrated excellent corrosion resistance due to the unique characteristics of coir fibers derived from coconut husk. Coir fibers offer good tensile strength, stiffness, and resistance to moisture absorption. The hydrophobic nature of coir fibers helps to reduce the moisture content within the composite, thereby minimizing the risk of corrosion. Coir FRP composites find applications in marine environments, such as reinforcement in boat hulls, off-shore structures, and coastal protection systems, where resistance to saltwater corrosion is crucial [29].

Sisal Fiber-Reinforced Polymer (FRP) Composites: Sisal fiber composites have been explored for their corrosion resistance properties. Sisal fibers possess high tensile strength, stiffness, and alkali resistance. These properties make sisal FRP composites suitable for anticorrosive applications in the construction industry. Sisal FRP composites can reinforce concrete structures, such as beams, columns, and panels, to improve their durability and mitigate corrosion caused by moisture and alkalis [30, 31].

Jute Fiber-Reinforced Polymer (FRP) Composites: Jute fibers, commonly employed to reinforce polyester composites through hand lay-up and compression molding, improve the physical characteristics of the polyester matrix. Both woven jute fabric and unidirectional jute fibers can serve as effective reinforcements. Jute's wood-like characteristics, high aspect ratio, impressive strength-to-weight ratio, and good insulation properties make it suitable for a wide array of applications, including doors, windows, furniture, corrugated sheets, I-shaped beams, trenchless rehabilitation of underground pipes, false roofing, floor tiles, and more [32–34].

Bamboo Fiber-Reinforced Polymer (FRP) Composites: Bamboo fibers, obtained from the leaves, branches, and trunks of bamboo plants, offer UV light absorbency and enhanced performance, making them attractive as reinforcement materials for polymer composites. These fibers have been integrated into various polymer matrices, including polyester, epoxy, phenolic resin, PLA, and PP, to enhance composite properties [35, 36].

While NFRPCs offer these advantages, they may have lower mechanical properties compared to synthetic fibers and can degrade under prolonged exposure to moisture or UV radiation. Ongoing research aims to optimize their properties for various anticorrosive applications [14].

Natural fibers such as cotton, jute, sisal, coir, hemp, and bamboo have garnered significant interest from engineers and scientists across various fields. Polyester composites reinforced with these fibers offer high impact and electrical resistance, excellent insulation, and a reduced environmental footprint compared to synthetic counterparts. Despite historical applications dating back to the nineteenth century, challenges persist in scaling up structural applications due to manufacturing processes and quality consistency [14]. Nonetheless, these natural fibers show great potential in reshaping industries with a focus on sustainability.

Environmental conditions significantly influence the performance of natural fiber-reinforced polymers (NFRPs). They can degrade in various environments, including biological, chemical, mechanical, thermal, photochemical, and aqueous conditions. Addressing these challenges is crucial for harnessing the full potential of NFRPs [37]. In concrete matrices, treatments can enhance their resistance to degradation and reduce detrimental effects, such as exposure to alkaline pore solutions [14].

2.2 Synthetic Fiber-Reinforced Polymer Composites

The most common synthetic fibers employed in the industry are glass fibers (GF), aramid fiber (AF), and carbon fiber (CF). Each of these fibers behaves differently depending on their application; for example, GFs are weak and easily degradable in the presence of water, acidic, and alkaline solutions. Also, they have shown degradation as reinforcement in concrete structures [38].

However, GFs are divided into many types such as E-glass, S-glass, and C-glass. Each of these variations has specific applications and overcomes some of the challenges that GFs present. Compared to GFs, CFs do not present moisture absorption and they offer improved corrosion resistance to different chemical solutions, as they employ different carbon allotropes, such as graphite. Moreover, aramid fibers (AFs) are known to have a high degree of moisture absorption and poor performance at elevated temperatures. The creep phenomena are observed in the AFs, leading to their strength degradation over time.

In order to address some of the drawbacks of the intrinsic properties of the fibers, thermosetting or thermoplastic polymers are employed as a matrix and binding agent for the fibers to increase and broaden the use of the composites in real-life applications. Synthetic fiber-reinforced polymers (SFRP) are generally made from glass (GFRP), carbon (CFRP), aramid (AFRP), and sometimes even basalt (BFRP) are also being used [39, 40]. SFRP offers improved corrosion resistance applications, high stiffness, strength-to-weight ratio, and improved strength compared to the NRFP. The SFRPCs have found themselves as a viable alternative to NFRPCs in many industrial fields, highlighting their use in concrete structures, high-voltage insulators, ship-hauls, decks, military applications such as bulletproof vests, car-body parts, etc.

In the subsequent sections, we will discuss in detail the types and uses of each one of the particular GFRPs, CFRPs, and AFRPs in the industry and some of their physical and electrochemical properties.

2.2.1 Fabrication Techniques for SFRPCs

Synthetic fiber-reinforced polymers (SFRP) are often manufactured by extrusion, spinning, polymerization, and filament processing methods. Figure 2 shows the schematics of some of the conventional manufacturing methods. The spinning method is subdivided into wet, dry, and melt spinning [41, 42]. The extrusion method passes molten material (i.e., resin and fiber material) through a nozzle which is moved by a worm screw and is pumped and forced into a closed mold or die-block under pressure to form the desired shape or component, which is cooled by cooling molds and the final product is subsequently ejected. This method is often used to manufacture aerospace and automotive industries, as producing small or intricate pieces is possible.

Melt spinning follows the same principle as in the extrusion method by pumping the fiber-polymer composite material. Fiber-polymer material in the form of pellets is fed through a hoover, melted, and transported by a worm screw. The material then passes through a nozzle and is pulled by a pump to form strands. The SFRP composite strands are subsequently cooled down in a cooling chamber and winded with the help of a mandrill. Moreover, polymerization is simply the by-product of the chemically bonded monomers, giving rise to a polymer with different physical properties. The obtained polymer holds the fiber together, which acts as a binding agent.

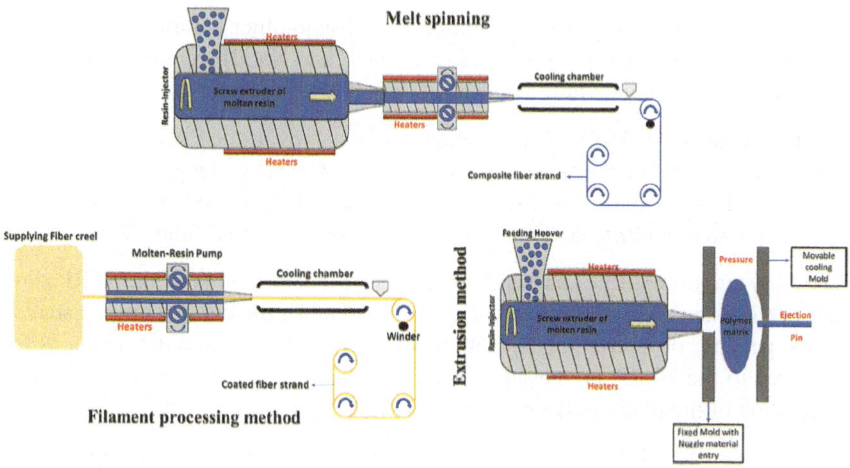

Fig. 2 Schematic of some conventional SFRP manufacturing processes

Furthermore, the filament processing method consists of thin strands or filaments pulled through a molten resin bath. The strand is cooled down in the thin filament method and not entrapped in a mold as in the extrusion method. These strands are subsequently winded for future uses with the help of a mandril. The pultrusion or filament processing method is a well-known continuous process for the industry's thermoplastics and epoxies fiber-reinforced manufacturing. These filaments are used in various applications and industries, such as the textile industry onshore and off-shore gas and oil pipelines.

2.2.2 Properties, Role, and Advantages of Synthetic Fibers in Terms of Their Corrosion Resistance

We have briefly discussed the GFs, CFs, and AFs, usages, drawbacks, and benefits. Moreover, there are different ways in which SFRPs provide corrosion resistance properties for a given designed component. SFRPs improve the corrosion resistance of a component or protect it from harmful environments are as follows:

Physical barrier protection: SFRPs act as a physical barrier against corrosive elements; for example, off-shore marine applications protect the structures by preventing the highly corrosive chloride salts from coming into contact with the protected component surface.

Galvanic protection: The application of a sacrificial metallic coating often carries out galvanic protection of metallic components. However, SFRPs, specifically GRP composites, have been known to be employed in sewage pipes, structural components in chemical plants, and off-shore applications. Glass fiber dissolution in the presence of alkaline media is controlled by the disruption of the siloxane bonds in the GF and subsequent leaching into the solution [43]. Some other authors have reported that for the particular case of GFs, adding a $CaCl_2$ and NaOH solution to the GFRP composites forms an insoluble protective layer of Ca–Si.

Chemical resistance: Most of the synthetic fibers show a good corrosion-resistant performance to chemical attacks. For example, GFs have been seen to perform better in acidic solutions than AFs. Moreover, BFs present decreased mass loss due to corrosion in hydrochloric acid and sodium hydroxide solutions than GFs.

Flexibility and durability: The excellent flexibility and durability of the SFRP allow their use in many applications and shapes. These make the composites resistant to cracks, peeling, or interfacial separation between the matrix and the fibers. SFRP, compared to the NFRP, are easier to handle, as they present a more homogeneous distribution of the fibers/polymer matrix.

Long-service life: SFRP possesses a longer service life than NFRP composites, as they do not suffer from bio-degradation. They require minimal maintenance, making them cost-effective and economically feasible.

Synthetic fiber-reinforced polymers offer a versatile and effective solution to many engineered components in terms of their corrosion protection. It is worth mentioning that the proper selection of the type of SF and its application ensures their performance as a physical barrier, galvanic protection provision, and chemical resistance.

2.2.3 Examples of SFRPCs Specifically Developed for Anticorrosive Applications

Synthetic fiber-reinforced polymer composites (SFRPCs) have been widely used in various industrial applications. The main drawbacks of using NFRPCs are biodegradability, water retention, and moisture absorption, limiting their use in certain applications. SFRPCs offer excellent electrical insulation properties, a high strength-to-weight ratio, excellent durability, and good tribological and mechanical properties. Moreover, SFRPCs are corrosion-resistant in most acidic and alkaline media (i.e., corrosive media). Therefore, SFRPCs have found application in many industrial fields, such as the oil and gas industry, construction, and off-shore marine applications. From the wide variety of SFRP (i.e., GF, BF, and AF), certain applications with their benefits and drawbacks are explained below.

Glass fiber reinforced polymer (GFRP): Glass fibers are one of the most used fiber-reinforced polymers (FRPs) due to their excellent electrical insulation properties, making them suitable for applications where electrical conductivity must be minimized. Printing board structural components is just one of the applications of GFRP. Some of the most known electrical glass fiber composites are electrical glass (E-glass), structural glass (S-glass), and chemical glass (C-glass), among others, which offer a wide range of applications in automotive parts, pipes, tanks, and construction industry for reinforcing concrete structures. The E-glass fiber offers a high-strength-to-weight ratio and improved corrosion resistance. At the same time, the

S-glass enhances structures' strength and durability. Glass fibers, therefore, offer strength and durability with excellent tribological and mechanical properties [42]. It is noteworthy that commercial glass fibers have a notable capacity to withstand loads reaching a magnitude of 2.8 GPa [6].

Carbon fiber reinforced polymer (CFRP): Carbon fiber-reinforced polymers are simply a composite material in which carbon fibers are embedded in a polymer matrix (resin or epoxy). This material possesses a good strength-to-weight ratio, high stiffness, burst, and corrosion resistance. However, they are inherently electrically conductive, making them prone to galvanic corrosion in certain conditions. Carbon fiber composites possess a low coefficient of thermal expansion, allowing their use at relatively high temperatures for a polymeric composite material. With carbon contents varying up to 99%, as a filler in the polymeric matrix, they show a tensile strength of a range from 2 to up to 7 GPa.

Moreover, its axial compressive strength is 15–60% of its tensile strength. The CFRP composites are typically made from a rich carbon precursor named polyacrylonitrile (PAN), rayon, or pitch. This type of FRP is less prone to thermal deformation and expansion under temperature changes; also, it is highly corrosion-resistant to environmental factors or chemicals.

Aramid fiber reinforced polymers (AFRP): AFRP, also known as Kevlar fibers reinforced polymers (KFRP), are made of PPTA (p-phenylene terephthalamide). Kevlar offers low weight, high strength, abrasion resistance, and impact resistance to penetration. Some of the applications of various Kevlar fiber reinforced composites (i.e., Kevlar 29, Kevlar 49, Kevlar 129, etc.) include body armor protective gear (i.e., bulletproof vest), automotive, and aerospace components. Moreover, Kevlar has been reported to have a glass-transition temperature as high as 270 °C, allowing it to be employed in relatively high-temperature applications for polymeric materials. Aramid fibers have higher elastic modulus than glass fibers and high resistance to corrosion up to temperatures of 500 °C [39]. However, despite the advantages shown by Kevlar fibers, they show a poor compression-to-strength ratio. It was reported that the tensile strength/compressive ratio is 5:1, the tensile strength to shear strength ratio was 17:1, and the tensile modulus/shear modulus was 70:1 [6]. The difference in the tensile strength values for SFRP composites (i.e., glass, carbon, and aramid fibers) is shown in Table 1.

Table 1 Tensile strength comparison for most known SFRP

Type of fiber	Denomination	Tensile strength (GPa)	References
GFRP	E-glass	2–3.5	[44]
GFRP	A-glass	3.30	[6]
GFRP	S-glass	4.88	[6]
ARFP	Kevlar 29	3.4	[45]
ARFP	Kevlar 49	1.4–2.3	[46]
CRFP	Carbon	2.5–6.1	[1]

Each SRFP from glass, carbon, or aramid materials has advantages and disadvantages. The production cost of GRFP is less expensive than CFRP; however, the FRP made from GFs presents lower strength, as observed in some of the references in Table 1. Moreover, GFRP presents a good performance while extended at high temperatures for long periods, making them an appealing material for high-temperature applications. One of the main applications of GFRP is high-voltage insulation. The ARFP have limited industrial uses, as they present a high degree of crystallinity, making their processing difficult. However, they present a high dielectric strength and excellent mechanical and heat stability properties. These ARFPs, known as PET fibers, are often employed in conveyor belts, cord wrap, magnetic tape, etc. [41].

2.3 *Applications of FRPCs as Anticorrosive Materials*

2.3.1 Applications in Infrastructure

FRPs have attracted attention in their use in structural components as they are characterized by their high tensile strength in reinforced concrete structures [40]. Any structural component is subjected to tensile loads, so NFRP and SFRP have been employed as strengthening components in concrete structures. Generally, the reinforcing fibers are arranged parallel to the matrix, in which the maximum tensile capacity is achieved. It is well known that concrete structures are prone to corrosion, especially when exposed to harsh environments such as in coastal regions where chloride-induced corrosion of the reinforcing steels (i.e., reinforced steel bars) occurs, shortening the useful life of the structures [47]. Using GFRP and CFRP composites in concrete structures tends to address the strength deterioration over time. The degradation of these composites was addressed by Bank et al. [38], in which they classified the failure mechanisms of these into three main groups: Fiber mechanism, matrix mechanism, and interphase mechanisms. It was found that KFRP composites presented matrix plastification in the presence of moisture and elevated temperatures, leading to fiber buckling and subsequent tensile failure. GFRP suffers from corrosion degradation due to their contact with the corrosive solution, which permeates the reinforced concrete structures. Moreover, GFRP is more prone to moisture absorption than CFRP, leading to hydrolysis and scission of the polymeric chains, increasing the corrosion rate degradation [48, 49]. Also, the appearance of micro-cracks due to matrix degradation and mismatch of the parent phase and the reinforcing fibers leads to rupture [38].

The authors Ahmed et al. [50], presented a comprehensive overview of the long-term performance of GFRP, CFRP, and BFRP in concrete structures [50]. It was found that there was no substantial degradation of the GFRP bars in concrete structures for six months in marine environmental conditions. The GFRP bars and reinforced concrete structures did not present chemical degradation and loss of tensile strength after accelerated aging. However, the long-term performance of these composites must be investigated. Using steel fiber-reinforced polymer sheets to repair existing cracks on US150 bridges in Nelson County, USA, is a perfect example of using FRP's high strength and stiffness in structural applications [51]. The SFRP sheets were applied over the bridge girders' affected surfaces by a special mixture of epoxy and concrete.

2.3.2 Applications in Off-Shore Marine Applications

One of the main challenges of SFRP for off-shore marine applications is the extended periods of exposure to chloride salts, winds, and elevated temperatures [52]. Therefore, for the SFRP to be reliable, it must show anticorrosive-related properties without compromising functionality. Water vessels, off-shore structures like oil platform rigs, and other marine structural components are sensitive to corrosion due to salts and corrosive seawater [53, 54]. We have already discussed using SFRP in concrete structural components and their behavior in off-shore environments. Other applications of SFRP are mechanical components and ship hull plates; GFRP, CFRP, and BFRP are widely used due to their unmatched mechanical, tribological, and electrochemical properties [53].

GFRP composites are used in marine crafts, naval ships, hulls, and even submarines [48]. The authors Kootsookos and Mouritz [48], studied the performance, moisture absorption, degradation, and mechanical and electrochemical behavior of GFRP and CFRP in a polyester matrix in seawater after being immersed for a prolonged time [48]. It was found that the initial thickness of ~1.5–1.6 mm of the composites after being immersed in seawater conditions led to a mass and thickness change in the composites. An initial mass change was observed, and they attributed it to the water absorption for both composites. Moreover, the authors attributed the increase in weight for the GFRP to the chemical reaction between the fibers and the solution and the leaching of alkali elements. However, the CFRP was found not to absorb as much water as the GFRP, making it more resistant to moisture and the degradation effects of water.

The US Navy performed a clear application of FRP for ships and crafts structures, where low-cost and high-performance composite ship structures were made [55]. Critchfield et al. [55], employed a vacuum-assisted resin transfer molding process (RTM) to fabricate deckhouse structural modules. The RTM process is a closed-mold process under a high-pressure vacuum. The deckhouse was made of E-glass and isophthalic polyester resin. Moreover, the authors discussed using a single open molding process known as SCRIMP, where the resin distribution and fiber content reach up to 70%, with a small void content. The subsequent studies led the authors to conclude that manufacturing deckhouse panels is more economically feasible and has higher imparted qualities using the SCRIMP instead of conventional room temperature cure methods.

2.3.3 Application in the Oil and Gas Industry

Researchers over the past decade have studied composites as a viable alternative to conventional materials with higher mechanical, tribological, and, more importantly, electrochemical properties. The gas and oil industry employs pipelines to transport their products, often in their gaseous or liquid state, making them prone to corrosion. The lead cause of pipeline failure is attributed to the degradation of the pipeline material by corrosion, stress, and atmospheric conditions, leading to cracks and subsequent failure of the pipes. Natural reinforced fiber polymers (NRFP) are not an ideal choice for the reinforcement of pipelines, as they are known to be susceptible to water retention due to their inherent properties and structures, making them not a viable alternative for gas and liquid transportation (i.e., oil, water, and gases). Moreover, SFRP is known to be resistant to most alkali and acid-corrosive environments compared to the NFRP [56]. Nowadays, pipelines are composed of a metallic body, which is widely used and accepted in industrial applications; however, metals are severely attacked by corrosion from inside the pipe and in its outer frame. Furthermore, the metallic pipe is usually subjected to corrosion and abrasion due to the dragging of particles inside the pipe. In contrast, the pipe is exposed to atmospheric conditions in its outer frame, leading to external corrosion.

One of the most common methods to prevent the atmospheric corrosion of the pipes is by galvanic protection by applying metallic coatings. However, metallic coatings limit their protection role to the pipelines due to their discontinuity or surface damage. A new approach to protect oil and gas pipes from external and internal corrosion and abrasion is using FRP, more precisely SFRP, as coatings for the metallic pipes, as they offer a cost-effective solution and are lightweight and resistant to corrosion. FRP is widely used in the chemical and petroleum industries [57]. Some works related to pipeline corrosion protection by employing SFRP were conducted by Alabtah et al. [57]. They comprehensively studied the electrochemical and corrosion behavior of GRFP and AGRP with an epoxy resin matrix. The use of CFs and AFs for the study was done based on their superior properties compared to CFs, as they showed galvanic corrosion in the mechanically joined steels and CFRP. The authors employed a hand lay-up process for the preparation of the specimens, and their respective electrochemical characterizations (i.e., potentiodynamic polarization and electrochemical impedance spectroscopy) were performed in 0.5 M NaCl, HCl, H_2SO_4 solutions, resembling mild and severe atmospheric conditions. The authors found that the GFRP performed better than the AFRP in acidic solutions, showing better corrosion resistance. Moreover, the authors concluded that the presence of pores in the steel/GFRP and steel/AFRP led to the anodic dissolution of the steel, subsequently forming iron hydroxide products (i.e., $Fe(OH)_2$) and subsequent dissociation into iron oxides (i.e., Fe_2O_3, and Fe_3O_4). The failure of the protective coating was attributed to the deterioration of the FRP with time and the pile-up of oxide products in the porous structure of the already weakened epoxy.

For any given composite material (i.e., GFRP, AFRP, and CFRP), preventing damage and degradation of the composites ensures the components' long-term stability and useful life in critical environments. The involved corrosion mechanism in the composite is still not fully understood. Several authors have established that the fibers offer improved corrosion resistance to the epoxy resin matrix by homogenizing the existing structure, minimizing the appearance of cracks and their propagation, and increasing the polymeric matrix's intrinsic properties [58]. Wei et al. [59], studied the electrochemical performance of basalt fiber (BFs) and GFs (E-fiber) in NaOH and HCL solutions [59]. The authors found that the mass loss in acidic solution by increasing the treatment time was higher for the GFs. However, a reverse trend was observed in the case of the alkaline solution, in which the GFs outperformed the BFs. They suggested that the composite corrosion resistance may be improved by controlling the iron ions (i.e., Fe^{2+}) content during the anodic dissolution of the steel, which will lead to an increase in the stability of the composite in corrosive media [43, 58]. Zhang et al. [60] studied the service life of a transportation oil pipe made from a polyvinyl chloride (PVC) inner layer reinforced by a GFR thermosetting pipe, which failed during service in the Tarim Oilfield in China. The authors performed DSC, FTIR, and TG, among other characterizations, to determine the cause of failure for the PVC/GFR. They attributed the failure to a lower degree of cure for the GFR, which led to a decrease in the glass-transition temperature for the composite. Moreover, the PVC presented a non-uniform additive distribution at the matrix, possibly due to insufficient mixture before the extrusion process [60].

We have discussed the benefits and advantages of GFs, CFs, and AFs and some of their applications in the oil and gas industries. However, any engineered component is often subjected to stresses, which can lead to fracture damage in the composite materials. GFRP under corrosive media and stress can lead to a phenomenon known as stress-corrosion cracking, which produces severe failure in real-life components. Qi et al. [56] prepared a sandwich structure of GFs and CFs and carried out stress-corrosion experiments on GFRP composites with the pre-existence of a crack and submerged in corrosive media of 0.1 mol of HCl. The authors concluded that by lowering the iron ions concentration that arises from the anodic dissolution of the steel, better stability and increased corrosion-related properties would be enhanced.

3 Challenges and Future Perspectives

As discussed above, FRPCs offer diverse applications across industries, providing lightweight, high-strength, corrosion-resistant, and durable solutions. While FRPCs offer numerous advantages, they face challenges, including interfacial mismatches, limited performance at elevated temperatures, and manufacturing constraints.

Addressing these challenges requires ongoing research and development efforts to reduce production costs, standardize manufacturing processes, and enhance environmental sustainability through improved recycling methods. Furthermore, developing FRPCs with electrical conductivity and fire resistance remains an active pursuit.

Hybrid synthetic fiber-reinforced polymer (HFRP) composites emerge as a promising avenue for future applications, offering multifunctional capabilities such as environmental sensing and self-healing. Additionally, ensuring the collaboration between industry and academia is crucial to overcome FRPCs' limitations and promote their continued growth in various industrial sectors.

The potential areas of future studies in FRPCs are:

Modern Manufacturing Methods: Advancements in FRPCs' production can be achieved through the development of innovative manufacturing techniques such as additive manufacturing, automated fiber placement, and nanotechnology. These approaches promise to enhance the efficiency, precision, and adaptability of FRPCs' manufacturing processes.

Multifunctional FRPs: FRPCs can be engineered to possess diverse features, including environmental sensing, self-healing capabilities, and energy harvesting. Future research can explore the multifunctional potential of FRPCs, investigating how these capabilities can be harnessed in various applications beyond their current uses.

Fire Retardancy: The suitability of FRPCs for use in construction and automotive applications is contingent on their fire resistance. Future research endeavors can concentrate on the development of FRPCs' formulations that offer improved fire resistance and greater flame retardant properties, enhancing their safety in fire-prone environments.

Biomaterials in FRPCs: To reduce the environmental impact of FRPCs, sustainable and recyclable materials can be incorporated into their composition. Future research may focus on the creation of FRPCs derived from natural fibers and biodegradable materials, aligning with sustainability objectives and reducing their carbon footprint.

These potential areas of future studies underline the continuous evolution and significance of FRPCs in addressing a wide range of challenges across industries. By exploring these avenues, researchers can unlock the full potential of FRPCs, ushering in new possibilities for construction, corrosion protection, environmental sustainability, and safety.

4 Conclusion

In conclusion, this comprehensive exploration of fiber-reinforced polymer composites (FRPCs) highlights their pivotal role in the realm of materials science and engineering, particularly in the context of corrosion resistance. Our journey through these composite materials has unveiled two distinct categories: Natural Fiber-Reinforced Polymer Composites (NFRPCs) and Synthetic Fiber-Reinforced Polymer Composites (SFRPCs), each offering unique attributes and applications.

NFRPCs, rooted in sustainability and harnessed from nature's own fibers, exemplify the eco-conscious evolution of composite materials. Fabricated through techniques such as extrusion and compression molding, NFRPCs demonstrate remarkable corrosion resistance. Natural fibers not only serve as formidable barriers against corrosive forces in diverse environments but also contribute to the structural integrity of the composites, showcasing the synergy between sustainability and functionality.

On the other hand, SFRPCs represent a realm of innovation, strength, and adaptability. With synthetic fibers like glass, carbon, and aramid at their core, these composites undergo manufacturing processes such as extrusion, spinning, and polymerization. Synthetic fibers emerge as corrosion-resistant champions, safeguarding against chemical attacks while enhancing the longevity and durability of engineered components. Their applications span across various industries, including off-shore marine environments, where they excel in enduring extreme conditions while maintaining their structural integrity and corrosion resistance.

As with any material, both NFRPCs and SFRPCs face unique challenges, such as interfacial mismatches and sustainability concerns. To overcome these obstacles and fully harness the potential of FRPCs, a collaborative effort between industry and academia is imperative. Together, we can pioneer advancements in materials engineering and ensure the continued growth of FRPCs as a sustainable, corrosion-resistant solution across diverse industrial sectors.

In this collective endeavor, we envision a future where FRPCs stand at the forefront of materials innovation, creating a world that not only prioritizes sustainability but also offers uncompromised resistance to corrosion. As we forge ahead, the journey continues, marked by a commitment to excellence and a dedication to shaping a more resilient and sustainable future through the power of fiber-reinforced polymer composites.

Declarations

Conflicts of Interest The authors confirm that there are no conflicts of interest associated with this work.

Ethical Compliance The authors affirm that this study adheres to all ethical guidelines.

Competing Interests No competing interests are declared by the authors.

Data and Material Availability Data and materials are not applicable to this study.

References

1. Arpitha GR, Sanjay MR, Yogesha B (2014) Review on comparative evaluation of fiber reinforced polymer matrix composites. Adv Eng Appl Sci An Int J 4(4):44–47
2. Malkapuram R, Kumar V, Singh Negi Y (2008) Recent development in natural fiber reinforced polypropylene composites. J Reinf Plast Compos 28(10):1169–1189. https://doi.org/10.1177/0731684407087759
3. Shukla DK, Kasisomayajula SV, Parameswaran V (2008) Epoxy composites using functionalized alumina platelets as reinforcements. Compos Sci Technol 68(14):3055–3063. https://doi.org/10.1016/J.COMPSCITECH.2008.06.025

4. Shokrieh MM, Kefayati AR, Chitsazzadeh M (2012) Fabrication and mechanical properties of clay/epoxy nanocomposite and its polymer concrete. Mater Des 40:443–452. https://doi.org/10.1016/J.MATDES.2012.03.008
5. Vigneshwaran S et al (2020) Recent advancement in the natural fiber polymer composites: a comprehensive review. J Clean Prod 277:124109. https://doi.org/10.1016/j.jclepro.2020.124109
6. Waghmare S, Shelare S, Aglawe K, Khope P (2022) A mini review on fibre reinforced polymer composites. Mater Today Proc 54:682–689. https://doi.org/10.1016/j.matpr.2021.10.379
7. Kerni L, Singh S, Patnaik A, Kumar N (2020) A review on natural fiber reinforced composites. Mater Today Proc 28:1616–1621. https://doi.org/10.1016/j.matpr.2020.04.851
8. Biagiotti J, Puglia D, Kenny JM (2004) A review on natural fibre-based composites-part I: structure, processing and properties of vegetable fibres. J Nat Fibers 1(2):37–68. https://doi.org/10.1300/J395V01N02_04
9. Joshi SV, Drzal LT, Mohanty AK, Arora S (2004) Are natural fiber composites environmentally superior to glass fiber reinforced composites? Compos Part A Appl Sci Manuf 35(3):371–376. https://doi.org/10.1016/J.COMPOSITESA.2003.09.016
10. Gupta A, Kumar A, Patnaik A, Biswas S (2011) Effect of different parameters on mechanical and erosion wear behavior of bamboo fiber reinforced epoxy composites. Int J Polym Sci 2011. https://doi.org/10.1155/2011/592906
11. Ratna Prasad AV, Mohana Rao K (2011) Mechanical properties of natural fibre reinforced polyester composites: Jowar, sisal and bamboo. Mater Des 32(8–9): 4658–4663. https://doi.org/10.1016/J.MATDES.2011.03.015
12. Pulikkalparambil H, Rangappa SM, Siengchin S, Parameswaranpillai J (2021) Introduction to epoxy composites. Epoxy Compos Fabr Charact Appl 1–21. https://doi.org/10.1002/9783527824083.ch1
13. Parameswaranpillai J et al (2022) Tribological behavior of natural fiber-reinforced polymeric composites. INC. https://doi.org/10.1016/B978-0-323-90748-4.00014-5
14. Mohanty AK, Misra M, Drzal LT (2002) Sustainable bio-composites from renewable resources: opportunities and challenges in the green materials world. J Polym Environ 10(1–2):19–26. https://doi.org/10.1023/A:1021013921916/METRICS
15. Zaleha M, Shahruddin M, Idris MI (2012) A review on the mechanical and physical properties of natural fiber composites. Appl Mech Mater 229–231:276–281. https://doi.org/10.4028/WWW.SCIENTIFIC.NET/AMM.229-231.276
16. George J, Sreekala MS, Thomas S (2001) A review on interface modification and characterization of natural fiber reinforced plastic composites. Polym Eng Sci 41(9):1471–1485. https://doi.org/10.1002/PEN.10846
17. Zhao X et al (2020) Fabrication of composite coatings with core-shell nanofibers and their mechanical properties, anticorrosive performance, and mechanism in seawater. Prog Org Coatings 149:105893. https://doi.org/10.1016/J.PORGCOAT.2020.105893
18. Hawary E et al (2023) An overview of natural fiber composites for marine applications. J Mar Sci Eng 11(5):1076. https://doi.org/10.3390/JMSE11051076
19. Gopu GN, Joseph SA (2022) Corrosion behavior of fiber-reinforced concrete—a review. Fibers 10(5):38. https://doi.org/10.3390/FIB10050038
20. Karim MA, Abdullah MZ, Deifalla AF, Azab M, Waqar A (2023) An assessment of the processing parameters and application of fibre-reinforced polymers (FRPs) in the petroleum and natural gas industries: a review. Results Eng 18:101091. https://doi.org/10.1016/J.RINENG.2023.101091
21. Thyavihalli Girijappa YG, Mavinkere Rangappa S, Parameswaranpillai J, Siengchin S (2019) Natural fibers as sustainable and renewable resource for development of eco-friendly composites: a comprehensive review. Front Mater 6:481024. https://doi.org/10.3389/FMATS.2019.00226/BIBTEX
22. Pickering KL, Efendy MGA, Le TM (2016) A review of recent developments in natural fibre composites and their mechanical performance. Compos Part A Appl Sci Manuf 83:98–112. https://doi.org/10.1016/J.COMPOSITESA.2015.08.038

23. Prasad V, Suresh D, Joseph MA, Sekar K, Ali M (2018) Development of flax fibre reinforced epoxy composite with nano Tio_2 addition into matrix to enhance mechanical properties. Mater Today Proc 5(5):11569–11575. https://doi.org/10.1016/J.MATPR.2018.02.125
24. Al-Maharma AY, Al-Huniti N (2019) Critical review of the parameters affecting the effectiveness of moisture absorption treatments used for natural composites. J Compos Sci 3(1):27. https://doi.org/10.3390/JCS3010027
25. Rashed Labanieh A, Placet V, Grzegorz Kossakowski P, Wci W (2022) Fiber-reinforced polymer composites in the construction of bridges: opportunities, problems and challenges. Fibers 10(4):37. https://doi.org/10.3390/FIB10040037
26. Qureshi J (2023) A review of fibre reinforced polymer bridges. Fibers 11(5):40. https://doi.org/10.3390/FIB11050040
27. Colomer-Romero V, Rogiest D, García-Manrique JA, Crespo JE (2020) Comparison of mechanical properties of hemp-fibre biocomposites fabricated with biobased and regular epoxy resins. Mater (Basel) 13(24):1–8. https://doi.org/10.3390/MA13245720
28. Shahzad A (2012) Hemp fiber and its composites-a review. J Compos Mater 46(8):973–986. https://doi.org/10.1177/0021998311413623
29. Adeniyi AG, Onifade DV, Ighalo JO, Adeoye AS (2019) A review of coir fiber reinforced polymer composites. Compos Part B Eng 176:107305. https://doi.org/10.1016/J.COMPOSITESB.2019.107305
30. Gudayu A, Steuernagel L, Meiners I, Rotich G (2019) Sisal fiber reinforced polymeric composites: a review. Int J Text Eng Process 5(3):15–21
31. Dinesh Kumar Raju S, Balasubramanian M, Arul Jeya Kumar A (2020) Experimentalstudies on replacement of steel stirrups by sisal fiber reinforced polymers. IOP Conf Ser Mater Sci Eng 912(6). https://doi.org/10.1088/1757-899X/912/6/062058
32. Farzana M, Maraz KM, Sonali SN, Hossain MM, Alom MZ, Khan RA (2022) Properties and application of jute fiber reinforced polymer-based composites. GSC Adv Res Rev 11(1):084–094. https://doi.org/10.30574/gscarr.2022.11.1.0095
33. Pradeepa B, Malathi M, Kiruthika AV (2023) Properties of jute fibre reinforced polymer composites–a review. Eur J Eng Sci Technol 6(1):10–29. https://doi.org/10.33422/EJEST.V6I1.1052
34. Shahinur S, Alamgir Sayeed MM, Hasan M, Sayem ASM, Haider J, Ura S (2022) Current development and future perspective on natural jute fibers and their biocomposites. Polym (Basel) 14(7). https://doi.org/10.3390/POLYM14071445
35. Deshpande AP, Bhaskar Rao M, Rao CL, Extraction of bamboo fibers and their use as reinforcement in polymeric composites. https://doi.org/10.1002/(SICI)1097-4628(20000404)76:1
36. Li M et al (2020) Recent advancements of plant-based natural fiber–reinforced composites and their applications. Compos Part B Eng 200:108254. https://doi.org/10.1016/J.COMPOSITESB.2020.108254
37. Vroman I, Tighzert L (2009) Biodegradable polymers. Materials 2(2):307–344. https://doi.org/10.3390/MA2020307
38. Bank LC, Gentry TR, Barkatt A (1995) Accelerated test methods to determine the long-term behavior of FRP composite structures: environmental effects. J Reinf Plast Compos 14(6):559–587. https://doi.org/10.1177/073168449501400602
39. Alabtah FG, Mahdi E, Eliyan FF (2021) The use of fiber reinforced polymeric composites in pipelines: a review. Compos Struct 276(August):114595. https://doi.org/10.1016/j.compstruct.2021.114595
40. Navaratnam S, Selvaranjan K, Jayasooriya D, Rajeev P, Sanjayan J (2023) Applications of natural and synthetic fiber reinforced polymer in infrastructure: a suitability assessment. J Build Eng 66(January):105835. https://doi.org/10.1016/j.jobe.2023.105835
41. Rajak DK, Wagh PH, Linul E (2022) A review on synthetic fibers for polymer matrix composites: performance, failure modes and applications. Mater (Basel) 15(14). https://doi.org/10.3390/ma15144790

42. Rajak DK, Pagar DD, Menezes PL, Linul E (2019) Fiber-reinforced polymer composites: manufacturing, properties, and applications. Polym (Basel) 11(10). https://doi.org/10.3390/polym11101667
43. Scheffler C, Förster T, Mäder E, Heinrich G, Hempel S, Mechtcherine V (2009) Aging of alkali-resistant glass and basalt fibers in alkaline solutions: evaluation of the failure stress by Weibull distribution function. J Non Cryst Solids 355(52–54):2588–2595. https://doi.org/10.1016/j.jnoncrysol.2009.09.018
44. Ku H, Wang H, Pattarachaiyakoop N, Trada M (2011) A review on the tensile properties of natural fiber reinforced polymer composites. Compos Part B Eng 42(4):856–873. https://doi.org/10.1016/j.compositesb.2011.01.010
45. Yue C, Sui G, Looi H (2000) Effects of heat treatment on the mechanical properties of Kevlar-29 fibre. Compos Sci Technol 60(3):421–427. https://doi.org/10.1016/S0266-3538(99)00137-2
46. Zhu D, Mobasher B, Rajan SD (2011) Dynamic tensile testing of Kevlar 49 fabrics. J Mater Civ Eng 23(3):230–239. https://doi.org/10.1061/(ASCE)MT.1943-5533.0000156
47. Gadve S, Mukherjee A, Malhotra SN (2010) Corrosion protection of fiber-reinforced polymer-wrapped reinforced concrete. ACI Mater J 107(4):349–356. https://doi.org/10.14359/51663860
48. Kootsookos A, Mouritz AP (2004) Seawater durability of glass- and carbon-polymer composites. Compos Sci Technol 64(10–11):1503–1511. https://doi.org/10.1016/j.compscitech.2003.10.019
49. Ogbonna VE, Popoola API, Popoola OM, Adeosun SO (2022) A review on corrosion, mechanical, and electrical properties of glass fiber-reinforced epoxy composites for high-voltage insulator core rod applications: challenges and recommendations. Polym Bull 79(9):6857–6884. https://doi.org/10.1007/s00289-021-03846-z
50. Ahmed A, Guo S, Zhang Z, Shi C, Zhu D (2020) A review on durability of fiber reinforced polymer (FRP) bars reinforced seawater sea sand concrete. Constr Build Mater 256:119484. https://doi.org/10.1016/j.conbuildmat.2020.119484
51. Choo CC, Peiris A, Harik IE (2013) Repair using steel fiber reinforced polymer on US 150 bridges
52. Gellert EP, Turley DM (1999) Seawater immersion ageing of glass-fibre reinforced polymer laminates for marine applications. Compos Part A Appl Sci Manuf 30(11):1259–1265. https://doi.org/10.1016/S1359-835X(99)00037-8
53. Rubino F, Nisticò A, Tucci F, Carlone P (2020) Marine application of fiber reinforced composites: a review. J Mar Sci Eng 8(1). https://doi.org/10.3390/JMSE8010026
54. Kiratli S (2023) Marine applications of fiber-reinforced polymer matrix composites sakine, pp 68–77
55. Critchfield MO, Judy TD, Kurzweil AD (1994) Low-cost design and fabrication of composite ship structures. Mar Struct 7(2–5):475–494. https://doi.org/10.1016/0951-8339(94)90036-1
56. Qi C, Lin S, Yan X (2023) Damage calculation method for stress corrosion crack of glass fiber reinforced polymer composites. Compos Struct 306(June 2022):116575. https://doi.org/10.1016/j.compstruct.2022.116575
57. Alabtah FG, Mahdi E, Eliyan FF, Eltai E, Khraisheh M (2021) Towards the development of novel hybrid composite steel pipes: Electrochemical evaluation of fiber-reinforced polymer layered steel against corrosion. Polym (Basel) 13(21). https://doi.org/10.3390/polym13213805
58. Abidin NIZ, Sabri MFM, Kalantari K, Afifi AM, Ahmad R (2018) Corrosion detection for natural/synthetic/textiles fiber polymer composites. Struct Health Monit Biocomposites, Fibre-Reinf Compos Hybrid Compos 93–112. https://doi.org/10.1016/B978-0-08-102291-7.00006-X
59. Wei B, Cao H, Song S (2010) Tensile behavior contrast of basalt and glass fibers after chemical treatment. Mater Des 31(9):4244–4250. https://doi.org/10.1016/j.matdes.2010.04.009
60. Zhang X, Li H, Qi D, Li Q, Ding N, Cai X (2013) Failure analysis of anticorrosion plastic alloy composite pipe used for oilfield gathering and transportation. Eng Fail Anal 32:35–43. https://doi.org/10.1016/j.engfailanal.2013.03.001

Chapter 10
Short and Micro Cellulose-Based Environmentally Friendly Polymer Composites

Ishrat J. Badruddin, Sameer S. Rahatekar, and Shivam Tiwari

1 Introduction

In order to achieve the sustainability goals, there is a need to replace the widely used, well-known petrochemical-based monomers (non-renewable) with bio-sourced monomers (renewable sources) [118]. These renewable resources can be of plant and animal origin [116]. These renewable resources would help achieve the sustainability goal without compromising the resources needed for future generations. Also, due to their natural origin the naturally derived raw material could be replenished and contribute to the global biodiversity [18]. Therefore, there is a need to develop biomaterials that are of low cost, and comparable to the quality of the existing synthetic raw materials. Considering the global threat of climate change, it is imperative to be environmentally conscious. This sustainable development can be achieved by unlocking the potential of biological resources in a sustainable manner by being conscious of the manufacturing processes and techniques across all industries to obtain products in a sustainable manner [79].

One step towards achieving the sustainability goal would be to replace conventionally used synthetic polymers with biopolymers during the manufacturing of composites. The nature of the matrix material can be synthetic, non-biodegradable or biodegradable [13, 55, 115]. Non-biodegradable matrix materials remain intact for a prolonged time and are resistant to degradation with long-lasting durability. Biodegradable matrix materials, on the other hand, are obtained from natural biomass sources such as chitosan, cellulose and starch [10, 71]. Hybrid composites can be manufactured by reinforcing natural fibre into synthetic matrix materials, such as epoxy resins [63]. Resources can be efficiently used, and waste streams can be

I. J. Badruddin · S. S. Rahatekar · S. Tiwari (✉)
Composites and Advanced Materials Centre, School of Aerospace, Transport, and Manufacturing, Cranfield University, Bedfordshire MK43 0AL, UK
e-mail: myselfshivamtiwari@gmail.com

S. K. Sethi et al. (eds.), *Polymer Composites: From Computational to Experimental Aspects*, Materials Horizons: From Nature to Nanomaterials,
https://doi.org/10.1007/978-981-97-0888-8_10

managed in a better way by replacing fossil fuel-based resources with biomass for biocomposites. Green biocomposites are blends of matrix natural fibres and utilise renewable source for the matrix or reinforced fibres during their manufacturing [44, 109, 132]. These green biocomposites have improved stiffness, and are environmentally friendly, biodegradable, low weight, low cost and also can capture carbon from CO_2.

Fibres can be natural, semi-synthetic or synthetic origin depending on the origin and mode of manufacturing. The natural fibres for biocomposites production can be obtained through various natural sources such as bamboo, sisal, jute, jute, flax and hemp [91]. Natural fibres encompass all natural cellulosic fibres such as jute, hemp, coir, flax, cotton, sisal, ramie and protein-based fibres such as silk and wool. Some cellulosic fibres do not occur naturally instead are produced through chemical procedures from cotton, bamboo or pulp wood termed as manmade cellulose fibres such as cellulose acetate and viscose-rayon [48, 98]. Towards the approach of making the environment safer, researchers have been investigating the use of cellulose fibres as reinforcements or fillers instead of synthetic fibres during the reinforcement phase in composites due to their higher stiffness and strength, lower weight, lower cost, renewability and biodegradability [45, 66, 84].

1.1 Cellulose

Cellulose is a polysaccharide and the most abundantly available biopolymer on earth forming the basic structure of plant cell walls and can also be synthesised by microorganisms such as algae, bacteria and fungi [89]. The percentage content of cellulose from various sources is shown in Table 1.

Cellulose is hydrophobic, chiral, odourless, tasteless, biodegradable and renewable. Cellulose is also derived from cotton linters, ramie, hemp, pulp and flax. Cellulose structure consists of repeated glucose units linked together through glycosidic linkage forming long, stiff molecular chains [74]. The hydrogen bonding between the polymer chains provides crystalline configuration and stiffness to the structure. Cellulose is a linear carbohydrate long chain polymer chain of repeating units of β-(1 → 4)-linked d- glucopyranose units covalently bonded through the OH group of C1 and C4 carbon atoms. The repeating unit is cellobiose which is a dimer of glucose [3].

Table 1 Cellulose content (%) present in various natural sources

Cellulose source	Content (%)	References
Bacterial cellulose	>90	[119]
Mature cotton fibres	88–96.5	[93]
Plants	30–80	[134, 139]
Tunicate	60	[153]
Algae	8–47	[22, 147]

Three carbons in the monomeric cellulose unit consist of the -OH group which plays a significant role in the ultrastructure of cellulose, and modulation of the cellulose physicochemical properties due to their ability to form hydrogen bonds. Cellulose is a polysaccharide due to the presence of sugar monomers. Therefore, when hydrolysed, breakdown in glucose and reducing sugar. The molecular size of cellulose is represented as degree of polymerization (DP) which represents the number of anhydroglucose units per cellulose chain [60]. Cellulose chain DP ranges from 100 for MCC to 15,000 for native depending upon the mode of extraction. On the basis of the nature of the hydrogen bonding between the cellulose hydroxyl groups, cellulose exists as crystalline or allomorphic forms (I_α and I_β). I_α is predominantly obtained from plants whereas I_β is obtained from bacterial cellulose. The solubility of allomorphs depends upon the pattern of hydrogen bonding. Bacterial cellulose being abundant is advantageous over plant cellulose due to relatively easier extraction, production, greater mechanical strength, purity, flexibility and biodegradability. The key challenge related to the upscaling of cellulose production to commercial scale is the high processing cost. Researchers are exploring bacterial substrates that are cheap and are derived from water biomass for cellulose production in order to reduce the production cost. Other strategies for reduced production costs involve designing new fermentation vessels that are cost-effective with a new approach for static culturing and agitation.

Range of cellulose size can be obtained effectively through extraction using bottom-up approach where bacterial cellulose is obtained through bacteria (e.g. *Acetobacter Xylinam*) utilising glucose as a source or top-down approach which utilises agricultural residues, plants, cotton and wood for cellulose synthesis [39]. The properties and amount of produced cellulose depend on the origin, source and mode of extraction of the biomass. These sources are primarily composed of cellulose, lignin, hemicellulose and trace elements. In the matrix of soft hemicellulose and lignin, spirally oriented cellulose acts as reinforcements in the cell walls. Pure cellulose can be obtained through the removal process of lignin, hemicellulose and other trace elements. Due to the supramolecular structure, cellulose is insoluble in water and other organic solvents. However, its dissolution requires solvents such as ionic liquids which are expensive. Due to high stability, ionic liquids persist in the environment for a longer time and are poorly biodegradable [12]. The degree of toxicity of ionic liquid depends on the length of its side chain, and anionic and cationic moiety, and its effect varies from bacteria to high organisms [53, 125]. Imidazolium and pyridinium-based ionic liquids inhibit several enzyme activities such as acetylcholinesterase inhibition causing danger by distorting many neurological processes [49, 133]. Cellulose dissolution using toxic ionic liquids limits the application of cellulose for various applications.

Natural cellulose fibres are being used as reinforcements in biodegradable composites to replace glass fibres, and talc. Xi et al. investigated the use of cement composites reinforced with bamboo derived cellulose fibres showing increased flexural properties and impact energy absorption by the composites [145]. Cellulose Micro Filler (CMF) extracted from peanut oil cake improved the thermal and mechanical response of pineapple/flax natural fibre epoxy composites [128]. However, these cellulosic fibres' composites show typically lower strength as compared to glass fibre and carbon fibre composites. The reason is the incompatibility between the hydrophobic polymer matrix and the hydrophilic natural fibres reducing the thermal stability and interfacial bonding of the resultant composites [105]. Surface treatment using physicochemical modifications or chemical grafting of cellulose fibres is needed as the mechanical properties of cellulose fibres tend to reduce with moisture absorption. Liu et al. suggest chemical modification by removal of non-cellulosic components from natural fibres to improve the mechanical properties and thermal stability [88].

Natural fibres contribute to the plant in a greener way due to their CO_2 neutral origin [94]. The end applications of the resultant composites depend on the fibre properties which are dependent on various parameters such as fibre origin, mode of extraction, fibre treatment and the manufacturing method for fibre production. Cellulose fibres are a group of microfibrils that are generated through 20–200 chains of cellulose. These chains are asymmetric comprising a non-reducing and reducing ends [58]. The non-reducing end at position C4 has a supplementary OH group whereas the non-reducing end features a carbonyl functional group. Microcrystalline cellulose is one prominent class of cellulosic reinforcing agents that is widely employed for application in medical, defence, food, suspension stabiliser and cosmetic industries. The life cycle of MCC (Microcrystalline Cellulose) based composites from biomass sources until the final product is shown in Fig. 1. However, the challenge associated with the use of MCC fibres is due to its process of separation from natural sources and the inherent dispersibility, and incompatibility between the hydrophobic and non-polar polymer matrix with the hydrophilic and highly polar cellulose fibres [138].

1.2 Extraction Process

Scientists often integrate mechanical treatment with pretreatment strategies for cellulose nanomaterials extraction from wood fibres [107]. This approach is referred to as a top-down approach. These pretreatments are done to make the hydroxyl group of cellulose accessible, increasing the inner surface, break cellulose hydrogen bonds and alter the crystallinity of the fibres, making the fibres more reactive. Mechanical treatment has been the primary disintegration technique used to produce nanofibrillar cellulose (NFC) and microfibrillated cellulose (MFC). In addition to the high-pressure homogenisation processing technique, several kinds of mechanical treatment can be used to reduce the particle size of fibres to micro- or nano-scale. Cellulose nanoparticles can be extracted using mechanical, chemical and biological methods. These treatments include processes such as high-intensity ultra-sonication (HIUS), super-grinding, cryo-crushing, microfluidisation and steam explosion.

High-pressure homogenization method is widely employed for large-scale CNP or NFC production where cellulose pulp suspension is forced past a narrow channel under high impact using a piston where the fibre-water suspension is subjected to

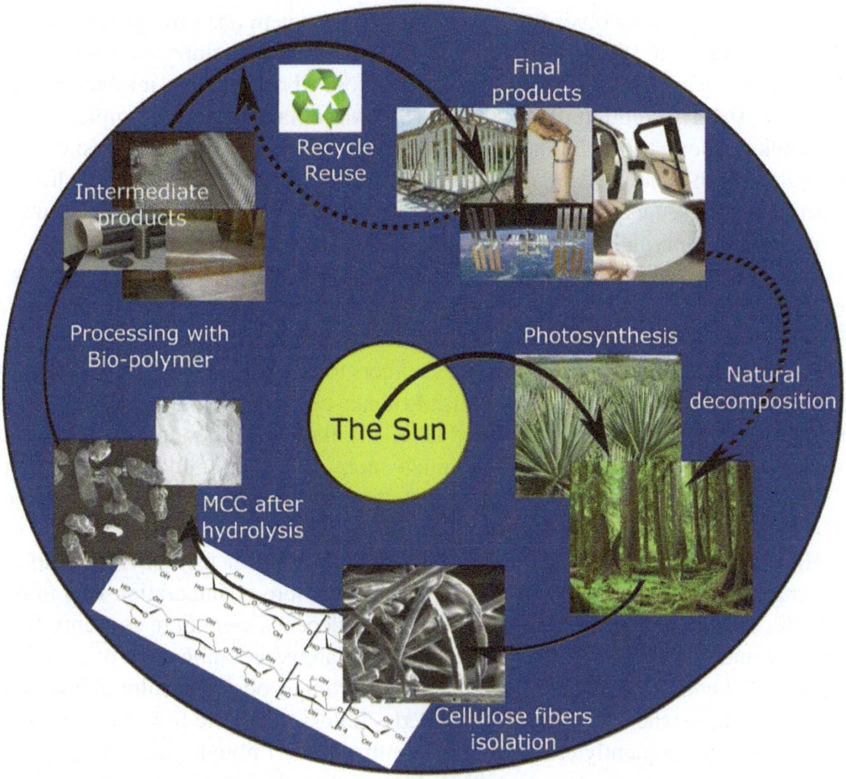

Fig. 1 Life cycle of microcrystalline cellulose-based biocomposites [138]

intense micronisation to disintegrate fibres into fibrils. Disintegration of fibres is promoted through frictional forces, turbulent flow and high mechanical shear forces. However, this process causes excess mechanical shearing forces to the CNP crystalline structure, consumes high energy and the disintegration of pulp fibres is incomplete [68, 107]. Removal of hemicellulose using hot water and lignin removal through green liquor followed by pretreatment of sugarcane bagasse and spruce using high-pressure homogenization improved the UV resistance and thermal stability of the cellulose nanofibres-based films as reported by [90].

Micro fluidization utilises a constant shear rate to produce homogenous CNF size distribution as compared to high-pressure homogenization [148]. However, the limitations associated with this technique are similar to high-pressure homogenisation. Fibre suspension is passed through a narrow Z-shape pressure treatment chamber where it is accelerated at high pressure with high impact and intense particle collision to separate fibrils off the fibre and sent through outlet [120]. Higher fibrillation degree can be obtained by reducing the size of the Z-shape constriction, which generates greater pressure in the chamber.

Grinding method involves the use of grindstones where one grinder is static and the other rotates at 1500 rpm through which the cellulose slurry is passed [80]. Due to the shear stress on the fibres, the cellulose fibres separate into nano-sized particles. A slurry solution of silk fibroin (SF) was obtained using a simple grinder from silk cocoons and cellulose nanofibres were used to prepare composite films through filtration and pressing methods. The CNF/SF nanofibres composite showed superior heat resistance, and mechanical properties in comparison to SF solution prepared by dissolving fibrions in concentrated neutral salt [103].

In the super-grinding process, the fibrous materials are thrust through the stator and rotor disc-gap in a rotating-grinding pattern [129]. "Supermass colloider" or ultra-fine friction grinder also known as Masuko grinder is an example of a super-grinder.

Refining is another approach that uses refiners to fibrillate pulp fibres into nanofibres [135]. Narrow gap-based specialised refiner can exhibit high-shearing forces on cellulosic fibre suspension, for efficient fibrillation. Processing of MFC and NFC through specialised refining is much more scalable for industrial applications as compared to steam and high-intensity ultra-sonication, microfluidisation, cryo-crushing and high-pressure homogenisation [73].

High-intensity ultra-sonication (HIUS) is another useful method for cellulose nanomaterials extraction utilising sound energy for fibre disintegration into fibrils [29, 96]. The fibrillation efficiency of the process depends upon fibre concentration, fibre size and distance, processing time and temperature. A combination of chemical pretreatment with HIUS disrupts cellulose fibres from wood or agricultural residues.

Cryocrushing where the moisture retained cellulosic fibres are frozen using liquid nitrogen and subsequently crushed into a combination of fibre fragments, fines and fibrils using high-impact grinding [27]. Making use of disintegrator, these cryo-crushed fibres are uniformly dispersed in water. Cryocrushing has infrequently been used for fibres disruption into fibrils. In a study conducted by [140], cryo-crushing using liquid nitrogen obtained nanofibrils of nanofibres with 30–100 nm diameter and several micrometres length [140]. Using the chemi-mechanical cryo-crushing technique, nanocellulosic material from agricultural residues such as soy hulls 20–120 nm width [140] and wheat straw 10–80 nm width can be used as strengthening additive in nanocomposite applications [9]. Diameter of nanofibrils extracted from kraft pulp fibres, flax bast fibres, hemp fibres and rutabaga fibres ranges from 5 to 60 nm when subjected to the cryo-crushing process [19].

2 Cellulose-Based Biocomposites Using Different Fillers/reinforcers

Biomass-derived cellulose be analysed on various levels of its structures, morphology such as porous, fibrous or macrofibrils; supramolecular level involving hydrogen bonds, microfibrils, amorphous and crystalline regions [3]. At the molecular level it can be based on the hydrogen bonds and glucan chains, and at dimensional scale

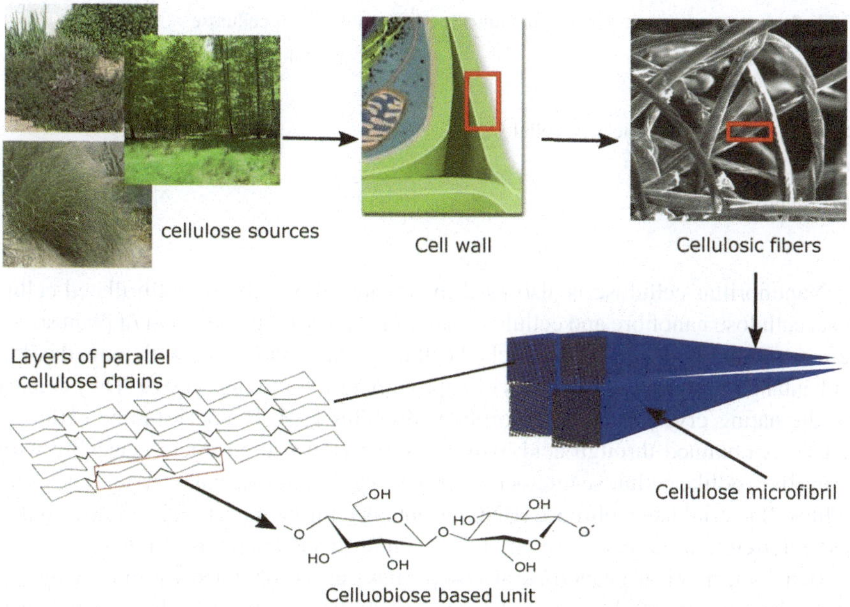

Fig. 2 Structural hierarchy of cellulose from biomass source to molecule [138]

range from nano-scale to macro-scale, allowing modifying the cellulose molecule through chemical modifications and making them amenable for potential application in filtration, paper making, films and coatings, nanocomposites, rheological modifiers, biomedical engineering and pharmaceuticals [3]. Nanocellulose shows the nano-scale range of 1–100 nm width or diameter of the cellulose fibril particles of a cellulosic material. Figure 2 shows a schematic hierarchical structure from cellulose biomass to molecular arrangements [138].

Based on the different hierarchical levels, nanocellulose can be cellulose nanostructured materials and cellulose nanofibres [65]. Cellulose nanostructured materials are further divided as microfibrillar cellulose (MFC) [or cellulose microfibril (CMF)] and microcrystalline cellulose (or cellulose microcrystal). Cellulose nanofibres are further divided as nanofibrillar cellulose (NFC) [or cellulose nanofibril (CNF)] and nanocrystalline cellulose (or cellulose nanocrystals or cellulose nanowhiskers). The structure of MFC and NFC appears to be spaghetti-like containing both amorphous and crystalline regions [40]. Significant differences appear across the particle size distribution of cellulose nanomaterials as evident through Image and Microscopy analysis [65]. Cellulose nanofibres have gained interest due to their high aspect ratio [100], good strengthening effect, mechanical properties, low thermal expansion [50]. Addition of cellulose nanofibres as biodegradable reinforcements in nanocomposites offers great potential for green nanocomposites materials [2]. Table 2 shows the young modulus value for the cellulose based on different kinds.

Table 2 A comparative of Young's modulus for different kinds of cellulose

Component	Young's modulus (GPa)	References
Biomass	20	[104]
Single fibre network of bacterial cellulose	78	[61]
Crystalline cellulose	167.5	[130]
Microfibril	100–130	[42]

Nanofibrillar cellulose is also used interchangeably with microfibrillated cellulose, cellulose nanofibre and cellulose nanofibril. It is the smallest unit of plant structure consisting of a bundle of stretched cellulose chain molecules with long, flexible and entangled cellulose nanofibres of approximately 1–100 nm in size. They consist of alternating crystalline and amorphous domains [23]. Nanocrystalline cellulose (NCC) is obtained through acid hydrolysis and is also used interchangeably with terms like rodlike cellulose microcrystals, whiskers, cellulose nanocrystals and crystallites. Bacterial nanocellulose (BNC) is obtained through bacterial cellulose and is also referred to as biocellulose, microbial cellulose and bacterial cellulose.

Better mechanical properties of composites can be obtained by improving the hydrophobicity of cellulose fibres through surface modification which would enhance the interface between fibres and matrix. Pre-treatments increase the roughness of the surface, prevent moisture absorption, chemically modify the surface and clean the fibre surface.

Appropriate selection or processing technique is required depending upon the desired product geometry, the cost and the ease of manufacture and performance needed. The fabrication method for natural fibre composites is similar to glass fibres composite manufacturing [66]. Natural fibre composites are prepared using preforms of nonwovens, fabrics or fibres [5, 36, 51, 70, 75]. Commonly composites are widely prepared using prepregs and manufacturing methods such as vacuum bagging, hand laminating, resin transfer moulding (RTM), compression moulding, injection moulding and pultrusion [20]. Preparation of composites is done using Eq. 1 where V_f is the volume fraction of the fibres, W_f and W_m is the weight of the fibres, and the weight of the matrix, ρ_f and ρ_m are the density of the fibre and the matrix [110].

$$V_f = W_f/\rho_f\left(W_f/\rho_f\right) + (W_m/\rho_m) \tag{1}$$

Due to the presence of lignocellulose, which contain strong polarised hydroxyl groups, natural fibres are hydrophilic in nature rendering the fibres incompatible with other popular well-known resins resulting in an ineffective interface between resin and fibres [124]. Only some thermosets such as the phenol–formaldehyde and related polymers are less hydrophilic and thus less problematic.

2.1 *Pre-treatments of Composites*

Physical treatments modify the surface and structure of the fibres and thereby improve the mechanical bonding to polymers [21]. Cold plasma treatment and corona treatment use electric discharge to change the surface energy of the cellulose fibres through surface oxidation activation to increase the fibre matrix adhesion. Chemical treatment involves silylation, mercerization, peroxide, benzoylation and graft copolymerizations. Mercerization is an alkali treatment that uses a strong base aqueous solution for fibre treatment through swelling which modifies the dimensions, morphology, structure and mechanical properties of the fibres. Amongst the most effective methods of chemical treatment is graft copolymerization where cellulose is treated with selected ions and aqueous solution under high energy radiation [69]. When exposed to radiation, radicals form as a result of cellulose molecule crack formation allowing the formation of copolymers with suitable compatible matrix solution. Better adhesion of matrix and fibres is achieved during graft copolymerisation of vinyl monomers with natural fibres [69]. Improved adhesion may be achieved by using alkaline solution or coupling agents like maleic anhydride to the fibres.

Coupling agents are materials with intermediate properties that help bring compatibility between two incompatible materials [83]. Coupling agents work by reacting to the functional groups of the matrix, and hydroxyl group of cellulose, by transferring the stress between matrix and fibre. Examples of coupling agents include triazine, maleic-anhydride (MAH) and organosilanes. Improved mechanical performance, flexural strength and increased interfacial adhesion were obtained for jute/epoxy composites using—silane as a coupling agents [21]. Other chemical modifications for fibre/matrix adhesion involve acetylation, isocyanate treatment and treatment with stearic acid.

Different surface properties of cellulose can be obtained through tailoring of the reactive hydroxyl (–OH) group of cellulose using an oxidation process. 2,2,6,6-tetramethypiperidine-1-oxyl (TEMPO)-mediated oxidation method is one of the most established chemical pretreatment methods for catalytic carbohydrate oxidation. This method is also used in organic synthesis including the oxidation of sulphides, alcohols and organometallic compounds [50, 126].

2.1.1 Alkaline Extraction

In order to degrade lignin and improve the fibrils individualisation, fibre slurries are carefully treated through alkaline pretreatment preventing cellulose degradation before mechanical processing [69, 83, 140]. Alkaline pretreatment technique involves three procedures [3]. Improve the fibre swelling properties by immersing biomass fibres in conc. NaOH solution (12–18 wt%) for approximately 120 min making the hydrolysis more feasible. Secondly, dissolving hemicellulose at 60–80 °C using 1 M hydrochloric acid solution. Finally, disruption of lignin at 60–80 °C for 120 min placing the biomass fibres in sodium hydroxide (NaOH) solution.

2.1.2 Carboxymethylation

Carboxymethylation involves the addition of polymers such as carboxymethylated cellulose to enhance anionic charges (carboxyl groups) and decrease frictional forces between the fibres to ease the processing. Carboxymethylation pretreatment enables fibre disintegration into fibrils by inducing electrostatic repulsion between the fibres [102].

2.1.3 Enzymatic Pretreatment

An alternate strategies for chemical pretreatment such as TEMPO is the use of enzymatic pretreatment which improves the reactivity and accessibility of the fibre wall. Sub-groups of enzymes cellulase, A and B termed cellohydrolases, attack the crystalline cellulose region, whilst sub-groups C and D are called as endoglucanase attack the amorphous (disordered) cellulose region [107].

2.1.4 NCC Stabilisation and Surface Modification

Cellulose nanocrystals can be surface modified to make them compatible with hydrophobic thermoplastic matrices or improve their stability in organic media [129]. These methods include surface hydrophobization by esterification, carboxylation, silylation esterification or acylation, poly(ethylene glycol) (PEG), FITC labelling (FITC = fluorescein-5′-isothiocyanate), cationic surface functionalization and polymer grafting [59, 97]. An alternative to chemical surface modification is the adsorption of surfactants at the colloid surface to improve nanoparticle stability in organic solvents. This type of stabilisation has been used to stabilise cellulose nanocrystals with a surfactant [67]. A potential drawback to surface functionalization is the possibility that the distinctive properties of the NCC may be lost upon modification. The type of material sources and reinforcements used for composites manufacturing is shown in Table 3 [47].

2.2 *Biocomposites Fabrication Methods*

Conventional methods such as hand lay-up, compression moulding, injection, pultrusion, and extrusion are used to manufacture biocomposites [15, 16]. These manufacturing methods have been thoroughly developed over the years. PLA [99], soy protein [28] based thermoplastic matrix, and PBA [149], PVC [52, 54], copolymer of $LiClO_4$-doped ethylene oxide–epichlorohydrin suspension and latex matrices [106] can be used to manufacture NCC composites using extrusion or compression moulding. Biopolymers such as chitosan [10, 25, 76, 127] and regenerated cellulose [81, 151] can also be used to manufacture NCC composites using solution casting. Furfuryl

Table 3 Type of materials and reinforcement used for composite manufacturing for different applications

Material type	Reinforced/with	Application
Cellulose	Calcium solutions [117, 121]	Biomedical and environmental
Cellulose nanofibres	Poly(N-isopropylacrylamide) [152]	Aerogels
Carboxymethyl cellulose	Cellulose nanofibers [4]	Coating and packaging
Hydroxypropyl cellulose	Poly(acrylic acid) [1]	Textile printing
	Microcrystalline cellulose [41]	Food packaging
Hydroxypropylmethyl cellulose	Silanol (Si) groups [142]	Oil–water separation
Hydroxyethyl cellulose	Hyaluronic acid [77]	Hydrogels for transdermal delivery systems
Bacterial cellulose	Alginate/gelatine [31]	Biomedical
	Polyvinyl alcohol [95]	
Xylan hemicellulose	Chitosan [24]	Bone tissue regeneration
Hemicellulose	Nanocellulose [86]	Wound healing
Hemicellulose galactoglucomannan	Polylactic acid [146]	3D printing
Cellulose	Poly(methylmethacrylate) PMMA [85]	Nanocomposites
Nanocrystalline cellulose	Polysulfone [14]	
Cellulose fibre	Polypropylene PP [113]	Lightweight aerospace and automotives
Cellulose nanocrystals	Carbon fibre/epoxy [112]	

alcohol can be used for NCC incorporation through in situ polymerization technique [111].

Researchers are focusing towards modifying existing techniques to improve the quality and make biocomposites more cost-effective. Through dip coating and spin coating, PAH-based NCC composites can be manufactured [34, 35, 64]. One of the novel composite production systems nowadays relies on additive manufacturing which is capable to manufacture a three-dimensional printed model of polymer through layer-by-layer fabrication [57, 72, 108]. The choice of fabrication techniques depends on the required reinforcement orientation, aspect ratio and fibre dispersion in specific applications These techniques do not require additional tools and allow for the creation of intricate, lightweight composite structures. Additionally, they enable the production of multi-material composites, which traditional manufacturing techniques are incapable of achieving. [56]. The selection of technique and composite manufacturing process is dependent on various factors such as process limitations, raw material properties, production costs, final design, shape and size. The various MFC sources, matrix materials, preparation method and Young's modulus are shown in Table 4.

Table 4 MFC sources, matrix materials, preparation method and Young's modulus of the final composite

Source of MFC	Matrix material	Mode of preparation	Young's modulus (GPa)	Ref
Sugar-beet cellulose	Pheno-formaldeyde resin	High-pressure homogeniser	9.5	[82]
Opuntia ficus-indica parenchyma cell	Copolymerization of styrene, butyl acrylate, acrylic acid and acrylamide,	Chemical treatment followed by homogenisation	0.034	[92]
Potato tuber	Potato starch and glycerol	Chemical and mechanical treatment	7	[43]
Sisal	Polycaprolactone	Chemical	0.6	[122]

3 Applications

The growing demands for the environment-friendly materials have been highly explored in various fields of applications. Figure 3 shows some of the potential applications of the cellulose-based composites. Some of the applications of cellulose biocomposites have been discussed briefly below.

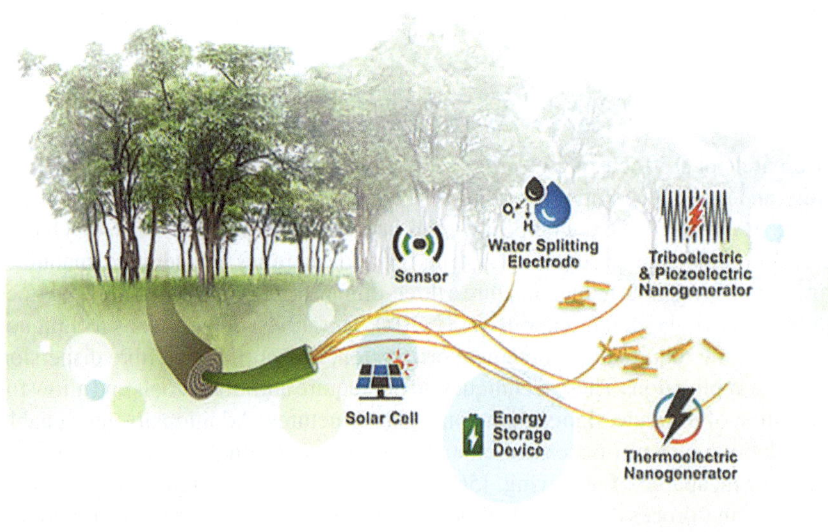

Fig. 3 The development of environment-friendly cellulose-based composites for various applications [131]

3.1 Automobile Application

There has been a growing trend for the use of biocomposites' materials as primary materials in automotive applications to manufacture new or future cars. Biocomposites as compared to non-renewable composites in the automotive industry offer several advantages such as lightweight, superior thermal and acoustic properties, reduced fuel consumption and reduced greenhouse gas emissions which makes them suitable for use in vehicle non-structural interior components such as dashboards, headliners, seat backs, seat fillers, interior panels, thermoacoustic insulation and wood trim [30, 37]. Incorporation of biopolymers in cars improves fuel efficiency due to reduced weight and provides increased resistance to external and heat impact [114]. Ford utilised polypropylene and agricultural residue wheat straw for the fabrication of cabin storage bin which helped generate extra income for farmers and reduce pollution [7]. Currently, Audi uses polyurethane and flax/sisal natural fibres for door trim panels, also some of its car models use cellulose-based biocomposites to manufacture door panels, seatbacks and boot lid finish panels [26]. Polyurethane foam reinforcement using kenaf and hemp fibre showed potential lightweight yet load-bearing applications in cars using sustainable materials.

The conventional automobile industry is dominated by the use of glass and carbon fibre as reinforcements in composites. To address environmental concerns, renewable alternatives are required to reduce petroleum-based composites' carbon footprints [101]. The use of biocomposites in automobiles could cut down production costs and reduce overall weight thereby improving the fuel efficiency. Furthermore, biocomposites are also known to reduce noise and vibrations through damping [6]. Various natural fibres such as coir, flax, hemp, sisal, kenaf and jute are being employed to manufacture biocomposites for automobiles. Different components such as seat pads, armrests, bumpers, door panels, cup holders, trunk covers and headrests can be made using biocomposites. Mercedes-Benz uses sisal-based composites for rear panel shelves, flax fibre composites for shelves and trunk covers and jute-based biocomposites for interior panels [78]. When compared to synthetic polymer, the use of biocomposites reduced the energy consumption by 80% and the weight by roughly 10%. Similarly, Volkswagen uses biocomposites to make door inserts, door panels, package trays and flap linings. Toyota used PP/PLA biocomposites inside toolbox areas, trims and package trays, soy foams for a vehicle seat and kenaf fibres in tyre covers [46, 137]. Ford uses hemp fibre composites in the front grills, bio-based cushions and soy foam seats in its various vehicle models [11].

3.2 Energy Applications

Utilisation of the natural polymer-based composites for energy-based applications is following an upward trend due to their biocompatibility and abundant renewable sources. Cellulose has marked its diverse features in different energy sectors such as energy harvesting and solar energy applications.

Energy harvesting from different sources which are renewable and abundant in nature and goes waste due to unfavourable attention towards it, is the beneficial sector to harness productive energy from it. Energy harvesting from abundant solar energy and waste mechanical sources is one of the major areas which is currently being progressed towards beneficial energy generation for a sustainable future.

3.2.1 Piezoelectric/Triboelectric Energy Harvesting

Cellulose provides a cheap and biodegradable option with substantial piezoelectric/triboelectric properties, which can be effective for energy harvesting applications. Alam et al. described a PENG made of multi-walled carbon nanotubes as conducting filler, native cellulose microfibres and PDMS. Under repetitive hand punching, the PENG produced an open-circuit voltage of 30 V and a power density of 9.0 W/cm^3, which was sufficient to power a large number of LEDs or other mobile electronics like a wristwatch, calculator or liquid–crystal display [8].

The crystallinity of cellulose is often greater than that of ordinary polymers. The hydroxyl groups on the surface of cellulose can create organised hydrogen bonds with other polymers, enhancing the directional dipole and performance of a piezoelectric system [123]. In recent work, cotton was used as a substrate to develop a polymer-based composite device for efficient piezoelectric energy harvesting applications. The cellulose containing cotton-PVDF composite hybrid showed better output as compared to pristine polymer and effectively generated substantial energy through different human motion modes like finger tapping, bending and foot tapping [136].

Strong, washable and biodegradable cellulose-based conductive macrofibres were created by wet-twisting and wet-stretching bacterial cellulose hydrogel combined with carbon nanotubes and polypyrrole (BC/CNT/PPy). The cellulose-based conductive macrofibres demonstrated good electrical conductivity of 5.32 S/cm and high tensile strength of 449 MPa, whilst still being able to preserve structural stability in water. Additionally, a degradation experiment showed that the cellulose-based macrofibres could be broken down in a cellulase solution in just 108 h. A fabric-based triboelectric nanogenerator (TENG) that was developed has a maximum output power of 352 W, which is sufficient to drive industrial electronics [62].

Zhang et al. created an all-cellulose energy harvesting device that used bacterial cellulose that was both pure and conductible for the friction layers and electrodes. By surface alteration using polypyrrole and carbon nanotubes, conductive bacterial cellulose might be created due to the widely dispersed -OH groups and the porous cellulose network topology. The created sandwich-structured device was capable of powering industrial electronics with a maximum output power of 3 W, open-circuit voltage of 29 V and a short-circuit current of 0.6 A. Their work played a sewable electronic piano and the online game "Greedy Snake" to show that precise fingertip control is possible. Additionally, the carbon nanotubes that were still present in the all-cellulose composites could be recycled after they had entirely broken down in 8 h [150].

3.2.2 Solar Energy/Organic Photovoltaics

Solar energy is one of the most abundant energy available to almost every section of the world. The energy generated from the solar rays is of high advantage when transformed into productive output. In order to effectively convert solar energy, a variety of materials, including silicon and group III–V elements, have been employed in solar cells. However, it is highly expensive to produce solar cells using inorganic materials. A successful conversion of solar energy to electrical output requires high surface area and better charge transport characteristics of the material. Being abundant, renewable and sustainable in nature, biomaterials are now being investigated for application in solar cells.

According to a recent study, organic solar cells can use cellulose nanopaper made from biomass—such as tobacco stalks or agroforestry residues—to trap light and capture it in a wide-angle fashion. Because cellulose has a different refractive index than other materials, the coexistence of its naturally crystalline and amorphous areas makes it easier to harvest obliquely incoming light and achieve wide-angle absorption. The device, which was made of cellulose nanopaper, displayed successful wide-angle harvesting and achieved a power-conversion efficiency of 16.17%. The combined effect of light-scattering and the cellulose's unique refractive index, along with the efficient light trapping, were attributed to the improved efficiency and capture angle [143].

Carboxymethyl cellulose sodium derived from rice straws of agroforestry residues can be applied as a cathode interlayer for organic solar cells in an inverted structure [144]. In this device, carboxymethyl cellulose sodium was engaged as an interfacial modifier with zinc oxide (ZnO) for the electrons transfer and collection. Benefiting from the co-interface layer that can reduce work function, increase absorption and better interfacial contact, the power-conversion efficiency of the device was remarkably improved to 12.01%, which is 9.4% higher than that of pristine ZnO-based organic solar cells. Importantly, studies have shown that cellulose-based solar cells can be separated and recycled into their major components, thus opening the door for a sustainable solar cell recycling industry [155].

With the use of CNC and silver, Zhou et al. developed a recyclable solar cell with a semi-transparent electrode that made a substantial contribution to the field of recyclable solar cells. A 2.7% conversion efficiency was discovered with these solar cells [154].

Bella et al. created DSSC with the goal of replacing petroleum-derived polymeric matrices with environmentally friendly jellifying agents like carboxymethylcellulose (CMC) and resolving the issue of moisture/water contamination in solar cells. The DSSC's exhibited a photovoltaic performance of 0.72% when 5.5 wt% CMC was employed as an electrolyte and high stability even under ageing circumstances [17].

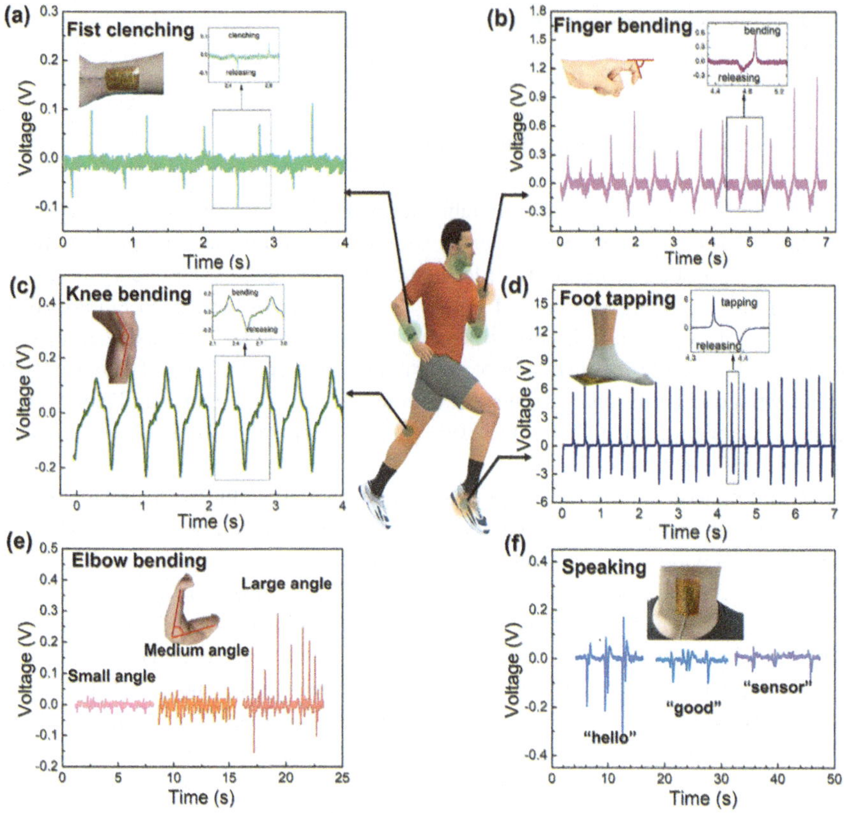

Fig. 4 Pictures displaying the sensing performance of the Bilayered pBT@CNF/PVDF-g-MA nanofibrous membrane for detecting human motions **a** fist clenching, **b** finger bending, **c** keen bending, **d** foot tapping during a walking movement, as well as **e** elbow bending at different angles and **f** speaking different words [141]

In a recent work, bilayered pBT@CNF/PVDF-g-MA nanofibrous membranes for detecting human motions were developed by Wang et al., which were used for energy generation using fist clenching, finger bending, keen bending, foot tapping during a walking movement, elbow bending at different angles and speaking different words as shown in Fig. 4 [141].

3.3 *Textiles/Wearables*

Cellulose has provided one of the better alternatives to the commercial textile material due to its good mechanical properties and biodegradability. In recent work,

Liu et al. (Fig. 5) carried out a new processing technique to develop cellulose-based multi-functional fibres for smart electronic-textiles applications. The approach involved the wet spinning of cellulose-based composite fibres using environmentally friendly alkali/urea solvents to breakdown cotton waste. Two co-continuous phases of polyaniline, a biocompatible intrinsically conductive polymer, were formed into fibres and placed side by side. A PANI-rich cellulose mix (50 wt. %) provided electrical conductivity, whereas neat cellulose (the substrate) provided high mechanical characteristics in the first phase. The side-by-side-drawn fibres have sufficient strength, flexibility and conductivities to be spun into yarns and work in a circuit to power an LED light. Additionally, they were created into a smart sensor for the detection of dangerous ammonia gas [87].

Natural dye has been incorporated in cellulose to develop composite fibres for potential textile applications. Coscia et al. manufactured cellulose-based fibres using the wet spinning method in the presence of ionic liquid. Natural curcumin was added as an additive to the cellulose solution in varying concentrations and its fibre properties were analysed. The dyed composite fibres (1 wt% curcumin) showed a tensile strength and modulus of 336 MPa and 16.2 Gpa, respectively, higher than the pristine cellulose fibres. The developed fibre showed its potential use in clothing/fashion applications [32]. Costa et al. prepared the dyed bacterial cellulose films using the plant-based dyes. In the work, natural pigments from *Clitoria ternatea L.* and *Hibiscus rosa-sinensis* were used with cellulose to develop composite films. Dyeing the cellulose films through an advanced process maintained the mechanical strength, crystallinity and thermal stability of the bacterial cellulose. The results suggested the usage of the developed material for textile and packaging applications [33]. A composite yarn developed by Darabi et al. from regenerated cellulose and PEDOT: PSS from ionic liquid showing better electrical conduction. The developed yarn material showed better washability and mechanical durability which points out to be a potential material for the e-textile applications [38].

4 Conclusions and Future Scopes

The consumption of the fossil fuels and petroleum-based materials has dominated over many decades but the alarming environmental effect due to their usage has led towards renewable and environment-friendly materials for a sustainable future. Cellulose is one of the most abundant natural materials obtained from different sources and possesses the characteristic features of renewability and biodegradability. The manufacturing of cellulose-based composites has been one of the major areas for researchers who find its usage in various fields of application. The cost-effective and biocompatible feature makes it most favourable in terms of sustainability and durability. These biocomposites have been well-explored in the automotive, energy and wearables applications providing better alternative to the conventional materials.

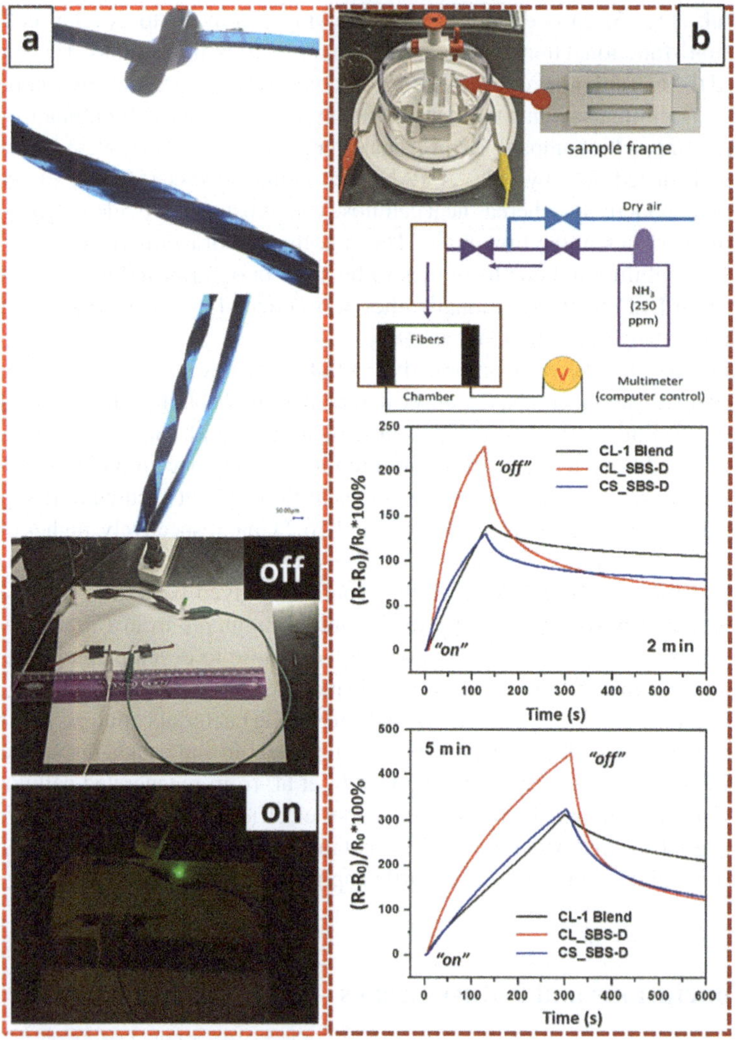

Fig. 5 **a** Representation of LED glowing through twisting and knotting of CS_SBS-D and its function as a circuit (twisted with the red polyester yarns); **b** The setting of ammonia sensing experiment (top) and the relative resistance (R-R0)/R0 of fibre samples in two different testing modes (bottom) [87]

Hence, cellulose-based composites are proving to be one of the prime materials for various engineering and manufacturing applications which can be a suitable alternative to traditional materials which are environmentally harmful in one or another way.

Declarations

Conflicts of Interest The authors confirm that there are no conflicts of interest associated with this work.

Ethical Compliance The authors affirm that this study adheres to all ethical guidelines.

Competing Interests No competing interests are declared by the authors.

Data and Material Availability Data and materials are not applicable to this study.

References

1. Abdel-Halim ES, Emam HE, El-Rafie MH (2008) Utilization of hydroxypropyl cellulose and poly(acrylic acid)-hydroxypropyl cellulose composite as thickeners for textile printing. Carbohydr Polym 74:938–941. https://doi.org/10.1016/j.carbpol.2008.05.013
2. Abdul Khalil HPS, Bhat AH, Ireana Yusra AF (2012) Green composites from sustainable cellulose nanofibrils: a review. Carbohydr Polym 87:963–979
3. Abdul Khalil HPS, Davoudpour Y, Islam MN et al (2014) Production and modification of nanofibrillated cellulose using various mechanical processes: a review. Carbohydr Polym 99:649–665
4. Abdulkhani A, Najd Mazhar A, Hedjazi S, Hamzeh Y (2020) Preparation of xylan biocomposite films reinforced with oxidized carboxymethyl cellulose and nanocellulose. Polym Bull 77:6227–6239. https://doi.org/10.1007/s00289-019-03075-5
5. Aisyah HA, Paridah MT, Sapuan SM et al (2021) A comprehensive review on advanced sustainable woven natural fibre polymer composites. Polymers (Basel) 13:1–45
6. Akampumuza O, Wambua PM, Ahmed A et al (2017) Review of the applications of biocomposites in the automotive industry. Polym Compos 38:2553–2569
7. Alam MA, Sapuan SM, Ya HH et al (2020) Application of biocomposites in automotive components: a review. In: Biocomposite and synthetic composites for automotive applications, pp 1–17
8. Alam MM, Mandal D (2016) Native cellulose microfiber-based hybrid piezoelectric generator for mechanical energy harvesting utility. ACS Appl Mater Interfaces 8:1555–1558. https://doi.org/10.1021/ACSAMI.5B08168/SUPPL_FILE/AM5B08168_SI_003.MPG
9. Alemdar A, Sain M (2008) Isolation and characterization of nanofibers from agricultural residues - wheat straw and soy hulls. Bioresour Technol 99:1664–1671. https://doi.org/10.1016/j.biortech.2007.04.029
10. Ambaye TG, Vaccari M, Prasad S et al (2022) Preparation and applications of chitosan and cellulose composite materials. J Environ Manag 301.https://doi.org/10.1016/j.jenvman.2021.113850
11. Andresen C, Demuth C, Lange A et al (2012) Biobased automobile parts investigation
12. Arendarczyk A, Zgórska A, Grabińska-Sota E (2011) Toxicity of 1-heksyl-3-methylimidazolium chloride according to selected marine and freshwater organisms
13. Ashothaman A, Sudha J, Senthilkumar N (2023) A comprehensive review on biodegradable polylactic acid polymer matrix composite material reinforced with synthetic and natural fibers. Mater Today Proc 80:2829–2839. https://doi.org/10.1016/j.matpr.2021.07.047
14. Bai H, Zhou Y, Zhang L (2015) Morphology and mechanical properties of a new nanocrystalline cellulose/polysulfone composite membrane. Adv Polym Technol 34.https://doi.org/10.1002/adv.21471
15. Bajpai PK, Ahmad F, Chaudhary V (2019) Processing and characterization of bio-composites. In: Handbook of ecomaterials, pp 123–139
16. Balla VK, Kate KH, Satyavolu J et al (2019) Additive manufacturing of natural fiber reinforced polymer composites: processing and prospects. Compos B Eng 174

17. Bella F, Galliano S, Falco M et al (2017) Approaching truly sustainable solar cells by the use of water and cellulose derivatives. Green Chem 19:1043–1051. https://doi.org/10.1039/C6GC02625G
18. Bhatia S (2016) Natural polymers vs synthetic polymer. In: Natural polymer drug delivery systems, pp 95–118
19. Bhatnagar A, Sain M (2005) Processing of cellulose nanofiber-reinforced composites. J Reinf Plast Compos 24:1259–1268. https://doi.org/10.1177/0731684405049864
20. Biswas S, Anurag J (2019) Fabrication of composite laminates. In: Reinforced polymer composites: processing, characterization and post life cycle assessment. John Wiley & Sons, Ltd., pp 39–53
21. Bledzki AK, Gassan J (1999) Composites reinforced with cellulose based fibres. Prog Polym Sci (Oxford) 24:221–274
22. Bogolitsyn K, Parshina A, Aleshina L (2020) Structural features of brown algae cellulose. Cellulose 27:9787–9800. https://doi.org/10.1007/s10570-020-03485-z
23. Brinchi L, Cotana F, Fortunati E, Kenny JM (2013) Production of nanocrystalline cellulose from lignocellulosic biomass: technology and applications. Carbohydr Polym 94:154–169. https://doi.org/10.1016/J.CARBPOL.2013.01.033
24. Bush JR, Liang H, Dickinson M, Botchwey EA (2016) Xylan hemicellulose improves chitosan hydrogel for bone tissue regeneration. Polym Adv Technol 27:1050–1055. https://doi.org/10.1002/pat.3767
25. Cabañas-Romero LV, Valls C, Valenzuela SV et al (2020) Bacterial cellulose-chitosan paper with antimicrobial and antioxidant activities. Biomacromol 21:1568–1577. https://doi.org/10.1021/acs.biomac.0c00127
26. Campilho RDSG (2017) Recent innovations in biocomposite products. In: biocomposites for high-performance applications: current barriers and future needs towards industrial development, pp 275–306
27. Chakraborty A, Sain M, Kortschot M (2005) Cellulose microfibrils: a novel method of preparation using high shear refining and cryocrushing. Holzforschung 59:102–107. https://doi.org/10.1515/HF.2005.016
28. Chan R, Lim LT, Barbut S, Marcone MF (2014) Extrusion and characterization of soy protein film incorporated with soy cellulose microfibers. Int Polym Proc 29:467–476. https://doi.org/10.3139/217.2866
29. Cheng Q, Wang S, Han Q (2010) Novel process for isolating fibrils from cellulose fibers by high-intensity ultrasonication. II. Fibril characterization. J Appl Polym Sci 115:2756–2762. https://doi.org/10.1002/app.30160
30. Cherradi Y, Rosca IC, Cerbu C, et al (2021) Acoustic properties for composite materials based on alfa and wood fibers. Appl Acoust 174. https://doi.org/10.1016/j.apacoust.2020.107759
31. Chiaoprakobkij N, Seetabhawang S, Sanchavanakit N, Phisalaphong M (2019) Fabrication and characterization of novel bacterial cellulose/alginate/gelatin biocomposite film. J Biomater Sci Polym Ed 30:961–982. https://doi.org/10.1080/09205063.2019.1613292
32. Coscia MG, Bhardwaj J, Singh N et al (2018) Manufacturing & characterization of regenerated cellulose/curcumin based sustainable composites fibers spun from environmentally benign solvents. Ind Crops Prod 111:536–543. https://doi.org/10.1016/J.INDCROP.2017.09.041
33. Costa AFdeS, de Amorim JDP, Almeida FCG et al (2019) Dyeing of bacterial cellulose films using plant-based natural dyes. Int J Biol Macromol 121:580–587.https://doi.org/10.1016/J.IJBIOMAC.2018.10.066
34. Cranston ED, Gray DG (2006) Morphological and optical characterization of polyelectrolyte multilayers incorporating nanocrystalline cellulose. Biomacromol 7:2522–2530. https://doi.org/10.1021/bm0602886
35. Cranston ED, Gray DG, Rutland MW (2010) Direct surface force measurements of polyelectrolyte multilayer films containing nanocrystalline cellulose. Langmuir 26:17190–17197. https://doi.org/10.1021/la1030729
36. Cristaldi G, Latteri A, Engineering GR-W Fabric (2010) Composites based on natural fibre fabrics. books.google.com

37. Dahal RK, Acharya B, Dutta A (2022) Mechanical, thermal, and acoustic properties of hemp and biocomposite materials: a review. J Compos Sci 6
38. Darabi S, Hummel M, Rantasalo S et al (2020) Green conducting cellulose yarns for machine-sewn electronic textiles. ACS Appl Mater Interfaces 12:56403–56412. https://doi.org/10.1021/ACSAMI.0C15399/ASSET/IMAGES/LARGE/AM0C15399_0008.JPEG
39. De Oliveira RL, Da Silva BH, De Assunção RMN et al (2011) Synthesis and characterization of microcrystalline cellulose produced from bacterial cellulose. J Therm Anal Calorim 106:703–709. https://doi.org/10.1007/s10973-011-1449-1
40. De Souza Lima MM, Borsali R (2004) Rodlike cellulose microcrystals: structure, properties, and applications. Macromol Rapid Commun 25:771–787
41. Dogan N, McHugh TH (2007) Effects of microcrystalline cellulose on functional properties of hydroxy propyl methyl cellulose microcomposite films. J Food Sci 72:E016–E022. https://doi.org/10.1111/J.1750-3841.2006.00237.X
42. Dufresne A (2013) Nanocellulose: a new ageless bionanomaterial. Mater Today 16:220–227
43. Dufresne A, Dupeyre D, Vignon MR (2000) Cellulose microfibrils from potato tuber cells: processing and characterization of starch-cellulose microfibril composites. J Appl Polym Sci 76:2080–2092. https://doi.org/10.1002/(SICI)1097-4628(20000628)76:14%3c2080::AID-APP12%3e3.0.CO;2-U
44. Edebali S (2020) Methods of engineering of biopolymers and biocomposites. In: Advanced green materials: fabrication, characterization and applications of biopolymers and biocomposites, pp 351–357
45. Eichhorn SJ, Dufresne A, Aranguren M et al (2010) Review: current international research into cellulose nanofibres and nanocomposites. J Mater Sci 45:1–33
46. Fan J, Nassiopoulos E, Brighton J et al (2011) New structural biocomposites for car applications. In: Society of plastics engineers - EUROTEC 2011 conference proceedings
47. Fernandes EM, Pires RA, Mano JF, Reis RL (2013) Bionanocomposites from lignocellulosic resources: properties, applications and future trends for their use in the biomedical field. Prog Polym Sci 38:1415–1441
48. Fischer S, Thümmler K, Volkert B et al (2008) Properties and applications of cellulose acetate. In: Macromolecular symposia, pp 89–96
49. Flieger J, Flieger M (2020) Ionic liquids toxicity—benefits and threats. Int J Mol Sci 21:1–41
50. Fukuzumi H, Saito T, Iwata T et al (2009) Transparent and high gas barrier films of cellulose nanofibers prepared by TEMPO-mediated oxidation. Biomacromol 10:162–165. https://doi.org/10.1021/BM801065U
51. Gairola SP, Tyagi YK, Gangil B, Sharma A (2020) Fabrication and mechanical property evaluation of non-woven banana fibre epoxy-based polymer composite. Mater Today: Proc 3990–3996
52. Ganster J, Fink HP (2006) Novel cellulose fibre reinforced thermoplastic materials. Cellulose 13:271–280. https://doi.org/10.1007/s10570-005-9045-9
53. Garcia MT, Gathergood N, Scammells PJ (2005) Biodegradable ionic liquids part II. Effect of the anion and toxicology. Green Chem 7:9–14
54. Gatenholm P, Bertilsson H, Mathiasson A (1993) The effect of chemical composition of interphase on dispersion of cellulose fibers in polymers. I. PVC-coated cellulose in polystyrene. J Appl Polym Sci 49:197–208. https://doi.org/10.1002/app.1993.070490202
55. Gatenholm P, Kubát J, Mathiasson A (1992) Biodegradable natural composites. I. Processing and properties. J Appl Polym Sci 45:1667–1677. https://doi.org/10.1002/app.1992.070450918
56. Gholampour A, Ozbakkaloglu T (2020) A review of natural fiber composites: properties, modification and processing techniques, characterization, applications. J Mater Sci 55:829–892
57. Giachini PAGS, Gupta SS, Wang W et al (2020) Additive manufacturing of cellulose-based materials with continuous, multidirectional stiffness gradients. Sci Adv 6.https://doi.org/10.1126/sciadv.aay0929

58. Giri J, Adhikari R (2012) A brief review on extraction of nanocellulose and its application. BIBECHANA 9:81–87. https://doi.org/10.3126/bibechana.v9i0.7179
59. Hakimi NMF, Hua LS, Chen LW et al (2021) Surface modified nanocellulose and its reinforcement in natural rubber matrix nanocomposites: a review. Polymers 13:3241. https://doi.org/10.3390/POLYM13193241
60. Heinze T, Petzold-Welcke K, van Dam JEG (2013) Polysaccharides: molecular and supramolecular structures. Terminology. In: The European polysaccharide network of excellence (EPNOE): research initiatives and results, pp 23–64
61. Hsieh YC, Yano H, Nogi M, Eichhorn SJ (2008) An estimation of the Young's modulus of bacterial cellulose filaments. Cellulose 15:507–513. https://doi.org/10.1007/s10570-008-9206-8
62. Hu S, Han J, Shi Z et al (2022) Biodegradable, super-strong, and conductive cellulose macrofibers for fabric-based triboelectric nanogenerator. Nanomicro Lett 14:1–20. https://doi.org/10.1007/S40820-022-00858-W/FIGURES/6
63. Jawaid M, Abdul Khalil HPS (2011) Cellulosic/synthetic fibre reinforced polymer hybrid composites: a review. Carbohydr Polym 86:1–18
64. Jean B (2014) Polymer–cellulose nanocrystals multilayer films, pp 105–119
65. Jonoobi M, Oladi R, Davoudpour Y et al (2015) Different preparation methods and properties of nanostructured cellulose from various natural resources and residues: a review. Cellulose 22:935–969
66. Joshi SV, Drzal LT, Mohanty AK, Arora S (2004) Are natural fiber composites environmentally superior to glass fiber reinforced composites? Compos Part A: Appl Sci Manuf 371–376
67. Kaboorani A, Riedl B (2015) Surface modification of cellulose nanocrystals (CNC) by a cationic surfactant. Ind Crops Prod 65:45–55. https://doi.org/10.1016/j.indcrop.2014.11.027
68. Kalia S, Boufi S, Celli A, Kango S (2014) Nanofibrillated cellulose: surface modification and potential applications. Colloid Polym Sci 292:5–31
69. Kalia S, Kaith BS, Kaur I (2009) Pretreatments of natural fibers and their application as reinforcing material in polymer composites-a review. Polym Eng Sci 49:1253–1272. https://doi.org/10.1002/pen.21328
70. Karahan M, Ozkan F, Yildirim K, Karahan N (2016) Investigation of the properties of natural fibre woven fabrics as a reinforcement materials for green composites. bibliotekanauki.pl. F& T in Eastern Europe
71. Karimi S, Dufresne AMd, Tahir P, et al (2014) Biodegradable starch-based composites: effect of micro and nanoreinforcements on composite properties. J Mater Sci 49:4513–4521.https://doi.org/10.1007/s10853-014-8151-1
72. Kaynak B, Spoerk M, Shirole A et al (2018) Polypropylene/cellulose composites for material extrusion additive manufacturing. Macromol Mater Eng 303.https://doi.org/10.1002/mame.201800037
73. Keijsers ERP, Yilmaz G, Van Dam JEG (2013) The cellulose resource matrix. In: Carbohydrate Polymers. pp 9–21
74. Klemm D, Heublein B, Fink H-P, Bohn A (2005) Cellulose: fascinating biopolymer and sustainable raw material. ChemInform 36.https://doi.org/10.1002/chin.200536238
75. Küçük M, Korkmaz Y (2012) The effect of physical parameters on sound absorption properties of natural fiber mixed nonwoven composites. Text Res J 82:2043–2053. https://doi.org/10.1177/0040517512441987
76. Kusmono WMW, Lubis FI (2021) Fabrication and characterization of chitosan/cellulose nanocrystal/glycerol bio-composite films. Polymers (Basel) 13.https://doi.org/10.3390/polym13071096
77. Kwon SS, Kong BJ, Park SN (2015) Physicochemical properties of pH-sensitive hydrogels based on hydroxyethyl cellulose–hyaluronic acid and for applications as transdermal delivery systems for skin lesions. Eur J Pharm Biopharm 92:146–154. https://doi.org/10.1016/J.EJPB.2015.02.025
78. La Mantia FP, Morreale M (2011) Green composites: a brief review. Compos Part A Appl Sci Manuf 42:579–588

79. Lange L, Connor KO, Arason S et al (2021) Developing a sustainable and circular bio-based economy in EU: by partnering across sectors, upscaling and using new knowledge faster, and for the benefit of climate, environment & biodiversity, and people & business. Front Bioeng Biotechnol 8
80. Lee H (2016) Preparation and chracterization of cellulose nanofibrils using various pretreament techniques 283
81. Lee TW, Jeong YG (2015) Regenerated cellulose/multiwalled carbon nanotube composite films with efficient electric heating performance. Carbohydr Polym 133:456–463. https://doi.org/10.1016/j.carbpol.2015.06.053
82. Leitner J, Hinterstoisser B, Wastyn M et al (2007) Sugar beet cellulose nanofibril-reinforced composites. Cellulose 14:419–425. https://doi.org/10.1007/s10570-007-9131-2
83. Li X, Tabil LG, Panigrahi S (2007) Chemical treatments of natural fiber for use in natural fiber-reinforced composites: a review. J Polym Environ 15:25–33
84. Liu D, Zhong T, Chang PR et al (2010) Starch composites reinforced by bamboo cellulosic crystals. Bioresour Technol 101:2529–2536. https://doi.org/10.1016/j.biortech.2009.11.058
85. Liu H, Liu D, Yao F, Wu Q (2010) Fabrication and properties of transparent polymethylmethacrylate/cellulose nanocrystals composites. Bioresour Technol 101:5685–5692. https://doi.org/10.1016/j.biortech.2010.02.045
86. Liu J, Chinga-Carrasco G, Cheng F et al (2016) Hemicellulose-reinforced nanocellulose hydrogels for wound healing application. Cellulose 23:3129–3143. https://doi.org/10.1007/s10570-016-1038-3
87. Liu W, Liu H, Zhao Z et al (2023) A novel structural design of cellulose-based conductive composite fibers for wearable e-textiles. Carbohydr Polym 321:121308. https://doi.org/10.1016/J.CARBPOL.2023.121308
88. Liu Y, Xie J, Wu N et al (2019) Characterization of natural cellulose fiber from corn stalk waste subjected to different surface treatments. Cellulose 26:4707–4719. https://doi.org/10.1007/s10570-019-02429-6
89. Loureiro dos Santos LA (2017) Natural polymeric biomaterials: processing and properties. In: Reference module in materials science and materials engineering
90. Luo L, Wang X, Zhang S et al (2021) Contribution of different pretreatments to the thermal stability and UV resistance performance of cellulose nanofiber films. Coatings 11:1–15. https://doi.org/10.3390/coatings11020247
91. Mahmud S, Hasan KMF, Jahid MA et al (2021) Comprehensive review on plant fiber-reinforced polymeric biocomposites. J Mater Sci 56:7231–7264
92. Malainine ME, Mahrouz M, Dufresne A (2005) Thermoplastic nanocomposites based on cellulose microfibrils from Opuntia ficus-indica parenchyma cell. Compos Sci Technol 65:1520–1526. https://doi.org/10.1016/j.compscitech.2005.01.003
93. Martínez-Sanz M, Pettolino F, Flanagan B et al (2017) Structure of cellulose microfibrils in mature cotton fibres. Carbohydr Polym 175:450–463. https://doi.org/10.1016/j.carbpol.2017.07.090
94. Miao M, Finn N (2008) Conversion of natural fibres into structural composites. J Text Eng 54:165–177
95. Millon LE, Wan WK (2006) The polyvinyl alcohol-bacterial cellulose system as a new nanocomposite for biomedical applications. J Biomed Mater Res B Appl Biomater 79:245–253. https://doi.org/10.1002/jbm.b.30535
96. Mishra SP, Manent AS, Chabot B, Daneault C (2012) Production of nanocellulose from native cellulose - Various options utilizing ultrasound. Bioresources 7:422–435. https://doi.org/10.15376/biores.7.1.422-436
97. Missoum K, Belgacem MN, Bras J (2013) Nanofibrillated cellulose surface modification: a review. Materials 6:1745–1766. https://doi.org/10.3390/ma6051745
98. Mitchell RL (1949) Viscose processing of cellulose. Ind Eng Chem 41:2197–2201. https://doi.org/10.1021/ie50478a033
99. Mokhena TC, Sefadi JS, Sadiku ER et al (2018) Thermoplastic processing of PLA/cellulose nanomaterials composites. Polymers (Basel) 10

100. Moon RJ, Martini A, Nairn J et al (2011) Cellulose nanomaterials review: structure, properties and nanocomposites. Chem Soc Rev 40:3941–3994. https://doi.org/10.1039/C0CS00108B
101. Mooney BP (2009) The second green revolution? Production of plant-based biodegradable plastics. Biochem J 418:219–232
102. Naderi A, Lindström T (2014) Carboxymethylated nanofibrillated cellulose: effect of monovalent electrolytes on the rheological properties. Cellulose 21:3507–3514. https://doi.org/10.1007/s10570-014-0394-0
103. Narita C, Okahisa Y, Yamada K (2019) A novel technique in the preparation of environmentally friendly cellulose nanofiber/silk fibroin fiber composite films with improved thermal and mechanical properties. J Clean Prod 234:200–207. https://doi.org/10.1016/j.jclepro.2019.06.215
104. Nishino T, Matsuda I, Hirao K (2004) All-cellulose composite. Macromolecules 37:7683–7687. https://doi.org/10.1021/ma049300h
105. Nurazzi NM, Asyraf MRM, Rayung M et al (2021) Thermogravimetric analysis properties of cellulosic natural fiber polymer composites: a review on influence of chemical treatments. Polymers (Basel) 13
106. Ofem M, Ene E, Ubi P et al (2020) Properties of cellulose reinforced composites: a review. Niger J Technol 39:386–402. https://doi.org/10.4314/njt.v39i2.9
107. Osong SH, Norgren S, Engstrand P (2016) Processing of wood-based microfibrillated cellulose and nanofibrillated cellulose, and applications relating to papermaking: a review. Cellulose 23:93–123. https://doi.org/10.1007/S10570-015-0798-5
108. Pattinson SW, Hart AJ (2017) Additive manufacturing of cellulosic materials with robust mechanics and antimicrobial functionality. Adv Mater Technol 2.https://doi.org/10.1002/admt.201600084
109. Phiri R, Mavinkere Rangappa S, Siengchin S et al (2023) Development of sustainable biopolymer-based composites for lightweight applications from agricultural waste biomass: a review. Adv Ind Eng Polym Res.https://doi.org/10.1016/j.aiepr.2023.04.004
110. Prakash KB, Fageehi YA, Saminathan R et al (2021) Influence of fiber volume and fiber length on thermal and flexural properties of a hybrid natural polymer composite prepared with banana stem, pineapple leaf, and S-glass. Adv Mater Sci Eng.https://doi.org/10.1155/2021/6329400
111. Pranger L, Tannenbaum R (2008) Biobased nanocomposites prepared by in situ polymerization of furfuryl alcohol with cellulose whiskers or montmorillonite clay. Macromolecules 41:8682–8687. https://doi.org/10.1021/ma8020213
112. Reale Batista MD, Drzal LT (2018) Carbon fiber/epoxy matrix composite interphases modified with cellulose nanocrystals. Compos Sci Technol 164:274–281. https://doi.org/10.1016/j.compscitech.2018.05.010
113. Reale Batista MD, Drzal LT, Kiziltas A, Mielewski D (2020) Hybrid cellulose-inorganic reinforcement polypropylene composites: lightweight materials for automotive applications. Polym Compos 41:1074–1089. https://doi.org/10.1002/pc.25439
114. Reddy TRK, Kim H-J, Park J-W (2016) Renewable biocomposite properties and their applications. In: Composites from renewable and sustainable materials
115. Rogovina SZ (2016) Biodegradable polymer composites based on synthetic and natural polymers of various classes. Polym Sci - Ser C 58:62–73. https://doi.org/10.1134/S1811238216010100
116. Kulkarni Vishakha S, Butte Kishor D, Rathod Sudha S (2012) Natural polymers – a comprehensive review. Int J Res Pharmaceut Biomed Sci 3:1597–1613
117. Salama A (2019) Cellulose/calcium phosphate hybrids: new materials for biomedical and environmental applications. Int J Biol Macromol 127:606–617
118. Schneiderman DK, Hillmyer MA (2017) 50th anniversary perspective: there is a great future in sustainable polymers. Macromolecules 50:3733–3749. https://doi.org/10.1021/acs.macromol.7b00293
119. Seddiqi H, Oliaei E, Honarkar H et al (2021) Cellulose and its derivatives: towards biomedical applications. Cellulose 28:1893–1931

120. Sethi S, Nayak SL, Arora B (2022) Microfluidization: a tool for improvement of techno-functional properties of foods. ACS Food Science and Technology 2:1433–1441. https://doi.org/10.1021/acsfoodscitech.2c00179
121. Shi RJ, Lang JQ, Wang T et al (2022) Fabrication, properties, and biomedical applications of calcium-containing cellulose-based composites. Front Bioeng Biotechnol 10:937266. https://doi.org/10.3389/FBIOE.2022.937266/BIBTEX
122. Siaueira G, Bras J, Dufresne A (2009) Cellulose whiskers versus microfibrils: influence of the nature of the nanoparticle and its surface functionalization on the thermal and mechanical properties of nanocomposites. Biomacromol 10:425–432. https://doi.org/10.1021/bm801193d
123. Song Y, Shi Z, Hu GH et al (2021) Recent advances in cellulose-based piezoelectric and triboelectric nanogenerators for energy harvesting: a review. J Mater Chem A Mater 9:1910–1937. https://doi.org/10.1039/D0TA08642H
124. Stanciu MD, Savin A, Nastac SM (2018) Mechanical and surface properties of lignocellulosic fibres reinforced composites. J Mech Eng 64:698–705. https://doi.org/10.5545/sv-jme.2018.5352
125. Stasiewicz M, Mulkiewicz E, Tomczak-Wandzel R et al (2008) Assessing toxicity and biodegradation of novel, environmentally benign ionic liquids (1-alkoxymethyl-3-hydroxypyridinium chloride, saccharinate and acesulfamates) on cellular and molecular level. Ecotoxicol Environ Saf 71:157–165. https://doi.org/10.1016/j.ecoenv.2007.08.011
126. Štolcová M, Šev A, Kaszonyi A (2009) Catalytic oxidation of bioglycerol to mono- and dicarboxylic acids. In: 44th international petroleum conference, pp 1–8
127. Strnad S, Zemljič LF (2023) Cellulose–chitosan functional biocomposites. Polymers (Basel) 15
128. Sumesh KR, Kanthavel K, Kavimani V (2020) Peanut oil cake-derived cellulose fiber: extraction, application of mechanical and thermal properties in pineapple/flax natural fiber composites. Int J Biol Macromol 150:775–785. https://doi.org/10.1016/j.ijbiomac.2020.02.118
129. Taniguchi T, Okamura K (1998) New films produced from microfibrillated natural fibres. Polym Int 47:291–294. https://doi.org/10.1002/(SICI)1097-0126(199811)47:3%3c291::AID-PI11%3e3.0.CO;2-1
130. Tashiro K, Kobayashi M (1991) Theoretical evaluation of three-dimensional elastic constants of native and regenerated celluloses: role of hydrogen bonds. Polymer (Guildford) 32:1516–1526. https://doi.org/10.1016/0032-3861(91)90435-L
131. Teng CP, Tan MY, Toh JPW et al (2023) Advances in cellulose-based composites for energy applications. Materials 16:3856
132. Thomas S, Balakrishnan P (2021) Green composites. Mater Horiz. https://doi.org/10.1007/978-981-15-9643-8
133. Thuy Pham TP, Cho CW, Yun YS (2010) Environmental fate and toxicity of ionic liquids: a review. Water Res 44:352–372
134. Thygesen A, Oddershede J, Lilholt H et al (2005) On the determination of crystallinity and cellulose content in plant fibres. Cellulose 12:563–576. https://doi.org/10.1007/s10570-005-9001-8
135. Tingaut P (2013) Potential applications of nanofibrillated cellulose from wood
136. Tiwari S, Devi A, Dubey DK, Maiti P (2023) Induced piezoelectricity in cotton-based composites for energy-harvesting applications. ACS Appl Bio Mater 6:1536–1545. https://doi.org/10.1021/ACSABM.2C01062/ASSET/IMAGES/LARGE/MT2C01062_0006.JPEG
137. Toyota (2023) Sustainability reporting. Discover Toyota, Toyota, UK. https://www.toyota.co.uk/discover-toyota/sustainability. Accessed 29 Oct 2023
138. Trache D, Hussin MH, Hui Chuin CT et al (2016) Microcrystalline cellulose: Isolation, characterization and bio-composites application—a review. Int J Biol Macromol 93:789–804
139. Ververis C, Georghiou K, Christodoulakis N et al (2004) Fiber dimensions, lignin and cellulose content of various plant materials and their suitability for paper production. Ind Crops Prod 19:245–254. https://doi.org/10.1016/j.indcrop.2003.10.006

140. Wang B, Sain M, Oksman K (2007) Study of structural morphology of hemp fiber from the micro to the nanoscale. Appl Compos Mater 14:89–103. https://doi.org/10.1007/s10443-006-9032-9
141. Wang L, Cheng T, Lian W et al (2022) Flexible layered cotton cellulose-based nanofibrous membranes for piezoelectric energy harvesting and self-powered sensing. Carbohydr Polym 275:118740. https://doi.org/10.1016/j.carbpol.2021.118740
142. Wang X, Xu S, Tan Y et al (2016) Synthesis and characterization of a porous and hydrophobic cellulose-based composite for efficient and fast oil-water separation. Carbohydr Polym 140:188–194. https://doi.org/10.1016/j.carbpol.2015.12.028
143. Wu J, Che X, Hu HC et al (2020) Organic solar cells based on cellulose nanopaper from agro-forestry residues with an efficiency of over 16% and effectively wide-angle light capturing. J Mater Chem A Mater 8:5442–5448. https://doi.org/10.1039/C9TA14039E
144. Wu J, Liu Y, Islam A et al (2020) From straw to device interface: carboxymethyl-cellulose-based modified interlayer for enhanced power conversion efficiency of organic solar cells. Adv Sci 7:1902269. https://doi.org/10.1002/ADVS.201902269
145. Xie X, Zhou Z, Yan Y (2019) Flexural properties and impact behaviour analysis of bamboo cellulosic fibers filled cement based composites. Constr Build Mater 220:403–414. https://doi.org/10.1016/j.conbuildmat.2019.06.029
146. Xu W, Pranovich A, Uppstu P et al (2018) Novel biorenewable composite of wood polysaccharide and polylactic acid for three dimensional printing. Carbohydr Polym 187:51–58. https://doi.org/10.1016/j.carbpol.2018.01.069
147. Zanchetta E, Damergi E, Patel B et al (2021) Algal cellulose, production and potential use in plastics: challenges and opportunities. Algal Res 56
148. Zeng J, Hu F, Cheng Z et al (2021) Isolation and rheological characterization of cellulose nanofibrils (CNFs) produced by microfluidic homogenization, ball-milling, grinding and refining. Cellulose 28:3389–3408. https://doi.org/10.1007/s10570-021-03702-3
149. Zepnik S, Hildebrand T, Kabasci S et al (2013) Cellulose acetate for thermoplastic foam extrusion. Cellul - Biomass Convers
150. Zhang J, Hu S, Shi Z et al (2021) Eco-friendly and recyclable all cellulose triboelectric nanogenerator and self-powered interactive interface. Nano Energy 89:106354. https://doi.org/10.1016/J.NANOEN.2021.106354
151. Zhang X, Liu X, Zheng W, Zhu J (2012) Regenerated cellulose/graphene nanocomposite films prepared in DMAC/LiCl solution. Carbohydr Polym 88:26–30. https://doi.org/10.1016/j.carbpol.2011.11.054
152. Zhang X, Wang Y, Zhao J et al (2016) Mechanically strong and thermally responsive cellulose nanofibers/poly(N-isopropylacrylamide) composite aerogels. ACS Sustain Chem Eng 4:4321–4327. https://doi.org/10.1021/ACSSUSCHEMENG.6B00814/SUPPL_FILE/SC6B00814_SI_001.PDF
153. Zhao Y, Li J (2014) Excellent chemical and material cellulose from tunicates: diversity in cellulose production yield and chemical and morphological structures from different tunicate species. Cellulose 21:3427–3441. https://doi.org/10.1007/s10570-014-0348-6
154. Zhou Y, Fuentes-Hernandez C, Khan TM et al (2013) Recyclable organic solar cells on cellulose nanocrystal substrates. Sci Reports 3:1–5. https://doi.org/10.1038/srep01536
155. Zhou Y, Khan TM, Liu JC et al (2014) Efficient recyclable organic solar cells on cellulose nanocrystal substrates with a conducting polymer top electrode deposited by film-transfer lamination. Org Electron 15:661–666. https://doi.org/10.1016/J.ORGEL.2013.12.018

Chapter 11
A Study on Machinability of High-Strength CFRP Composites: Influence of Coated Cutting Tools on Cutting Performance and Surface Quality

Rajneesh Raghav and Rahul S. Mulik

1 Introduction

The exceptional mechanical properties and lightweight nature of carbon fiber reinforced polymer composites, or CFRP composites, make them highly significant in a variety of industries. Because these materials are composed of carbon fibers incorporated in a polymer matrix, they are significantly stronger and lighter than traditional materials like steel or aluminum [1]. Industries like construction, sports equipment manufacturer, automotive, and aerospace all make extensive use of CFRP composites [2, 3]. Furthermore, the application of CFRP composites in seismic-resistant building and structural reinforcements demonstrates its versatility and importance in modern engineering and technology. Additionally, because of their adaptability and resistance to corrosion, they are helpful in marine and renewable energy applications. Overall, carbon fiber reinforced composites have revolutionized a variety of industries by offering state-of-the-art materials that optimize performance and sustainability.

Carbon fiber reinforced composites (CFRP) are challenging to machine due to a number of reasons. The main source of difficulty is the material's heterogeneous nature as the carbon fibers are embedded in the polymer matrix. As the carbon fibers are abrasive in nature, they can cause tool to wear out rapidly, reducing their lifetime and increasing the cost of machining. Furthermore, a phenomenon of separation of layers while machining known as delamination occurs in CFRPs, which can affect

R. Raghav (✉) · R. S. Mulik
Department of Mechanical and Industrial Engineering, Indian Institute of Technology Roorkee, Roorkee, Uttrakhand 247667, India
e-mail: rraghavtamta@me.iitr.ac.in

S. K. Sethi et al. (eds.), *Polymer Composites: From Computational to Experimental Aspects*, Materials Horizons: From Nature to Nanomaterials,
https://doi.org/10.1007/978-981-97-0888-8_11

the structural integrity of finished goods. Additionally, variations in material properties and orientation of fibers in the polymer also cause difficulties in achieving precise tolerances. To overcome these problems, specialized cutting tools, cutting-edge tactics, and adaptive machining processes are used to improve machinability and lower defects. Although CFRP composites have excellent mechanical characteristics, it's important to handle the challenges of their machining with caution and specialist techniques.

Utilizing tools coated by CVD or PVD, diamond-like carbon (DLC) or polycrystalline diamond (PCD) coatings is one of the key strategies for enhancing the machinability of CFRP composites. These coatings provide crucial barrier layers that significantly enhance the tool's performance and longevity. They prolong the life of the tool and guarantee consistent cutting quality by efficiently reducing friction, heat generation, and cutting-edge wear. Furthermore, these coatings can also improve the workpiece's surface finish and lessen the chance of burr formation. In summary, the use of coated cutting tools is necessary to minimize wear and prevent damage to the tool and the workpiece during the efficient and high-quality machining of CFRP composites.

This book chapter is a useful resource for researchers, engineers, and practitioners working on the machining of high-strength CFRP composites. Along with discussing the difficulties in machining CFRP, it also offers practical guidance on how to choose and enhance coated cutting tools to achieve higher cutting performance and surface quality. Ultimately, this chapter advances the application of CFRP-based solutions across numerous industries.

2 High-Strength CFRP Composites

Carbon fiber reinforced polymer (CFRP) composites with high strength are state-of-the-art materials that are recognized for their exceptional strength-to-weight ratio. Because these advanced composites may substantially reduce total weight while maintaining structural integrity, they are being used widely in a variety of industries. Because of their exceptional strength and lightweight nature, high-strength CFRP composites are essential for maximizing performance and fuel efficiency in a variety of industries. In this section, we will provide a brief description of CFRP's characteristics, composition, multitude of uses, and challenges with machining.

2.1 Composition and Properties

As the name implies, Carbon fibers are incorporated into a polymer resin to create carbon fiber reinforced polymer, or CFRP [4]. Carbon fibers make up the reinforcing portion of the composite. These fibers, which are renowned for their low weight and high tensile strength, are what give CFRP its remarkable strength characteristics.

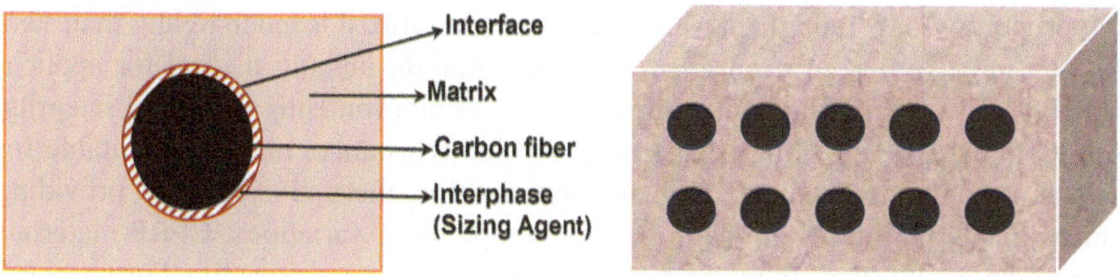

Fig. 1 Illustration of CFRP

The polymer matrix, often epoxy resin, acts as a binding agent, distributing loads among the carbon fibers and keeping them together. This matrix helps to protect the fibers from outside forces and influences while also giving the composite structure a great deal of flexibility. When these two components are properly blended, a material with exceptional mechanical properties, corrosion resistance, and lightweight characteristics is created. Because of this, CFRP is a good material to utilize in modern engineering applications. Figure 1 shown an illustration of CFRP composites.

The composition of CFRP composites consists of many necessary elements. The main reinforcing material is carbon fiber, which is composed of strong, very thin strands of carbon atoms. Typically, precursor compounds like pitch or polyacrylonitrile (PAN) are utilized to create these fibers. High-strength carbon fibers are created by aligning the carbon atoms in these materials and removing impurities through a variety of high-temperature operations. After that, a polymer matrix—which acts as a binding agent and facilitates the transmission of loads between the fibers—is mixed with the carbon fibers. The excellent mechanical strength, low weight, and exceptional adhesive properties of epoxy resins make them a popular option for matrix material. The process of impregnation, sometimes referred to as wet layup, is widely employed to create CFRP composites. Using this technique, liquid epoxy glue is saturating the carbon fibers. The resin undergoes a chemical transformation during the following stage, curing, which strengthens it and binds the fibers together.

A number of factors, including the fiber orientation, fiber volume fraction, and kind of polymer matrix employed, can be used to modify the composition of CFRP composite materials. Engineers may create CFRP materials with certain characteristics, such high tensile strength, low weight, and superior fatigue resistance, thanks to this customization. Moreover, different fillers and additives, such fire retardants, or nanoparticles, may be added to CFRP composites to improve particular qualities or satisfy criteria. Strength, stiffness, and other mechanical qualities can be optimized by adjusting the orientation and content of the carbon fibers inside the matrix, which makes CFRP composites an adaptable material with a multitude of uses. Strong, light carbon fibers combined with a resilient polymer matrix produce a material that is remarkably mechanically performant while still being relatively light, which makes it the material of choice for applications requiring high-performance materials.

High demand for CFRP composites in various sectors is because of their outstanding properties. CFRP combines the lightweight and high-strength characteristics of carbon fibers with the versatility and durability of polymers. One of its key properties is its remarkable strength-to-weight ratio, it is much lighter than steel but can have similar or even greater tensile strength, making it ideal for applications where weight reduction is crucial without compromising structural integrity. Additionally, CFRP exhibits excellent corrosion resistance, making it suitable for harsh environments. It boasts high stiffness and low thermal expansion, providing dimensional stability even under extreme temperature variations. CFRP materials are also known for their fatigue resistance, ensuring a long service life. Furthermore, CFRP's versatility allows for precise tailoring of its properties to meet specific application requirements, making it an indispensable material in modern engineering and technology.

2.2 Applications in Various Industries

The utilization of CFRP materials has a well-established history, and in the present century, their widespread adoption is evident across various structural domains, spanning from land to sky applications, encompassing everything from compact sports equipment to enormous spacecrafts [5]. Over the past 10 years, there has been a remarkable surge in the worldwide demand for CFRP (Carbon Fiber Reinforced Polymer). It is anticipated that from 2010 to 2022, this demand will exhibit a compound annual growth rate of approximately 11.98% [6]. Figure 2 depicts a simple overview of the structural applications of CFRP in different sectors. The most important applications are automobiles, aerospace, energy production, sporting equipments, textile and paper industry, civil engineering, acoustics, and portable power sources. CFRP finds extensive application in the automotive industry to enhance fuel efficiency, optimize vehicle performance, and above all, guarantee passenger safety. A diverse range of contemporary automobile parts, such as the chassis, suspension system, engine components, gearbox, driveshaft, hood, roof panels, side doors, and even wheels, are manufactured using CFRP materials [7]. A study [8] demonstrates that a 10% weight decrease can improve fuel efficiency by 6–8%.

The aerospace industry has increasingly turned to the utilization of CFRPs due to the urgent need to tackle various technical challenges, including but not limited to fatigue resistance, fracture resistance, static resistance, corrosion resistance, engine noise reduction, and fuel consumption reduction [9]. Aerospace components, such as aircraft doors, brakes, fasteners, fuel tanks, fuselages, cockpits, ribs, wings, landing gear systems, spoilers, keel beams, as well as tail assemblies, encompassing both horizontal and vertical stabilizers are developed by CFRP. Furthermore, CFRPs enhanced with nanofillers like graphene and carbon nanotubes are extensively employed in the production of wind turbine blades. These materials are favored for their exceptional durability against fractures and fatigue failure. Additionally, they offer the advantage of efficiently converting wind energy by altering their configuration without sacrificing their rigidity and structural integrity [9]. CFRPs also found their application in a wide range of sports equipment, including skiboards, rackets, fishing rods, golf clubs, bows and arrows, bicycles, baseball bats, yachts, ice hockey gear, and rowing

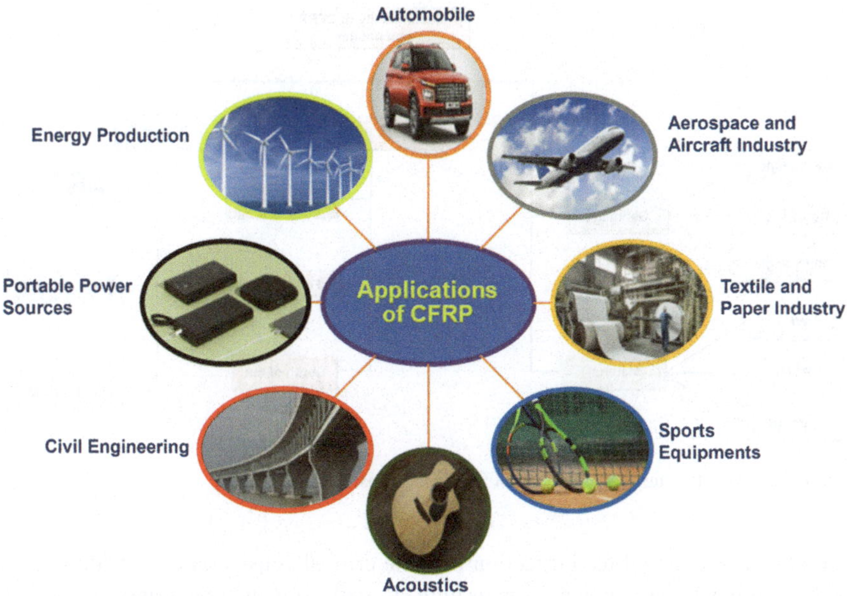

Fig. 2 Application areas of CFRPs

equipment [10]. Other than this CFRPs can be used in civil engineering structures, portable power sources, musical instruments, and several manufacturing industries.

2.3 Machining Challenges

Machining carbon fiber reinforced composites (CFRPs) presents distinct challenges compared to traditional metals and alloys machining [11]. This is primarily due to the unique characteristics of CFRP composites, including their anisotropic nature, non-uniform composition, textured structure, and limited thermal conductivity, which can pose significant difficulties in the machining process. Figure 3 lists the different factors affecting machinability of CFRP composites. Anisotropy and non-uniformity have an impact on the machining of FRP composites by influencing phenomena such as fiber delamination, pull-outs, peel-ups, and machining parameters [12]. CFRP composites tend to delaminate easily when subjected to machining forces, making it crucial to optimize cutting parameters such as feed rate, cutting speed, and tool geometry to minimize this issue. The presence of delamination in polymer composites resulted in a decrease in both their stiffness and strength, ultimately contributing to material failure. Delamination also amplifies the challenges associated with machining processes such as turning, drilling, and milling when working with polymer composites. Also, the anisotropic nature of CFRP composites, where the material properties vary in different directions, adds complexity to machining

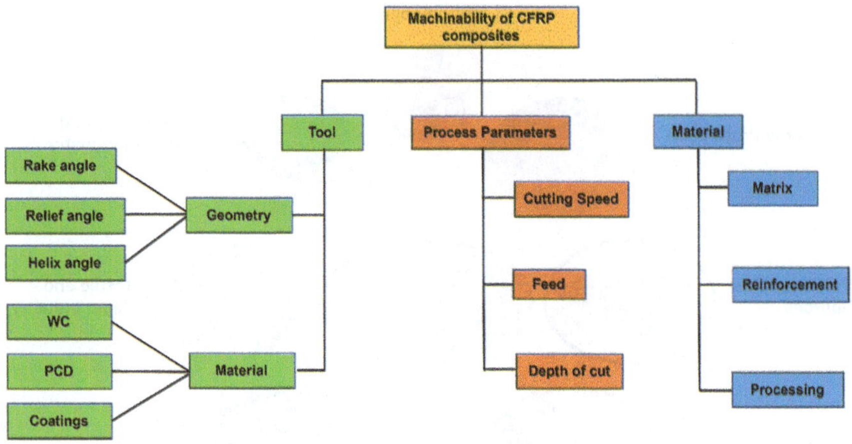

Fig. 3 Factors affecting machinability of CFRP composites

as tool orientation and feed direction become critical considerations. Additionally, CFRP composites are known for their high strength-to-weight ratio and exceptional stiffness, which can lead to rapid tool wear and increased cutting forces during machining. This necessitates the use of specialized cutting tools made from materials like polycrystalline diamond (PCD) or cubic boron nitride (CBN), which are expensive and require careful maintenance. Moreover, achieving tight tolerances and maintaining a high-quality surface finish can be particularly challenging while machining CERPs.

Dust and airborne particles generated during CFRP machining pose health hazards to operators due to the release of fine carbon fibers, which can cause respiratory issues if not adequately controlled [13]. Furthermore, achieving precise tolerances can be challenging because CFRP composites are prone to thermal expansion and dimensional instability during machining processes, necessitating careful monitoring and control of environmental conditions. The challenges in machining can be addressed by modifying the cutting tool's geometry, applying thin-film coatings to the cutting tool, and adjusting cutting parameters. In this book chapter thin-film coating effects on improving the machinability of CFRP composites have been discussed in order to effectively address these challenges and harness the benefits of CFRP composites in various industries, including aerospace and automotive.

3 Coated Cutting Tools: Types and Properties

Machining carbon fiber reinforced polymer (CFRP) composites can be challenging due to various factors discussed in the previous section. There are several ways in which machining challenges can be overcome. One of the techniques is to use coated cutting tools that can help improve the machining process by increasing tool life and reducing the risk of damage to the workpiece. This section discusses the role of coated cutting tools in machining CFRPs.

3.1 Overview of Cutting Tool Materials

Tool materials used for machining Carbon Fiber Reinforced Polymers (CFRPs) play a critical role in achieving precise and efficient machining processes. CFRPs are known for their high strength and lightweight properties, but they can be extremely abrasive and challenging to work with due to the abrasive nature of carbon fibers. As a result, tool materials need to be carefully selected to withstand the abrasive wear and provide good thermal conductivity. The cutting tools required for machining these materials should provide adequate levels of strength, toughness, and hardness to combat the high cutting loads. Also, thermal conductivity and thermal shock resistance become of great importance when cutting intermittently or at high cutting speeds, where the thermal load applied is significant. The hardness of a tool material plays a crucial role in its suitability as a cutting tool for machining processes. Figure 4 presents a comparative value of the relative hardness of different tool materials when compared to single-crystal diamond, which is renowned for its status as the hardest known material.

Commonly used tool materials for machining CFRPs include carbide, ceramics, polycrystalline diamond (PCD), and polycrystalline Cubic Boron Nitride (PCBN). Carbide tools are popular due to their toughness and cost effectiveness, but they may experience more significant wear in prolonged machining operations. Ceramics, such as silicon nitride, offer excellent resistance to heat and abrasion, making them suitable

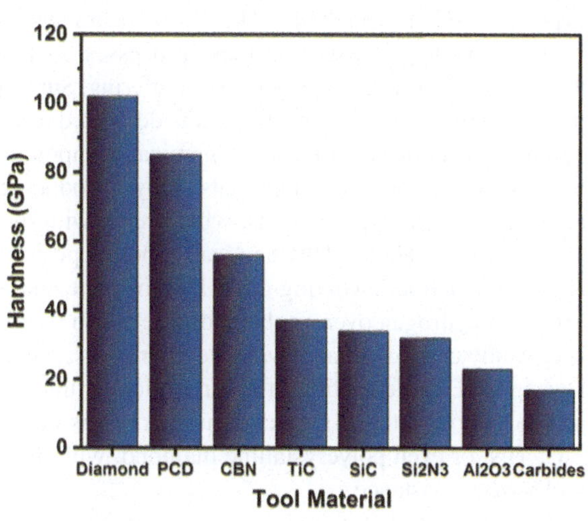

Fig. 4 Hardness values associated with materials used for cutting tools [14]

for high-temperature machining of CFRPs. PCD tools are favored for their exceptional hardness and wear resistance, making them suitable for high-speed machining of CFRPs. Choosing the appropriate tool material depends on factors like cutting speed, feed rate, and the specific CFRP composition to ensure efficient and precise machining while extending tool life. Polycrystalline diamond and cubic boron nitride tools are especially effective in achieving precision and minimizing delamination in CFRP machining, making them indispensable in industries like aerospace and automotive where CFRPs are increasingly used to reduce weight while maintaining structural integrity.

3.2 Role of Coatings in Enhancing Tool Performance

Coatings play a crucial role in enhancing tool performance across a wide range of industries and applications. These thin, protective layers applied to the surface of cutting tools, machine components, and various other tools offer several significant advantages. First and foremost, coatings can significantly increase tool durability and lifespan by providing resistance to wear, abrasion, and corrosion. This helps tools to maintain their sharpness and precision over a more extended time period, reducing the frequency of tool replacements and downtime. Furthermore, coatings can improve the efficiency of cutting and machining processes. By reducing friction and heat generation during cutting or drilling, they contribute to lower energy consumption and reduce the risk of overheating, which can lead to tool failure. Coatings also enhance the surface finish of the workpiece, resulting in better quality products and reduced post-processing requirements.

At present, tool coatings are categorized based on the technology used for their deposition, which includes physical vapor deposition (PVD) and chemical vapor deposition (CVD) methods. The PVD technique encompasses a wide range of vacuum coating processes. In these processes, material is extracted from a source through methods like evaporation or sputtering. Subsequently, this material is carried by the energy of vapor particles and is deposited as a film onto designated surfaces within a vacuum environment. Chemical compounds can be deposited by either employing similar source materials or by introducing reactive gases like nitrogen, oxygen, or basic hydrocarbons, which react with the metal from the PVD source [15]. Whereas, The CVD (Chemical Vapor Deposition) process entails a gas-phase chemical reaction, involving gases like methane and nitrogen, in the presence of an excess of hydrogen over a solid substrate, often a carbide cutting tool. This process is typically carried out at elevated temperatures, ranging from 1073 to 1373 K [16]. During CVD, a deposition of several microns thick single or multilayer coating is applied onto the tool's surface. These coatings may consist of hard materials like TiC, TiN, or even polycrystalline diamond, with the primary aim of enhancing the tool's wear resistance.

Coatings can be tailored to specific applications, providing advantages such as high-temperature resistance, improved lubricity, or enhanced chemical resistance.

They can also help reduce the adhesion of materials like chips or debris to the tool, which can further improve performance and reduce maintenance needs. As CFRP composites gained widespread popularity, a variety of coatings, both in terms of their types and deposition methods, have been extensively utilized to enhance the wear resistance of cutting tools used in CFRP machining. Many scientific studies [17–20] have been done to analyze the effect of coatings on cutting tool performance. All these studies show that coatings on cutting tools not only extend their lifespan but also contribute to higher precision and quality in machining operations, making them indispensable in modern manufacturing. In summary, coatings are a fundamental component in the world of tools and machining, enhancing tool performance by increasing durability, reducing friction, and tailoring properties to suit various applications. As technology continues to advance, the development of innovative coatings continues to play a pivotal role in pushing the boundaries of what tools can achieve.

3.3 Coated Tool Selection for CFRP Machining

Selecting the right coated tools for machining Carbon Fiber Reinforced Polymer (CFRP) is essential to achieve high-quality results and prolong tool life. Depositing hard coatings onto the tool surface serves the following goals: enhancing resistance to wear, decreasing friction, reducing the need for lubrication, improving thermal hardness, and boosting fracture strength. As a result, this process extends the lifespan of the tool and enhances productivity [21]. CFRP is a composite material known for its strength and lightweight properties, but it can be abrasive and challenging to machine. There are some key considerations when choosing coated tools for CFRP machining. First and most important is the material compatibility, select tools that are specifically designed for machining composite materials like CFRP. These tools are typically made from solid carbide, ceramics, or polycrystalline diamond (PCD) materials, which offer high wear resistance. The second important thing to consider is a selection of coating, choose a suitable coating for your tools. Common coatings for CFRP machining include diamond-like carbon (DLC), polycrystalline diamond (PCD), or polycrystalline cubic boron nitride (PCBN). These coatings provide excellent wear resistance and reduce the likelihood of tool wear and delamination.

4 Cutting Performance Evaluation

In every machining and manufacturing operations most crucial factor is the evaluation and assessment of the cutting tool performance to optimize the efficiency of cutting, tool life, and cost effectiveness. In this section, we will discuss a few crucial elements that are thought to be significant for assessing cutting performance while machining CFRPs. When machining CFRP, regular monitoring and assessment of these variables will help optimize cutting performance, improving part quality, and lowering production costs. Optimizing cutting parameters, tool selection, and machining techniques in response to these evaluations results in Better outcomes in CFRP machining processes.

4.1 Tool Wear Analysis

Monitoring and comparing the performance of various cutting tools used becomes very important while machining CFRPs as these tools are exposed to abrasive wear, delamination, and thermal degradation. By comparing the wear characteristics of different tool materials, coatings, and cutting parameters, researchers can make decisions about tool selection and machining strategies, ultimately improving the efficiency and quality of CFRP machining processes. Tool wear in machining CFRP is mainly due to abrasion and micro-chipping. Abrasion results from tool surfaces sliding against abrasive fibers, causing rounding and wear lands. Micro-chipping is due to oscillating forces from fiber fracture, leading to voids on the cutting edge. To resist these wear types, high hardness and toughness are needed. Uncoated tools, while cost-effective, tend to exhibit faster tool wear when machining CFRP. The reason for the quick deterioration of the cutting edge is the abrasive nature of CFRPs which accelerates tool wear. As a result of the abrasive wear mechanism, the surface quality and dimensional accuracy of machined products may decline. Several research studies have concluded that cutting tool without any coating demonstrates higher values of flank wear and chipping, resulting in the requirement of frequent tool changes and longer production downtime [22, 23]. Figure 5 shows the SEM images of tool comparing the wear tool undergoes after machining process.

In addition, it has been found that when cutting CFRP, special coated tools—such as those with polycrystalline diamond (PCD) or diamond-like carbon (DLC) coatings—perform better. By lowering friction and increasing hardness, these coatings

Fig. 5 SEM micrograph of **a** uncoated and **b** CVD diamond coated tips after machining test [22] (Adapted from)

can prolong the life of tools. The coatings serve as a protective barrier, reducing the direct contact between the abrasive fibers and the tool substrate, thus minimizing wear. Moreover, coated tools can lessen the requirement for frequent tool changes and the resulting production disruptions by helping to retain consistent cutting performance over extended periods of time. Although coated tools may initially cost more, due to their superior machining performance and longer lifespan long-term cost effectiveness is achieved by coated tools. The significance of taking coated tool alternatives into account when cutting CFRP materials is shown by these findings.

In conclusion, when cutting CFRPs, coated tools have a number of benefits over uncoated tools. They offer enhanced surface polish, decreased wear, longer tool life, and possibly cheaper total machining costs. However, to get the best results, it's crucial to select the right coating material and cutting parameters based on the particular CFRP material and machining circumstances. To get the most out of coated tools and optimize the machining process, regular tool wear analysis and maintenance are necessary.

4.2 Cutting Forces

Basically, there are three components of cutting force involved while machining CFRPs which are cutting force (F_c), thrust force (F_t), and feed force (F_f). The force needed to remove material is denoted by F_c, the force applied axially to the tool is shown by F_t, and the force needed to move the cutting tool along the workpiece by F_f. All these forces depend on several factors, which include tool geometry, cutting parameters, and the properties of CFRPs materials. One significant component that influences the cutting forces used in CFRP machining is the fiber orientation in the composite. The orientation of the fibers in CFRP composites, which are embedded in a matrix material, has a big influence on the cutting forces. Cutting in a direction perpendicular to the fiber orientation, i.e., across the fibers usually results in higher cutting force values in comparison to cutting in the direction parallel to the fiber orientation which is cutting along the fibers. In addition, cutting parameters such as cutting speed, feed rate, and depth of cut have a significant impact on cutting forces. Higher cutting speeds can minimize cutting forces but may also raise the danger of tool wear, while adjusting the feed rate and depth of cut allows for fine-tuning the forces based on the specific requirements of the machining operation. According to a number of studies, the feed rate has the greatest physical and statistical impact on cutting force [24–26]. An et al. suggest using a high cutting speed and a larger depth of cut to accomplish stable machining of CFRPs because doing so lowers the specific cutting force [27].

Higher values of cutting forces during machining have a huge impact on the process and quality of the finished product. Reduced tool life and increased tool wear are two of the most noticeable immediate consequences. Coated cutting tools might be used to overcome this problem. Using coated tools to machine CFRP has various benefits. Because of wear and increased friction, uncoated tools may require more cutting forces, which could make the machining process less predictable. On the other hand, coated tools provide more consistent and controllable cutting forces due to their increased longevity, decreased friction, and greater heat dissipation. While dealing with CFRPs, especially in high-precision applications, coated tools are frequently the recommended option to guarantee better control over the machining process and yield superior results.

4.3 Surface Finish Assessment

When machining hard to cut and challenging materials like carbon fiber reinforced polymer composites, assessment of surface roughness becomes a critical factor for evaluating the quality of the final machined product. It is affected by a number of factors, such as the cutting parameters and coating material selection. Coatings are thin layers that are deposited to the surface of cutting tools to enhance their surface properties like wear resistance and reduce friction during machining operations. Because of the abrasive nature of CFRPs, choosing the right coating material is crucial when machining it. Cutting tools are often protected with hard coatings such as titanium nitride (TiN) or diamond-like carbon (DLC). These coatings are often used to reduce and minimize tool wear and extend the life of cutting tool, which can have an indirect impact on surface roughness by maintaining constant cutting conditions over an extended length of time. Also, these protective coatings serve as a shield against abrasive wear and lower the possibility of tool-chip adhesion while cutting CFRP with coated tools. The outcome of this protection is lower cutting forces, less heat production, and less tool vibration. Coated tools are therefore better suited to keep a smoother and more consistent surface finish while machining.

Cutting parameters, including cutting speed, feed rate, and depth of cut, have a direct and immediate impact on the surface roughness of CFRP components. In a study Gao et al. [28] examined machined CFRP surfaces, finding that fiber cutting angle had the most significant impact on surface roughness, followed by cutting speed, DoC, and rake angle, as indicated by ANOVA results. In another study Liu et al. [29] found that higher cutting speeds improve the surface quality of machined CFRP composites in FEM simulations. Through these research it can be seen that adjusting these parameters allows machinists to control the material removal process and, consequently, the quality of the machined surface. When machining CFRP, it is crucial to strike a balance between material removal rates and surface finish. Higher cutting speeds and lower feed rates can result in smoother surfaces, but they may also increase the risk of delamination and fiber pullout, which can degrade the quality of the final product. On the other hand, lower cutting speeds and higher feed rates can reduce these issues but may lead to increased surface roughness due to insufficient material removal. A thorough investigation and assessment of the particular CFRP material to be machined are necessary to determine the ideal cutting parameters.

5 Conclusions

Although the demand for CFRP materials is very high in several industries because of their significant mechanical and lightweight properties, their challenging machining requires effective solutions. One such potential solution for addressing these machining challenges is to utilize coated cutting tools. This chapter has systematically discussed how surface finishes can be enhanced and machining efficiency can be significantly increased by carefully choosing tool coatings. Coatings that decrease delamination and withstand abrasive wear have shown tremendous promise in achieving these aims. With more sectors turning to CFRP composites for their benefits, the information in this chapter is a helpful resource for researchers, engineers, and practitioners.

By proper selection of coated cutting tools and utilizing optimized parameters for machining, industries can enhance the machining efficiency and attain improved surface finishes while cutting CFRP materials. As the demand for high strength and lightweight materials continues to grow in the market, the methods, knowledge, and approaches discussed in this chapter will be essential in promoting the efficient use of CFRP composites while reducing machining difficulties. In the end, this study advances the more general objective of making it possible for CFRP materials to be widely adopted in a variety of applications, resulting in more effective and sustainable production processes.

Declarations

Conflicts of Interest The authors confirm that there are no conflicts of interest associated with this work.

Ethical Compliance The authors affirm that this study adheres to all ethical guidelines.

Competing Interests No competing interests are declared by the authors.

Data and Material Availability Data and materials are not applicable to this study.

References

1. Praveenkumara J, Madhu P, Yashas Gowda TG, Sanjay MR, Siengchin S (2022) A comprehensive review on the effect of synthetic filler materials on fiber-reinforced hybrid polymer composites. J Text Inst (Taylor & Francis) 113(7):1231–1239. https://doi.org/10.1080/00405000.2021.1920151
2. Slamani M, Gauthier S, Chatelain JF (2015) A study of the combined effects of machining parameters on cutting force components during high speed robotic trimming of CFRPs. Meas J Int Meas Confed (Elsevier Ltd.) 59:268–283. https://doi.org/10.1016/j.measurement.2014.09.052
3. Makhdum F, Jennings LT, Roy A, Silberschmidt VV (2012) Cutting forces in ultrasonically assisted drilling of carbon fibre-reinforced plastics. J Phys Conf Ser 382(1)

4. Liu Y, Zwingmann B, Schlaich M (2015) Carbon fiber reinforced polymer for cable structures-a review. Polymers (Basel) 7(10):2078–2099
5. Dydek K, Boczkowska A, Kozera R, Durałek P, Sarniak Ł, Wilk M, Łogin W (2021) Effect of Swcnt-Tuball paper on the lightning strike protection of CFRPS and their selected mechanical properties. Materials (Basel) 14(11)
6. Zhang J, Chevali VS, Wang H, Wang CH (2020) Current status of carbon fibre and carbon fibre composites recycling. Compos Part B Eng (Elsevier Ltd.) 193:108053. https://doi.org/10.1016/j.compositesb.2020.108053
7. Das TK, Ghosh P, Das NC (2019) Preparation, development, outcomes, and application versatility of carbon fiber-based polymer composites: a review. Adv Compos Hybrid Mater 2(2):214–233
8. Joost WJ (2012) Reducing vehicle weight and improving U.S. energy efficiency using integrated computational materials engineering. JOM 64(9):1032–1038
9. Jureczko M, Pawlak M, Mężyk A (2005) Optimisation of wind turbine blades. J Mater Process Technol 167(2–3):463–471
10. Zhang Q (2013) The "black revolution" of sports equipment: application of carbon fiber reinforced plastics (CFRP). Appl Mech Mater 440:69–73
11. Islam F, Ramkumar J, Milani AS (2015) A simplified damage prediction framework for milling of unidirectional carbon fiber-reinforced plastics. Adv Manuf Polym Compos Sci (Taylor & Francis) 1(4):175–184. https://doi.org/10.1080/20550340.2015.1117747
12. Wang GD, Melly SK (2018) Three-dimensional finite element modeling of drilling CFRP composites using Abaqus/CAE: a review. Int J Adv Manuf Technol 94(1–4):599–614
13. Nguyen-Dinh N, Hejjaji A, Zitoune R, Bouvet C, Salem M (2020) New tool for reduction of harmful particulate dispersion and to improve machining quality when trimming carbon/epoxy composites. Compos Part A Appl Sci Manuf (Elsevier) 131:105806. https://doi.org/10.1016/j.compositesa.2020.105806.
14. Sheikh-Ahmad J, Davim JP (2012) Tool wear in machining processes for composites. Mach Technol Compos Mater 116–153
15. Bouzakis KD, Michailidis N, Skordaris G, Bouzakis E, Biermann D, M'Saoubi R (2012) Cutting with coated tools: coating technologies, characterization methods and performance optimization. CIRP Ann - Manuf Technol CIRP 61(2):703–723. https://doi.org/10.1016/j.cirp.2012.05.006
16. Arenas Cabello FJ (2015) "Eco Sostenible," Ecosostenible N°. 41, 2008, 30–37. ISSN-e 1699-3942
17. Mkaddem A, Ben Soussia A, El Mansori M (2013) Wear resistance of CVD and PVD multilayer coatings when dry cutting fiber reinforced polymers (FRP). Wear (Elsevier B.V.) 302(1–2):946–954. https://doi.org/10.1016/j.wear.2013.03.017
18. Wang X, Kwon PY, Sturtevant C, Kim DDW, Lantrip J (2013) Tool wear of coated drills in drilling CFRP. J Manuf Process (The Society of Manufacturing Engineers) 15(1):127–135. https://doi.org/10.1016/j.jmapro.2012.09.019
19. Davim JP, Mata F (2008) Chemical vapour deposition (CVD) diamond coated tools performance in machining of PEEK composites. Mater Des 29(8):1568–1574
20. Wang X, Shen X, Yang G, Sun F (2017) Evaluation of boron-doped-microcrystalline/nanocrystalline diamond composite coatings in drilling of CFRP. Surf Coat Technol (Elsevier) 330:149–162. https://doi.org/10.1016/j.surfcoat.2017.10.002
21. Raghav R, Kumar A, Mulik RS (2023) Comparison of CrN and CrTiN hard coatings developed on SiAlON tool inserts via magnetron sputtering process. J Mater Eng Perform 11665
22. Belmonte M, Oliveira FJ, Lanna MA, Silva CRM, Corat EJ, Silva RF (2004) Turning of CFRC composites using Si_3N_4 and thin CVD diamond coated Si_3N_4 tools. Mater Sci Forum 455–456:609–613
23. Asaithambi S, Gowri S (2013) Study on machinability of CFRP composite using tin coated PCBN and uncoated PCBN cutting tool. IOSR J Mech Civ Eng 2320–2334
24. Davim JP, Reis P (2003) Drilling carbon fiber reinforced plastics manufactured by autoclave-experimental and statistical study. Mater Des 24(5):315–324

25. Eneyew ED, Ramulu M (2014) Experimental study of surface quality and damage when drilling unidirectional CFRP composites. J Mater Res Technol (Korea Institute of Oriental Medicine) 3(4):354–362. https://doi.org/10.1016/j.jmrt.2014.10.003
26. Karpat Y, Bahtiyar O (2015) Tool geometry based prediction of critical thrust force while drilling carbon fiber reinforced polymers. Adv Manuf (Shanghai University) 3(4):300–308
27. An Q, Ming W, Cai X, Chen M (2015) Study on the cutting mechanics characteristics of high-strength UD-CFRP laminates based on orthogonal cutting method. Compos Struct (Elsevier Ltd.) 131:374–383. https://doi.org/10.1016/j.compstruct.2015.05.035
28. Gao C, Xiao J, Xu J, Ke Y (2016) Factor analysis of machining parameters of fiber-reinforced polymer composites based on finite element simulation with experimental investigation. Int J Adv Manuf Technol 83(5–8):1113–1125
29. Liu H, Xie W, Sun Y, Zhang J, Chen N (2018) Investigations on micro-cutting mechanism and surface quality of carbon fiber-reinforced plastic composites. Int J Adv Manuf Technol 94(9–12):3655–3664

Chapter 12
Reactive Extrusion: Filled Polymer Compounds and Its Applications

Faique Shakil

1 Reactive Extrusion Fundamentals

Introduction: Reactive extrusion stands at the forefront of materials engineering, offering a transformative approach to create filled polymer compounds with enhanced properties. Reactive extrusion is a sophisticated processing technique that involves the simultaneous application of mechanical energy and chemical reactions within an extrusion system. It combines the benefits of polymer processing with in-situ chemical reactions, allowing for the synthesis of complex materials with controlled structures.

Types of Reactive Extrusion: There are various types of reactive extrusion, including grafting, crosslinking, reactive blending, etc. Each type offers distinct advantages for different applications and composite designs.

These types are listed here.

1. Grafting Reactive Extrusion.
2. Cross-linking Reactive Extrusion.
3. Reactive Blending Extrusion.
4. Polymerization in Extrusion.
5. Degradation and Chain Scission.
6. Functionalization and Modification.
7. Masterbatch Compounding.

One such example of Reactive Blending Extrusion and Grafting Reactive Extrusion could be the compatibilization of PP and PBT with the GMA, in which devolatilization is involved. It is a unit operation designed to eliminate volatile molecules from an extruder while operating at reduced pressure. Utilizing a modular

F. Shakil (✉)
Materials Engineering, ETC-IN, Robert Bosch India, Bengaluru, India
e-mail: faiqueshakil@gmail.com

S. K. Sethi et al. (eds.), *Polymer Composites: From Computational to Experimental Aspects*, Materials Horizons: From Nature to Nanomaterials,
https://doi.org/10.1007/978-981-97-0888-8_12

twin-screw extruder offers the flexibility to position the devolatilization process at various points along the barrel. The significance of removing any remaining monomers becomes evident when considering the in-situ compatibilization of polypropylene (PP) and polybutylene terephthalate (PBT). There are two potential approaches: a two-step reactive extrusion and a one-step reactive extrusion. In the former, the initial extrusion step involves grafting glycidyl methacrylate (GMA) onto PP to produce PP-g-GMA, followed by the second step, which entails using PP-g-GMA as a reactive compatibilizer to blend PP and PBT. In the latter approach, both the free radical grafting of GMA onto PP and the compatibilization of PP and PBT with PP-g-GMA occur within a single reactive extrusion step. [1]

Reactive extrusion offers several advantages due to its unique combination of extrusion and chemical reactions. Some of the key advantages include:

1. **Efficiency:** Reactive extrusion allows for the integration of multiple processes into a single step, reducing the need for separate reaction vessels and processing steps. This leads to more efficient use of time, energy, and resources.
2. **Continuous Processing:** Process is inherently continuous, which makes reactive extrusion suitable for large-scale production. This continuous nature ensures consistent product quality and reduces batch-to-batch variations.
3. **Enhanced Product Properties:** Chemical reactions that occur during extrusion can lead to improvements in material properties such as mechanical strength, thermal stability, chemical resistance, and barrier properties. This enables the creation of polymers with enhanced performance characteristics.
4. **Tailored Materials:** Reactive extrusion allows for precise control over the polymer structure and composition. This means that polymers with specific functionalities, molecular weights, and distributions can be tailored to meet specific application requirements.
5. **Reduced Wastes:** By integrating reactions directly into the extrusion process, there's often less waste generated compared to traditional methods where separate reaction steps might be involved.
6. **Cost Savings:** The consolidation of processes and the reduction in waste contribute to cost savings in terms of raw materials, energy consumption, and labor.
7. **Shorter Reaction Times:** Reactive extrusion can accelerate reaction kinetics due to the combination of high shear forces and elevated temperatures. This can lead to faster reaction times compared to conventional methods.
8. **Processing of Difficult-to-React Polymers:** Reactive extrusion can make it possible to process polymers that are traditionally difficult to react due to their high viscosity or heat sensitivity. The extrusion process can provide better mixing and dispersion of reactants.
9. **Scalability:** Reactive extrusion processes developed on a laboratory or pilot scale can often be easily scaled up to industrial production levels, ensuring consistency in product quality and properties.

10. **Versatility:** Reactive extrusion can be applied to a wide range of polymers and chemical reactions, making it a versatile platform for modifying and functionalizing polymers for various industries.
11. **Reduced Environmental Impact:** The efficiency of reactive extrusion, along with the potential for using eco-friendly reactants and reducing waste, can contribute to a lower environmental footprint compared to traditional chemical processes.

These advantages make reactive extrusion a valuable tool in the field of polymer processing and materials development, with applications spanning from packaging and automotive industries to electronics, coatings, and biomedical materials.

In one of the instances of reactive extrusion is in which extrusion pretreatment is involved as it offers the benefit of having a wide range of processing options including functionalization and modification of materials. These options encompass several factors such as the size of the biomass, liquid infusion, the configuration of the extruder screw, temperature, screw rotation speed, the duration of residence, as well as the cleaning and dehydration of the biomass following extrusion. These factors can be classified into three main categories: pre-processing actions, extruder-related parameters, and post-processing activities [2].

Jaeyoung Jang and colleagues conducted a research exploring the in-situ reactive extrusion which is the type of crosslinking and polymerization process in which the modification of PET flakes were achieved using MDI methylenedi(phenyl isocyanate), aiming to enhance the properties of recycled PET (rPET). The results revealed notable improvements, including increased molecular weight, enhanced thermal properties, and improved oxygen barrier properties with higher MDI content. The intrinsic viscosity (IV) of rPET exhibited a positive correlation with MDI content, indicating the formation of new amide bonds as confirmed by FTIR analysis. Despite the positive outcomes, the study noted a reduction in crystallinity and oxygen barrier properties in rPET as MDI content increased. However, the overall migration of rPET-MDI remained within acceptable limits for food-contact applications. The higher molecular weight achieved in rPET-Ms positions them as suitable alternatives for applications requiring elevated IV compared to virgin PET (vPET). The study also highlighted color deterioration in rPET-Ms due to the low UV stability of MDI, suggesting potential strategies such as incorporating optical brighteners or optimizing the solid-state polymerization (SSP) process [3].

In another scenario, one study conducted by Bánhegyi and colleagues delved into the challenges of additive manufacturing technologies, specifically fused filament fabrication (FFF) or fused deposition modeling (FDM). Standard polymer grades often face issues like interlayer adhesion, shrinkage, and warpage. The study explored solutions such as blending engineering polymers, high-performance polymers, multilayer and coextruded structures, and biodegradable polymers to overcome these challenges in extrusion-based additive manufacturing. The conclusions emphasized the need to consider differences in thermal behavior, compatibility, interfacial adhesion, and mixing when selecting polymeric components and technologies for multicomponent 3D printing.

In the first scenario, an innovative in-situ mixing process, utilizing a double extruder head, was employed to produce polyamide printing filaments reinforced with thermotropic liquid crystalline (TLCP) fibers. To accommodate the significantly different processing temperatures of the TLCP (330 °C) and polyamide (PA) components (250 °C), two independent gear pumps were utilized. The PA component served as the primary melt flow, while the molten TLCP was introduced perpendicularly. A static mixer cooled the melt to 270 °C before combining the two streams, ensuring good dispersion. The resulting strands were re-granulated and extruded into 3D printing filaments. During filament formation and 3D printing, the solidified TLCP phase, present in varying quantities (20, 40, and 60%), maintained elongation and orientation, enhancing reinforcement in the printing direction. The optimum printing temperature was 280 °C, coinciding with the TLCP component's melting point, contributing to gradually increasing interlayer strength. Despite the absence of a compatibilizer, the composite exhibited competitive mechanical properties comparable to short glass-fiber and carbon-fiber-reinforced composites.

In the second scenario, an experiment explored the combination of mechanical properties of polycarbonate (PC) and acrylonitrile butadiene styrene (ABS) through core–shell printing filaments manufactured using coaxial co-extrusion. The stiffer, more brittle, and stronger PC component served as the core, while ABS formed the shell layer. Over a wide composition range (25/75, 50/50, and 75/25%), the core–shell structure remained intact during filament manufacturing and 3D printing. This approach, distinct from blending, showcased the compatibility of the PC-ABS polymer pair. A designed experiment investigated the impact of composition, extrusion temperature, nozzle input geometry, and layer thickness on mechanical properties (yield stress and elongation). The results demonstrated that the yield stress and elongation increased with PC content, with marked increases in yield strain. The measured values fell within the range determined by the parameters of pure components, remaining comparable to those of a benchmarked melt-blended grade. This innovative approach provides a novel means of combining the properties of two polymers in a single printing filament, particularly when direct melt blending proves challenging. The study highlighted the experimental nature of developing extrusion-based 3D printing technologies, acknowledging the importance of well-designed solutions to expand additive manufacturing possibilities. This type of extrusion is also a type of functionalization and modification type of reactive extrusion [4].

The success of reactive extrusion depends on a combination of process parameters, material properties, and reaction kinetics. Here are some key factors that influence the success of reactive extrusion:

1. **Material Compatibility:** The compatibility of the polymers and reactants is crucial. Polymers should have functional groups that can react with the introduced reagents, ensuring a successful chemical reaction during the extrusion process.

2. **Reactivity of Functional Groups:** The reactivity of functional groups in the polymers or reactants determines how quickly and efficiently the chemical reactions take place. High reactivity can lead to faster reaction kinetics and shorter processing times.
3. **Temperature:** Temperature plays a significant role in reaction rates and product quality. It affects the mobility of polymer chains and reactants, as well as the activation energy of the chemical reactions. The chosen temperature should be conducive to the desired reactions while avoiding degradation.
4. **Shear Rate and Mixing:** Shear forces generated during extrusion play a critical role in promoting mixing between reactants, enhancing mass transfer, and improving dispersion. Proper mixing ensures that reactants come into contact and react effectively.
5. **Residence Time:** The time that polymer materials spend within the extruder, known as residence time, affects the extent of the reaction. Balancing residence time with reaction kinetics is essential to achieve desired reaction outcomes without overreacting or degrading the material.
6. **Extruder Configuration:** The design and geometry of the extruder, including screw profile and barrel length, impact the level of shear and mixing. Choosing the right extruder configuration for the specific reactions and materials is important.
7. **Additive or Catalyst Concentration:** The concentration of additives, catalysts, or initiators can significantly influence the reaction rate and the final properties of the product. Optimal concentration levels need to be determined through experimentation.
8. **Reaction Inhibition or Side Reactions:** Undesired reactions or inhibition of reactions can occur due to competing reactions or the presence of impurities. Understanding the potential for side reactions and their impact is crucial for product quality.
9. **Purity of Materials:** Impurities, moisture, or other contaminants in the starting materials can interfere with reactions or cause degradation. High-quality starting materials are essential for achieving consistent and predictable outcomes.
10. **Process Monitoring and Control:** Monitoring and controlling process parameters such as temperature, pressure, torque, and material flow rate are crucial to ensure reproducibility and consistency in product quality.
11. **Scale-Up Considerations:** Successfully scaling up from laboratory scale to production-scale extrusion requires careful consideration of how process parameters and material behaviors change with larger volumes and different equipment.
12. **Safety:** Reactive extrusion involving chemically reactive species may pose safety risks. Adequate precautions, such as proper ventilation, protective equipment, and emergency procedures, should be in place.
13. **Characterization and Analysis:** Analytical techniques such as spectroscopy, thermal analysis, and rheology can provide insights into the progress of reactions and the properties of the modified polymers.

14. **Trial and Error:** Reactive extrusion often involves experimentation and optimization. Iterative testing and adjustments are important for identifying the best process conditions and achieving the desired product properties.

Considering these factors and tailoring the reactive extrusion process to the specific polymers and reactions are key to achieving successful outcomes and desired modifications in polymer materials.

Grafting and reactive extrusion methods are advanced approaches for modifying the surface of ground tire rubber (GTR) to enhance compatibility with matrices. Grafting involves attaching longer polymer chains onto GTR particles through methods like surface etching, oxidation, hydroxylation, or irradiation, followed by growing or attaching longer chains onto initiating sites. However, this method is complex, involving multiple chemical synthesis steps and the use of expensive or toxic solvents, making it time-consuming and economically challenging for industrial applications. On the other hand, in-situ grafting by reactive extrusion of GTR is a promising and industrially feasible approach. It allows simultaneous devulcanization/oxidation and modification/functionalization of waste rubber. This continuous, solvent-free process enables cost-effective chemical reactions using existing industry equipment. Products obtained through reactive extrusion exhibit good quality and repeatability. Grafting and reactive extrusion technologies have been applied to improve the tensile strength and thermal stability of GTR-modified polyester and LLDPE/HDPE composites. These methods enhance crack resistance in rubber and modified GTR composites due to improved interfacial adhesion between the modified GTR and the matrix. Reactive extrusion processes are particularly promising for industrial processing of GTR composites due to their simplicity, solvent-free nature, and shorter processing times. These methods hold great potential for developing high-quality, cost-effective GTR-based materials for various applications [5].

2 Filled Polymer Compound Design and Formulation Considerations

In this section, we explore the crucial aspects of designing and formulating filled polymer compounds using reactive extrusion. The judicious selection of polymer matrices, fillers, and compatibilizers plays a pivotal role in tailoring the material properties to meet specific application requirements.

In the realm of polymer science and engineering, the design and formulation of filled polymer compounds stand as a pivotal domain. Filled polymers, also known as polymer composites, offer a unique way to tailor material properties by incorporating various additives, fillers, and reinforcements into a polymer matrix.

In one of the research, Polyamide 12-multiwalled carbon nanotube (MWCNT) nanocomposites were effectively synthesized through in-situ polymerization in a micro-compounder, representing a novel approach. The study systematically explored the impact of molar mass, screw speed, and MWCNT content on the dispersion and electrical properties of the nanocomposites. Optimal MWCNT dispersion was achieved with high molar mass PA12 (20,000 g/mol) at a screw speed of 200 rpm, although these samples exhibited the highest electrical percolation thresholds. Notably, in-situ polymerization nanocomposites demonstrated superior MWCNT dispersion compared to bulk PA12 composites prepared through direct compounding. However, the bulk PA12 composites displayed a lower electrical percolation threshold. The findings underscore the effectiveness of in-situ polymerization of lauryl lactam in achieving well-dispersed MWCNTs in PA12 nanocomposites, offering a promising processing route for enhanced material properties. The factors which have been effective for better properties of nanocomposites were: (1) molar mass, (2) screw speed, and (3) nanoparticles content on dispersion [6].

Some major factors which need to be in consideration when conducting the reactive extrusion are:

1. **Additive Selection and Role:** The heart of filled polymer compound design lies in the selection of additives. Additives encompass a broad spectrum, ranging from reinforcing agents like fibers and particles to functional modifiers such as stabilizers, flame retardants, and colorants. Each additive serves a distinct purpose, affecting mechanical, thermal, electrical, and aesthetic properties. Understanding the synergistic effects of these additives is essential to achieving the desired composite characteristics.
2. **Filler Effects on Properties:** Fillers, including particulate materials like nanoparticles, microspheres, and fibers, have a profound impact on the properties of polymer compounds. Mechanical reinforcement, dimensional stability, thermal conductivity, and electrical behavior can all be tailored through proper filler selection and dispersion. The interaction between the filler and polymer matrix critically influences these outcomes.
3. **Dispersion and Compatibility:** Achieving uniform dispersion of fillers within the polymer matrix is paramount. Poor dispersion can lead to weak interfaces, compromised mechanical properties, and decreased material performance. Strategies such as surface modification, compatibilizers, and masterbatch techniques enhance filler–polymer interactions and ensure proper dispersion.
4. **Manufacturing Processes:** Filled polymer compounds can be produced through various manufacturing processes, including extrusion, injection molding, compression molding, and more. The choice of process profoundly affects the distribution of additives and fillers, influencing the final material properties and the potential for defects or inconsistencies.
5. **Property Enhancement and Tailoring:** The beauty of filled polymer compounds lies in their versatility to enhance specific properties. Reinforcements like glass fibers can bolster strength and stiffness, while conductive fillers enable electrical conductivity. Flame retardants enhance fire resistance, and nanofillers impart unique properties at the nanoscale. By judiciously selecting additives and fine-tuning their ratios, materials can be tailored for optimal performance in diverse applications.

6. **Challenges and Considerations:** While filled polymer compounds offer exceptional customization, they also present challenges. Achieving a balance between enhancing properties and avoiding compromises in other aspects can be intricate. Factors like filler-matrix compatibility, shrinkage, and anisotropy require careful attention during formulation and processing.
7. **Emerging Trends and Future Prospects:** As materials science continues to evolve, so do the possibilities for filled polymer compounds. Advancements in nanotechnology, biocompatible additives, and sustainable fillers are reshaping the landscape. Additionally, computational modeling and predictive tools are enabling more efficient compound design by providing insights into the interactions between additives and polymers.

In summary, the design and formulation of filled polymer compounds epitomize the synergy between chemistry, physics, and engineering. By strategically incorporating additives and fillers, engineers and scientists can create materials with tailored properties for a myriad of applications, from automotive components to medical devices. This section sheds light on the art and science of crafting compounds that transcend the properties of individual components, opening doors to innovation and unlocking new dimensions of material performance.

In another study conducted by Mathieu et. al., a comprehensive investigation was carried out to explore three different methods for blending bioresorbable ceramic fillers (hydroxyapatite or β-tricalcium phosphate) with a poly L-lactic acid polymer matrix, a crucial step in the preparation of porous composite structures for biomedical applications. The study compared the dry mixing approach, which resulted in a ceramic particle network enveloping polymer pellet, with solvent and melt methods that achieved a more uniform filler dispersion. Comparative analysis demonstrated that solvent and melt processing techniques yielded superior homogeneity and mechanical properties. Consequently, the solvent-free melt extrusion method was selected for preparing ceramic/polymer blends before supercritical CO2 foaming. This approach ensured homogeneous blends without the potential risk of toxic solvent residues, laying the foundation for the development of tailored porous structures in the field of bone tissue engineering [7].

In the study conducted by Eichner et al., a detailed process route was explored for fabricating highly filled composites with a polymer matrix using the spouted bed process and warm compaction. The objective was to achieve uniformly polymer-coated micron-sized copper particles through the spouted bed technique. The research demonstrated that elutriation of the finest particle fractions was inevitable when using particles with a broad size range, even in a specially designed spouted bed with a large relaxation zone. Despite particle agglomeration during the coating process, the final composite's properties remained unaffected. Electrical resistivity measurements were employed to assess coating quality, revealing that the resistivity increased significantly, indicating good coating quality, even at high copper filling degrees. Moreover, mechanical property analysis indicated a rise in modulus of elasticity and bending strength with increasing copper fraction up to 74 vol.%. However, exceeding this optimal filling degree of copper particles, specifically at 78.1 vol.%, resulted in a drastic reduction in mechanical properties, likely due to surpassing the percolation threshold of the copper particles [8].

3 Reactive Extrusion Process

Polymer blends involve the combination of two or more distinct materials, which can be polymers or copolymers, through physical mixing, with or without chemical interactions between them. This method of polymer blending is an economical and promising technique for creating novel materials using existing polymers, a practice as old as the polymer industry itself. This is evident from their widespread use in consumer products worldwide.

Blends are categorized as either miscible or immiscible, with the former referring to a homogeneous mixture at the molecular level. Most polymer blends tend to be immiscible and require compatibility adjustments to modify their interfaces and morphologies. Compatibilization is a critical process that involves the modification of polymer blends that would otherwise be immiscible on an interfacial scale. This process results in the creation of an interphase, which can be understood as a transitional region marked by a composition gradient, as well as the stabilization of the blend's overall morphology. Blending biodegradable polymers is one strategy for reducing overall material costs and tailoring specific properties and degradation rates. In contrast to the copolymerization approach, blending offers a simpler and quicker means of achieving desired characteristics. Moreover, blending allows for the incorporation of less expensive polymers with one another. The blend's miscibility is a critical factor that influences the final properties of the polymer.

The key benefit of compatibilization lies in its ability to either significantly delay or eliminate phase separation within polymer blends. Several methods can be employed to achieve compatibilization, as outlined in:

(a) Incorporating non-reactive agents like statistical, graft, or block copolymers whose units are miscible with different phases of the blend.
(b) Cross-linking the components within the blend.
(c) Introducing reactive or functionalized homopolymers that contain groups capable of forming bonds with other blend components, a process known as reactive compatibilization.
(d) The introduction of double-functionalized, low molecular weight compounds that can link the two polymer phases between the polymers, termed as in-situ reactive compatibilization.
(e) The application of mechanochemical methods during thermoplastic blend processing, which generates macro-radicals through temperature and shear forces. These macro-radicals can combine to form block copolymers that act as compatibilizers.

It's important to note that the effectiveness of additives in enhancing compatibility within a blend is closely tied to the specific types of polymers being blended, as well as the nature of the additives themselves. Among various types of organic peroxides, dicumyl peroxide (DCP) stands out as the most used initiators other than compatibilizing agents in practice [4].

Reactive extrusion is a dynamic process that melds polymer processing and chemical reactions, offering a pathway to engineer novel materials with enhanced properties. However, the successful execution of reactive extrusion hinges on a meticulous optimization process. This section delves into the intricacies of optimizing the reactive extrusion process, from understanding the underlying principles to fine-tuning parameters for achieving desired material modifications.

1. **Fundamental Understanding:** Before embarking on the optimization journey, a solid grasp of the fundamental principles of reactive extrusion is essential. This includes comprehending the chemical reactions involved, reaction kinetics, and the interplay between material properties and processing conditions.
2. **Defining Objectives:** The optimization process begins by defining clear objectives. Are you aiming to enhance mechanical strength, improve thermal stability, or introduce new functionalities? Clearly articulating the desired outcomes sets the stage for informed decision-making throughout the optimization process.
3. **Selection of Polymers and Additives:** The choice of polymers and additives is pivotal. Understanding the compatibility of different polymers, functional groups, and additives ensures that the reaction chemistry aligns with the desired modifications. Selecting appropriate materials paves the way for successful optimization.
4. **Process Variables:** Temperature, residence time, shear rate, and mixing intensity are some of the critical process variables that influence reaction kinetics and material properties. Altering these parameters can lead to varying degrees of reaction, affecting both the extent of modification and the potential for degradation.
5. **Reaction Monitoring and Analysis:** Optimization necessitates continuous monitoring and analysis of the reaction progress. Techniques such as spectroscopy, thermal analysis, and rheology provide insights into reaction kinetics, degree of conversion, and changes in material properties.
6. **Experimental Design:** Employing experimental design methodologies streamlines the optimization process. Techniques like Design of Experiments (DOE) allow for systematic exploration of process variables and their interactions, helping identify optimal conditions efficiently.
7. **Balancing Reaction Kinetics and Material Integrity:** Achieving the desired level of reaction is a delicate balance. Pushing the reaction to completion might compromise material integrity, leading to undesired side reactions or degradation. Optimization involves finding the sweet spot where reaction kinetics align with material performance.
8. **Scaling Up:** Successful optimization at the laboratory scale doesn't guarantee a seamless transition to production. The scalability of optimized conditions should be evaluated, considering potential changes in equipment, material behavior, and processing dynamics.

9. **Iterative Approach:** Reactive extrusion optimization is rarely a linear path. It often involves iterations, wherein insights gained from initial experiments inform subsequent adjustments. Flexibility and adaptability are key attributes in navigating the optimization journey.
10. **Data Analysis and Modeling:** Data collected during optimization can be harnessed for predictive modeling. Computational simulations and modeling techniques provide insights into the effects of process variables and aid in identifying optimal conditions before experimental trials.
11. **Multi-Objective Optimization:** In complex scenarios, where multiple properties are targeted simultaneously, multi-objective optimization methods help strike a balance between competing objectives. Pareto optimization and trade-off analyses assist in decision-making.
12. **Characterization of Optimized Compounds:** The final step of the optimization process involves comprehensive characterization of the optimized polymer compounds. Mechanical, thermal, and other relevant properties are evaluated to ensure that the modified materials align with the defined objectives.

In another review report authored by Banhegyi et al., the focus is on fused filament fabrication (FFF) or fused deposition modeling (FDM), a widely used additive manufacturing technology employing polymer melts to construct intricate structures. However, standard polymer grades often face challenges such as interlayer adhesion, shrinkage, and warpage, hindering them from meeting specific requirements. To tackle these issues unique to extrusion-based additive manufacturing, various techniques are employed. The study explores solutions tailored for extrusion-based additive manufacturing, distinct from traditional multicomponent materials used in injection molding. Techniques such as blending polymers, incorporating high-performance polymers, developing multilayer and coextruded structures, and exploring biodegradable polymers are discussed. These innovative approaches address challenges and enhance the capabilities of 3D printing using polymer extrusion. Some crucial factors, including differences in thermal behavior, compatibility, interfacial adhesion, and mixing technologies, significantly impact the success of multicomponent 3D printing. Routine compatibilization techniques aim to adjust properties like viscosity and mechanical strength, while specialty technologies involve cutting-edge methods such as modified mixing/printing heads, co-extrusion, and multilayer technologies. These innovative approaches offer unexpected advantages and pave the way for advancements in the realm of additive manufacturing [4].

In the in-depth review authored by Tiandong Zhang et al., polymer-based film capacitors have emerged as significant components in various applications, owing to their attributes such as high-power density, flexibility, and cost-effectiveness. The study explores the significant attention garnered by the doping of organic fillers into polymer films, driven by the advantages of easy processing, compatibility with the polymer matrix, and excellent film-forming properties. The distinction between doping organic fillers and blending two phases lies in the visible interface resulting from the doping of a secondary phase due to the immiscibility of fillers and the matrix.

Organic fillers, broadly categorized into conductive, semiconductive, and insulating types, offer diverse properties. Conductive fillers, exemplified by polyaniline (PANI), enhance the dielectric constant of polymer films. However, the blind pursuit of high dielectric constant using conductive fillers may negatively impact dielectric loss and conductivity. Conversely, introducing high-electron-affinity organic semiconductive fillers, like p-type semiconductive (PHT) or n-type semiconductive (PTCBI), captures injected and excited electrons, suppressing leakage current and improving breakdown strength.

Insulating organic fillers distinguish themselves by influencing the polarization character and enhancing electric breakdown strength. For instance, a novel ferroconcrete-like structure was introduced by embedding PVDF ferroelectric fibers into a PMMA linear matrix. This resulted in PMMA/PVDF all-organic films with a higher discharge energy storage density of 17.7 J/cm^3 and improved efficiency compared to blends and sandwiched films. Additionally, the study explores the use of low dielectric loss and good insulation properties of organic fillers like poly(acrylonitrile butadiene styrene) (ABS) and acrylonitrile butadiene rubber (NBR) in ferroelectric polymer matrices to reduce polarization loss and alleviate local electric field distortion [9].

In another study conducted by Ch Rani and colleagues, novel poly(vinyl alcohol + zinc oxide) [(PVA + ZnO)] polymer films doped with varying Chromium ion concentrations were prepared at room temperature using the slow evaporation method. The research employed a thorough analysis involving EPR, optical, FTIR, XRD, and SEM spectroscopic techniques to investigate the electronic mechanisms within these films. XRD analysis revealed the semi-crystalline nature of both PVAZn and Chromium-doped PVAZn films, with SEM confirming surface homogeneity in the Chromium-doped variants. EPR spectroscopy identified isolated Chromium ions in rhombic symmetry and distorted sites within the films. Optical absorption spectra indicated the presence of Cr3þ ions in octahedral positions, allowing for the calculation of optical band gap and Urbach energy. FTIR analysis confirmed the successful incorporation of Chromium ions into the PVAZn polymer matrix. These detailed findings provide crucial insights into the structural and electronic properties of Chromium-doped PVAZn polymer films, opening new possibilities for applications in real-time holography, nanophotonics, optical sensors, and microlasers [6].

In the realm of nature and sustainability, a study has been conducted by Sandhya Alice Varghese and colleagues, in which innovative biocomposite films were introduced, comprising poly(3-hydroxybutyrate-co-3-hydroxyvalerate) (PHBV) polymer incorporated with Ceiba pentandra bark fibers. These composites were meticulously prepared using the casting technique, and their properties were comprehensively examined. The incorporation of Ceiba pentandra bark fibers resulted in significant enhancements in various aspects of the composite films compared to pure PHBV. These enhancements included improved tensile properties, percentage crystallinity, and biodegradability. Furthermore, the films displayed commendable thermal stability and enhanced hydrophilicity, facilitating biodegradation in soil environments. Antimicrobial tests demonstrated inhibition against S. aureus, particularly at higher fiber loadings. In practical applications, specifically in packaging fresh strawberries, the composite films with a 10% fiber loading exhibited optimal performance. These films successfully maintained tensile strength, biodegradability, and effectively preserved the strawberries, preventing mold attacks. These promising results indicate that the developed biodegradable films, incorporating Ceiba pentandra fibers, have the potential to serve as eco-friendly active packaging solutions. These solutions can significantly contribute to extending the shelf life of perishable fruits while offering environmentally sustainable alternatives in the packaging industry [10].

The study conducted by Marvin Sauter and colleagues delves into the potential of filament spraying as a thermal spraying technique, with a focus on depositing fine and ultra-fine ceramic powders. In contrast to the limitations of traditional high-velocity oxy-fuel flame spraying (HVOF), filament spraying introduces a novel approach where a polymer-ceramic powder filament is combusted, releasing individual ceramic particles into the flame, ultimately deposited onto a substrate. The authors fabricated filaments utilizing low-density polyethylene (LDPE) and polybutylene adipate-terephthalate (PBAT) with alumina (Al_2O_3) and alumina-titania (Al_2O_3-TiO_2 87/13) ceramic powders. Their investigation revealed successful deposition of fine ceramic powders, showcasing particle sizes of approximately 1–5 μm for Al_2O_3 and 1–8 μm for Al_2O_3-TiO_2. The microstructure of the filament-sprayed coatings displayed density, indicating complete polymer combustion and individual release of molten powder particles. Moreover, the phase composition, deposition efficiency, microhardness, and surface roughness of the coatings were analyzed and compared favorably with traditional suspension spraying methods. The authors conclude that filament spraying holds promise as a thermal spraying technique for fine and ultra-fine ceramic powders, while acknowledging the need for further optimization and exploration of its feasibility across a broader spectrum of materials. The advantages of filament spraying, including ease of handling, reduced likelihood of clogging, and minimized agglomeration, position it as a noteworthy advancement in thermal spraying technology. However, addressing the incorporation of defects into the coatings remains a crucial area for further refinement [11].

In conclusion, reactive extrusion process optimization marries scientific acumen with engineering finesse. It demands a thorough understanding of both the chemistry and mechanics involved. By strategically navigating the variables and parameters, engineers and scientists can unlock the full potential of reactive extrusion, yielding materials with tailored properties that cater to the demands of diverse applications.

4 Processing of Filler Materials in Reactive Extrusion

Reactive extrusion, the convergence of chemical reactions and polymer processing, provides a versatile platform for incorporating filler materials into polymers. Fillers, ranging from nanoparticles to fibers, can significantly enhance the properties of polymer compounds. This section delves into the intricacies of processing filler materials within the realm of reactive extrusion, highlighting the challenges, methods, and benefits of achieving optimal filler dispersion and reinforcement.

1. **Filler Selection and Modification:** Choosing the right filler material is the first step. Fillers can be mineral-based (e.g., clays, silica), metal-based (e.g., nanotubes), or organic (e.g., cellulose). Some fillers might require surface modification to improve compatibility with the polymer matrix or to facilitate chemical reactions.
2. **Dispersion Challenges:** Achieving uniform filler dispersion is a common challenge in polymer processing. Agglomerates or poor dispersion can lead to reduced material properties and compromised performance. Reactive extrusion offers mechanisms to overcome these challenges through intense mixing and shear forces.
3. **Enhancing Filler-Matrix Interaction:** Incorporating fillers often requires promoting interactions with the polymer matrix. Functionalization of fillers or the use of compatibilizers can enhance filler-matrix adhesion, leading to improved mechanical properties and stability.
4. **Mixing Strategies:** Reactive extrusion provides an effective mixing environment due to the shear forces and turbulence generated by the screw design. Introducing mixing elements and optimizing screw configuration further enhance the dispersion of filler materials.
5. **Shear-Induced Alignment:** Fillers such as fibers or nanoparticles can exhibit alignment along the flow direction due to shear forces during extrusion. This alignment influences material anisotropy and mechanical properties, offering tailored reinforcement.
6. **Chemical Reactions with Fillers:** Reactive extrusion can facilitate chemical reactions between fillers and the polymer matrix. Silane coupling agents, for example, can react with both filler surfaces and polymer chains, enhancing filler dispersion and adhesion.
7. **In Situ Filler Synthesis:** In some cases, filler materials can be synthesized in-situ within the extruder. This can lead to unique filler morphologies and enhanced compatibility with the polymer matrix.
8. **Monitoring Filler Dispersion:** Real-time monitoring techniques such as spectroscopy, microscopy, and rheology help assess filler dispersion and the progress of filler–polymer interactions during processing.
9. **Benefits of Filler Incorporation:** Processing fillers through reactive extrusion yields numerous benefits. Mechanical reinforcement improved thermal conductivity, reduced shrinkage, enhanced flame retardancy, and reduced material cost are among the advantages.
10. **Optimization for Desired Properties:** Optimizing filler content, processing conditions, and filler-matrix interactions is essential to achieve desired material properties. The balance between enhancing properties and maintaining material integrity must be struck.

11. **Application-Specific Considerations:** The selection of fillers and processing parameters is highly dependent on the intended application. Conductive fillers are suitable for electronics, while reinforcement with fibers finds applications in structural materials.
12. **Future Directions:** The field of filler processing in reactive extrusion is continuously evolving. Nanocomposites, bio-based fillers, and sustainable reinforcement are emerging areas that offer new opportunities for material enhancement.

One of the studies conducted by Orevaoghene Eterigho-Ikelegbe et al. is the development of polymer composites utilizing coal as a filler. The study delves into the polymeric nature of coal and its interaction with polymer matrices, focusing on factors such as coal content, coal structure, and interfacial bonding with the polymer matrix, all of which influence the mechanical performance of the composites. Challenges associated with hydrophilic coal, such as moisture absorption and poor dispersion in polymers, are discussed. Potential solutions, including the use of coupling agents and coal-like polymers to enhance compatibility, are explored. Despite the potential for cost-effective coal/polymer composites, further research on optimal coupling agents and comprehensive findings is deemed essential for their widespread applications beyond energy-related fields [12].

In summary, the processing of filler materials within the realm of reactive extrusion is a delicate dance between material science and process engineering. The judicious selection of fillers, coupled with innovative processing strategies, unlocks the potential to engineer polymer compounds with enhanced properties tailored to diverse applications. Through a synergy of chemistry, mechanics, and meticulous optimization, reactive extrusion becomes a gateway to materials that surpass conventional expectations.

5 Fillers in Reactive Extrusion

The infusion of nanofillers into polymer matrices via reactive extrusion has revolutionized the field of polymer nanocomposites, opening avenues for enhanced material properties and novel applications. This section delves into the importance of fillers and nanofillers in reactive extrusion, exploring their potential, processing intricacies, and the resulting material advancements.

1. **Automotive Industry:** Nanofillers have found a significant role in the automotive sector, where lightweight and improved mechanical properties are paramount. For instance, nanoclay incorporated into polymer matrices enhances the mechanical strength and thermal stability of automotive components, contributing to lighter and more fuel-efficient vehicles.
2. **Reactive Extrusion as a Versatile Platform:** Reactive extrusion serves as an ideal platform for the incorporation of nanofillers. The high shear forces, intense mixing, and controlled temperature conditions within the extruder facilitate

uniform dispersion of nanofillers and enhance their interaction with the polymer matrix.

3. **Dispersion Challenges and Solutions:** Uniform dispersion of nanofillers within the polymer matrix is a paramount challenge. The high surface energy of nanofillers can lead to agglomeration. Innovative strategies such as in-situ polymerization, compatibilizers, and surface modifications enable improved dispersion and prevent agglomeration.
4. **Interactions and Interfaces:** Effective nanofiller dispersion is influenced by the interactions between the nanofiller surface and the polymer matrix. Surface functionalization, coupling agents, and interfacial engineering enhance filler-matrix adhesion, minimizing interfacial gaps and improving overall material performance.
5. **Exfoliation and Alignment:** Incorporating layered nanofillers, such as clay, presents the opportunity for exfoliation—separating individual layers—and alignment along the extrusion flow direction. Both phenomena contribute to enhanced material properties and tailored structural characteristics.
6. **Shear-Induced Enhancement:** The shear forces generated during reactive extrusion contribute to the exfoliation and alignment of nanofillers. These shear-induced effects aid in achieving desired filler dispersion and orientation, leading to improved material properties.
7. **Tailoring Material Properties:** The unique properties of nanofillers allow for precise tailoring of material properties. By selecting appropriate nanofiller types, concentrations, and processing conditions, engineers can achieve materials optimized for specific applications.
8. **Challenges in Process Optimization:** Optimizing nanofiller incorporation through reactive extrusion requires a delicate balance between achieving desired property enhancements and maintaining processability. Factors such as filler content, processing parameters, and polymer-nanofiller interactions must be harmonized.

In a new study conducted by V. Martin, Jean-François Witz, Frédéric Gillon, et al., researchers successfully manufactured parts and conducted mechanical characterizations on sintered 316L steel, validating the quality of filling and the capabilities of the setup. Several challenges were addressed, such as controlling temperature gradients and preventing granule blockages. The study also tackled the complexity of printing intricate geometries from pellets by employing extrusion control and multi-material printing of support structures. By utilizing pellets filled with a compatible binder polymer matrix, the process's applicability was extended to a wide range of materials. Essentially, mixing the appropriate polymer matrix with the powder of the target material enables this versatility. Notably, companies like PolyMIM and BASF offer feedstock for MIM (Metal Injection Molding) applications in various metal alloys and ceramic materials, which can be repurposed for 3D printing, broadening the scope of materials that can be utilized. Moreover, the study highlighted the flexibility of the E3D Tool Changer gantry, enabling the complexity of the process to be enhanced. FFF, one of the earliest additive manufacturing processes, involves melting solid feedstock in a moving print head to deposit it as a viscous liquid. The part's geometry is built layer by layer, with each layer formed by extruding material filaments in a complex pattern. Filaments bond during application, ensuring part cohesion. Conventional material feeding uses a solid filament stored as a spool, relying on filament quality for proper functioning.

For 3D printing with filled materials, such as polymers with metal or ceramic particles, adaptation of the extrusion head is necessary. The structure comprises a heating unit, extrusion nozzle, Archimedean extrusion screw, and heat exchanger with a conveying pipe. This adaptation allows the use of small pellets, ensuring the dissipation of heat to prevent premature melting and plug formation. Examples demonstrate the use of polymer filaments filled with metal or ceramic powders, intended for printing with a conventional Fused Filament Fabrication (FFF) printer. Extrusion screws, also known as Archimedean screws, play a crucial role in thermoplastic polymer applications. They are the primary tool for creating large formulations from blends of polymers with different properties. Schematics illustrating these processes are shown in Figs. 1 and 2 [13].

Regarding the importance of dispersion and mixing of nanoparticles, in a study conducted by Vanchiappan Aravindan et. Al., the impact of $Li_2O.SiO_2.TiO_2.P_2O_5$ (LSTP) particle size on polyethylene oxide (PEO)-based composite polymer electrolytes (CPE) for Li-metal batteries with $LiFePO_4$ cathodes was investigated. The

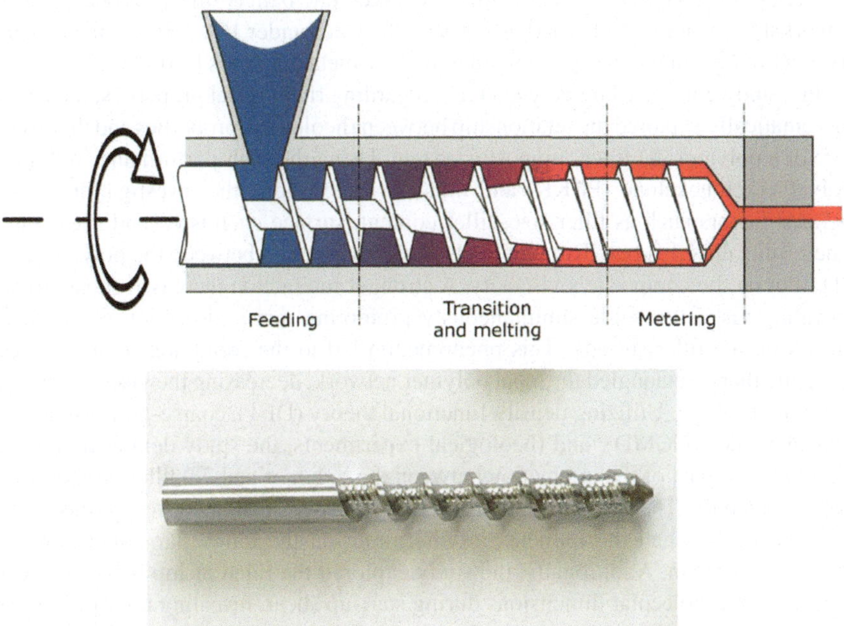

Fig. 1 Extrusion screw typical design

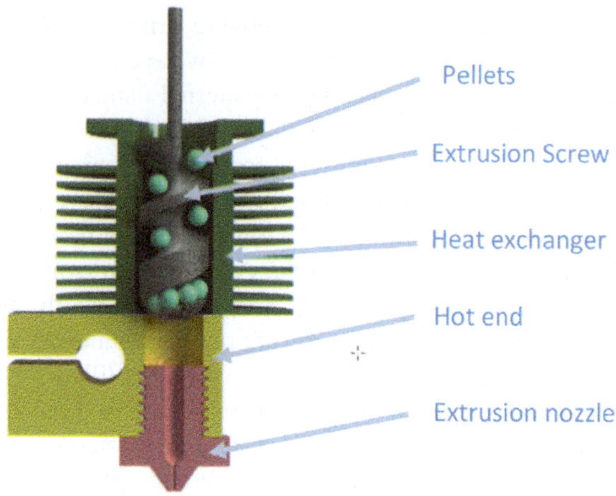

Fig. 2 Schematic view of pellet extruder

research found that nano-sized LSTP particles significantly improve electrochemical stability at 0.1 mA cm^{-2}, leading to enhanced ionic conductivity (1.09×10^{-3} S cm^{-1} at 80 °C) and stability at higher voltages (>4.5 V). The PEO-based CPE loaded with nano-LSTP demonstrates high Li-ionic conductivity (5.44×10^{-4} S cm^{-1}) and improved electrochemical stability even at higher current densities. When applied in the $LiFePO_4$-nm-LSTP CPE/Li composite solid-state battery, the system exhibits a remarkable capacity of 110 mAh g^{-1} after 100 cycles under 1C at 60 °C, showcasing its potential for high-energy next-generation Li-metal batteries [14].

In a study conducted by Peiyuan Gao, regarding rheological properties, the study systematically explored the relationship between rheological properties and the structure of a polymer-2D filler composite system. Using the high-performance polymer polyetherketoneketone (PEKK) and mica as an example, the investigation incorporated factors such as filler size, filler content, surface chemistry, and shear rate. The results unveiled two distinct effects of the interaction between the polymer and 2D filler on the composite's viscosity. A stronger interaction increased shear stress, elevating viscosity, while simultaneously promoting more adsorption of polymer chains on the filler surface. This phenomenon led to the formation of an ordered structure that disentangled the local polymer network, decreasing the viscosity of the continuous phase. Utilizing density functional theory (DFT), coarse-grained molecular dynamics (CGMD), and rheological experiments, the study demonstrated the significant impact of the interaction between the polymer and 2D filler on the rheological behavior. The findings emphasized the polymer-2D filler system's sensitivity to normal stress and the influence of shear rate on the molecular dimensions of the polymer chain. Additionally, the study explored the relationship between shear viscosity and molecular dimensions during start-up shear, revealing multiple stages of behavior consistent with transient shear viscosity trends. This study offers valuable insights into the fine-tuning of rheological properties in polymer-2D filler composites by manipulating filler surface chemistry. These insights are pivotal for designing new feedstocks applicable to extrusion-based additive manufacturing processes [15].

In the research conducted by Kalaimani Markandan et al., significant advancements in additive manufacturing (AM) of polymers have reshaped the landscape of 3D-printed polymer composites. This study explores innovative techniques dedicated to enhancing these composites by strategically incorporating and aligning fillers. Utilizing methods such as shear, ultrasound, electric and magnetic fields, and self-assembly, researchers have endeavored to exert control over filler alignment, resulting in substantial improvements in microstructure and overall properties. The analysis delves into various analytical models, establishing connections between microstructures and macroscopic mechanical properties, offering invaluable insights into the field. While filler alignment has demonstrated enhancements in properties such as electrical conductivity, the review underscores the persistent challenges, notably clumping and other defects. It highlights the necessity for continuous research and development efforts to fully unleash the potential of perfectly ordered fillers in 3D-printed polymer composites. These endeavors hold the promise of delivering a wide array of structural and functional benefits, catering to diverse applications in the realm of polymer composites [16].

6 Fillers and Nanofillers: Applications Across Industries

In the study conducted by Gao et al., significant advancements were made in the understanding and enhancement of BST ($Ba_{0.6}Sr_{0.4}TiO_3$)/polymer composites for various applications. BST, a ferroelectric solid–solution material, was investigated due to its excellent dielectric properties, including a high dielectric constant and tunable behavior related to the composition-dependent Curie temperature (Tc). However, these materials faced challenges such as poor flexibility, high dielectric loss at microwave frequencies, and limited energy storage density, hindering their potential applications in electric devices. To overcome these limitations, researchers explored BST/polymer composites, carefully examining their microstructure and the interface between BST fillers and polymers. Better compatibility between fillers and polymers proved essential, leading to improved composite microstructure stability. Various techniques, including filler modification, functionalization, and hot processing, were employed to enhance the microstructure and interface of these composites. Surface modification of inorganic fillers played a crucial role, promoting interfacial interactions and efficient filler dispersion within the polymer matrix. This approach aimed to minimize defects or voids in the composite structure, often achieved through processes like hot pressing. The dielectric properties of BST/polymer composites were found to be significantly influenced by the size, shape, and concentration of BST fillers. Increasing BST filler content generally enhanced dielectric properties. However, an excessive loading of ceramic particles led to increased defects, adversely affecting the dielectric performance.

Researchers successfully predicted the dielectric constant of BST/polymer composites using the Yamada model, making necessary adjustments to the ceramic filler's shape factor for accurate estimations. Furthermore, the investigation extended to the realm of composite dielectric tunability, focusing primarily on 0–3-type BST/polyvinylidene fluoride (PVDF)-based composites. Researchers explored the influence of BST filler concentration and shape on dielectric tunability [17].

Nanofillers have ushered in a new era of material innovation, revolutionizing a wide array of sectors across industries. Their unique properties and capabilities have led to transformative advancements, addressing challenges, and enhancing performance. Let's delve into the diverse sectors where nanofillers are making a profound impact, with illustrative examples showcasing their applications.

1. **Automotive Industry:** Nanofillers have found a significant role in the automotive sector, where lightweight and improved mechanical properties are paramount. For instance, nanoclay incorporated into polymer matrices enhances the mechanical strength and thermal stability of automotive components, contributing to lighter and more fuel-efficient vehicles.
2. **Aerospace and Defense:** Nanofillers are altering the landscape of aerospace materials. Carbon nanotubes, integrated into composites, provide exceptional strength and conductivity. An example is carbon nanotube-reinforced polymer composites used in aircraft components to enhance structural integrity while reducing weight.
3. **Electronics and Optoelectronics:** The electronics industry benefits from nanofillers for thermal management. Silicon carbide nanoparticles, dispersed in polymers, improve heat dissipation in electronic devices. This advancement allows for more compact and efficient electronics, as seen in LED lighting systems.
4. **Packaging and Consumer Goods:** Nanofillers revolutionize packaging materials by enhancing barrier properties. Clay nanoparticles are employed in polymer films used for food packaging, creating a barrier against oxygen and moisture, extending product shelf life, and ensuring food safety.
5. **Energy and Renewable Resources:** Nanofillers are transforming the energy sector. Graphene nanosheets, integrated into battery electrodes, enhance electrical conductivity, enabling faster charging and longer-lasting batteries. This technology is evident in electric vehicles that benefit from improved battery performance.
6. **Healthcare and Biomedicine:** Nanofillers offer novel solutions in healthcare. Silver nanoparticles embedded in wound dressings exhibit antimicrobial properties, reducing the risk of infections. Similarly, nanoparticles loaded with drug molecules enable targeted and controlled drug delivery in cancer therapies.
7. **Construction and Infrastructure:** Nanofillers are redefining construction materials. Carbon nanotubes mixed with concrete enhance tensile strength, leading to more resilient and durable structures. This is showcased in bridges and buildings that withstand environmental stressors with improved longevity.
8. **Textiles and Apparel:** Nanofillers are making textiles smarter. Silver nanoparticles integrated into fabrics create antimicrobial textiles, reducing odor and promoting hygiene. Nano-enhanced fabrics are used in sports apparel and medical garments.

9. **Environmental Remediation:** Nanofillers are harnessed for environmental sustainability. Titanium dioxide nanoparticles on surfaces break down pollutants when exposed to sunlight, purifying air and water. Such technology is used in self-cleaning coatings for buildings and water treatment systems.
10. **Cosmetics and Personal Care:** Nanofillers are transforming the beauty industry. Nanoscale pigments in sunscreens offer transparent UV protection. Silica nanoparticles in skincare products provide enhanced moisturization through improved water retention on the skin.

In each of these sectors, nanofillers are driving innovation and shaping the future of materials science. Their applications transcend traditional limitations, offering enhanced properties, improved performance, and novel functionalities that cater to the evolving needs of industries worldwide.

7 Characterization Techniques for Nanofilled Reactive Extrusion Compounds

Characterizing nanofilled compounds resulting from reactive extrusion is a complex endeavor due to the intricate interactions between nanofillers and polymers. A suite of advanced techniques is essential to unravel the structure, morphology, and properties of these materials. This section delves into the key characterization techniques employed to understand nanofilled reactive extrusion compounds.

1. **Scanning Electron Microscopy (SEM):** SEM provides high-resolution images of material surfaces, offering insights into nanofiller dispersion, agglomeration, and interface morphology. It aids in visualizing the spatial arrangement of nanofillers within the polymer matrix.
2. **Transmission Electron Microscopy (TEM):** TEM delves deeper, providing nanoscale details of nanofiller dispersion and interactions. It allows researchers to visualize individual nanoparticles, measure particle sizes, and assess the extent of nanofiller exfoliation.
3. **X-ray Diffraction (XRD):** XRD identifies the crystalline structure of nanofillers and polymers, aiding in assessing filler exfoliation or intercalation. It offers information about the degree of order and alignment within the material.
4. **Fourier Transform Infrared Spectroscopy (FTIR):** FTIR helps identify chemical interactions between nanofillers and polymers. Shifts in characteristic absorption peaks can indicate changes in molecular bonding, demonstrating filler–polymer interactions.
5. **Dynamic Mechanical Analysis (DMA):** DMA measures the mechanical properties of nanofilled compounds across a range of temperatures and frequencies. It highlights changes in stiffness, glass transition temperature, and damping behavior resulting from nanofiller incorporation.

6. **Thermogravimetric Analysis (TGA):** TGA evaluates the thermal stability and degradation behavior of nanocomposites. It aids in understanding how nanofillers influence thermal stability and decomposition pathways.
7. **Differential Scanning Calorimetry (DSC):** DSC assesses the melting and crystallization behavior of nanofilled compounds. It reveals changes in crystallinity, thermal transitions, and potential interactions between the polymer matrix and nanofillers.
8. **Rheological Analysis:** Rheological measurements provide insights into the flow behavior and viscosity of nanofilled compounds. Changes in viscosity can indicate nanofiller dispersion, interaction, and processing dynamics.
9. **Mechanical Testing:** Mechanical tests, including tensile, flexural, and impact testing, evaluate the impact of nanofillers on material strength, stiffness, and toughness. Such tests assess how nanofillers reinforce or alter mechanical properties.
10. **Electrical and Thermal Conductivity Measurements:** Electrical and thermal conductivity measurements determine how nanofillers impact these properties. Nanofilled compounds often exhibit enhanced conductive properties due to the presence of conductive nanofillers.
11. **Atomic Force Microscopy (AFM):** AFM images surfaces with nanometer-scale resolution, aiding in visualizing nanofiller topography and distribution on material surfaces.

In conclusion, characterizing nanofilled reactive extrusion compounds necessitates a multidisciplinary approach. Employing a combination of advanced techniques helps unravel the complexities of nanofiller dispersion, interactions, and their impact on material properties. By leveraging these characterization tools, researchers gain insights that guide the optimization of nanofilled compounds for targeted applications.

Vanchiappan Aravindan et. al. utilized FE-SEM, TEM, and AFM techniques to investigate the impact of different-sized LSTP particles in PEO-based composite polymer electrolytes (CPEs). Microscopic analysis confirmed that both micron and nano-sized ceramic fillers were well-distributed in the PEO matrix, with nano-sized LSTP particles showing a higher surface area, enhancing effective Lewis acid–base interactions. The incorporation of nanoparticles improved the ionic conductivity of CPEs compared to those with micron-sized fillers. Atomic force microscopy (AFM) images revealed a uniform distribution of LSTP particles on PEO-based CPEs, indicating strong interaction between the ceramic filler, Li salt, and PEO polymer, essential for achieving good ionic conductivity and electrochemical stability. Thermal analysis through TGA and DSC studies demonstrated improved thermal stability in CPEs with nano-sized LSTP particles, with higher onset temperatures, suggesting enhanced thermal resistance due to shielding effects [14].

8 Unlocking the Potential: Reactive Extrusion and Nanofillers in Materials Engineering

In the vast landscape of materials engineering, the synergy between reactive extrusion and fillers stands as a beacon of innovation, offering limitless possibilities. The journey we embarked on, from the fundamental principles of reactive extrusion to the transformative power of nanofillers, has revealed a world where science, engineering, and imagination converge to shape the materials of tomorrow. Reactive extrusion, with its capacity to blend mechanical processing and chemical reactions seamlessly, has redefined how we craft materials with tailored properties. Its efficiency, scalability, and versatility have paved the way for groundbreaking advancements across industries. From aerospace to healthcare, automotive to electronics, the marriage of chemistry and engineering in reactive extrusion has birthed materials that exceed our expectations. The art of formulating filled polymer compounds, a symphony of material compatibility, dispersion, and performance enhancement, has brought us materials that meet the most demanding application requirements. Whether it's the pursuit of stronger, lighter automotive components or the quest for sustainable packaging solutions, the versatility of filled polymer compounds has left an indelible mark.

9 Conclusion

The introduction of nanofillers into the realm of materials engineering has rewritten the rules. At the nanoscale, we've discovered a treasure trove of materials with properties beyond imagination. Reactive extrusion serves as the gateway to harnessing these nanofillers, enabling precise control over material characteristics and the creation of materials with enhanced strength, conductivity, and functionality. As we look to the future, the possibilities are boundless. Materials engineered through reactive extrusion and nanofillers are poised to drive innovation in ways we can scarcely predict. Smarter electronics, sustainable infrastructure, breakthroughs in healthcare, and eco-friendly solutions are all within our reach.

In closing, the path of discovery and innovation in materials engineering never ceases. We invite you to embrace this path, to be part of the journey that pushes the boundaries of what's possible. The fusion of chemistry, physics, and engineering continues to yield remarkable materials that address the challenges of our ever-advancing world. Step forward with curiosity, with a spirit of exploration, and with the knowledge that the future of materials engineering is in your hands. Together, we unlock the potential that propels us toward a brighter and more sustainable tomorrow.

Declarations

Conflicts of Interest The authors confirm that there are no conflicts of interest associated with this work.

Ethical Compliance The authors affirm that this study adheres to all ethical guidelines.

Competing Interests No competing interests are declared by the authors.

Data and Material Availability Data and materials are not applicable to this study.

References

1. Li T-T, Cheng S-B, Feng L-F et al (2023) A review on the free radical grafting of vinyl monomers onto polyethylene and polypropylene by reactive extrusion. Chem Eng Sci 278:118916
2. Guiao KS, Gupta A, Tzoganakis C et al (2022) Reactive extrusion as a sustainable alternative for the processing and valorization of biomass components. J Clean Prod 355:131840
3. Jang J, Shin H, Seo J (2023) In-situ chain extension of polyethylene terephthalate flakes using reactive extrusion as an upcycling approach. Chem Eng Sci 282:119289
4. Bánhegyi G (2023) Polymer compatibility and interfaces in extrusion-based multicomponent additive manufacturing – a mini-review. Adv Ind Eng Polym Res
5. Phiri MM, Phiri MJ, Formela K et al (2022) Grafting and reactive extrusion technologies for compatibilization of ground tyre rubber composites: compounding, properties, and applications. J Clean Prod 369:133084
6. Rani C, Linga Raju Ch (2023) Spectroscopic properties and effect of chromium in (PVA + ZnO) polymer films. Mater Today: Proc
7. Mathieu L, Bourban P, Manson J (2006) Processing of homogeneous ceramic/polymer blends for bioresorbable composites. Compos Sci Technol 66(11–12):1606–1614
8. Eichner E, Salikov V, Bassen P et al (2017) Using dilute spouting for fabrication of highly filled metal-polymer composite materials. Powder Technol 316:426–433
9. Zhang T, Sun H, Yin C et al (2023) Recent progress in polymer dielectric energy storage: from film fabrication and modification to capacitor performance and application. Prog Mater Sci 140:101207
10. Varghese SA, Pulikkalparambil H, Rangappa SM et al (2020) Novel biodegradable polymer films based on poly(3-hydroxybutyrate-co-3-hydroxyvalerate) and Ceiba pentandra natural fibers for packaging applications. Food Packag Shelf Life 25:100538
11. Sauter M, Roth A, Grebhardt A et al (2023) High velocity flame spraying of highly-filled ceramic—polymer filaments (F-HVOF). Surf Coat Technol 458:129324
12. Eterigho-Ikelegbe O, Yoro KO, Bada S (2021) Coal as a filler in polymer composites: a review. Resour Conserv Recycl 174:105756
13. Vincent M, Jean-François W, Frédéric G et al (2021) Low cost 3D printing of metals using filled polymer pellets
14. Ramkumar B, Aravindan V, Ramasamy H et al (2022) Ternary metal oxide filled PEO-based polymer electrolyte for solid-state lithium metal battery: The role of filler particle size. Solid State Sci 132:106958
15. Gao P, Muller SE, Chun J et al (2023) Effects of 2D filler on rheology of additive manufacturing polymers: Simulation and experiment on polyetherketoneketone-mica composites. Polymer 269:125722
16. Markandan K, Lai CQ (2023) Fabrication, properties and applications of polymer composites additively manufactured with filler alignment control: A review. Compos B Eng 256:110661
17. Gao F, Zhang K, Guo Y et al (2021) (Ba, Sr)TiO3/polymer dielectric composites–progress and perspective. Prog Mater Sci 121:100813

Chapter 13
Drilling Behavior of Glass Fibre Reinforced Polymer Composites

Pawan Kumar Rakesh, Rajesh Kumar, and Jayant Kumar

1 Introduction

The use of composite materials was initiated in the early 1940s, and the subsequent advancement of polymer composite materials was primarily motivated by the requirements of the space, defense, and aviation sectors. During this period, the primary focus was on achieving optimal performance rather than cost-effectiveness [1–3]. The utilization of polymer matrix composites has seen significant growth in the last decade. Consequently, researchers and scientists have directed their attention to pertinent factors related to the machinability of these materials [4–6]. The study examined four distinct drill types in order to analyze their effects on tool wear pattern. The extent of wear seen in the case of the high-speed steel (HSS) drill was greater compared to that of the carbide drill. The term "combing wear" was used to describe the slender belt-like grooves that were generated by chipping during the first phase of drilling. The study saw a progressive rise in torque as the number of drilled holes increased, eventually reaching a practically constant value after drilling 100 holes of composites [7].

In their experimental investigation, Tagliaferri et al. [8] conducted a study on woven glass fibre reinforced polymer (GFRP) composites. A larger ratio was associated with improved hole quality. The ratio exhibited independence with respect to variables such as resin type, fibre shape, laminate thickness, and fabrication technique. The experimental results indicate that the tensile strength of the drilled GFRP

P. K. Rakesh (✉) · R. Kumar
Department of Mechanical Engineering, National Institute of Technology Uttarakhand, Srinagar 246174, Uttarakhand, India
e-mail: pawankumarrakesh@gmail.com

J. Kumar
Department of Mechanical Engineering, Katihar Engineering College, Katihar 854109, Bihar, India

S. K. Sethi et al. (eds.), *Polymer Composites: From Computational to Experimental Aspects*, Materials Horizons: From Nature to Nanomaterials,
https://doi.org/10.1007/978-981-97-0888-8_13

laminates is unaffected by the level of damage. The bearing strength, in this case, was found to be impacted by the level of damage. It was observed that as the damage decreased, there was a corresponding rise in the bearing strength [9–11]. Hocheng and Dharan [12] conducted an observation indicating that delamination phenomena manifest throughout the drilling process, namely in the form of pushdown at the exit and peel-up at the entry. The drill consistently applies a compressive force to the laminate material located at its center. The laminate in the vicinity of the drill tends to experience a displacement away from the connection between adjacent layers surrounding the hole. As the drill point nears its termination, the remaining uncut thickness diminishes and thus, the resistance to deformation diminishes as well. When the peel-up phenomenon occurs, the first contact between the drill's cutting edge and the laminate will result in abrasion. Subsequently, by progressive motion, it exhibits a tendency to displace the abraded substance towards the direction of the flute. This movement applies an upward peeling force to detach the top lamina from the uncut section, which is kept in place by a downward acting thrust force [13, 14].

In their study, Di-Paolo et al. [15] found three distinct damage processes that contribute to the development of delamination in materials. These mechanisms include plate bulging, crack opening, and fibre tearing/twisting. A bulge refers to a fracture mode, namely Mode-I, which is induced by a dispersed load that directly applies pressure from the chisel edge onto the laminate. Upon the departure of the chisel edge from the laminate, fragments of fibre originating from the surface of the laminate at the point of exit undergo detachment. Crack initiation takes place when a concentrated load is applied to the fibres by the drilling lip. Fibre ripping and twisting is the third kind of damage. The fibres that have been cut undergo tearing fracture in Mode-III. This occurs owing to the torque and twisting of the fibres caused by the combined downward thrust force. According to Chen [16], the influence of the cutting speed on the cutting forces is deemed to be negligible. In contrast, it was observed that the cutting force decreased as the feed rates were reduced. To enhance the quality of the hole at its exit point, it is necessary to reduce the feed rate during the drilling procedure. The analysis focused on many characteristics related to tool geometry, including the point angle, helix angle, chisel edge rake angle, and web thickness. The study revealed that there was a drop in the tangential force, also known as torque, as the point angle grew. Conversely, there was a rise in the thrust force. When the helix and chisel edge rake angles were increased, there was a concurrent drop in both thrust and torque. The thrust force and torque exhibit a positive correlation with the augmentation of both the web thickness and chisel edge length [17]. According to the findings of Lin and Shen [18], it was determined that the wear rate of multi-faceted drills is higher than that of regular twist drills when the cutting speed is increased. This phenomenon may be ascribed to the acute external edge of the multi-faceted drill, which is prone to rapid erosion or fracturing.

According to Piquet et al. [19], the drilling caused damage is greatly influenced by the drill tip shape. The study examined two distinct drill tip geometry of K20 rated "micrograin" tungsten carbide. The drill tips were equipped with three cutting blades, including a rake angle ranging from 0 to 6° and a clearance angle of 6°. The study's findings indicate that in order to get a high-quality hole, pre-drilling is a necessary step when using standard drill points. This pre-drilling process helps to counteract the negative impact of the chisel edge effect. The altered geometric configuration yielded very favorable outcomes without the need for pre-drilling. In addition to the improved shape, the variable feed rate was used, resulting in the production of high-quality holes. According to Lachaud et al. [20], the use of a twist drill during the drilling process leads to the occurrence of several sorts of defects inside the laminate material. There are four primary categories into which they may be classified: hole entrance, circular flaw, heat damage, and hole exit. The occurrence of hole entrance defects may be attributed to the presence of unidirectional plies in the laminate. This phenomenon results in the rupture of the first lamina that comes into contact with the drilling. The presence of a circular flaw may be attributed to the angular alignment of the cutting edge in relation to the orientation of the fibres. The asymmetrical behavior of the fibres therefore results in the formation of an elliptical aperture. The minor axis of the ellipse is of lesser magnitude compared to the drill diameter and aligns with the orientation of the fibres inside the ply. Thermal damage arises as a consequence of the friction occurring between the fibres and the two smaller cutting blades of the drill. The phenomenon under consideration has the potential to induce matrix burning and crater formation in the vicinity of the aperture perimeter. The elimination of fibres results in a surface irregularity on the lateral surface of the aperture. The occurrence of hole exit is attributed to the inability of the chisel edge of the twist drill to effectively penetrate the material. The impact of drilling settings on the amount of delamination was investigated by Capello [21]. The extent of the damage was assessed via the use of digital image processing techniques. The study's findings indicate that the peel-up delamination phenomenon is affected by the feed rate and may be mitigated by reducing the helix angle of the twist drill. On the contrary, pushdown delamination is contingent upon several conditions, including backup support, drill temperature, and feed rate. The application of pre-heating to the drill tip results in a decrease in pushdown delamination and an increase in thermal damage. It was observed that the power and specific cutting pressure exhibited lower values while using the "Straight Shank" drill as compared to the "Brad & Spur" drill under identical cutting circumstances. The feed rate has significant physical and statistical relevance in relation to both power and specific cutting pressure. The 'Brad & Spur' drill point demonstrated superior hole quality in comparison to the 'Straight Shank' drill point. The feed rate has the most significant physical and statistical impact on delamination at the entry for both drills [22]. Similarly, the cutting speed demonstrates a comparable effect on delamination at the exit.

In their study, Hocheng and Tsao [23] introduced an analytical methodology for the estimation of the critical thrust force associated with several drill point geometries, including the saw drill, candle stick drill, core drill, and step drill. It was hypothesized that the stress applied by various drill locations on the laminate to be drilled would vary. The analytical models used were only based on axial loading, with no consideration given to the influence of torque. The impact of chisel edge length and pilot hole diameter on delamination was investigated by Tsao et al. [24]. The use of a pilot hole serves to prevent the chisel edge from making direct contact with the laminate material, hence resulting in a reduction of the force applied by the drill during the drilling process. The notion of pre-drilling has been shown to effectively mitigate the essential thrust force responsible for delamination. The magnitude of the thrust force is substantially reduced by the elimination of the chisel edge's influence. The study's findings indicated that the optimization of the chisel edge length resulted in the ability to use greater feed rates for drilling medium to large size holes, hence mitigating the occurrence of delamination. In their study, El-Sonbaty et al. [25] examined the impact of different drilling parameters on the thrust force, torque, and surface roughness associated with the process of drilling a hole. The initiation of the torque cycle exhibits a temporal delay of several seconds relative to the onset of the thrust force cycle. The chisel's sharp edge effectively pierces the workpiece, resulting in a forceful shove. The cutting forces experienced during the drilling of GFRP laminate were found to decrease as the cutting speed rose. Conversely, the cutting forces were shown to rise with larger drill diameters and higher fibre volume fractions. The surface roughness of drilled holes was found to be increased when greater cutting speeds and fibre volume fractions were used. The surface roughness of the drilled holes was found to be greater when lower feed rates were used, in comparison to higher feed rates.

The oscillatory vibration aided (OVA) drilling technique for GFRP composites was introduced by Ramkumar et al. [26, 27]. The analysis of drilling performance was conducted by online measurement of cutting forces, power, and acoustic emission (AE) monitoring. The OVA drilling technique exhibited superior thrust force and torque responsiveness. The number of holes bored using traditional drilling was determined to be 30, whereas the use of OVA drilling resulted in a doubling of this quantity to 60. The study's findings indicate that OVA drilling is a viable method for conducting drilling operations without causing damage to GFRP composites. In their research, Wang et al. [28, 29] performed an experimental investigation focusing on the use of vibration drilling techniques to GFRP and CFRP laminate materials. The study specifically examined the use of high-speed steel and carbide drills in this context. The vibration amplitudes and frequencies were manipulated in the context of vibration drilling. The use of high-speed steel (HSS) drills for drilling carbon fibre reinforced polymer (CFRP) and GFRP materials leads to a higher degree of tool wear as compared to the usage of carbide drills. The experimental findings revealed that the magnitude of the thrust force generated during vibration drilling was comparatively lower in comparison to that of conventional drilling. The fibre is prone to easy breakage due to the tool's sudden cutting action, which operates in a pulse intermittent cutting mode. This cutting mode alters the mechanics of chip creation, resulting in less chip deformation and decreased thrust force.

The process of removing material from fibre reinforced polymers (FRPs) is much more intricate in comparison to metals, mostly owing to the intrinsic characteristics of FRPs, such as their non-uniformity and anisotropy [30]. As a consequence, there is a high occurrence of fibre pullout, delamination, matrix burning, and substandard surface quality. The researchers created a finite element model to quantitatively assess the material damage. This model was then confirmed by non-destructive dye penetrant testing, as per experimental procedures. The occurrence of damage resulting from drilling in [(0/90)/0]s GFRP laminates was seen to be distributed in the vicinity of the hole. The study conducted by Mohan et al. [31] aimed to investigate the impact of feed rate and drill diameter on various machining output parameters. The investigation of machining parameters, including drill size, feed rate, and cutting speed, has been conducted to ensure the absence of damage during the drilling process of GFRP composites. The link between thrust force and torque, as influenced by feed rate and drill diameter, has been analyzed in order to construct a semi-empirical correlation with an existing model based on cutting parameters.

In a study, Paul et al. [32] conducted optimization techniques on three distinct drill point geometries, namely conical, racon, and helical, with the objective of minimizing both thrust force and torque during drilling operations. The optimization results demonstrate a decrease in both thrust force and torque by around 40% compared to the conventional conical drill. The optimized racon drill exhibits a significant decrease in thrust, while demonstrating just a little enhancement in torque compared to the benchmark racon drill. The helical drill that has been optimized demonstrates a reduction in both torque and thrust force, surpassing the benchmark helical drill by more than 40% in each aspect. The optimized drill geometry exhibits higher rake angles, resulting in a reduction in thrust forces attributed to significant negative rake angles. The model for estimating thrust force and torque during drilling of composite materials was developed by Langella et al. [33]. The examination of the overall force magnitude produced by the cutting lips and chisel edge when drilling GFRP composites reveals that both the thrust force and the potential for material damage escalate in direct correlation with the feed rate and point angle. The study conducted by Varatharajan et al. [34] examined the machinability attributes, including thrust force, torque, tool wear, surface roughness, and chip morphology, of both thermoplastic and thermoset materials. Thermoplastics have superior impact resistance in comparison to thermosets. In contrast to the brittle fracture mode often seen in thermosets, thermoplastic composites exhibit a ductile mode of failure. The results obtained from the dynamic mechanical study indicate that the incorporation of reinforcing thermoplastics leads to an improvement in both the glass transition temperature and the damping qualities. In their study, Shaikh and Channiwala [35] proposed a theoretical framework aimed at forecasting the mechanical properties of fibre composites that include porous reinforcements. The model they devised was based on hexagonal fibre packing. The concept of the interphase zone between the fibre and matrix was created in order to account for impurities prevalent in composites. The evaluation of interphase volume and characteristics was conducted using the rule of mixing [36]. In their study, Sardinas et al. [37] used a micro-genetic algorithm to optimize the procedure. The use of multi-objective optimization, using an a posteriori methodology, enhances the versatility in the selection of optimum cutting parameters for the drilling process of composite materials. The used evolutionary algorithm successfully generated a collection of solutions that were evenly distributed, enabling the organization of Pareto's front with a relatively minimal computing burden. The cutting forces may be determined by empirical formulae derived from multivariable linear regression analysis, which takes into account factors such as fibre volume fractions, feed rate, and cutting speed. The occurrence of delamination showed a positive correlation with the increase in feed rate [38]. In their study, Rao et al. [39] conducted a numerical simulation to investigate the chip formation process during orthogonal machining of UD-GFRP and UD-CFRP composites. The researchers used quasi-static analysis for their investigation. A two-phase micromechanical model was developed using an explicit finite element algorithm. The model successfully predicted the occurrence of fibre failure, matrix damage, and interfacial debonding between the fibre and matrix at certain locations throughout the machining process [40].

The study conducted by Faraz et al. [41] examined the drilling load profiles throughout the drilling process and conducted a quantitative analysis of the resulting delamination at the hole exit in the drilling of GFRP composites. This research was done utilizing several types of uncoated carbide drill bits, including two specialized bits and one ordinary bit. The traditional drill bit exhibited more thrust force and resulted in increased delamination at the hole exit in comparison to alternative drill designs. The mechanical loads experienced throughout the drilling process provide valuable insights into the characteristics of a certain drill bit and facilitate comprehension of the dynamic relationship between the tool and the workpiece. The study conducted by Rahamathullah and Shunmugam [42] investigated the process of micro-drilling of GFRP composites using a carbide drill with a diameter of 0.32 mm. The empirical investigations conducted on blind hole drilling demonstrate a discernible decrease in both the maximum thrust force and torque throughout the process of peck drilling. The thrust force in microdrilling exhibits an upward trend as the feed rate increases, mostly attributed to the rise in chip load. Conversely, the thrust force experiences a decline as the speed increases, primarily owing to the phenomenon of thermal softening, similar to what is seen in the traditional drilling of composites. The peck drilling method exhibits reduced thrust force and torque in comparison to direct drilling when applied to laminates of same thickness. In their study, Wang et al. [28] introduced an examination of optimization and methods for drilling operations on CNC machine tools. The focus was on the use of plane rake faced twist drills, and a deterministic optimization methodology was used. The use of graphical representations in the study of optimization on the feed-speed domain facilitated a more comprehensive comprehension of the economic attributes. The optimization technique was explicitly outlined, ensuring the attainment of the global optimal solution, as a consequence of the effect exerted by the restrictions and machining performance data [43]. Table 1 presents a comprehensive overview of prior research endeavors pertaining to the drilling of polymer matrix composite materials. These investigations have primarily examined various aspects such as operating conditions (specifically cutting speed and feed rate), drill point geometries, tool wear, drilling forces, drilling-induced damage, theoretical approaches, and finite element models of drilled laminates.

Table 1 Previous investigations focusing on the drilling of PMCs

Investigators	Inference
Tsao and Hocheng [24] Bhatnagar et al. [30] Rahamathullah and Shunmugam [42]	**Thrust force and torque** • On increasing the feed rate, thrust force and torque increased
Velayudham [44] Palanikumar et al. [45] Mishra et al. [46] Durao et al. [47] Kumar et al. [48] Singh et al. [49]	**Drill point geometries** • Different types of drill point geometries are twist drill, jo drill, parabolic drill, saw drill, core drill, trepanning tool, candle stick drill, step drill, brad and spur, 8-facet drill, stub length, and straight shank • Different types of drilling-induced damage are fibre pull out, matrix burning, spalling, peel-up and push down delamination, matrix cracking
Rubio et al. [50] Kishore et al. [5] Albuquerque et al. [51] Latha et al. [43]	**Drilling-induced damage** • X-ray, computerized tomography (CT) scan, and non-destructive dye penetrant test can be used for the visualization of drilling-induced damage
Sakuma et al. [7] Durao et al. [47] Faraz et al. [41] Mishra et al. [46]	**Tool wear** • Different types of tool wear can be observed during drilling such as flank wear, crater wear, and combing wear
Shaikh and Channiwala [35] Sardinas et al. [37] Singh et al. [36]	**Analytical model** • Predictive models were developed based on artificial neural network and fuzzy logic
Latha et al. [43] Singh et al. [49] Rakesh et al. [52] Albuquerque et al. [51]	**Finite element approach** • Simple drill point geometries have been used to predict the thrust force • Drill point geometries influence the drilling-induced damage

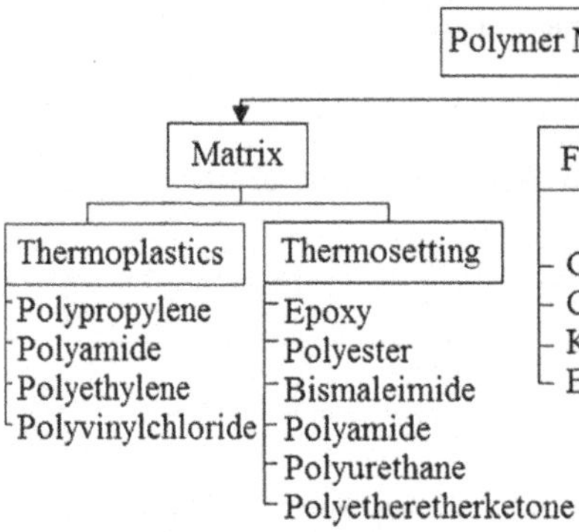

Fig. 1 Classifications of PMCs

1.1 *Matrix Materials and Fibres*

Various matrix materials, including thermosets and thermoplastics, together with various kinds of fibres such as glass, carbon, and kevlar, have been used in the production of polymer matrix composites, as seen in Fig. 1. Thermoplastic matrix materials exhibit greater toughness in comparison to thermosets; nevertheless, their thermal resistance is comparatively weaker, limited to the temperature range of 100–200 °C. Epoxy and polyesters are the matrix materials that are often used in various applications. Polyamides have a high level of thermal resistance, withstanding temperatures above 300 °C. The investigation of matrix materials including fillers such as silicon carbide (SiC), graphite, and alumina has also been conducted.

1.2 *Drill Point Geometry*

The study of drill point geometry has been a key area of research. The implementation of alterations to the drill geometry has yielded improvements in hole quality, an increased quantity of drilled holes, and a decrease in drilling forces. The avoidance of damage has been taken into consideration via the optimization of twist drill geometry, including adjustments to the helix angle, rake angle, and web thickness.

1.3 *Operating Variables*

The majority of researchers have conducted optimization studies on the operational factors, namely the cutting speed and feed rate. In the drilling process of composite materials, the feed rate has a greater impact on the thrust force, torque, and damage compared to the cutting speed. An increase in the feed rate resulted in an escalation of drilling-induced damage. A decrease in the feed rate leads to an increase in the duration of engagement, which subsequently causes damage as a consequence of the heat generated. The development of a thrust force controller was undertaken in order to regulate the feed rate during the drilling process of laminates.

1.4 Drilling-Induced Damage

The process of drilling composite laminates entails unique conditions and needs that are often distinct from those found in metal machining. The distinct behavior of composite materials during machining operations may be attributed to their anisotropic character. The machining process is significantly influenced by the combined effects of the different characteristics of the fibre and the matrix, as well as the orientation of the fibre. The estimate and characterization of damage generated by drilling has been a topic of significant interest in the field. The extent of damage incurred during the drilling process for an anisotropic or orthotropic laminated material is contingent upon the inherent properties and mechanical qualities of the elongated and uninterrupted fibres used inside the lamina. The stacking sequence and mechanical properties inside the interfaces of laminae are oriented in various directions (matrix) over the thickness of the laminate. The topic of interest is the geometry and material properties of drills. The operating conditions refer to the specific circumstances and parameters under which a system or process functions. Process variables refer to the factors or elements that might influence or impact a certain process. These variables can include various parameters, conditions, or inputs that are involved in the execution.

2 Manufacturing of Polymer Matrix Composites

The composite material with a polymer matrix is composed of two main components: the matrix and the reinforcement. The matrix materials used in this context are often thermosets, namely epoxy. The reinforcing component, on the other hand, might take the shape of fibres or textiles.

2.1 Matrix Material

Epoxy is classified as a copolymer in the field of polymer chemistry. The formation of the substance occurred by the combination of two distinct chemical compounds. The terms 'resin' and 'hardener' are often used to designate these substances. The resin is composed of monomers or short chain polymers that possess an epoxy group at both ends. The term "monomer" refers to the fundamental unit of a polymer that is capable of repeating in a chain-like fashion. Upon combining these compounds, the amine groups undergo a reaction with the epoxy groups, resulting in the formation of a covalent bond. The reaction between each amino (NH) group and an epoxy group leads to the formation of a polymer with a high degree of cross-linking, resulting in a stiff and robust material. Epoxy resins may undergo rapid and efficient curing within a temperature range of 20–180 °C, without any emission of volatile components. The resin and hardener have to be thoroughly mixed together until a uniform and consistent blend is achieved. The term 'epoxy' denotes a chemical moiety comprising an oxygen atom that is covalently bound to two carbon atoms, which are themselves already connected in some manner. Given that amines exhibit a propensity to co-react with epoxy groups in a fixed ratio, it is essential to attain the appropriate mix ratio between resin and hardener to guarantee the occurrence of a comprehensive reaction. If the amine and epoxy components are not properly combined in the appropriate proportions, there will be residual resin or hardener present in the matrix. This presence of unreacted materials will have an impact on the ultimate qualities of the material after it undergoes curing. Manufacturers often devise a mix ratio that facilitates precise blending of the resin and hardener, often accomplished by the measurement of weight or volume. The resin and hardener are combined in a 1:1 ratio. In this study, the resin Araldite 'AW 106' is used in conjunction with the hardener 'HV 953'.

Epoxy exhibits advantageous properties due to the presence of two ring groups at its core, enabling it to effectively absorb mechanical and thermal stresses. Consequently, epoxy demonstrates exceptional characteristics such as stiffness, toughness, heat resistance, high strength, tenacity, elasticity, resistance to dynamic loads, compatibility with various structural materials, prolonged operational lifespan, minimal shrinkage, high electrical insulation, and commendable chemical resistance. Additionally, epoxy serves as laminating resin in diverse industrial contexts.

2.2 *Reinforcement*

Glass fibres are renowned for their exceptional strength and rigidity, inherent fire resistance, ability to withstand corrosion and cost-effectiveness. The E-glass is widely used and derives its nomenclature from its favorable electrical characteristics. The S-glass has exceptional strength, rigidity, and thermal stability. The GFRP laminates were reinforced using woven E-glass fibre tape. Glass fibre mostly consists of silicon oxide (SiO_2). The use of woven glass fibre mat as a kind of reinforcement is seen in Fig. 2.

The manufacturing process of GFRP laminates included the use of the hand lay-up method. The mold is comprised of two components. Figure 3 illustrates a somewhat flattened section of the mold. The purpose of creating two half plates was to enhance the efficiency of the compression molding technique used in laminate production.

The mold was constructed using mild steel. The mold was specifically designed to produce GFRP laminates with dimensions of 300×300 mm^2, while allowing for

Fig. 2 Woven glass fabrics

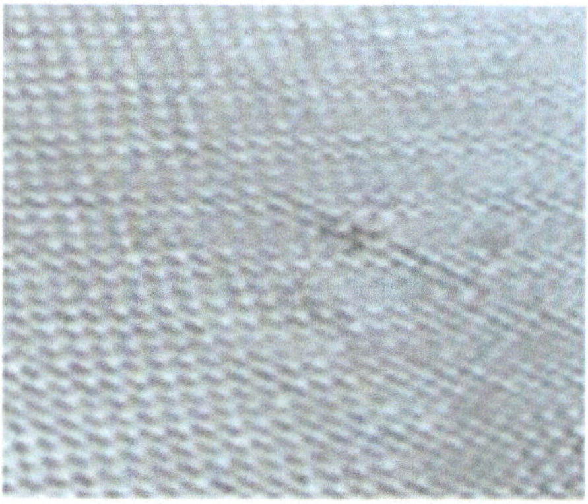

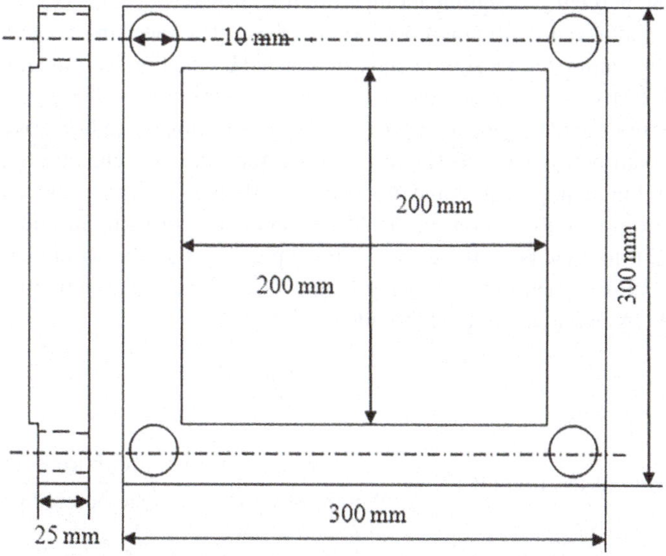

Fig. 3 A half flat part for making GFRP laminates

the production of laminates with different thicknesses. The matrix material used in this study was Araldite 'AW 106' in combination with hardener 'HV 953'. The use of woven glass fibre as the reinforcing material inside an epoxy matrix was seen. The resin and hardener were combined in appropriate ratios, resulting in a homogeneous mixture achieved via continuous agitation of the resin inside the flask to ensure the absence of any voids. In order to aid the simple removal of the laminate, a gel coat was placed onto the mold prior to the lay-up procedure. The woven glass fibre mats were precisely trimmed to match the dimensions of the flat laminate mold and thereafter placed onto the surface of the mold. The composite solution was administered onto a substrate of woven glass fibre using manual application techniques using a hand brush and roller. A sequence of four layers of glass fibre mats was consecutively positioned on top of each other. To minimize the presence of voids and trapped air bubbles in the uncured laminate system, the surplus resin was expelled using compression. A wooden roller was used for the purpose of eliminating surplus resin. The upper plate of the mold was positioned onto the laminate, and afterwards, the bolts were securely fastened in order to exert pressure. The mold was subjected to ambient temperature conditions for about 24 h to facilitate cure. A set of four GFRP laminates, each with a thickness of 4 mm, were fabricated.

2.3 Determination of Fibre Volume Fraction

The volume percent is a significant measure that provides insight into the extent of reinforcing achieved by the incorporation of fibres. The mechanical characteristics of composite laminates are significantly influenced by the fibre volume fraction (vf). A burn-off test was conducted in order to determine the fibre volume percentage of glass fibre reinforced polymer composites. The combustion procedure was conducted inside the confines of a Muffle furnace. The samples were weighed using a computerized weighing equipment with a precision of 0.01 mg. The burn-off samples were extracted from various sections of the GFRP laminate in order to introduce randomization and minimize potential errors. Table 2 presents the fibre volume percentage of the Glass Fibre Reinforced Polymer laminate. The fibre volume fraction (v_f) was determined by using the method provided in Eq. (1).

$$vf = \frac{\rho c.Wf}{\rho f} \tag{1}$$

$$\text{Fiber weight fraction } (Wf) = \frac{\text{Weight of fiber after burn - out}}{\text{Weight of GFRP laminate before burn - out}}$$

whereas, ρ_c is the composite laminate density, ρ_f is the fibre density; W_f is the fibre weight fraction.

Table 2 Fibre volume fraction estimation

S. No.	Volume (cm^3)	Initial Wt. (gm)	Fibre Wt. (gm)	ρ_c (gm/cm^3)	W_f	v_f
1	1.764	4.06961	2.45459	2.3070	0.6031	0.5477
2	1.68	3.96824	2.44313	2.3620	0.6156	0.5724
3	1.6	3.89021	2.32649	2.4313	0.5980	0.5724

3 Experimental Set-Up

The drilling force signals were recorded and analyzed using specialized data collecting and Dyno-Ware assessment software. The composite laminate specimen was secured inside a stiff fixture that was affixed to the dynamometer, which in turn was positioned on the machine table. The dynamometer was securely affixed to the table by use of square-headed bolts that were inserted into the T-slots of the drilling machine table. The composite laminate was affixed to the dynamometer by using washers (spacers) to create a gap between the two, ensuring that the drill does not come into contact with the dynamometer. Figure 4 illustrates the drilling experimental set-up, including the dynamometer.

3.1 Measurement of Drilling Forces

The drilling forces during the drilling of GFRP laminates were measured using a Kistler Type 9272 piezoelectric drill dynamometer, which consists of four components. The dynamometer is comprised of four components, namely piezoelectric sensors that are installed under significant preload conditions, positioned between a base plate and a top plate. The device has a small design, exhibits durability, and demonstrates the ability to quantify three perpendicular thrust forces (Fx, Fy, Fz) as well as torque (Mz) generated during milling operations. In the context of drilling, the use of measurement techniques is limited to capturing signals related to thrust force (Fz) or torque (Mz) alone. The dynamometer has notable characteristics such

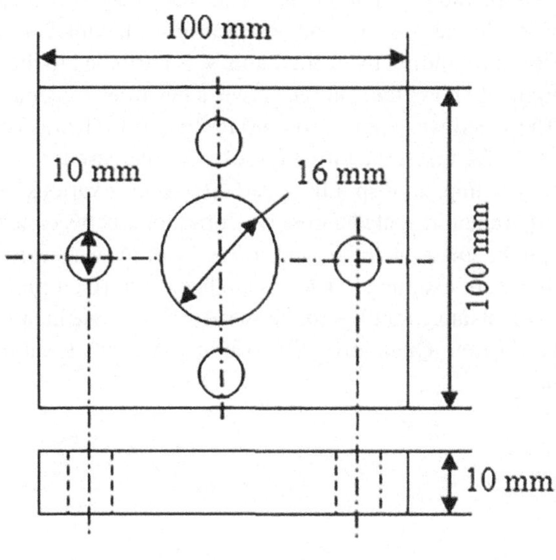

Fig. 4 Laminate holding fixtures

as high stiffness and natural frequency, enabling it to accurately detect and quantify even minute variations in thrust force and torque. The operational mechanism of a dynamometer is as follows: the force of thrust and the moment exerted on the dynamometer are communicated from the top plate to a load washer consisting of four components. The sensor in question consists of four components, each of which includes a pair of quartz rings. One of these pairs is specifically designed to detect compression, while the other two pairs are designed to detect shear forces in either the X or Y direction. The supplementary collection of quartz plates, known as the sensor, is organized in a manner that produces an electric charge that is directly proportional to the moment Mz exerted around the sensor's axis. The dynamometer experiences a thrust force that may be divided into three distinct components. The electrical charge generated by quartz plates is gathered by electrodes that are linked to the sensor.

The multi-channel charge amplifier is responsible for acquiring the output voltage signals that are proportionate to the forces exerted on the dynamometer. The data obtained by the charge amplifier is subjected to analysis via the use of DynoWare software, which is facilitated by the data collecting system. The programme facilitates the storage of individual force measurement data and its corresponding set-up. Signal analysis also allows for the mitigation of undesired signals that arise from the excessive impact of temperature, sometimes referred to as drift compensation [53, 54].

3.2 Drilling Forces

The spindle speed of 355 RPM and feed rate of 0.05 mm/rev. for the drilling process were chosen (Fig. 5). The observation of maximal drilling forces and damage occurring surrounding the drilled hole was attributed to the impact action during drilling. Figure 6 shows the relation between the thrust forces and feed rate at different speeds. The speed was varied from 90 to 2800 rpm. It was observed that with the increase of the feed rate, the thrust forces were increased.

The impact of spindle speed on torque at various feed rates is seen in Fig. 7. The experiment revealed a positive correlation between torque and feed rates. When the spindle speed was increased, a drop in torque was seen across all the various feed rates that were examined. The researchers performed preliminary tests to determine an optimal range for the spindle speed and feed rate in order to achieve high-quality hole production. Qualitative observations were made on the damage around the drilled hole.

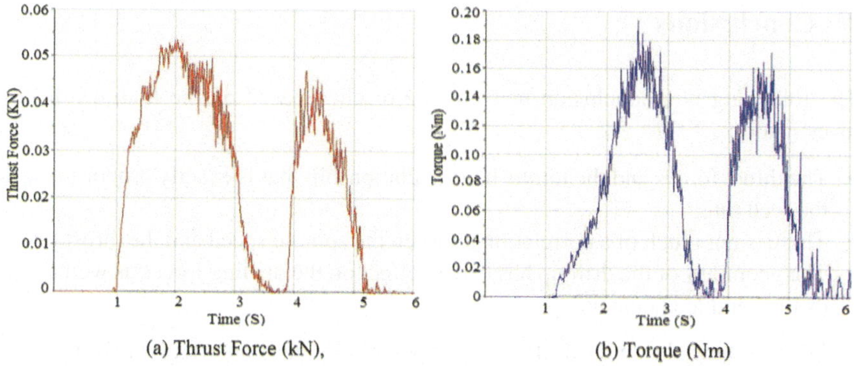

(a) Thrust Force (kN), (b) Torque (Nm)

Fig. 5 Drilling forces with Jo drill geometry

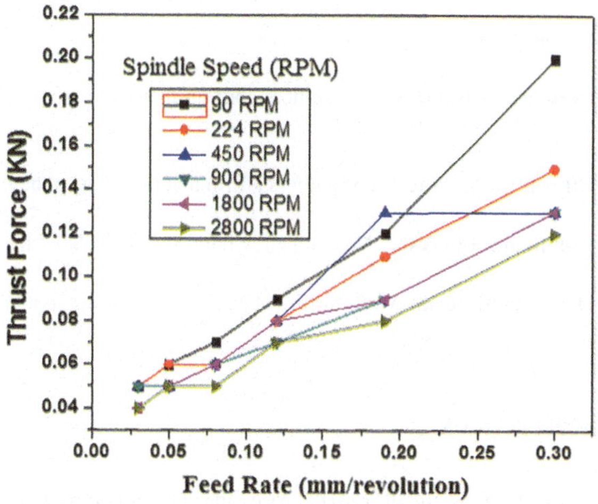

Fig. 6 Thrust force versus feed rate at different spindle speed

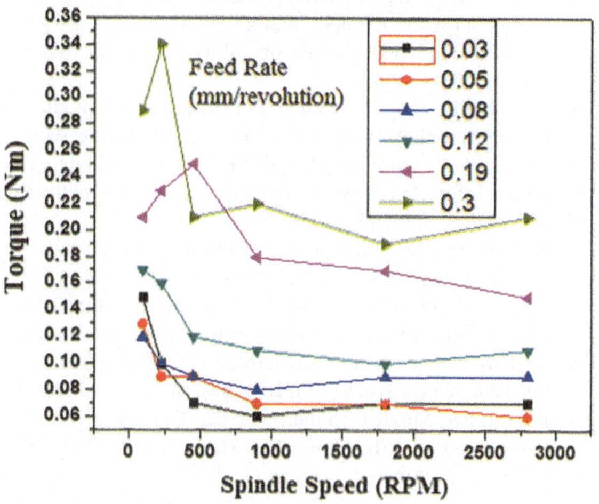

Fig. 7 Torque versus spindle speed at different feed rates

4 Conclusions

The following is a conclusion that may be reached based on the findings of this investigation.

- The thrust forces and the torque have a relationship that is exactly proportional to the feed rate.
- There is not much of a correlation between the spindle speed and the thrust force.
- The geometry of the drill tip has had an effect on the drilling forces as well as the drilling-induced damage.

Damage brought on by drilling might potentially have an effect on the residual strength of drilled composite laminates. Therefore, further research has to be done before employing drilled laminates for any structural purposes.

Declarations

Conflicts of Interest The authors confirm that there are no conflicts of interest associated with this work.

Ethical Compliance The authors affirm that this study adheres to all ethical guidelines.

Competing Interests No competing interests are declared by the authors.

Data and Material Availability Data and materials are not applicable to this study.

References

1. Rakesh PK, Singh I (2018) Processing of green composites. Springer. ISBN 978-981-13-6019-0
2. Naik TJ, Singh I, Sharma AK (2022) Processing of polymer matrix composites using microwave energy: a review. Compos A Appl Sci Manuf 156:106870
3. Chung DDL (2019) A review of multifunctional polymer-matrix structural composites. Compos B Eng 160:644–660
4. Rakesh PK, Singh I, Kumar D (2010) Failure prediction in glass fiber reinforced plastics laminates with drilled hole under uni-axial loading. Mater Des 31(6):3002–3007
5. Kishore RA, Tiwari R, Rakesh PK, Singh I, Bhatnagar N (2011) Investigation of drilling in fiber reinforced plastics using response surface methodology. Proc Inst Mech Eng Part B: J Eng Manuf 225(3):453–457
6. Rakesh PK, Davim JP (2022) Cellulose composites: processing and characterization. DE GRUYTER, Germany. ISBN: 978-3-11-076869-5
7. Sakuma K, Yokoo Y, Seto M (1984) Study on drilling of reinforced plastics (GFRP and CFRP)-relation between tool material and wear behavior. Bulletin of JSME. 27(228):1237–1244
8. Tagliaferri V, Caprino G, Diterlizzi A (1990) Effect of drilling parameters on the finish and mechanical properties of GFRP composites. Int J Mach Tools Manuf 30(1):77–84
9. Rakesh PK, Singh I, Kumar D (2012) Compressive behavior of glass fiber reinforced plastic laminates with drilled hole. Adv Mater Res 410:349–352
10. Rakesh PK, Singh I, Kumar D (2012) Flexural behavior of glass fiber reinforced plastic laminates with drilled hole. Proc Inst Mech Eng Part L: J Mater Des Appl 226(2):149–158

11. Gairola S, Gairola S, Sharma H, Rakesh PK (2019) Impact behavior of pine needle fiber/pistachio shell filler based epoxy composite. J Phys: Conf Ser 1240(012096):1–7
12. Hocheng H, Dharan CKH (1990) Delamination during drilling in composite laminates. J Eng Ind Trans ASME 112:236–239
13. Rakesh PK, Singh I, Kumar D (2012) Drilling of composite laminates with solid and hollow drill point geometries. J Compos Mater 46:3173–3180
14. Caprino G, Tagliaferri V (1995) Damage development in drilling glass fiber reinforced plastics. Int J Mach Tools Manuf 35(6):817–829
15. Di-Paolo G, Kapoor SG, Devor RE (1996) An experimental investigation of the crack growth phenomenon for drilling of fiber reinforced composite materials. J Eng Ind Trans ASME 118:104–110
16. Chen W-C (1997) Some experimental investigations in the drilling of carbon fiber reinforced plastic composite laminates. Int J Mach Tools Manuf 37(8):1097–1108
17. Tsao CC (2008) Experimental study of drilling composite materials with step-core drill. Mater Des 29:1740–1744
18. Lin SC, Shen J-M (1999) Drilling unidirectional glass fiber reinforced composite materials at high speed. J Compos Mater 33(9):827–851
19. Piquet R, Ferret B, Lachaud F, Swider P (2000) Experimental analysis of drilling damage in thin carbon/epoxy plate using special drills. Compos A 31:1107–1115
20. Lachaud F, Piquet R, Collombet F, Surcin L (2001) Drilling of composite structures. Compos Struct 52:511–516
21. Capello E (2004) Workpiece damping and its effects on delamination damage in drilling thin composite laminates. J Mater Process Technol 148(2):186–195
22. Capello E, Tagliaferri V (2001) Drilling damage of GFRP and residual mechanical behavior - part II: static and cyclic bearing loads. J Compos Tech Res 23(2):131–137
23. Hocheng H, Tsao CC (2006) Effects of special drill bits on drilling induced delamination of composite materials. Int J Mach Tools Manuf 46:1403–1416
24. Tsao CC, Hocheng H (2003) The effect of chisel length and associated pilot hole on delamination when drilling composite materials. Int J Mach Tools Manuf 43(11):1087–1092
25. El-Sonbaty I, Khashaba UA, Machaly T (2004) Factors affecting the machinability of GFR/epoxy composites. Compos Struct 63:329–338
26. Ramkumar J, Aravindan S, Malhotra SK, Krishnamurthy R (2004) An enhancement of the machining performance of GFRP by oscillatory assisted drilling. Int J Adv Manuf Technol 23:240–244
27. Ramkumar J, Malhotra SK, Krishnamurthy R (2004) Effect of work piece vibration on drilling of GFRP laminates. J Mater Process Technol 152:329–332
28. Wang J, Zhang Q, Mathew P (2011) Optimization of cutting conditions in drilling operations with plane rake faced twist drills. Mach Sci Technol 15(1):91–109
29. Wang X, Wang LJ, Tao JP (2004) Investigation on thrust in vibration drilling of fiber reinforced plastics. J Mater Process Technol 148:239–244
30. Bhatnagar N, Jalutharia MK, Singh I (2008) Prediction of thrust force and torque when drilling composite materials. Int J Mater Prod Technol 32(2/3):213–224
31. Mohan NS, Kulkarni SM, Ramachandra A (2007) Delamination analysis in drilling process of glass fiber reinforced plastic (GFRP) composite materials. J Mater Process Technol 186:265–271
32. Paul A, Kapoor SG, Devor RE (2005) Chisel edge and cutting lip shape optimization for improved twist drill point design. Int J Mach Tools Manuf 45:421–431
33. Langella A, Nele L, Maio A (2005) A torque and thrust prediction model for drilling of composite materials. Compos A Appl Sci Manuf 36:83–93
34. Varatharajan R, Malhotra SK, Vijayaraghavan L, Krishnamurthy R, (2006) Mechanical and machining characteristics of GF/PP and GF/Polyester composites. Mater Sci Eng 132B:134–137
35. Shaikh AA, Channiwala (2006) Experimental and analytical investigation of jute polyester composite for long continuous fiber reinforcement. J Reinf Plast Compos 25(8):863–873

36. Singh I, Bhatnagar N (2006) Drilling of uni-directional glass fiber reinforced plastic (UD-GFRP) composite laminates. Int J Adv Manuf Technol 27:870–876
37. Sardinas RQ, Reis P, Davim JP (2006) Multi-objective optimization of cutting parameters for drilling laminate composite materials by using genetic algorithms. Compos Sci Technol 66:3083–3088
38. Khashaba UA, Seif MA, Elhamid MA (2007) Drilling analysis of chopped composites. Compos A 38:61–70
39. Rao GVG, Mahajan P, Bhatnagar N (2007) Machining of UD-GFRP composites chip formation mechanism. Compos Sci Technol 67:2271–2281
40. Isik B, Ekici E (2010) Experimental investigations of damage analysis in drilling of woven glass fiber-reinforced. Int J Adv Manuf Technol 49:861–869
41. Faraz A, Heymann T, Biermann D (2011) Experimental investigation on drilling GFRP epoxy composite laminates using specialized and conventional uncoated cemented carbide drills. Mater Manuf Processes 26(4):609–617
42. Rahamathullah I, Shunmugam MS (2011) Thrust and torque analyses for different strategies adapted in microdrilling of glass fiber reinforced plastic. Proc Inst Mech Eng Part B: J Eng Manuf 225:505–518
43. Latha B, Senthilkumar VS, Palanikumar K (2011) Modeling and optimization of process parameters for delamination in drilling glass fiber reinforced plastic (GFRP) composites. Mach Sci Technol 15(2):172–191
44. Velayudham A, Krishnamurthy R (2007) Effect of point geometry and their influence on thrust and delamination in drilling of polymeric composites. J Mater Process Technol 185:204–209
45. Palanikumar K, Rubio JCC, Abrao AM, Correia AE, Davim JP (2008) Influence of drill point angle in high-speed drilling of glass fiber reinforced plastics. J Compos Mater 42(24):2585–2597
46. Mishra R, Malik J, Singh I (2010) Prediction of drilling-induced damage in unidirectional glass fiber reinforced plastic laminates using an artificial neural network. Proc Inst Mech Eng Part B: J Eng Manuf 224(5):733–738
47. Durão LMP, Tavares JMRS, Albuquerque VHC, Marques AT, Magalhães AG, Vieira AA (2010) Tool effects on hybrid laminates drilling. Mater Manuf Processes 25(6):476–481
48. Kumar S, Chauhan SR, Rakesh PK, Singh I, Davim JP (2012) Drilling of glass fiber/vinyl ester composites with fillers. Mater Manuf Processes 27(3):314–319. Impact Factor: 4.616
49. Inderdeep A, Rakesh PK, Malik J (2012) Soft Computing techniques for predicting forces and damage in drilling of polymer composites. In: Mechatronics and manufacturing engineering: research and development. Woodhead Publishing, UK, pp 227–255. ISBN: 978 0 85709 150 5
50. Rubio JC, Abrao AM, Faria PE, Correia AE, Davim JP (2008) Effects of high speed in the drilling of glass fiber reinforced plastic: Evaluation of the delamination factor. Int J Mach Tools Manuf 48:715–720
51. Albuquerque VHCD, Tavares JMRS, Durao LMP (2010) Evaluation of delamination damage on composite plates using an artificial neural network for the radiographic image analysis. J Compos Mater 44(9):1139–1159
52. Rakesh PK, Sharma V, Singh I, Kumar D (2011) Delamination in fiber reinforced plastics: a finite element approach. Engineering 3:249–254
53. Kishore RA, Tiwari R, Dvivedi A, Singh I (2009) Taguchi analysis of the residual tensile strength after drilling in glass fiber reinforced epoxy composites. Mater Des 30:2186–2190
54. Kumar J, Kumar V, Singh I, Rakesh PK (2021) Joining behavior of polymeric composites fabricated using agricultural waste as fillers. J Adhes Sci Technol 35(15):1652–1663

Chapter 14
Finite Element Modelling and Dynamic Behavior Analysis of Sandwich Panel with Cutout

Ravi Kumar, Rajesh Kumar, Sandeep Tiwari, and Chetan Kumar Hirwani

1 Introduction

Composite materials are engineered materials that are formed by combining two or more different constituents to create a material with enhanced properties and performance that cannot be achieved by any single constituent on its own. These constituents, referred to as phases, can be materials with different physical or chemical properties, such as polymers, metals, ceramics, or natural fibers. The combination of these materials in a composite creates a material with improved characteristics, including strength, stiffness, durability, and more, making composites highly versatile and widely used in various industries.

Key components of composite materials include:

- Reinforcement: The reinforcement material in the form of fibers, particles, or sheets, is often added/used in the composite to enhance the composite's mechanical properties. Reinforcements can be made of materials like carbon fibers, glass fibers, aramid fibers, natural fibers (e.g., bamboo or hemp), or metallic particles.
- Matrix: The matrix is the material that surrounds and binds the reinforcement materials. It typically acts as a support structure and provides protection to the reinforcement. Common matrix materials include polymers (e.g., epoxy, polyester, or thermoplastics), metals (e.g., aluminum), and ceramics.

Composites are categorized based on the type of matrix and reinforcement used and can be seen in Fig. 1. The most common types include:

R. Kumar · R. Kumar · S. Tiwari · C. K. Hirwani (✉)
Department of Mechanical Engineering, National Institute of Technology Patna, Patna 800005, Bihar, India
e-mail: chetanhirwani111@gmail.com

S. K. Sethi et al. (eds.), *Polymer Composites: From Computational to Experimental Aspects*, Materials Horizons: From Nature to Nanomaterials,
https://doi.org/10.1007/978-981-97-0888-8_14

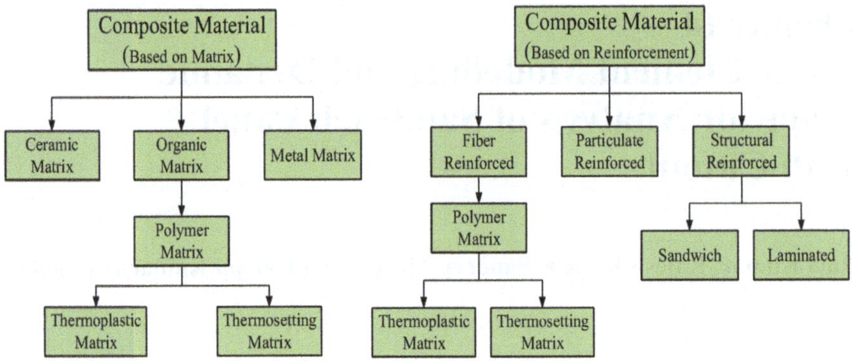

Fig. 1 Types of composite material

- Fiber-Reinforced Composites: These composites use fibers (e.g., carbon, glass, or aramid) as the reinforcement and a matrix material (usually a polymer or resin). They are known for their excellent strength-to-weight ratio and are widely used in aerospace, automotive, sporting goods, and construction.
- Particulate Composites: In these composites, small solid particles (e.g., ceramic or metal particles) are added to a matrix material. Particulate composites can enhance properties like wear resistance, thermal conductivity, and electrical conductivity. They are used in applications such as brake pads, metal matrix composites, and some aerospace components.
- Laminar Composites: These composites consist of multiple layers of materials, often in the form of sheets or laminates. The layers are stacked and bonded together to create a structure with specific properties. Laminar composites are widely used in aerospace for their lightweight and high-strength characteristics.
- Sandwich Composites: Sandwich composites combine two thin, stiff face sheets with a lightweight core material. This design provides exceptional strength and stiffness while minimizing weight. They are used in applications like aircraft panels, boat hulls, and architectural components.

As the properties could be tailored with different combinations and proportions of the reinforcement and matrices materials, there are various advantages. A few important of them are discussed below:

- **High Strength and Stiffness**: Composites can be designed to be stronger and stiffer than their individual components, making them ideal for structural applications.
- **Lightweight**: Due to their high strength-to-weight ratio, composites are significantly lighter than many traditional materials, making them suitable for weight-sensitive applications.
- **Corrosion Resistance**: Many composite materials are naturally corrosion-resistant, reducing maintenance and extending the lifespan of structures and components.

- **Tailored Properties**: Engineers can customize composites to meet specific requirements by selecting the type of reinforcement, matrix material, and manufacturing processes.
- **Design Flexibility**: Composite materials allow for complex shapes and designs, enabling innovative and efficient solutions in various industries.
- **Reduced Energy Consumption**: In transportation and aerospace, lightweight composites contribute to reduced fuel consumption and emissions.
- **Electrical and Thermal Properties**: Composites can be engineered to exhibit electrical insulating or conductive properties, as well as various levels of thermal conductivity.

Composite materials have a wide range of applications across various industries due to their unique combination of properties. Here are some common applications of composite materials in various fields which are available in Fig. 2 and a brief discussion is listed below:

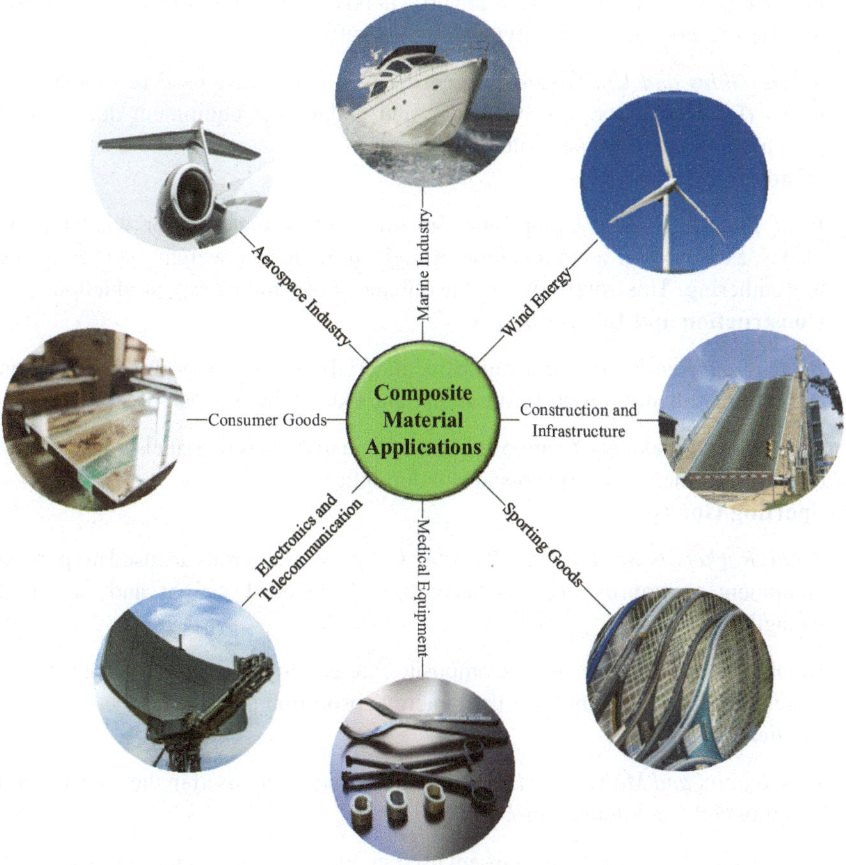

Fig. 2 Application of composite materials in various field

- **Aerospace Industry**:

  Aircraft Components: Composite materials are used in aircraft components, including wings, fuselage sections, and tail surfaces, to reduce weight and improve fuel efficiency.

  Spacecraft: Composites are used in spacecraft components and satellite structures, where weight savings are critical for launching and maintaining payloads in space.
- **Automotive Industry**:

  Vehicle Body Parts: Composite materials are used in automotive body panels, such as hoods, fenders, and doors, to reduce vehicle weight and improve fuel economy.

  Chassis and Suspension Components: Composites are employed in components like suspension arms and drive shafts for their strength and weight reduction benefits.
- **Marine Industry**:

  Boat Hulls: Composite materials are used in boat hulls to provide lightweight and durable solutions for various types of watercrafts.

  Submersibles and Underwater Structures: Composites are used in submarines, ROVs (Remotely Operated Vehicles), and underwater equipment due to their resistance to corrosion and water damage.
- **Wind Energy**:

  Wind Turbine Blades: Composites are the primary material for wind turbine blades, as they offer a combination of high strength, low weight, and resistance to weathering. This contributes to the efficiency of wind energy production.
- **Construction and Infrastructure**:

  Bridges and Reinforcement: Composite materials are used in bridge construction and reinforcement to improve durability and extend the lifespan of structures.

  Architectural Elements: Composites are used in architectural panels, facades, and cladding for their aesthetic appeal and durability.
- **Sporting Goods**:

  Tennis Rackets, Golf Clubs, and Bicycles: Composite materials are used in sporting equipment to improve performance through reduced weight and increased strength.

  Helmets and Protective Gear: Composites are employed in protective gear such as helmets and body armor for their energy absorption properties.
- **Medical Equipment**:

  Wheelchairs and Mobility Aids: Composite materials are used in the construction of lightweight and durable mobility aids.

  Orthopedic Devices: Composites are used in orthopedic implants and support for their biocompatibility and strength.

- **Electronics and Telecommunication**:

  Antennas and Radomes: Composites are used in telecommunications equipment for their electromagnetic transparency and durability.

  Printed Circuit Boards (PCBs): Composites are used in high-performance PCBs for their thermal and electrical properties.
- **Defense and Aerospace**:

  Armored Vehicles: Composites are used in the construction of armored vehicles to enhance protection while keeping weight manageable.

  Military Aircraft: Composites play a crucial role in the construction of military aircraft and UAVs (Unmanned Aerial Vehicles) for their lightweight and stealth properties.
- **Consumer Goods**:

  Luggage: Composite materials are used in lightweight and durable luggage and bags.

  Furniture: Composites are used for modern and stylish furniture pieces that are lightweight and robust.

The above discussed are just a few examples, and composite materials continue to find applications in new and innovative ways across various industries. The versatility of composites allows engineers and designers to tailor the material properties to meet specific requirements, making them a critical part of modern engineering and manufacturing.

The present work mainly focused on the sandwich, which is one of the composite materials. It is also extensively used in aerospace for various applications due to its lightweight and high-strength properties. The structure of a sandwich composite typically consists of a lightweight core material, such as foam or honeycomb, sandwiched between two face sheets, which are usually made of high-strength composite materials like carbon fiber-reinforced composites or fiberglass-reinforced composites. The adhesive bonding process, in conjunction with the material properties employed in fabrication, plays a pivotal role in determining the performance and stability of sandwich structures.

Presently, structures featuring cutouts are experiencing widespread utilization in various industries. In the context of a sandwich composite plate, a cutout denotes a deliberately designed void or aperture. These cutouts serve a multitude of purposes, encompassing the facilitation of housing components such as fasteners, creating access points, or accommodating windows. Additionally, they play a significant role in contributing to the overall reduction in the weight of the structure, which is of paramount importance for enhanced efficiency and performance. The presence of a cutout can significantly affect the mechanical behavior of the plate [1]. Sandwich composite plates with cutouts are widely used in aerospace, automotive, marine, and other industries where they are employed in challenging environments, enduring significant axial mechanical loads during their operational service [2].

Sandwich structures with the capability to deform under shear loads have a tendency to undergo buckling when exposed to various mechanical loads. Buckling is a structural instability phenomenon in which a slender structural component, such as a column, beam, or shell, fails suddenly and uncontrollably due to the application of compressive loads. It occurs when the critical load or stress level is exceeded, causing the structural member to deform laterally or buckle out of its original shape. Buckling is a mode of failure that can lead to structural collapse or loss of load-carrying capacity. This buckling behavior is a critical consideration in the analysis and design of such structures. However, buckling does not signify the ultimate failure of the structure, as it retains the ability to withstand additional loads even after experiencing geometric distortion from buckling, a property referred to as its post-buckling strength. Therefore, from a designer's perspective, it is crucial to determine the precise critical buckling load parameter of layered sandwich structural components. This knowledge is essential for achieving optimal designs for future applications. Material property investigation is the essential step while selecting the material for structural applications and assessing the structural behavior (bending/buckling/vibration) of composite materials [3–6]. Mechanical properties with microstructure characterization using an experimental approach are primary studies reported in open literatures [7–10]. Molecular dynamics-based study is a new approach to assess the performance of polymer composites and new age materials [11–13]. Extensive research has been conducted in the academic literature concerning the buckling behavior of composite plates under uniform in-plane loading condition.

Sriwasan and Chellapandi [14] investigated the dynamical stability of symmetrical, anti-symmetric, and asymmetric plates with clamped edges using a finite strip deformations model that did not take transverse deformation caused by shear or rotating inertia into account. The state-space technique was used by Khdeir and Reddy [15] to tackle the buckling issue of cross-ply layered beams with varied end conditions. The buckling and free vibration analysis of stepped layered composite beams has been conducted by Song and Waas [16] using the simple higher-order model. Gilat et al. [17] investigated the elastic buckling load of a layered composite flat plate, employing global–local plate theory as their analytical framework. They compared the outcomes of a zig-zag theory with those of a discrete layer theory in various other scenarios. Wang and Chen [18] investigated the axisymmetric dynamical stability of sandwiched circular sheets with a limited damper layer. The sheets were subjected to a repeated equal radial force across the outside edge, with the viscous material assumed to be frequency dependent. The combined layer wise hypothesis for sandwich panels was expanded by Carrera and Demasi [19]. The total amount of unknowns in such models is proportional to the number of layered composite layers, and it grows as the quantity of layers rises. Based on lower-order kinematic model, Park et al. [20] described the thermally buckled of a flat composite structure using SMA fiber. It is shown that employing SMA fibers reduces thermal big deformation and the crucial temperature. Shukla et al. [21] concentrated on analyzing the buckling behavior of a rectangular laminated flat panel employing the First-Order Shear Deformation Theory (FSDT). They explored various combinations of boundary conditions, including clamped, simply supported, and free, to investigate how they influenced the critical loads. Their study also delved into the impact of factors such as number of layers, plate aspect ratio, lamination scheme, and material parameters on these critical loads. Ungbhakorn and Singhatanadgid [22] examined the critical buckling load of the laminated structure by utilizing the Kantorovuch method. The accuracy of the developed method was validated by comparing it to existing Levy and Rayleigh–Ritz solutions. The results provided strong evidence that this semi-analytical approach can effectively analyze the buckling behavior of laminated unidirectional and cross-ply symmetrical plates, regardless of the combination of simple, clamped, or free supports used. Woo et al. [23] studied the post-buckling reaction of an FGM panel and a narrow panel under side compressible stresses and an elevated temperature field using a third order kinematic model. Liang and Chen [24] conducted research on the buckling behavior of square cell honeycomb structures and observed that various types of cells exhibit distinct buckling modes, each characterized by different governing equations. Additionally, they devised an innovative iterative optimization approach specifically tailored for square cell honeycomb panels. Using an improved HSDT finite elements approach, Chakrabarti and Sheikh [25] investigated the dynamic unstable of layered sandwich sheets exposed to in-plane edge stress. The effect of diverse panel designs, boundary circumstances, layering sequences, and thickness proportions on the dynamic stability behavior of these systems was investigated. Shen [26] investigated both the post-buckling and buckling reactions of FG panels using the piezoelectric fibers reinforcing composite actuator. They employed a higher-order kinematic model and governing equation that took into account piezoelectric effects. For the purpose of static, buckling, and frequency response of usually layered composite beams with different end conditions, Boay and Wee [27] devised a closed form formulation utilizing Euler–Bernoulli theory in combination with the classical laminate system. Under axial compression, the dimensional stability of multilayer composite beams reinforced with piezoelectric materials was investigated by Fridman and Abramovich [28]. Shariyat [29] investigated the longitudinal elasticity of sandwiched panels using a high-order global–local model that assures the stable condition of all shifts and transverse stress constituents. Prakash et al. [30] used the shear deformed finite elements technique to investigate the thermally post-buckling of FGM skewed panels. The Mori–Tanaka technique was utilized in their investigation to evaluate the most efficient material qualities. FSDT in solutions was employed, together with linear piezoelectric constitutive equations. In their investigation, Yang et al. [31] delved into the buckling characteristics of paper honeycomb sandwich panels featuring composite skins, subjecting them to dynamic and compressive forces using ANSYS. Their exploration of wall structure using finite element codes, including MARC, ANSYS, and ABAQUS, aimed to demonstrate the consistency between the numerical results and the corresponding experimental findings. Ferreira et al. [32] investigated the buckling behavior of flat panel laminated composite structures using the lower-order shear deformation theory and a collection approach. They validated their numerical model by comparing it to previously published results and found a close agreement between the computed outcomes and the existing data. Neves et al. [33] pioneered the exploration of the buckling behavior within sandwich structures featuring graded functionally oriented face skins. Their research employed a sine shear deformation model and incorporated radial basis functions as essential components. Ouinas and Achour [34] conducted a comprehensive analysis of a square laminated composite plate featuring an elliptical notch subjected to axial load. They utilized the commercial finite element analysis software ABAQUS to evaluate the plate's stability conditions. Jeyakrishnan et al. [35] carried out a comprehensive analysis of eigenvalues for a sandwich structure featuring a honeycomb core. Their study encompassed the application of a theoretical approach and meticulous examination of the structural response to static loads while maintaining simply supported conditions. Jabbari et al. [36] carried out a research study aimed at exploring the buckling behavior of a circular sandwich plate equipped with a piezoelectric actuator. Their investigation took temperature variations into account and utilized the FSDT model, with a clamped boundary condition as a key component of their analysis. Heydari et al. [37] employed the generalized differential quadrature technique (GDQM) to investigate the bending reactions of carbon nanotube functionally graded plates exposed to uniform pressure and surrounding in elastomeric media using FSDT. Yufeng and Wei [38] employed the FSDT approach in their research to examine the buckling behavior of a symmetric laminated plate. They focused on analyzing the plate's response under simply supported conditions. Kheirikhah and Babaghasabha [39] presented a numerical model for the study of buckling and bending characteristics in a sandwich plate. Their approach involved utilizing the finite element analysis software ANSYS for a comprehensive analysis of the plate's structural behavior. Kahya [40] investigated the stability of layered and sandwich composite beams. Their study involved the utilization of a lower-order kinematic model in combination with finite element analysis to assess the structural stability of these beams. Fan and Wang [41] investigated the bending responses of laminated hybridized plates with CNT-reinforced layers placed in elastic foundation under the impact of matrix fractures. Hajlaoui et al. [42] conducted a study to explore the buckling properties of a composite shell reinforced with carbon nanotubes. Their analysis involved the use of the FSDT to investigate the structural behavior of the composite shell. Amiri et al. [43] conducted an investigation into the stress and stability behavior of a flat sandwich panel. This particular sandwich structure was characterized by a soft core made of foam and face skins embedded with carbon nanotubes. Higginson et al. [44] undertook a study to examine the buckling behavior of a non-rectangular plate. Their approach involved the utilization of the Rayleigh–Ritz methodology, and the results were validated through finite element analysis steps. Nima and Ganesan [45] focused their research on the buckling analysis of a symmetrically layered flat panel. They applied the FSDT and the Ritz method, considering a specific set of boundary conditions in their analysis. Vinh et al. [46] used FSDT and finite element stages to investigate the buckling and bending reactions of a functionally graded plate. To meet the free-conditions on both the top and bottom surfaces, they customize the transverse shear stresses. Chen et al. [47] utilized the Galerkin method based on the FSDT to investigate the stability characteristics of a sandwich structure featuring a corrugated core. Their analysis was conducted under the simply supported edge conditions. Ye et al. [48] concentrated their research on the stability testing of a layered sandwich plate with functionally graded material. They employed the layerwise theory to analyze and assess the plate's stability under various conditions. Pavan et al. [49–51] introduced a numerical method based on Isogeometric collocation. They utilized this method to conduct an extensive investigation of the static, dynamic, and stability characteristics of a laminated structure. Afonso et al. [52] conducted a study focused on techniques for controlling the static and dynamic buckling behavior of a layered sandwich flat panel. Their investigation included the implementation of a numerical approach that incorporated the utilization of piezoelectric patches.

Based on the literature review discussed, it is evident that the majority of research efforts have focused on analyzing the buckling behavior of laminated composite flat plates. There has been relatively limited research dedicated to the study of sandwich structures. As of now, to the best of the authors' awareness, there exists no open literature addressing the buckling behavior of sandwich composite plates featuring cutouts, specifically analyzed using the FSDT, when subjected to uniaxial mechanical loading conditions. In order to achieve this objective, a comprehensive numerical model has been developed in simulation tool ABAQUS, utilizing lower-order kinematics principles, specifically tailored for analyzing sandwich composite plates featuring cutouts. To accomplish this, the sandwich structure, consisting of face sheets and a core with cutouts, was modeled using two-dimensional elements in ABAQUS, specifically employing eight-noded elements of type S8R5. Furthermore, the current model's validity has been confirmed by comparing the predicted results with the published data. Following a successful validation study, the developed numerical model has been employed for in-depth investigations into the buckling behavior of flat sandwich panels, taking into account the effects of cutouts, including their size, shape, position, the ratio of core thickness to face thickness, and end conditions.

2 Mathematical Formulation

A sandwich composite structure, as depicted in Fig. 3, is a type of material assembly of two outer layers, known as face skin, and a core material placed between them. The dimensions of the sandwich composite panel are represented as $a \times b \times t$, where: 'a' refers to a length, 'b' refers to a width, and 't' refers to a uniform thickness. The thickness of the face skin and core are denoted as 't_f 'and 't_c'. A cutout of arbitrary shape is positioned at the center of the composite structure.

2.1 Displacement Field

In the FSDT, it is assumed that the in-plane displacements exhibit linear variation within the thickness of a composite structure, while the out-of-plane displacement is considered to be solely dependent on the in-plane coordinates. The mathematical

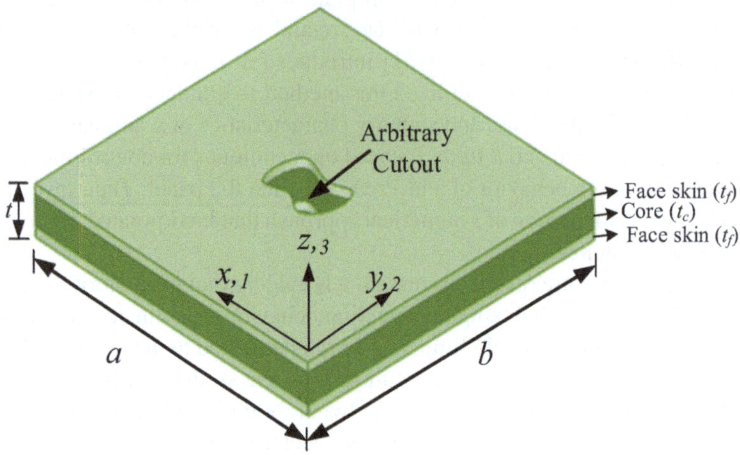

Fig. 3 Sandwich plate with arbitrary cutout

expression representing the displacement field is as follows:

$$\left.\begin{aligned} u(x_1, y_2, z_3) &= \overline{u}(x_1, y_2) + z_3\phi_{x_1}(x_1, y_2) \\ v(x_1, y_2, z_3) &= \overline{v}(x_1, y_2) + z_3\phi_{y_2}(x_1, y_2) \\ w(x_1, y_2, z_3) &= \overline{w}(x_1, y_2) + z_3\phi_{z_3}(x_1, y_2) \end{aligned}\right\} \tag{1}$$

where, $\overline{u}$ and $\overline{v}$ denote in-plane displacements, typically along the $x,_1$ and $y,_2$ axes, and $\overline{w}$ represents the out-of-plane displacement. The axes of rotation within the mid-plane are indicated by the symbols ϕ_{x_1} and ϕ_{y_2}.

2.2 Constitutive Relation

The constitutive relation for the kth lamina, describing the stress–strain relationship at an arbitrary angle θ about any arbitrary axis, is expressed as follows:

$$\{\sigma\} = \left[\overline{Q}\right]\{\varepsilon\} \tag{2}$$

here, the symbols $\{\sigma\}$, $\left[\overline{Q}\right]$ and $\{\varepsilon\}$ are used to represent stress, reduced transformation stiffness, and strain, respectively.

2.3 Finite Element Formulation

The finite element method (FEM) is widely recognized as a valuable numerical tool for conducting structural analyses involving diverse geometric shapes and complex material properties. In this current study, an eight-noded isoparametric element with six degrees of freedom (DOFs) per node is utilized for the purpose of discretization. The displacement vector, denoted as 'λ', at any point within the mid-plane can be represented as follows:

$$\{\lambda\} = \sum_{i=1}^{n} [N]\{\lambda_i\} \tag{3}$$

where, $\{\lambda_i\} = \left\{\overline{u}, \overline{v}, \overline{w}, \phi_{x_1}, \phi_{y_2}, \phi_{z_3}\right\}$ is expressed as the nodal displacement vector, and $[N]$ represents the nodal interpolation function.

2.4 Governing Equation

The final formulation of the governing equation for the buckling analysis of sandwich flat panel is presented as follows:

$$\delta\Pi = 0 \tag{4}$$

here, $\Pi = U + U_{in}$

$$[K] - P_{cr}[K_G] = 0 \tag{5}$$

where, K is the global stiffness matrix, K_{G} is the geometrical stiffness matrix, and P_{cr} denotes the eigenvalue or critical buckling load. Now, the end condition is applied in Eq. (5) to solve it and to obtain the critical buckling load or eigenvalue.

2.5 Boundary Condition

The following boundary conditions have been used while solving the problems.

Clamped condition: $\overline{u} = \overline{v} = \overline{w} = \phi_{x_1} = \phi_{y_2} = \phi_{z_3} = 0$ at $x = 0$ and a

Simply supported: $\overline{v} = \overline{w} = \phi_{x_1} = \phi_{z_3} = 0$ at $x = 0$ and a

Pinned condition: $\overline{u} = \overline{v} = \overline{w} = 0$ at $x = 0$ and a

3 Steps for Buckling Analysis of Sandwich Panel

Buckling analysis in ABAQUS involves determining the critical load at which a structure or component becomes unstable and experiences a sudden deflection or failure due to buckling. Here are the general steps for performing a buckling analysis in ABAQUS:

- **Geometry Creation (PART Module)**: The steps start by creating the geometry of the sandwich panel within the PART module. This involves defining the size, shape, and dimensions of the panel.
- **Material Property Specification (PROPERTIES Module)**: Under this module, the material properties for both the face sheet and the core of the sandwich are being defined. This includes parameters such as elastic modulus, Poisson's ratio, and density. Using this module, the material property from the inbuilt material library as well as user defined material could be added to the panels.
- **Lay-Up Assembly (ASSEMBLY Module)**: The parts created in the previous process are assembled to obtain the desired/complete lay-up of the sandwich using the ASSEMBLY module. Also, for the face sheet, the number of plies and their orientation/ stacking sequence is also been defined.
- **Setup (SETUP Module)**: The SETUP module, initiates the setup process for the analysis. Under the "Linear Perturbation" category, the "Buckle" option is used to set up a buckling analysis. Also, the desired number of eigenvalues for the buckling analysis has been specified.
- **Mesh Generation (MESH Module)**: The mesh generation for the final assembled part is performed within the MESH module. The creation of mesh for the sandwich panel should be appropriately refined so that it ensures investigation of the buckling responses of the sandwich panel with high/sufficient accuracy.
- **Boundary Conditions (LOAD Module)**: Using this module, the boundary conditions to constrain the panel's degrees of freedom are being imposed. These conditions restrict the panel's movement and simulate its interaction with the surrounding environment.
- **Load Application (LOAD Module)**: This module facilitates to apply the required loads to the sandwich panel. In particular, define how and where the load will be applied, such as specifying the panel edge or other relevant locations. Using this module, the required in-plane mechanical load at the appropriate edge of the sandwich panel has been applied.
- **Job Creation (JOB Module)**: JOB module allows to create a final job. The job with the defined analysis parameters, such as load magnitude, boundary conditions, and analysis type (buckling in this case) has been set up using this module.
- **Job Submission (JOB Module)**: After configuring the job, it is submitted to run the analysis. The software solves the equations and computes the buckling modes and corresponding eigenvalues.

- **Results Visualization (VISUALIZATION Module)**: Finally, the evaluated responses could be seen using the VISUALIZATION module. One can view the buckling modes, eigenvalues, stress distributions, and other relevant information to assess the behavior of the sandwich panel.

The described steps offer a methodical framework for conducting a buckling analysis of a composite/sandwich panel within the commonly used finite element analysis (FEA) software ABAQUS. Additionally, a visual representation of the buckling analysis process is provided in Fig. 4, showing the flow diagram of the steps involved.

4 Results and Discussion

4.1 Validation Study

In this section, a comprehensive comparative analysis has been conducted to evaluate the accuracy and dependability of the simulation model. To do so, a symmetric cross-ply laminated plate with a simply supported edge subjected to uniaxial mechanical loading, as described in reference [53], was chosen as the basis for the study. The laminated plate under consideration is configured with two distinct lay-up schemes, namely 0°/90°/0° and 0°/90°/90°/0°, both with the thickness ratio of 10. The material parameters employed in the analysis are consistent with those outlined in the reference and are also provided below for reference.

$$\frac{E_1}{E_2} = 3, 10\, G_{12} = G_{13} = 0.6E_2$$
$$G_{23} = 0.5E_2\, \upsilon_{12} = 0.25$$

The dimensionless critical load factor of layered composite plate is calculated by using the following equation:

$$\lambda_{P_{cr}} = \frac{P_{cr}a^2}{E_2 t^3} \tag{6}$$

where P_{cr} indicates the critical load or eigenvalue.

The comparison of the presently calculated and corresponding published results is presented in Table 1, exhibits a good agreement.

4.2 New Numerical Examples

After the verification of the current numerical model's accuracy, it has now been extended to explore a parametric effect on the buckling behavior of sandwich panel

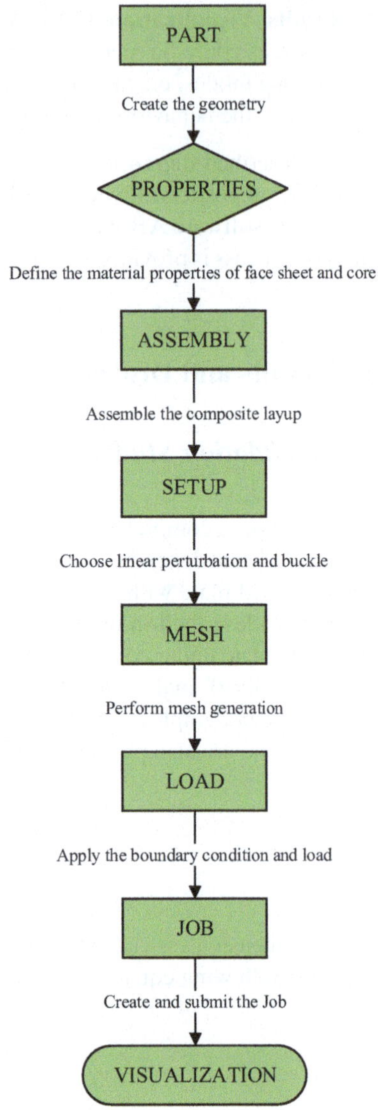

Fig. 4 Steps for buckling analysis of sandwich panel

with cutout. For the analysis, a square ($a/b = 1$) sandwich flat panel with lay-up (0°/90°/core/ 90°/0°) as shown in Fig. 5, square cutout ($c/a = 0.2$), thickness ratio ($a/t = 40$), core-face thickness ratio of ($t_c/t_f = 16$) subjected to uniaxial axial compressive load (at $x = a$) and one end clamped boundary condition, unless otherwise specified. The material properties of sandwich structures used in further ingestion are provided in Table 2. The critical buckling load of sandwich structures is presented in dimensionless form by using Eq. (6).

Table 1 Critical load factor (λP_{cr}) for symmetric cross-ply plate with ($a/t = 10$) subjected to uniaxial loading

Laminate	Solution method	Orthotropy ratio (E_1/E_2)	
		3	10
[0°/90°/0°]	Present	6.86	8.11
	Reddy and Phan [54]	5.39	9.94
	Singh et al. [53]	5.410	9.89
[0°/90°]	Present	7.17	9.64
	Liu et al. [55]	5.41	10.01
	Singh et al. [53]	5.41	10.02

Fig. 5 Lay-up for sandwich panel

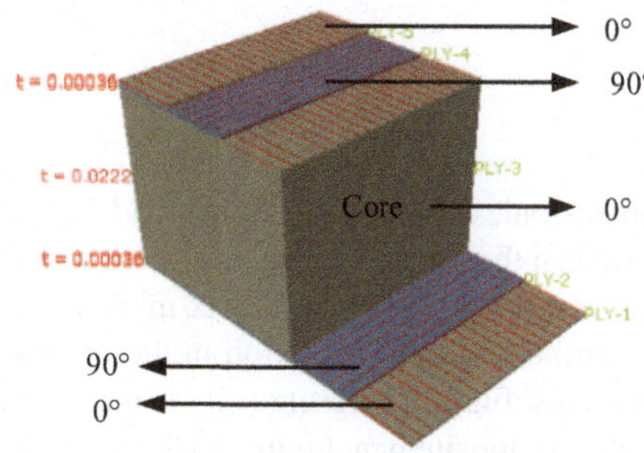

Table 2 Materials properties for the sandwich panel

Material properties	Face sheet	Core
$E_1(Pa)$	25×10^9	0.04×10^9
$E_2(Pa)$	1×10^9	0.04×10^9
$G_{12}(Pa)$	0.5×10^9	0.016×10^9
$G_{13}(Pa)$	0.5×10^9	0.06×10^9
$G_{23}(Pa)$	0.2×10^9	0.06×10^9
$\upsilon_{12} = \upsilon_{13}$	0.25	0.25
υ_{23}	0.25	0.25
$\rho(Kg/m^3)$	1000	94.195

4.2.1 Effect of Cutout Ratio on Buckling Behavior of Sandwich Plate

In this particular example, the cutout ratio effect on the buckling behavior of the sandwich panel has been investigated. The term "cutout ratio" typically refers to the ratio of the cutout area (the portion of the panel that has been removed or cutout) to the total surface area of the panel. The sandwich structure has been examined using five distinct cutout ratios, precisely at values of ($c/a = 0$, 0.2, 0.3, 0.4, and 0.5). The

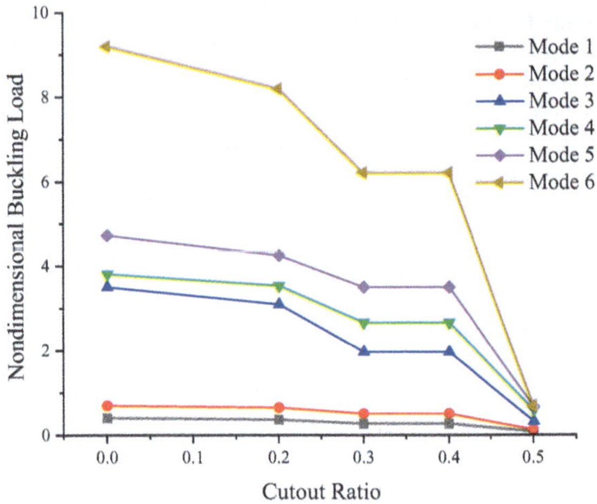

Fig. 6 Impact of cutout ratio on buckling behavior of sandwich flat panel

dimensionless critical buckling load has been computed and is depicted in Fig. 6. Based on the plotted figure, it is clear that the critical buckling load decreases as the cutout ratio increases for all six modes. This decrease in critical buckling load may be attributed to the reduction in the overall stiffness of the plate as the cutout ratio increases. Figure 7 presents all six mode shapes for the computed buckling behavior of the composite panel with a 0.2 cutout ratio.

4.2.2 Effect of Cutout Shape on Buckling Behavior of Sandwich Plate

In this particular case, the investigation has focused on assessing how the shape of the cutout affects the stability of a sandwich flat panel. Four distinct cutout shapes (square, rectangular, circular, and elliptical), as depicted in Fig. 8, were considered to assess their impact on the panel's buckling behavior. It is noteworthy that the areas of all the cutout shapes are kept identical for the analysis. The non-dimensional buckling loads calculated for these different cutout shapes have been graphed in Fig. 9. Notably, the square cutout shape exhibits the maximum critical buckling load, while the circular cutout shape yields the minimum. The rectangular and elliptical cutout shapes fall in between, demonstrating moderate values concerning their critical buckling loads. This analysis emphasizes the significance of cutout shape when evaluating the stability of sandwich flat panels and highlights the potential impact on their structural performance.

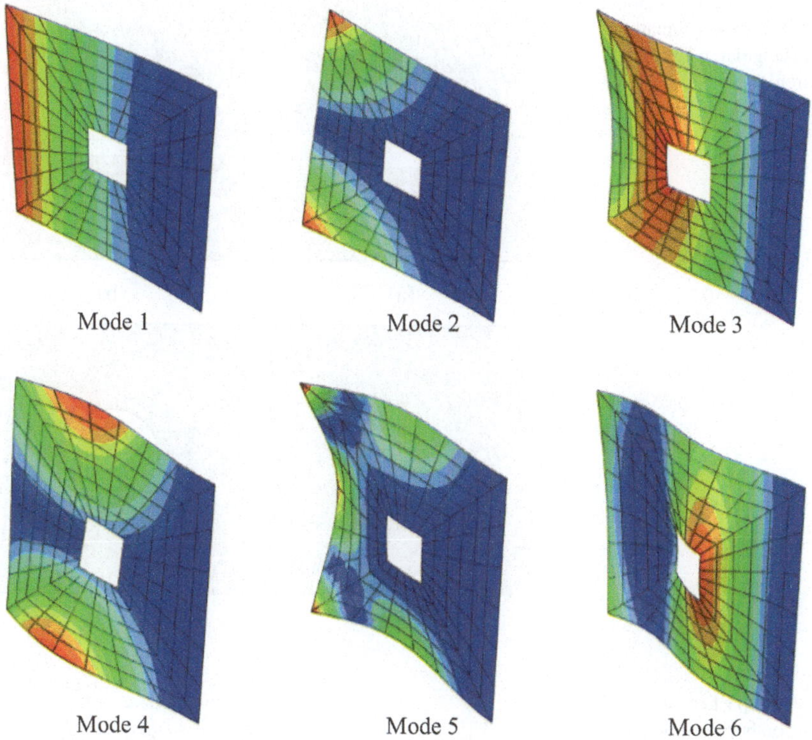

Fig. 7 Mode shape of stability analysis of sandwich panel

4.2.3 Effect of Cutout Position on Buckling Behavior of Sandwich Plate

In this example, the position of the cutout is a key consideration when studying the buckling behavior of the sandwich structure. Initially, the cutout is placed concentrically, and then it is shifted towards the corner of the panel for analysis, as illustrated in Fig. 10. Following the investigation, the calculated non-dimensional critical buckling loads are presented in Table 3. A notable observation from the table is that when the cutout is shifted towards the corner, the critical buckling load decreases for all modes. This behavior can be attributed to the reduction in the overall stiffness of the panel as the cutout is positioned eccentrically. Essentially, the displacement of the cutout from the center leads to a decrease in the panel's resistance to buckling, and this observation has implications for the structural design and stability analysis of sandwich structures.

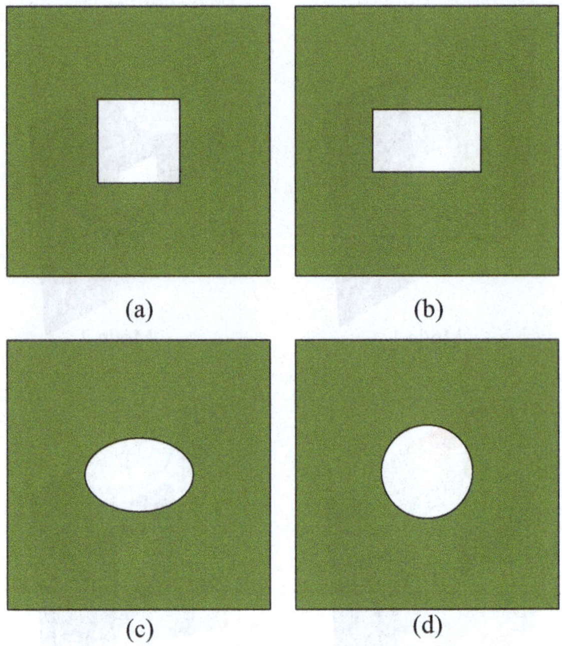

Fig. 8 Various types of Cutout shapes. **a** Square. **b** Rectangular. **c** Elliptical. **d** Circular

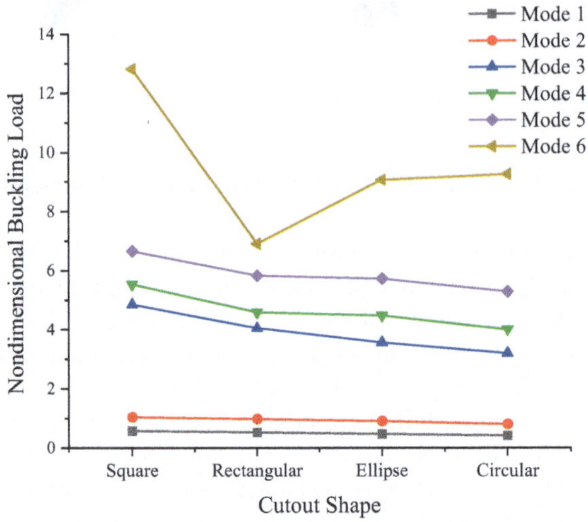

Fig. 9 Impact of cutout shape on buckling behavior of sandwich flat panel

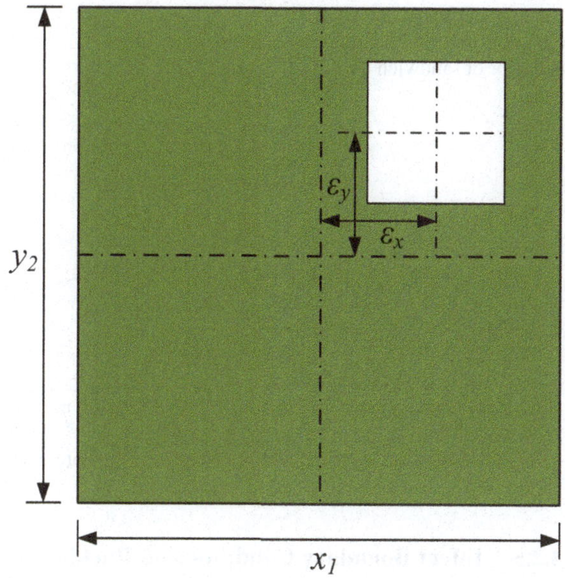

Fig. 10 Position of the cutout within the sandwich plate

Table 3 Nondimensional critical buckling load for sandwich panel with concentric and eccentric cutout

Eccentricity		Mode					
ε_x	ε_y	1	2	3	4	5	6
0	0	0.404	0.722	3.523	3.830	4.743	9.219
0.1	0.1	0.341	0.669	3.231	3.579	4.511	7.680
0.2	0.2	0.330	0.661	3.085	3.662	4.630	7.040
0.25	0.25	0.325	0.660	3.003	3.747	4.733	6.720

4.2.4 Effect of Core to Face Thickness on Buckling Behavior of Sandwich Plate

This illustration evaluates the influence of the core-face thickness ratio on the buckling characteristics of sandwich panel with cutout. The impact of the core-face thickness ratio on the buckling behavior of the sandwich panel is prominently evident in Fig. 11. As the core-face thickness ratio increases, it becomes apparent that the critical buckling load decreases. This decrease in the critical buckling load can be attributed to the fact that a higher core-face thickness ratio leads to a reduction in the overall stiffness of the panel. In essence, the panel becomes less resistant to buckling as the core-face thickness ratio is increased, which is a critical consideration when designing and assessing the stability of such sandwich structures.

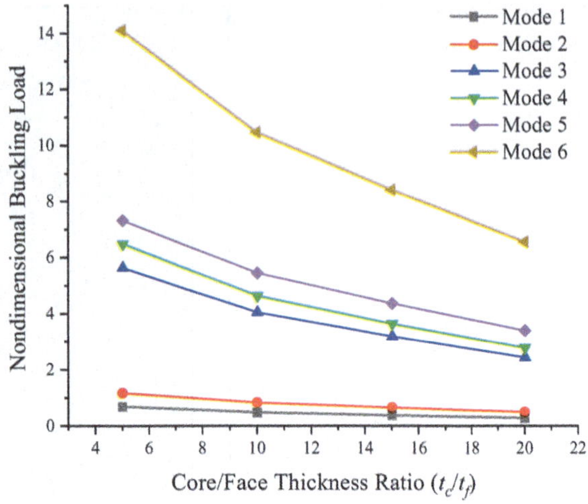

Fig. 11 Impact of core-face thickness ratio on buckling behavior of sandwich flat panel

4.2.5 Effect Boundary Conditions on Buckling Behavior of Sandwich Plate

The boundary conditions imposed on a sandwich composite plate could have a significant impact on its buckling behavior. In this case, various boundary conditions have been carefully assessed to examine their effects on the stability responses of the sandwich panel. Three distinct edge conditions—clamped, simply supported, and pinned were implemented at one end, to compute the critical buckling load, which is visualized in Fig. 12. Analysis of the plotted figure leads to the conclusion that the clamped end conditions (representing fixed boundary conditions) result in the highest critical load, while the pin end condition yields the lowest critical load. The simply supported end condition (allowing rotation but not translation) demonstrates an intermediate critical load value. It could have happened due to progressive stiffness levels of the composite material, following the order: clamped > simply supported > pinned.

5 Conclusions

The present comprehensive study focused on buckling behavior analysis of sandwich composite plates with predefined damage/cutout subjected to uniaxial mechanical load numerically. The core of this analysis relies on the FSDT and FEM. To execute this investigation, the widely adopted FE software, ABAQUS, is employed. To assure the validity of the numerical model, an initial verification process has been undertaken, involving a meticulous comparison of computed results with available data from published literature. Following this validation phase, the study expanded to explore the various structural parameter effects on buckling responses via a series of

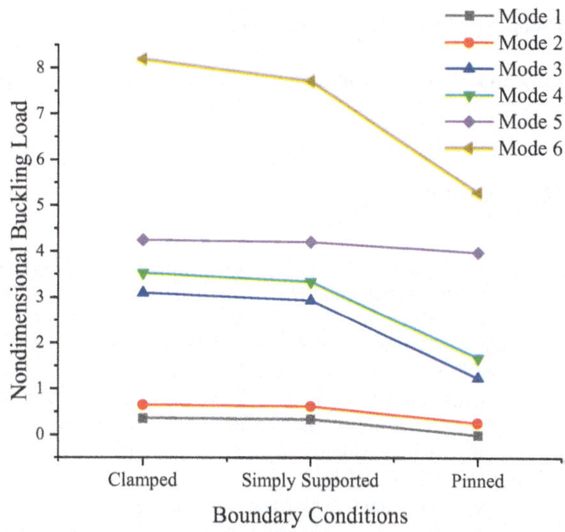

Fig. 12 Impact of boundary conditions on buckling behavior of sandwich flat panel

new examples. The results obtained from these new cases/illustrations are presented and discussed in the subsequent lines.

- The subsequent results provide insights into the non-dimensional buckling behavior of a flat sandwich panel, taking into consideration the impact of cutouts, as analyzed through the finite element method.
- The findings highlight the significant influence of cutout size on the critical buckling load. An increase in cutout size correlates with a decrease in the critical load.
- Moreover, the shape of the cutout also plays a role in determining the buckling load of the structure. Square cutouts yield the maximum critical load, while circular cutouts result in the minimum critical load.
- Additionally, when the cutout is positioned eccentrically, the critical buckling load may decrease across all modes.
- Critical buckling load is also influenced by the thickness of core to face sheet. As core-face thickness ratio increases, the critical load may decrease.
- The maximum eigenvalue is observed in the case of the clamped edge restriction, while the pinned condition results in the minimum eigenvalue. The simply supported conditions exhibit a moderate eigenvalue.

Declarations

Conflicts of Interest The authors confirm that there are no conflicts of interest associated with this work.

Ethical Compliance The authors affirm that this study adheres to all ethical guidelines.

Competing Interests No competing interests are declared by the authors.

Data and Material Availability Data and materials are not applicable to this study.

References

1. Kumar R, Hirwani CK (2022) Investigation on nonlinear bending behavior of sandwich shell panel with cutout using HSDT and FE approach. Mech Based Des Struct Mach. https://doi.org/10.1080/15397734.2022.2158868
2. Kumar R, Tiwari S, Kumar HC (2023) On transient responses of sandwich plate with cutout using FEM. Mater Today Proc. https://doi.org/10.1016/J.MATPR.2023.05.409
3. Marichelvam MK, Manimaran P, Verma A, Sanjay MR, Siengchin S, Kandakodeeswaran K et al (2021) A novel palm sheath and sugarcane bagasse fiber based hybrid composites for automotive applications: an experimental approach. Polym Compos 42:512–521. https://doi.org/10.1002/PC.25843
4. Tiwari S, Hirwani CK, Barman AG (2023) Fabrication and mechanical property investigation of banana/glass/epoxy hybrid composites. J Inst Eng (India): Ser C 104:521–527. https://doi.org/10.1007/S40032-023-00936-X/METRICS
5. Tiwari S, Hirwani CK, Barman AG (2022) Eigenfrequency behavior of banana/glass/epoxy hybrid composite-a numerical and experimental investigation. J Nat Fibers 19:9514–9530. https://doi.org/10.1080/15440478.2021.1982838
6. Tiwari S, Gopal Barman A, Kumar HC (2023) Dynamic deflection behaviour of plant fibre reinforced laminated composite plate structure using simulation model. Mater Today Proc 91:39–43. https://doi.org/10.1016/J.MATPR.2023.04.333
7. Verma A, Baurai K, Sanjay MR, Siengchin S (2020) Mechanical, microstructural, and thermal characterization insights of pyrolyzed carbon black from waste tires reinforced epoxy nanocomposites for coating application. Polym Compos 41:338–349. https://doi.org/10.1002/PC.25373
8. Verma A, Singh VK (2019) Mechanical, microstructural and thermal characterization of epoxy-based human hair-reinforced composites. J Test Eval 47:1193–1215. https://doi.org/10.1520/JTE20170063
9. Tiwari S, Barman AG, Pandey SM, Hirwani CK (2023) Experimental measurement of structural static and dynamic characteristics of glass/banana hybrid composite. Mapan - J Metrol Soc India 38:481–488. https://doi.org/10.1007/S12647-023-00630-4/METRICS
10. Verma A, Negi P, Singh VK (2018) Experimental investigation of chicken feather fiber and crumb rubber reformed epoxy resin hybrid composite: mechanical and microstructural characterization. J Mech Behav Mater 27:1–24. https://doi.org/10.1515/jmbm-2018-0014
11. Verma A, Parashar A (2018) Molecular dynamics based simulations to study the fracture strength of monolayer graphene oxide. Nanotechnology 29:115706. https://doi.org/10.1088/1361-6528/AAA8BB
12. Kumar R, Parashar A (2023) Atomistic simulations of pristine and nanoparticle reinforced hydrogels: a review. Wiley Interdiscip Rev Comput Mol Sci 13:e1655. https://doi.org/10.1002/WCMS.1655
13. Kumar R, Parashar A (2023) Effect of the degree of polymerization and water content on the thermal transport phenomena in PEGDA hydrogel: a molecular-dynamics-based study. Phys Chem Chem Phys 25:18960–18972. https://doi.org/10.1039/D3CP00667K
14. Srinivasan RS, Chellapandi P (1986) Dynamic stability of rectangular laminated composite plates. Comput Struct 24:233–238. https://doi.org/10.1016/0045-7949(86)90282-8
15. Khdeir AA, Redd JN (1997) Buckling of cross-ply laminated beams with arbitrary boundary conditions. Compos Struct 37:1–3. https://doi.org/10.1016/S0263-8223(97)00048-2

16. Song SJ, Waas AM (1997) Effects of shear deformation on buckling and free vibration of laminated composite beams. Compos Struct 37:33–43. https://doi.org/10.1016/S0263-8223(97)00067-6
17. Gilat R, Williams T, Aboudi J (2001) Buckling of composite plates by global–local plate theory. Compos B Eng 32:229–236. https://doi.org/10.1016/S1359-8368(00)00059-7
18. Wang H-J, Chen L-W (2003) Axisymmetric dynamic stability of sandwich circular plates. Compos Struct 59:99–107. https://doi.org/10.1016/S0263-8223(02)00187-3
19. Carrera E, Demasi L (2003) Two benchmarks to assess two-dimensional theories of sandwich, composite plates. AIAA J 41:1356–1362. https://doi.org/10.2514/2.2081
20. Park JS, Kim JH, Moon SH (2004) Vibration of thermally post-buckled composite plates embedded with shape memory alloy fibers. Compos Struct 63:179–188. https://doi.org/10.1016/S0263-8223(03)00146-6
21. Shukla KK, Nath Y, Kreuzer E, Kumar KV (2005) Buckling of laminated composite rectangular plates. J Aerosp Eng 18:215–223. https://doi.org/10.1061/(ASCE)0893-1321(2005)18:4(215)
22. Ungbhakorn V, Singhatanadgid P (2006) Buckling analysis of symmetrically laminated composite plates by the extended Kantorovich method. Compos Struct 73:120–128. https://doi.org/10.1016/j.compstruct.2005.02.007
23. Woo J, Meguid SA, Stranart JC, Liew KM (2005) Thermomechanical postbuckling analysis of moderately thick functionally graded plates and shallow shells. Int J Mech Sci 47:1147–1171. https://doi.org/10.1016/j.ijmecsci.2005.04.008
24. Liang S, Chen HL (2006) Investigation on the square cell honeycomb structures under axial loading. Compos Struct 72:446–454. https://doi.org/10.1016/j.compstruct.2005.01.022
25. Chakrabarti A, Sheikh AH (2006) Dynamic instability of laminated sandwich plates using an efficient finite element model. Thin-Walled Struct 44:57–68. https://doi.org/10.1016/j.tws.2005.09.003
26. Shen HS (2009) A comparison of buckling and postbuckling behavior of FGM plates with piezoelectric fiber reinforced composite actuators. Compos Struct 91:375–384. https://doi.org/10.1016/j.compstruct.2009.06.005
27. Chai GB, Yap CW (2008) Coupling effects in bending, buckling and free vibration of generally laminated composite beams. Compos Sci Technol 68:1664–1670. https://doi.org/10.1016/j.compscitech.2008.02.014
28. Fridman Y, Abramovich H (2008) Enhanced structural behavior of flexible laminated composite beams. Compos Struct 82:140–154. https://doi.org/10.1016/j.compstruct.2007.05.007
29. Shariyat M (2010) A generalized global–local high-order theory for bending and vibration analyses of sandwich plates subjected to thermo-mechanical loads. Int J Mech Sci 52:495–514. https://doi.org/10.1016/j.ijmecsci.2009.11.010
30. Prakash T, Singha MK, Ganapathi M (2008) Thermal postbuckling analysis of FGM skew plates. Eng Struct 30:22–32. https://doi.org/10.1016/j.engstruct.2007.02.012
31. Yang X, Lan Q, Zhong Y (2011) Buckling analysis and experiment of fiber-paper honeycomb sandwich structure composites. Adv Mat Res 314–316:566–570. https://doi.org/10.4028/WWW.SCIENTIFIC.NET/AMR.314-316.566
32. Ferreira A, Castro L, Roque C, Reddy JN, Bertoluzza S (2011) Buckling analysis of laminated plates by wavelets. Comput Struct 89:626–630. https://doi.org/10.1016/j.compstruc.2011.01.007
33. Neves AMA, Ferreira AJM, Carrera E, Cinefra M, Jorge RMN, Soares CMM (2012) Buckling analysis of sandwich plates with functionally graded skins using a new quasi-3D hyperbolic sine shear deformation theory and collocation with radial basis functions. ZAMM - J Appl Math Mech/Z Angew Math Mech 92:749–766. https://doi.org/10.1002/zamm.201100186
34. Ouinas D, Achour B (2013) Buckling analysis of laminated composite plates [(θ/−θ)] containing an elliptical notch. Compos B Eng 55:575–579. https://doi.org/10.1016/j.compositesb.2013.07.011
35. Jeyakrishnan PR, Chockalingam KKSK, Narayanasamy R (2013) Studies on buckling behavior of honeycomb sandwich panel. Int J Adv Manuf Technol 65:803–815. https://doi.org/10.1007/s00170-012-4218-9

36. Jabbari M, Farzaneh Joubaneh E, Mojahedin A (2014) Thermal buckling analysis of porous circular plate with piezoelectric actuators based on first order shear deformation theory. Int J Mech Sci 83:57–64. https://doi.org/10.1016/j.ijmecsci.2014.03.024
37. Heydari MM, Hafizi Bidgoli A, Golshani HR, Beygipoor G, Davoodi A (2015) Nonlinear bending analysis of functionally graded CNT-reinforced composite Mindlin polymeric temperature-dependent plate resting on orthotropic elastomeric medium using GDQM. Nonlinear Dyn 79:1425–1441. https://doi.org/10.1007/s11071-014-1751-0
38. Xing Y, Xiang W (2017) Analytical solution methods for eigenbuckling of symmetric cross-ply composite laminates. Chin J Aeronaut 30:282–291. https://doi.org/10.1016/j.cja.2016.12.027
39. Kheirikhah MM, Babaghasabha V (2016) Bending and buckling analysis of corrugated composite sandwich plates. J Braz Soc Mech Sci Eng 38:2571–2588. https://doi.org/10.1007/s40430-016-0498-6
40. Kahya V (2016) Buckling analysis of laminated composite and sandwich beams by the finite element method. Compos B Eng 91:126–134. https://doi.org/10.1016/j.compositesb.2016.01.031
41. Fan Y, Wang H (2017) Nonlinear low-velocity impact on damped and matrix-cracked hybrid laminated beams containing carbon nanotube reinforced composite layers. Nonlinear Dyn 89:1863–1876. https://doi.org/10.1007/s11071-017-3557-3
42. Hajlaoui A, Chebbi E, Dammak F (2019) Buckling analysis of carbon nanotube reinforced FG shells using an efficient solid-shell element based on a modified FSDT. Thin-Walled Struct 144. https://doi.org/10.1016/j.tws.2019.106254
43. Amiri A, Mohammadimehr M, Anvari M (2020) Stress and buckling analysis of a thick-walled micro sandwich panel with a flexible foam core and carbon nanotube reinforced composite (CNTRC) face sheets. Appl Math Mech 41:1027–1038. https://doi.org/10.1007/s10483-020-2627-7
44. Higginson K, Fernando D, Veidt M, Burnton P, You Z, Heitzmann M (2021) Local buckling of FRP thin-walled plates, shells and hollow sections with curved edges and arbitrary lamination. Thin-Walled Structures 168:108242. https://doi.org/10.1016/j.tws.2021.108242
45. Nima SJ, Ganesan R (2021) Buckling analysis of symmetrically laminated composite plates including the effect of variable pre-stress field using the Ritz method. Eur J Mech A Solids 90:104323. https://doi.org/10.1016/j.euromechsol.2021.104323
46. Van Vinh P, Tounsi A (2022) The role of spatial variation of the nonlocal parameter on the free vibration of functionally graded sandwich nanoplates. Eng Comput 38:4301–4319. https://doi.org/10.1007/s00366-021-01475-8
47. Chen W, Yang JS, Wei DY, Yan ST, Peng LX (2022) Buckling analysis of corrugated-core sandwich plates using a FSDT and a meshfree Galerkin method. Thin-Walled Struct 180:109846. https://doi.org/10.1016/j.tws.2022.109846
48. Ye W, Liu J, Zang Q, Yin Z, Lin G (2022) Buckling analysis of three-dimensional functionally graded sandwich plates using two- dimensional scaled boundary finite element method. Mech Adv Mater Struct 29:2468–2483. https://doi.org/10.1080/15376494.2020.1866125
49. Pavan GS, Muppidi H, Dixit J (2022) Static, free vibrational and buckling analysis of laminated composite beams using isogeometric collocation method. Eur J Mech A Solids 96:104758. https://doi.org/10.1016/j.euromechsol.2022.104758
50. Tiwari S, Hirwani CK, Barman AG (2023) Dynamic behavior of hybrid composite shallow shell panel-FEM HOSDT and experimental approach. Eur J Mech A Solids 97:104772. https://doi.org/10.1016/J.EUROMECHSOL.2022.104772
51. Kumar P, Arya R, Sharma N, Hirwani CK, Panda SK (2023) Curved fiber-reinforced laminated composite panel and variable stiffness influence on eigenfrequency responses: a higher-order FE approach. J Vib Eng Technol 11:2349–2359. https://doi.org/10.1007/S42417-022-00706-6/METRICS
52. Afonso A, Tomé M, Moleiro F, Araújo AL (2023) Dynamic buckling of active sandwich panels. Compos Struct 322:117355. https://doi.org/10.1016/j.compstruct.2023.117355
53. Singh S, Singh J, Shukla KK (2013) Buckling of laminated composite plates subjected to mechanical and thermal loads using meshless collocations. J Mech Sci Technol 27:327–336. https://doi.org/10.1007/s12206-012-1249-y

54. Reddy J, Phan ND (1985) Stability and vibration of isotropic, orthotropic and laminated plates according to a higher-order shear deformation theory. J Sound Vib 98:157–170. https://doi.org/10.1016/0022-460X(85)90383-9
55. Liu L, Chua L, Ghista DN (2007) Mesh-free radial basis function method for static, free vibration and buckling analysis of shear deformable composite laminates. Compos Struct 78:58–69. https://doi.org/10.1016/j.compstruct.2005.08.010

Chapter 15
Chemically Treated Pinus Roxburghii Fibre Reinforced PLA Composites: An Overview

Rajesh Kumar, Pawan Kumar Rakesh, and Dungali Sreehari

1 Introduction

The increasing demand for composite materials, there is an emerging need for high-performance and ecologically sustainable alternatives. Extensive research has been conducted by scholars to investigate and develop innovative natural fibres in order to address these requirements, with each fibre exhibiting unique and notable attributes. Plant-based fibres are the primary source of natural fibres used in composite fabrications. Several extraction methods, such as water retting, mechanical extraction, steam explosion, acid hydrolysis, and mechanochemical extraction, have been used. The water retting process is both environmentally sustainable and economically efficient. The water retting procedure is considered advantageous in comparison to other fibre extraction methods due to its straightforward methodology and cost-effectiveness in delivering high-quality fibres. In their study, Gu et al. [1] used coconut husks which were submerged in a water container for a duration of 5 months. The separation of the fibres from the husk was helped by the partial degradation of the pulp present on the shell. Following the retting procedure, the coconut husks were subjected to mechanical beating with a hammer. The process of separating the individual coir fibres included combing them with the use of a carding frame. The study conducted by Reddy et al. [2] examined coconut leaf sheaths obtained from the Arecaceae family. The coconut leaf sheath mats were obtained from coconut trees and thereafter immersed in water for a duration of 7 days. In order to eliminate any contaminants, the fibres underwent a purification process including the use of distilled water. Subsequently, the fibre mats were subjected to thermal treatment in a high-temperature oven, with temperatures ranging from 105 to 110 °C, for a duration of one day, in

R. Kumar · P. K. Rakesh (✉) · D. Sreehari
Mechanical Engineering Department & Design Innovation Center, National Institute of Technology Uttarakhand, Pauri Garhwal, Srinagar 246174, India
e-mail: pawankumarrakesh@gmail.com

S. K. Sethi et al. (eds.), *Polymer Composites: From Computational to Experimental Aspects*, Materials Horizons: From Nature to Nanomaterials,
https://doi.org/10.1007/978-981-97-0888-8_15

order to completely remove any residual moisture. In their study, Saravanakumar et al. [3] performed an experimental investigation including the immersion of the bark of the Prosopis juliflora plant in water for a duration of up to two weeks. The process of immersion facilitated the microbial decomposition of the bark, leading to a significant level of degradation that made the extraction of fibres viable. The researchers employed a conventional combing technique including a thin, elongated brush with metal teeth to separate the fibres from the deteriorated bark. In their study, Shanmugasundaram et al. [4] conducted the extraction of fibres from the stalks of Areca palm leaf (APLSF). Following the process of retting, the stalks were subjected to the action of a wooden mallet. Subsequently, the extracted fibres were subjected to a comprehensive washing and cleaning process using a continuous flow of water. This was followed by a drying time of two days at an ambient temperature to remove any residual moisture from the fibres. Vijay et al. [5] detected the presence of Parthenium Hysterophorus plants and proceeded to gather their stems. The stems underwent a 10-day water immersion before to being treated to the conventional retting technique, which included the hand application of a wire brush. The retted fibres underwent a washing process using conventional water and were then subjected to a 72-h drying period under sunshine. The water retting process was used by Manimaran et al. [6] to extract fibres from Furcraea Foetida (FF). In the first stage, the leaves, which were encircled by pointed spines, were meticulously removed, and thereafter discarded. Subsequently, the leaves were completely immersed in a receptacle filled with water and left to undergo the retting process for a duration of 20 days, until decomposition took place, and the subsequent step was the separation of fibres. The leaves were carefully arranged in a flat position, and the outer layer was meticulously eliminated by a process of scouring. The fibres were subjected to a purification process using water in order to remove any undesirable odour that may have arisen during the retting process. Subsequently, the fibres were subjected to a drying process in a shaded environment to prevent direct exposure to sunlight. Sreenivasan et al. [7] conducted a study in which they observed that mechanical decortication methods were used to remove fibres from sansevieria leaves, namely the sansevieria cylindrical fibres. The fibres that were obtained were subjected to a washing process using water and afterwards dried for a duration of 24 h under sunshine. Thamae and Baillie [8] used elongated blades for the purpose of extracting foliage from the Agave Americana botanical specimen. The leaves underwent a process of boiling in water for a duration of two hours, resulting in their attainment of a softened condition. Subsequently, the leaves that had been softened were delicately compressed between stones in order to extract the fibres. Following this, the fibres were subjected to a thorough rinsing process using running water in order to remove any residual soft tissue present. The fibres that were obtained were allowed to undergo a natural drying process under ambient conditions, exposed to both air and sunshine.

Fig. 1 Steps of extracted Pinus roxburghii fibre

2 Surface Treatment of Natural Fibres

Cellulose stands out as the most prevalent constituent among the components, characterised by a sophisticated composition consisting of both crystalline and non-crystalline phases. The cellulose's crystalline phase has a considerable number of hydroxyl groups. Consequently, cellulose exhibits hydrophilic properties [9, 10]. This characteristic results in diminished interfacial adhesion with diverse hydrophobic matrix materials. Cellulosic fibres' characteristics may be greatly affected by environmental variables. Moreover, untreated natural fibres have a greater proportion of amorphous constituents, including pectin, wax, and lignin, which may be mitigated via the use of several treatment methods. Materials with insufficient interfacial adhesion are prone to occurrences such as fibre pullouts, debonding, and premature failure [11]. It is commonly acknowledged in academic circles that natural fibres possess hydroxyl groups that exhibit a high degree of hydrophilicity [12, 13]. Figure 1 shows the steps involved in the extraction of Pinus roxburghii fibre.

3 Characterisation of Untreated Natural Fibres

A detailed investigation was undertaken by Rosa et al. [14] on the okra fibre originating from the Malvaceae family. The fibres were discovered to have originated in Egypt and are derived from the stem of the okra plant. The investigation of its thermal stability was conducted utilising TGA. The comprehensive investigation conducted by Sreenivasan et al. [15] extensively examined the physical, chemical, mechanical, and microstructural characteristics of Sansevieria cylindrical fibre (SCF). The SCF has favourable characteristics for applications requiring low weight due to its relatively low density and unique properties. The supercritical fluids (SCFs) exhibit a semi-crystalline nature and include Ib cellulose within their structure. The study conducted by Baskaran et al. [16] examined a natural fibre derived from the bark of Dichrostachys cinerea. The fibre was found to have 72.4% cellulose by weight, with a comparatively lower hemicellulose concentration of 13.08%. The calculated confidence interval (CI) value is 57.82%. The degradation temperature of the fibre was determined to be 226 °C according to the TGA study. The natural fibre obtained from the bark of the Vachellia Farnesiana (VF) tree was described by Vijay et al. [17]. The VF fibre demonstrated characteristics that are very suitable for lightweight applications. In their study, Elenga et al. [18] conducted an analysis on the physico-mechanical characteristics of fibres derived from wild raffia palm plants. The study conducted by Belouadah et al. [19] focused on the extraction of fibres from the stem of the Lygeum spartum plant (LSF). The LSF material demonstrated a brittle behaviour when subjected to tensile stress, resulting in structural collapse. The researchers used SEM to examine the morphology of the samples and make predictions about their surface qualities. The stems of LSF exhibit a substantial presence of fibre cells that are arranged in an aligned manner and are bound together by non-cellulosic constituents.

In their study, Kathiresan et al. [20] conducted an analysis of the properties of fibres obtained from the leaf of Artistida hystrix. The fibres that were extracted had the following characteristics: a CI (crystallinity index) of 44.85%, a cellulose content of 59.54% by weight, a hemicellulose content of 11.35% by weight, a lignin content of 8.42% by weight, and a density of 540 kg/m^3. In the research done by Moshi et al. [21], fibres were isolated from the root of the Ficus religiosa tree. The fibers chemical composition study indicated that it consisted of 55.6% cellulose, 13.9% hemicellulose, 10.1% lignin, 0.72% wax, 9.33% moisture, and 4.86% ash. In order to forecast surface characteristics, the researchers used SEM for the purpose of analysing morphology. The TGA findings indicated that the fibre exhibited thermal stability up to a temperature of 325 °C. Additionally, the highest temperature at which degradation occurred was found to be 400 °C.

The fibres obtained from the bark of the Melia Azedarach tree (MAF) were described by Pandiarajan et al. [22]. Moreover, the morphological examination of MAF revealed a consistent smoothness seen on the external surfaces of the fibres. Furthermore, the investigators conducted measurements on the fibre diameter, yielding a value of 44.6 ± 0.78 μm. The physicomechanical and chemical properties of the FF plant fibre were examined by Manimaran et al. [23]. The density of FF fibres was found to be 778 kg/m^3, which was notably lower in comparison to synthetic fibres. This particular attribute offers distinct benefits in the development of lightweight applications. The FF demonstrated a notably elevated confidence interval of 56.6%, hence enhancing its mechanical qualities in comparison to other alternatives. The remarkable tensile qualities of FF fibres make them a very promising and optimal alternative to traditional synthetic fibres.

The work undertaken by Indran et al. [24] included the collection of root fibres from the Cissus quadrangularis plant. The study revealed that the spatial arrangement of chemical components inside the fibres had a direct impact. The extraction of fibres often employs a conventional methodology including microbial degradation and subsequent combing. The basic features of Cardiospermum Halicababum Fibre (CHF) generated from the Cardiospermum Halicababum plant were investigated by Vinod et al. [25]. The plant fibres showed reduced quantities of wax and moisture content. The chemical study conducted indicated a marginal increase in the proportion of lignin, resulting in the occurrence of brittle fractures during the tensile test. In addition, it is noteworthy that the CHF demonstrated exceptional thermal stability, with the capacity to withstand temperatures of up to 336 °C, surpassing the temperature at which thermoplastics undergo polymerization. In their work, Hyness et al. [26] performed an investigation whereby they collected fibres from the plant species Heteropogon contortus. The cellulose percentage of the plant fibre derived from Heteropogon contortus fibres (HCFs) was found to be very high at 64.875%, resulting in enhanced mechanical strength. Furthermore, the lower density of HCFs (602 kg/m^3) renders them very suitable for lightweight applications.

The study conducted by Seki et al. [27] focused on the extraction of fibres from the Ferula Communis plant. The Archimedes technique was used to determine the fibre density, with ethanol being utilised as the measurement liquid. In their study, Maheshwaran et al. [28] collected natural fibre from the Epipremnum aureum plant and performed a series of assays to investigate its qualities. The X-ray diffraction and functional group analysis outcomes have provided confirmation of the significant occurrence of cellulose inside the fibres of Epipremnum aureum. The cellulose content has been determined to be 49.33% with a confidence interval (CI). Furthermore, the thermal study findings indicate that the extracted fibre has a notable resistance to elevated temperatures, with the capacity to withstand temperatures up to 328.9 °C. The study conducted by Ahmed et al. [29] focused on the extraction of fibres from the stem of the Elettaria cardamomum plant. These attributes include a crystallinity index (CI) value of 36.84% and a crystal size (CS) of 5.82 nm.

The study conducted by Khan et al. [30] focused on the manual extraction of fibres from the stem of Eleusine indica grass. The extracted fibre has a considerable cellulose content, amounting to 61.3 wt%. Additionally, the fibre exhibits a comparatively low density, measuring 1143 kg/m^3. The confidence interval (CI) of the fibre was found to be 45%. Furthermore, the fibre's tensile strength exhibited promising characteristics for utilisation in semi-structural scenarios. In their study, Nascimento et al. [31] conducted an extraction of lignocellulosic fibres from the leaves of Yucca aloifolia L. Subsequently, they conducted a detailed investigation of the chemical, morphological, thermal, and mechanical characteristics of these fibres. The chemical study conducted indicates that the fibres consist of a substantial proportion of cellulose, accounting for roughly 52.5% of the overall composition. Furthermore, the XRD study resulted in a confidence interval (CI) of 69.43% for the cellulose. The results of the SEM examinations conducted on the fibre surfaces revealed the presence of substantial non-cellulosic materials, which were found to be statistically significant.

4 Characterisation of Alkali-Treated Natural Fibres

Scientists consistently investigate novel natural fibres as possible substitutes or additions to manmade materials. The use of a chemical technique aimed at eliminating wax from the fibres leads to the formation of a porous surface, hence substantially enhancing the adhesive strength inside composites. Specifically using sodium hydroxide (NaOH), has shown notable advantages when compared to other chemical treatments for the purpose of surface modification. This very effective chemical technique induces modifications to the surface of the fibre, resulting in a reduction in its moisture content. In addition, the use of alkali treatment has been shown to be very successful in the removal of non-cellulosic chemical polymers from the fibres. Various studies have been conducted by researchers to explore the impact of varying dosages of NaOH solution in order to get the desired results. The chemical treatment of sisal, hemp, and jute fibres was examined by Mwaikambo and Ansell [32]. However, after the alkalisation process, all the fibres showed surfaces that were characterised by unevenness. The efficacy of this approach in the identification and verification of chemical group elimination during the process of alkalization was also revealed.

In the research done by Shanmugasundaram et al. [33], fibres were isolated from Areca palm plants. The findings indicated a rise in fibre density and a reduction in fibre diameter as a consequence of the therapy. In their study, Cai et al. [34] conducted an experiment involving the immersion of abaca fibres in NaOH solutions of varying concentrations (5%, 10%, or 15% by weight) for a duration of 2 h. The original tensile strength of the abaca fibre that had not undergone any treatment was determined to be 717 MPa, while its Young's modulus was measured to be 18.6 GPa. After undergoing treatment with a NaOH solution with a weight percentage of 5%, the abaca fibre showed a notable enhancement in both tensile strength, which saw a rise of 8%, and Young's modulus, which revealed a substantial improvement of 36% as compared to the untreated state.

In their research, Lassoued et al. [35] investigated the physicochemical property of natural fibres obtained from palm trees in Tunisia. Furthermore, the treatment successfully eradicated contaminants and non-cellulosic constituents from the surface of the fibre. The study conducted by Babu et al. [36] examined the effects of NaOH treatment on several characteristics of Phaseolus vulgaris fibres (PVFs), including their chemical composition, thermal behaviour, and surface attributes. It was determined that the optimal duration for treatment with the 5% NaOH solution was 45 min. In their research, Ganapathy et al. [37] performed a study focused on the fibre obtained from the root of the banyan tree. Subsequently, the isolated fibres were subjected to alkali treatment by the researchers. In addition, the fibre subjected to treatment had reduced levels of silicon (Si) and calcium (Ca), indicating the effective elimination of wax, lignin.

The study conducted by Kommula et al. [38] examined the impact of alkali treatment on the mechanical, physical, thermal, and chemical characteristics of Napier Grass Fibre (NGF) that was extracted. Various concentrations of alkali (5, 10, and

15%) were used for the treatment of NGF, with a consistent soaking time of 2 h. This enhancement may be attributed to the improved structure of the fibres, with a particularly significant improvement seen when subjected to a 10% alkali treatment. In their study, Kathirselvam et al. [39] documented significant enhancements in both the surface and structural properties of Thespesia populnea bark fibre after exposure to different concentrations of NaOH and varying soaking durations. The optimal chemical composition and tensile properties were achieved by subjecting the fibres to a 5% NaOH concentration treatment for a duration of 60 min. The implemented treatment effectively eradicated formless substances and impurities from the fibre matrix, resulting in a decrease in fibre diameter and cross-sectional area.

The crystal structure and thermal stability of untreated and alkali-treated cotton shell fibres were investigated by Rajkumar et al. [40]. The researchers have made a discovery that the use of ideal approaches resulted in CI values (22.28%) that are similar to those obtained from raw cotton shell fibres. The TG/DTG study showed that the cellulose degradation temperature was higher in the optimally treated cotton shell fibres, suggesting an improvement in thermal stability. The optimal treatment resulted in a decrease in the quantity of amorphous constituents, such as lignin and hemicelluloses, as assessed by the use of FTIR spectroscopy and chemical analysis. In their study, Mylsamy and Rajendran [41] conducted an investigation into the mechanical and physio-chemical characteristics of Agave americana fibre. The researchers investigated the fibre in both its untreated form and after undergoing alkali treatment. The analysis concluded that the use of a 5% NaOH solution yielded the most optimal outcomes in terms of alkaline therapy effectiveness. Based on the findings of the study, it was determined that these fibres exhibited suitability in enhancing the structural integrity of ecologically sustainable composites. The use of chemical analysis and FTIR examinations shown that the application of alkali treatment effectively removed the presence of amorphous hemicellulose inside the fibres.

In their study, Amroune et al. [42] investigated on the chemical and thermo-physical properties of date palm fibre. The date palm fibres were subjected to treatment using alkaline solutions containing varying concentrations of NaOH, along with different immersion durations at ambient temperature. The water-retting procedure was used by Pandiarajan et al. [43] to obtain raw fibres from the rinds of the Thespesia Populnea plant. The fibre that underwent treatment exhibited a drop in its amorphous constituents, namely hemicellulose (decreasing from 20.5 wt.% to 16.1 wt.%) and lignin (decreasing from 20 wt.% to 15 wt.%).

The study conducted by Pokhriyal et al. [44] focused on the extraction of fibres from the culm stems of Himalayacalamus falconeri. It was shown that the application of NaOH alkali treatment effectively eradicated the OH groups present on the surface of the fibre, resulting in a significant reduction in non-cellulosic constituents. Moreover, it was observed that the surface roughness measurement of the fibres treated with a 5% alkali solution was 1.57 times greater compared to the untreated fibres treated with hydrofluoric acid (HF).

Singh and Rout [45] conducted a study on the analysis of cellulosic fibres derived from the leaves of Borassus flabellifer L. The findings of the study indicated that the application of alkali treatment resulted in a rise in cellulose content, accompanied by a decrease in hemicellulose, lignin, and wax content. The alterations resulted in an increase in tensile strength and density, accompanied by a decrease in the weight of the fibres. According to XRD research, there was an increase in the CI of both untreated and treated fibres, ranging from 46.58 to 50.34%. Additionally, the crystal size (CS) of the fibres also increased, ranging from 2.36 to 2.72 nm. The study conducted by Jaramillo-Quiceno et al. [46] examined the impact of alkali treatment on the mechanical, physical, thermal, and chemical properties of extracted Pineapple Leaf Fibres. The process of mercerization resulted in notable alterations in the lignocellulosic structure, as shown by infrared analysis.

5 Natural Fiber reinforced Polymer Composite

In recent years, the widespread use of synthetic fibre composites across diverse sectors has been attributed to their exceptional characteristics and qualities. Nevertheless, the escalating negative effects on the environment have resulted in a heightened need for natural fibre composites. The use of natural fibre reinforcement in polymer matrix composites offers significant ecological benefits, such as convenient disposal methods and effective utilisation of raw resources. Numerous scholars have highlighted the advantages associated with natural fibre composites in comparison to their synthetic counterparts, including their reduced density, extensive accessibility, capacity for recycling, biodegradability, and cost-effectiveness. However, a notable obstacle persists in the comparatively diminished strength of natural fibre composites when compared to composites composed of synthetic fibres [47–51]. The physical and mechanical characteristics of short-sisal/polypropylene composites were investigated by Joseph et al. [52]. The fibre composites that underwent chemical treatment exhibited improved interfacial bonding, leading to reduced water absorption. Nevertheless, the tensile property exhibited a decline when both the fibre concentration and water absorption levels rose. Moreover, the manner in which the fibres are treated and oriented has a substantial influence on their tensile performance. According to the findings of the study, it has been shown that the tensile properties of composites may experience a decline when subjected to ultraviolet (UV) radiation. The thermomechanical properties of composites including kenaf fibre and PLLA were examined by Nishino et al. [53]. The mechanical properties of kenaf/PLLA composites exhibited significant improvement due to the establishment of a robust interfacial bond between the constituent materials, resulting in a fibre volume percentage of 70%. Kousaalya et al. [54] developed bio-based laminate composites by using PLA sheets in conjunction with randomly oriented mats composed of areca fibres. The composites were created using the film stacking technique. The flexural, impact, and tensile strength of sisal fibre and polypropylene composites were examined by Mukhopadhyay and Srikanta [55]. The manufacturing process of composite laminates with fresh and aged sisal fibre reinforcement included the use of a compression moulding machine using the film stacking technique. The behaviour of aged and newly manufactured fibres inside the composites was examined by the researchers. The decline in strength found in aged fibres may be attributed to the process of cellulose oxidation. Interestingly, it was found that composites with aged fibres exhibited superior mechanical properties compared to those incorporating fresh fibres.

Ranakoti and Rakesh [56] used compression moulding as a manufacturing technique to produce composites using Tasar Silk Waste (TSW), jute fibre, and epoxy. The researchers observed a positive correlation between the percentage of TSW and the void fraction, with the highest void fraction occurring at 14 wt% of TSW. The incorporation of a 12 wt% TSW augmentation resulted in a notable enhancement of the composite's mechanical characteristics.

Oksman et al. [57] produced composite materials by incorporating flax fibre reinforcement into PLA. A manufacturing process was used to produce an extruder, wherein the flax fibre content was varied within the weight range of 30–40%. In their study, Islam et al. [58] used a film-stacking technique to produce composites by combining alkaline-treated long hemp fibres with PLA. In addition, the composite's CI and TGA results indicated that including alkali-treated fibres yielded superior performance compared to composites made with untreated fibres. The objective of the study conducted by Indran et al. [59] was to enhance the mechanical characteristics of a composite material consisting of Cissus quadrangularis Stem Fibre (CQSF) and unsaturated polyester. This was achieved by the manipulation of fibre length and quantity used. Enhanced mechanical qualities were seen when the fibre content and length were augmented, reaching the highest level of improvement at 30% and 40 mm, respectively. Nevertheless, upon further increasing the fibre %, the researchers observed a notable disintegration of the composite material. The researchers examined SEM images of the fractured specimens in order to ascertain the cause of the failure. The analysis conducted revealed that the most prevalent forms of failure seen were debonding and fibre pullout, particularly in cases where there was a higher concentration of fibres. This phenomenon may be attributed to the proximity of the fibres, leading to direct contact between them and inadequate dispersion of the matrix material inside the narrow spaces between the fibres.

The composite materials were created by Rosa et al. [60] by the combination of phormium Tenax leaf fibres with epoxy resin. The fibres were oriented in both random and unidirectional arrangements. The researchers made a significant finding that composites enhanced with quasi-unidirectional fibres exhibited superior performance in terms of modulus and strength compared to pure epoxy resin, in both tensile and flexural loading scenarios. The disintegration of the cellulose component of the fibre was seen to occur at a temperature of 337 °C, whereas the decomposition of the epoxy resin took place at a slightly lower temperature of 335 °C. Nevertheless, the composites that were generated exhibited decomposition at an elevated temperature of 347 °C, a result that was unforeseen. In their study, Yoganandam et al. [61] used epoxy resin as a binding agent to fabricate composites using natural fibres derived from Calotropis procera. The composites exhibited the best tensile and impact strength at a fibre loading of 30%. The mechanical properties of polyester composites using Cissus quadrangularis fibre were assessed by Mayandi et al. [62].

The researchers made a discovery indicating that the optimal composite configuration consisted of fibres with a length of 40 mm and a loading of 40%. This particular combination exhibited enhanced mechanical strength and toughness. The study used SEM morphological analysis to examine the prevalence of interface failures. The study conducted by Omar et al. [63] focused on the technique of film stacking as a means of producing a bio-composite material using Agave Tequilana Fibre (ATF). The samples were created by the researchers using three different fibre loadings, namely 20, 40, and 60% by weight. The flexural strength values of the treated samples ranged from 76.10 to 77 MPa, while the flexural modulus values ranged from 3.36 to 3.76 GPa. These samples had a fibre content of 60%. The mechanical characteristics of polyester composites reinforced with untreated and treated palmyra fibres were investigated by Sivakumar et al. [64]. The use of Palmyra fibres as a reinforcing material, in conjunction with polyester acting as the matrix, is seen. The findings indicate that the palmyra fibre composites subjected to treatment have favourable tensile and flexural properties. The tensile tests yielded a maximum strength of about 48 MPa and a fracture elongation of 6%. The compression approach was used by the researchers to fabricate the composite material, while a hydraulic press was utilised in the manufacturing process of the composite. The modulus of rupture, elasticity, and proportional limit exhibited a substantial increase in composites that included treated fibres.

6 Conclusions and Future Scope

Given the escalating environmental apprehensions and the surging need for sustainable resources, scholars have been actively investigating inventive resolutions within the domain of composite materials. These endeavours strive to discover novel approaches in tackling these difficulties. Numerous experiments have been conducted to ascertain the optimal natural fibre composites for certain technological purposes. The use of natural fibres has been more favoured owing to their wide range of advantageous properties, including but not limited to low density, cost-effectiveness, convenience of use, durability, and capacity to decompose naturally.

The needles of Pinus roxburghii, which are found in abundance in the hilly northern area of Uttarakhand, India, provide a distinct and noteworthy perspective. The use of PRN fibres into bio-composite reinforcement holds promise for decreasing dependence on synthetic fibres and prioritising the utilisation of renewable natural resources. Given this context, the current chapter explores the potential use of Pinus roxburghii fibres (PRF) as a reinforcement material for PLA composites, in line with the principles of sustainability. The research involves several aspects, including the extraction of PRF, the implementation of alkali treatment, and the subsequent integration of these fibres in the production of composites based on PLA.

Declarations

Conflicts of Interest The authors confirm that there are no conflicts of interest associated with this work.

Ethical Compliance The authors affirm that this study adheres to all ethical guidelines.

Competing Interests No competing interests are declared by the authors.

Data and Material Availability Data and materials are not applicable to this study.

References

1. Gu H (2009) Tensile behaviours of the coir fibre and related composites after NaOH treatment. Mater Des 30:3931–3934. https://doi.org/10.1016/j.matdes.2009.01.035
2. Reddy Obi K, Sivamohan Reddy G, Uma Maheswari C et al (2010) Structural characterization of coconut tree leaf sheath fiber reinforcement. J For Res 21:53–58. https://doi.org/10.1007/s11676-010-0008-0
3. Saravanakumar SS, Kumaravel A, Nagarajan T, Ganesh I (2014) Investigation of physico-chemical properties of alkali-treated Prosopis juliflora fibers 37–41. https://doi.org/10.1080/1023666X.2014.902527
4. Shanmugasundaram N, Rajendran I, Ramkumar T (2018) Characterization of untreated and alkali treated new cellulosic fiber from an Areca palm leaf stalk as potential reinforcement in polymer composites. Carbohydr Polym 195:566–575. https://doi.org/10.1016/j.carbpol.2018.04.127
5. Vijay R, Singaravelu DL, Vinod A et al (2021) Characterization of alkali-treated and untreated natural fibers from the stem of Parthenium hysterophorus. J Nat Fibers 18:80–90. https://doi.org/10.1080/15440478.2019.1612308
6. Manimaran P, Senthamaraikannan P, Sanjay MR et al (2018) Study on characterization of Furcraea foetida new natural fiber as composite reinforcement for lightweight applications. Carbohydr Polym 181:650–658. https://doi.org/10.1016/j.carbpol.2017.11.099
7. Sreenivasan VS, Somasundaram S, Ravindran D et al (2011) Microstructural, physico-chemical and mechanical characterisation of Sansevieria cylindrica fibres - an exploratory investigation. Mater Des 32:453–461. https://doi.org/10.1016/j.matdes.2010.06.004
8. Thamae T, Baillie C (2007) Influence of fibre extraction method, alkali and silane treatment on the interface of Agave Americana waste HDPE composites as possible roof ceilings in Lesotho. Compos Interfaces 14:821–836. https://doi.org/10.1163/156855407782106483
9. Kumar R, Rakesh PK, Sreehari D, Kumar D, Naik TP (2023) Experimental investigations on material properties of alkali retted Pinus roxburghii fiber. Biomass Convers Biorefinery 23:1–17
10. Chaitanya S, Singh I (2016) Novel aloe vera fiber reinforced biodegradable composites - development and characterization. J Reinf Plast Compos 35:1411–1423. https://doi.org/10.1177/0731684416652739
11. Pokhriyal M, Rakesh PK (2023) Processing and characterization of novel Himalayacalamus falconeri fiber reinforced biodegradable composites. Biomass Convers Biorefinery 23:1–16
12. Kumar R, Rakesh PK, Sreehari D (2023) Investigation on physico-chemical, mechanical and thermal properties of extracted novel Pinus roxburghii fiber. J Nat Fibers 20(1):2157924
13. Ranakoti L, Rakesh PK, Gangil B (2021) Effect of Tasar silk waste on the mechanical properties of Jute/Grewia optiva fibers reinforced epoxy laminates. J Nat Fibers 19(15):10462–10474
14. De Rosa IM, Kenny JM, Puglia D et al (2010) Morphological, thermal and mechanical characterization of okra (Abelmoschus esculentus) fibres as potential reinforcement in polymer composites. Compos Sci Technol 70:116–122
15. Sreenivasan VS, Ravindran D, Manikandan V, Narayanasamy R (2011) Mechanical properties of randomly oriented short Sansevieria cylindrica fibre/polyester composites. Mater Des 32:2444–2455. https://doi.org/10.1016/j.matdes.2010.11.042

16. Baskaran PG, Kathiresan M, Senthamaraikannan P, Saravanakumar SS (2018) Characterization of new natural cellulosic fiber from the bark of Dichrostachys Cinerea. J Nat Fibers 15:62–68. https://doi.org/10.1080/15440478.2017.1304314
17. Vijay R, Singaravelu DL, Vinod A et al (2018) Characterization of raw and alkali treated new natural cellulosic fibers from Tridax procumbens. Int J Biol Macromol. https://doi.org/10.1016/j.ijbiomac.2018.12.056
18. Elenga RG, Dirras GF, Goma Maniongui J et al (2009) On the microstructure and physical properties of untreated raffia textilis fiber. Compos Part A Appl Sci Manuf 40:418–422. https://doi.org/10.1016/j.compositesa.2009.01.001
19. Belouadah Z, Ati A, Rokbi M (2015) Characterization of new natural cellulosic fiber from Lygeum spartum L. Carbohydr Polym 134:429–437. https://doi.org/10.1016/j.carbpol.2015.08.024
20. Kathiresan M, Pandiarajan P, Senthamaraikannan P, Saravanakumar SS (2016) Physico-chemical properties of new cellulosic Artisdita hystrix leaf fiber. Int J Polym Anal Charact 21:663–668. https://doi.org/10.1080/1023666X.2016.1194636
21. Moshi AAM, Ravindran D, Bharathi SRS et al (2020) Characterization of a new cellulosic natural fiber extracted from the root of Ficus religiosa tree. Int J Biol Macromol 142:212–221. https://doi.org/10.1016/j.ijbiomac.2019.09.094
22. Pandiarajan P, Baskaran PG, Kathiresan M, Kanth S (2020) Physico chemical characterization of fiber from Melia azedarach barks as an effective reinforcement in polymer matrices. J Nat Fibers. https://doi.org/10.1080/15440478.2020.1798846
23. Manimaran P, Sanjay MR, Senthamaraikannan P et al (2019) Synthesis and characterization of cellulosic fiber from red banana peduncle as reinforcement for potential applications. J Nat Fibers 16:768–780. https://doi.org/10.1080/15440478.2018.1434851
24. Indran S, Edwin Raj R, Sreenivasan VS (2014) Characterization of new natural cellulosic fiber from Cissus quadrangularis root. Carbohydr Polym 110:423–429. https://doi.org/10.1016/j.carbpol.2014.04.051
25. Vinod A, Vijay R, Singaravelu DL et al (2021) Extraction and characterization of natural fiber from stem of Cardiospermum halicababum. J Nat Fibers 18:898–908. https://doi.org/10.1080/15440478.2019.1669514
26. Hyness NRJ, Vignesh NJ, Senthamaraikannan P et al (2018) Characterization of new natural cellulosic fiber from Heteropogon contortus plant. J Nat Fibers 15:146–153. https://doi.org/10.1080/15440478.2017.1321516
27. Seki Y, Sarikanat M, Sever K, Durmuşkahya C (2013) Extraction and properties of Ferula communis (chakshir) fibers as novel reinforcement for composites materials. Compos Part B Eng 44:517–523. https://doi.org/10.1016/j.compositesb.2012.03.013
28. Maheshwaran MV, Hyness NRJ, Senthamaraikannan P, Sanjay MR (2017) Characterization of natural cellulosic fiber from Epipremnum aureum stem characterization of natural cellulosic fiber from Epipremnum aureum stem 0478. https://doi.org/10.1080/15440478.2017.1364205
29. Ahmed J, Balaji MA, Saravanakumar SS, Senthamaraikannan P (2021) A comprehensive physical, chemical and morphological characterization of novel cellulosic fiber extracted from the stem of Elettaria Cardamomum plant. J Nat Fibers 18:1460–1471. https://doi.org/10.1080/15440478.2019.1691121
30. Khan A, Vijay R, Singaravelu DL et al (2021) Extraction and characterization of natural fiber from Eleusine indica grass as reinforcement of sustainable fiber reinforced polymer composites. J Nat Fibers 18:1742–1750. https://doi.org/10.1080/15440478.2019.1697993
31. Nascimento do HM, dos Santos A, Duarte VA et al (2021) Characterization of natural cellulosic fibers from Yucca aloifolia L. leaf as potential reinforcement of polymer composites. Cellulose 28:5477–5492. https://doi.org/10.1007/s10570-021-03866-y
32. Mwaikambo LY, Ansell MP (2002) Chemical modification of hemp, sisal, jute, and kapok fibers by alkalization. J Appl Polym Sci 84:2222–2234. https://doi.org/10.1002/app.10460
33. Shanmugasundaram N, Rajendran I (2016) Characterization of raw and alkali-treated mulberry fibers as potential reinforcement in polymer composites. J Reinf Plast Compos 35:601–614. https://doi.org/10.1177/0731684415625822

34. Cai M, Takagi H, Nakagaito AN et al (2016) Effect of alkali treatment on interfacial bonding in abaca fiber-reinforced composites. Compos Part A Appl Sci Manuf 90:589–597. https://doi.org/10.1016/j.compositesa.2016.08.025
35. Lassoued M, Mnasri T, Hidouri A, Ben Younes R (2018) Thermomechanical behavior of Tunisian palm fibers before and after alkalization. Constr Build Mater 170:121–128. https://doi.org/10.1016/j.conbuildmat.2018.03.070
36. Babu BG, Princewinston D, Saravanakumar SS et al (2022) Investigation on the physicochemical and mechanical properties of novel alkali-treated Phaseolus vulgaris fibers. J Nat Fibers 19:770–781. https://doi.org/10.1080/15440478.2020.1761930
37. Ganapathy T, Sathiskumar R, Senthamaraikannan P et al (2019) Characterization of raw and alkali treated new natural cellulosic fibres extracted from the aerial roots of banyan tree. Int J Biol Macromol 138:573–581. https://doi.org/10.1016/j.ijbiomac.2019.07.136
38. Kommula VP, Reddy KO, Shukla M et al (2013) Physico-chemical, tensile, and thermal characterization of napier grass (native African) fiber strands. Int J Polym Anal Charact 18:303–314. https://doi.org/10.1080/1023666X.2013.784935
39. Kathirselvam M, Kumaravel A, Arthanarieswaran VP, Saravanakumar SS (2019) Characterization of cellulose fibers in Thespesia populnea barks : influence of alkali treatment 217:178–189. https://doi.org/10.1016/j.carbpol.2019.04.063
40. Rajkumar R, Manikandan A, Saravanakumar SS (2016) Physicochemical properties of alkali-treated new cellulosic fiber from cotton shell. Int J Polym Anal Charact 21:359–364. https://doi.org/10.1080/1023666X.2016.1160509
41. Mylsamy K, Rajendran I (2010) Investigation on physio-chemical and mechanical properties of raw and alkali-treated Agave Americana fiber. J Reinf Plast Compos 29:2925–2935. https://doi.org/10.1177/0731684410362817
42. Amroune S, Bezazi A, Dufresne A et al (2021) Investigation of the date palm fiber for green composites reinforcement: thermo-physical and mechanical properties of the fiber. J Nat Fibers 18:717–734. https://doi.org/10.1080/15440478.2019.1645791
43. Pandiarajan P, Kathiresan M, Baskaran PG, Jeyakanth (2020) Characterization of raw and alkali treated new cellulosic fiber from the rinds of Thespesia populnea plant. J Nat Fibers. https://doi.org/10.1080/15440478.2020.1852996
44. Pokhriyal M, Rakesh PK, Sanjay, et al (2023) Effect of alkali treatment on novel natural fiber extracted from Himalayacalamus falconeri culms for polymer composite applications. Biomass Convers Biorefinery.
45. Singh JK, Rout AK (2022) Characterization of raw and alkali-treated cellulosic fibers extracted from Borassus flabellifer L. Biomass Convers Biorefinery. https://doi.org/10.1007/s13399-022-03238-x
46. Jaramillo-Quiceno N, Vélez R JM, Cadena ChEM et al (2018) Improvement of mechanical properties of pineapple leaf fibers by mercerization process. Fibers Polym 19:2604–2611. https://doi.org/10.1007/s12221-018-8522-3
47. Chaturvedi A, Ranakoti L, Rakesh PK, Mishra NK (2021) Experimental investigations on mechanical properties of walnut shell and pine needle ash polylactic acid biocomposites. Compos Theory Pract 21(3):114–120. Scopus
48. Kumar J, Kumar V, Singh I, Rakesh PK (2021) Joining behavior of polymeric composites fabricated using agricultural waste as fillers. J Adhes Sci Technol 35(15):1652–1663
49. Ranakoti L, Gupta M, Rakesh PK (2019) Analysis of mechanical and tribological behavior of wood flour filled glass fiber reinforced epoxy composite. Mater Res Express 6:085327
50. Gupta MK, Rakesh PK (2019) Characterization of pine needle ash particulates reinforced surface composite fabricated by friction stir process. Mater Res Express 06(4):046539
51. Gairola S, Gairola S, Sharma H, Rakesh PK (2019) Impact behavior of pine needle fiber/pistachio shell filler based epoxy composite. J Phys: Conf Ser 1240(012096):1–7
52. Joseph PV, Rabello MS, Mattoso LHC et al (2002) Environmental effects on the degradation behaviour of sisal fibre reinforced polypropylene composites. Compos Sci Technol 62:1357–1372. https://doi.org/10.1016/S0266-3538(02)00080-5

53. Nishino T, Hirao K, Kotera M et al (2003) Kenaf reinforced biodegradable composite. Compos Sci Technol 63:1281–1286. https://doi.org/10.1016/S0266-3538(03)00099-X
54. Kousaalya AB, Biddappa BI, Krumm K et al (2018) Poly(lactic acid)/areca fiber laminate composites processed via film stacking technique. J Appl Polym Sci 135:1–12. https://doi.org/10.1002/app.45795
55. Mukhopadhyay S, Srikanta R (2008) Effect of ageing of sisal fibres on properties of sisal - polypropylene composites. Polym Degrad Stab 93:2048–2051. https://doi.org/10.1016/j.polymdegradstab.2008.02.018
56. Ranakoti L, Rakesh PK (2020) Physio-mechanical characterization of tasar silk waste/jute fiber hybrid composite. Compos Commun 22:100526. https://doi.org/10.1016/j.coco.2020.100526
57. Oksman K, Skrifvars M, Selin JF (2003) Natural fibres as reinforcement in polylactic acid (PLA) composites. Compos Sci Technol 63:1317–1324. https://doi.org/10.1016/S0266-3538(03)00103-9
58. Islam MS, Pickering KL, Foreman NJ (2010) Influence of alkali treatment on the interfacial and physico-mechanical properties of industrial hemp fibre reinforced polylactic acid composites. Compos Part A Appl Sci Manuf 41:596–603. https://doi.org/10.1016/j.compositesa.2010.01.006
59. Indran S, Edwin Raj RD, Daniel BSS, Binoj JS (2018) Comprehensive characterization of natural Cissus quadrangularis stem fiber composites as an alternate for conventional FRP composites. J Bionic Eng 15:914–923. https://doi.org/10.1007/s42235-018-0078-9
60. De Rosa IM, Santulli C, Sarasini F (2010) Mechanical and thermal characterization of epoxy composites reinforced with random and quasi-unidirectional untreated Phormium tenax leaf fibers. Mater Des 31:2397–2405. https://doi.org/10.1016/j.matdes.2009.11.059
61. Yoganandam K, Ganeshan P, NagarajaGanesh B, Raja K (2020) Characterization studies on Calotropis procera fibers and their performance as reinforcements in epoxy matrix. J Nat Fibers 17:1706–1718. https://doi.org/10.1080/15440478.2019.1588831
62. Mayandi K, Rajini N, Pitchipoo P et al (2016) Extraction and characterization of new natural lignocellulosic fiber Cyperus pangorei. Int J Polym Anal Charact 21:175–183. https://doi.org/10.1080/1023666X.2016.1132064
63. Omar HC, Isidro DC, Philip LJ, Adriana EO (2022) Film-stacking method as an alternative Agave tequilana fibre/PLA composite fabrication. Mater Today Commun 31:103853. https://doi.org/10.1016/j.mtcomm.2022.103853
64. Sivakumar S, Vignesh V, Arasu IV et al (2021) Experimental investigation on tensile and flexural properties of randomly oriented treated palmyra fibre reinforced polyester composites. Mater Today Proc 46:7556–7560. https://doi.org/10.1016/j.matpr.2021.01.511

Chapter 16
Machine Learning-Enabled Data-Driven Research on Paper-Reinforced Composite Materials

Saureng Kumar, Sanjeev Kumar, and S. C. Sharma

1 Introduction

A composite material is a mixture of two or more components, one of which functions as a strengthening agent, such as fibers or sheets, and the other as the matrix phase. Composite materials have surfaced as a revolutionary innovation in the realm of materials science, providing distinct combinations of properties that remain beyond reach when considering individual constituent materials. One particularly encouraging avenue within this dynamic field centers on the utilization of paper-reinforced composites, presenting a renewable and biodegradable substitute for conventional non-biodegradable materials. The increasing global concern over environmental degradation and the accumulation of non-biodegradable waste has intensified the search for sustainable materials. Paper-reinforced composites, being a naturally occurring and abundant resource, have garnered attention as a potential reinforcing agent for composite materials. Because of its inherent strength, low weight, and environmental friendliness [1], it is an excellent contender for improving the mechanical characteristics of composite constructions. The research work [2] was motivated by the urgent need to discover eco-friendly materials that can reduce the negative environmental effect of traditional materials. The inception of composite materials marked a pivotal moment in the evolution of materials research, enabling the fusion of attributes that were previously inaccessible through individual constituent materials. Amid the extensive body of research in this domain, the integration of

S. Kumar (✉) · S. Kumar · S. C. Sharma
Electronics and Computer Discipline, Indian Institute of Technology Roorkee, Roorkee, India
e-mail: skumar@pp.iitr.ac.in

S. Kumar
e-mail: skumar1@pp.iitr.ac.in

S. C. Sharma
e-mail: scs60fpt@iitr.ac.in

S. K. Sethi et al. (eds.), *Polymer Composites: From Computational to Experimental Aspects*, Materials Horizons: From Nature to Nanomaterials,
https://doi.org/10.1007/978-981-97-0888-8_16

paper-reinforced composites emerges as a notable avenue to explore. These composites not only provide a renewable and biodegradable alternative to non-biodegradable materials but they also bring in a new era of inventive possibilities. The drive driving this research stems from the need to create materials that promote environmental responsibility. For example, Moshood et al. [3] investigated the environmental effects for the sustainable plastic economy as well as associated factors that influence their long-term viability. Song et al. [4] have discussed the plastic waste and other biowaste compost materials suitable for plant growth. In order to increase the global environmental concerns, the demand for eco-friendly, sustainable, and biodegradable natural fiber-reinforced composites (NFRCs) [5] has been used in place of non-renewable synthetic fibers. Green et al. [6] have investigated the tensile strength frequently decreased as the size of the specimen decreased, reaching a maximum reduction of 64% of notched composites. A similar type of study has been conducted by Liu et al. [7]. The author has performed an open-hole tensile tests on T700/SYE20005 carbon fiber.

Besides the experimental investigation of the mechanical properties, some of the studies focused on the machine learning-based prediction, Wang et al. [8] have used machine learning techniques to predict compression property of braided-textile tubes based on that energy absorbing properties were analyzed. Saureng et al. [9] proposed machine learning algorithms for energy content prediction in packaged food. These foods are supplied from source to destination using trucks. Then, the author [10] has used machine learning technique for the supply chain risk assessment during the transportation. Furthermore, a deep neural network method has been applied on the composite hydrogel [11] from the microstructural images and measured the performance in terms of MAE, MSE, and R-value. The author claims better performance as the result error prediction is less than 7%. Choosing the best fiber types and layup stacking arrangements for mechanical and non-mechanical properties, a Generative Adversarial Network (CGAN) has developed to design target strength, maximum deformation, and minimum thickness [12]. Balasundaram et al. [13] collected 26 different natural fiber composites of Coir, Jute, Banana, etc. and investigated using the machine learning along with Apriori algorithm to measure the density, tensile strength, hardness, etc. A comprehensive investigation is required for the prediction of compressive strength, compressive strain at rupture, strain energy density for resilience, strain energy, vent hole. In this research, we have performed a data-driven research on paper-reinforced composite material. This study supplements the application of machine learning technique in the field of paper-reinforced composite material. Furthermore, the author also discussed the mechanical properties, challenges and potential application areas of the NFRCs. However, the use of paper-reinforced composites has the potential to reduce our reliance on non-biodegradable alternatives, which contribute to the burden of pollution and waste buildup.

The rest of the paper is organized as follows. In Sect. 2, a literature review has illustrated, Sect. 3 represents the discussion of the roadmap for the mentioned machine learning techniques such as Support Vector Machine, Random Forest, Decision Tree, and k-Nearest Neighbor that are briefly introduced and some basic concepts are clarified. Section 4, data collection and data preprocessing concept for training and testing of the network. In addition, we evaluate the performance in terms of accuracy, recall, precision, F1 score, and error rate in Sect. 5. Finally, conclusion and future direction of the research work are presented.

2 Literature Review

In the field of artificial intelligence (AI), machine learning enables computers to learn and gain experience automatically, eliminating the need for explicit programming. The primary objective of machine learning is to develop models and algorithms that can autonomously access and utilize data for various purposes, such as generating insights, making informed decisions, and drawing meaningful conclusions. Table 1 shows the data-driven-based research in the field of polymer composites.

3 Roadmap of the Machine Learning Framework

The roadmap serves as a comprehensive and structured approach to developing, evaluating, and deploying machine learning models while emphasizing the importance of rigorous data preprocessing and performance evaluation. In this pursuit, various critical factors such as compressive strength, compressive strain at rupture, strain energy density for resilience, and vent hole considerations come into play. In light of these factors, our research delves deeply into the examination of paper-reinforced composite materials. Our research has two main objectives, concentrating on these materials' mechanical properties and environmental benefits. In this regard, we have also used machine learning for predictive modeling. We hope that these initiatives will advance material research while advancing environmentally friendly production methods. For predictive modeling, we have included machine learning. In our opinion, these activities advance both material research and the use of environmentally friendly industrial techniques. In the end, our main goal is to aid in the creation of high-performance, sustainable materials. Our work is to advance material science and environmentally responsible manufacturing techniques, thereby forging a more solid and harmonious link between technical advancement and environmental protection. Developing an effective machine learning solution involves several essential steps to ensure accurate predictions and reliable model performance. We have adapted the dataset [12] for the mechanical properties of damaged fiber composite. Further, we preprocess the data by handling missing values, outliers, and inconsistencies. Additionally, machine learning models often possess hyperparameters that influence their behavior during training. To determine the optimal configuration for these hyperparameters and achieve the best model performance, we leverage the random search technique (Fig. 1).

Machine learning model offers a variety of algorithms suitable for accurate prediction such as including Support Vector Machine (SVM), Decision Tree (DT), Random

Table 1 Overview of data-driven-based research in the field of polymer composite

Serial number	Polymer composites	Characterization	Methodology/ Techniques	Key parameters	References
1	S-type woven glass fiber-reinforced polymer matrix composites (PMCs)	Erosion behavior	Artificial neural network (ANN)	Slurry pressure, impingement angle and nozzle diameter	[14]
2	Hybrid nano-filled glass/epoxy composites	Mechanical behaviour	SEM analysis	Stiffness, strength	[15]
3	Aluminum metal matrix composites (AMCs)	Mechanical behavior	Artificial neural network (ANN)	Tensile and microhardness	[16]
4	Carbon fiber-reinforced polymer nanocomposite	Machining behavior	Response surface methodology (RSM), particle swarm optimization (PSO)	Surface delamination, tearing, abrasion, denting, cutting of fibers, formation of voids, fiber bending and pullout	[17]
5	Hybrid fiber-reinforced polymer composites (HFRPC)	Interfacial and tensile properties	Deep neural network (DNN), Oppositional-based firefly optimization (OFFO)	Strength-to-weight, stiffness-to-weight, corrosion resistance	[18]
6	Red brick dust (RBD)	Wear mechanisms behavior	Artificial neural network (ANN)	Erosion characteristics	[19]
7	Paper-reinforced composites (proposed)	Strength	Artificial intelligence, machine learning	Compressive strength, compressive strain at rupture, strain energy density for resilience, strain energy, vent hole	

Forest (RF), and k-Nearest Neighbor (KNN), and finally, we evaluate the performance based on accuracy, precision, recall, error rate, and F1 score.

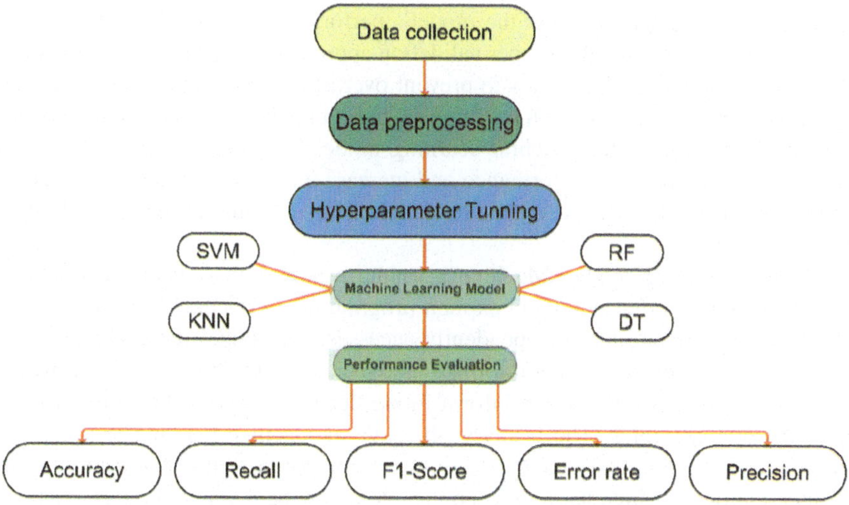

Fig. 1 Schematic representation of roadmap

4 Data Collection and Preprocessing

Data collection is a crucial process that involves gathering clear, precise, and concise data from various sources. The objective is to use this data for analysis and decision-making purposes. The collected data may initially be in a raw form, which can include repetitive, incomplete, and disorganized information. In the context of identifying and collecting relevant data on paper-reinforced composite materials, including mechanical properties and environmental data, data preprocessing techniques come into play. Data cleaning is an essential part of this process, as it involves converting the raw data into a format that is easily understandable and usable.

Hyperparameter Tuning: In order to maximize the performance of the model, the ideal combination of hyperparameters must be carefully sought. The objective is to identify the values that produce the highest levels of accuracy, precision, recall, or any other evaluation metric that is pertinent to the current challenge.

There are numerous methods and plans for tweaking hyperparameters. Grid search is a popular technique that involves training and evaluating the model for each combination of predetermined hyperparameter values. Although this thorough search can be computationally expensive, it ensures that the optimum combination will be found inside the designated search space. Another well-liked method is random search, which draws values for the hyperparameters at random from a predetermined distribution. Although less computationally costly than grid search, this method still has a reasonable chance of locating the ideal hyperparameters. For hyperparameter tweaking, more sophisticated techniques like Bayesian optimization or evolutionary algorithms can be used. These methods effectively explore the hyperparameter space by using statistical models or evolutionary principles to direct the search process. It is essential to assess the model's performance during the hyperparameter tuning process using the proper validation approaches, such as cross-validation or hold-out validation. This helps to prevent overfitting and ensure that the chosen hyperparameters generalize well to new data. In general, hyperparameter adjustment is essential for improving machine learning models' capacity for prediction. We can improve the model's performance and increase its capacity to generate precise predictions on fresh, unforeseen data by methodically looking for the optimal set of hyperparameters.

Machine learning model: Machine learning, a branch that enables systems to learn and gain experience without explicit programming. Machine learning focuses on computer software that independently accesses and utilizes data. The learning process begins with feedback or information associated with illustrations, real-world instances, or suggestions, which inform future decisions based on patterns in data and provided examples. The primary objective of machine learning is to enable autonomous learning and adaptive actions on computers without human support or intervention. Machine learning is applicable to any field where there is a dependence on an input–output link. Data are utilized to train machine learning algorithms, making it crucial to gather and compile accurate data. Thus, it is crucial to meticulously select and adequately preprocess the pertinent data to ensure the effective management of the problem. Forecasting relies on a variety of machine learning techniques, which can be classified according to the input data they employ. In our evaluation of machine learning models (SVM, RF, DT, KNN), we have assessed their performance across accuracy, precision, recall, and F1 score. The summary of our findings is presented in Table 2.

These categories include the unsupervised learning and supervised learning. Figure 2 illustrates prediction process of the mentioned machine learning algorithms. The RF model performs the superior accuracy, i.e., 96%. Hence. we select the Random Forest model for further analysis of the performance.

The prediction process commences with training set, test set, and validation set which are the three main subsets that make up the dataset used to start the prediction process. Initializing the model's parameters is crucial before training it. In this step, the weights and biases of the model are given initial values. The learning process and final performance of the model can be considerably influenced by appropriate initialization. The machine learning algorithm creates a trained model after the training phase. This model is able to predict outcomes based on fresh, unforeseen data since it has learned from the training set of data.

Table 2 Summary of findings

Parameter	SVM	RF	DT	KNN
Accuracy	0.91	0.96	0.91	0.66
Precision	0.93	0.88	0.92	0.98
Recall	0.99	0.72	0.75	0.65
F1 score	0.96	0.8	0.8	0.78

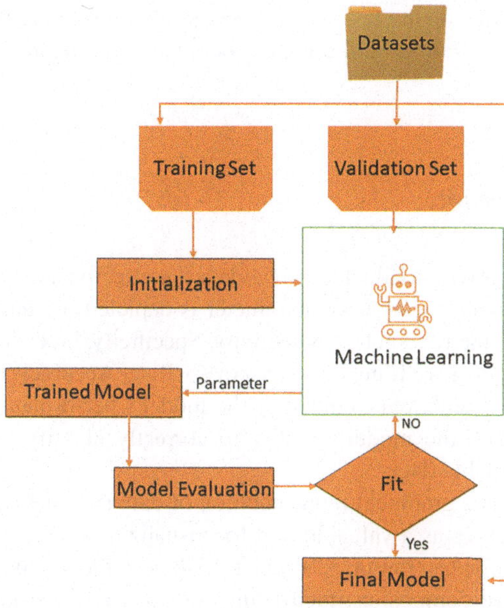

Fig. 2 Typical prediction process through machine learning algorithms

I. Training Set: The training set is used to train the machine learning algorithm. It contains a large amount of labeled data, where the input features are paired with the corresponding output labels. The algorithm learns patterns and relationships within the data during this training phase.
II. Test Set: Once the algorithm is trained, it is evaluated using the test set. The test set contains data that the algorithm has not seen before. By applying the trained model to the test set, its predictive performance can be assessed. This step helps to measure the algorithm's ability to generalize and make accurate predictions on unseen data.
III. Validation Set: In some cases, a validation set is used during the training process. This set is separate from the test set and is used to fine-tune the model's hyperparameters. Hyperparameters are settings that control the learning process, such as the learning rate or the number of hidden layers in a neural network. The validation set helps to optimize these hyperparameters and improve the model's performance.

The final step involves evaluating the performance of the trained model. Various evaluation metrics, such as accuracy, precision, recall, or mean squared error, can be used to assess how well the model performs on the test set. This evaluation helps to determine the effectiveness and reliability of the machine learning algorithm in making predictions.

Overall, the typical prediction process through machine learning algorithms involves preparing and dividing the dataset, training the model using the training set, evaluating its performance on the test set, and potentially fine-tuning the model using a validation set. The initialized and trained model is then evaluated to assess its predictive capabilities.

5 Results and Discussion

The performance evaluation was conducted to validate the effectiveness and accuracy of model. The performance evaluation parameter is depicted in Table 3. This involves measuring various metrics such as Sensitivity, Specificity, Precision, Negative Predictive Value, False Positive Rate, False Discovery Rate, Accuracy, F1 Score, and Matthews Correlation Coefficient to quantify the model's performance. These metrics provide insights into the model's ability to correctly classify or predict outcomes based on the available data.

Heat Map: A heat map is a graphical representation of data in which values are depicted using colors. It serves as a valuable tool for visualizing relationships and patterns among multiple variables. In Fig. 3, it is used to display the relative strength or strain of paper-reinforced composites, facilitating comparisons of strain energy density for resilience and strain energy across different materials, or the compressive strength and strain at rupture of various materials. Typically, heat maps employ a spectrum of colors to represent the intensity of the data. Darker hues indicate stronger correlations while lighter hues suggest weaker ones.

Table 3 Performance evaluation analysis

Performance evaluation	Value	Derivation
Sensitivity	0.7273	TPR = TP/(TP + FN)
Specificity	0.989	SPC = TN/(FP + TN)
Precision	0.8889	PPV = TP/(TP + FP)
Negative predictive value	0.9677	NPV = TN/(TN + FN)
False positive rate	0.011	FPR = FP/(FP + TN)
False discovery rate	0.1111	FDR = FP/(FP + TP)
False negative rate	0.2727	FNR = FN/(FN + TP)
Accuracy	0.9608	ACC = (TP + TN)/(P + N)
F1 score	0.8	F1 = 2TP/(2TP + FP + FN)
Matthews correlation coefficient	0.7833	TP*TN − FP*FN/sqrt((TP + FP)*(TP + FN)*(TN + FP)*(TN + FN))

TP-True Positive P-Condition Positive PPV-Positive Predictive Value ACC-Accuracy
TN-True Negative N-Condition Negative NPV-Negative Predictive Value
FP-False Positive TPR- True Positive Rate FPR-False Positive Rate
FN-False Negative SPC-Specificity FDR-False Discovery Rate

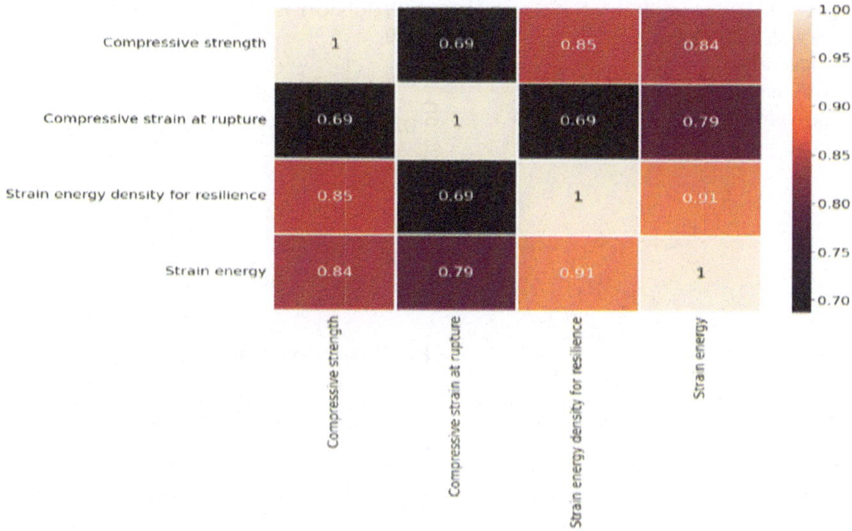

Fig. 3 Heat map

Precision-recall curve: A precision-recall curve is a graphical representation of a classification model's performance in terms of precision and recall. The comparison graph is plotted with the various threshold values, as depicted in Fig. 4a. This curve illustrates the trade-off between recall and precision for different threshold values. The Area Under the Curve (AUC) is calculated by measuring the area beneath the precision-recall curve, which yields values of 0.67 and 0.57, respectively. Additionally, the Average Precision (AP) is shown to be 94%, as indicated in Fig. 4b.

Receiver Operating Characteristic (ROC): Receiver Operating Characteristic analysis is used to evaluate the performance of classification models. The ROC curve summarizes the trade-off between true positive rate (sensitivity) and false positive rate (1 − specificity) across different classification thresholds. The received ROC curve is shown in Fig. 5, the ROC curve area of 92% indicates that the model has a high discriminatory power, with a strong ability to correctly classify positive cases while minimizing false positives. It suggests that our model is effective in making accurate predictions. Furthermore, training and testing set scatter plots are used for model development and evaluation as shown in Fig. 6a, b. It visualizes relationships between two continuous variables. The focus of scatter plots is on visualizing relationships within the data, not necessarily comparing the training and testing datasets. The model performing the given task is shown in Fig. 7.

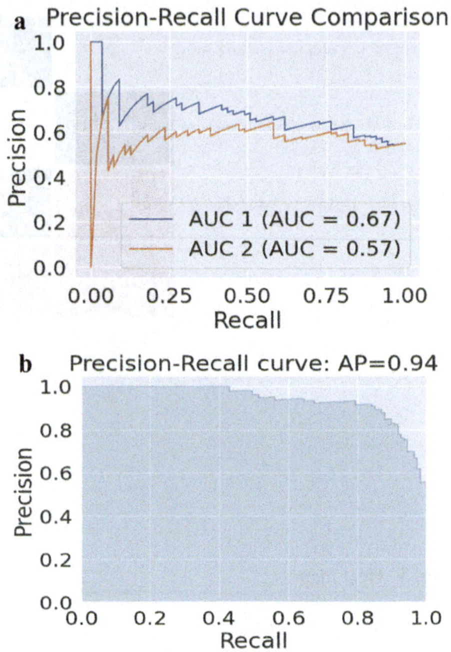

Fig. 4 **a** Precision-recall curve with the different threshold values. **b** Precision-recall curve average precision

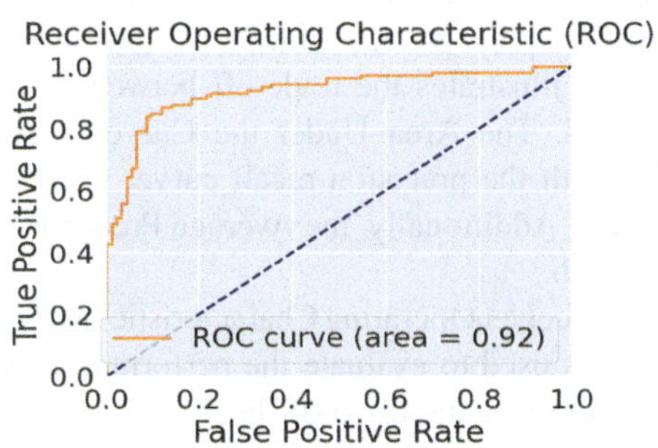

Fig. 5 Receiver operating characteristic

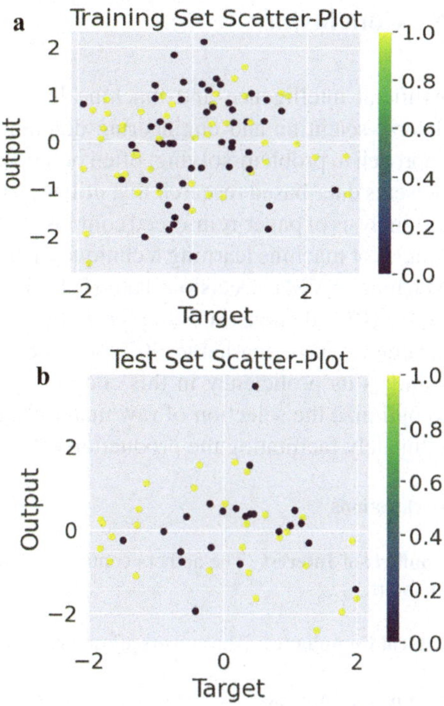

Fig. 6 **a** Training set scatter plot. **b** Testing set scatter plot

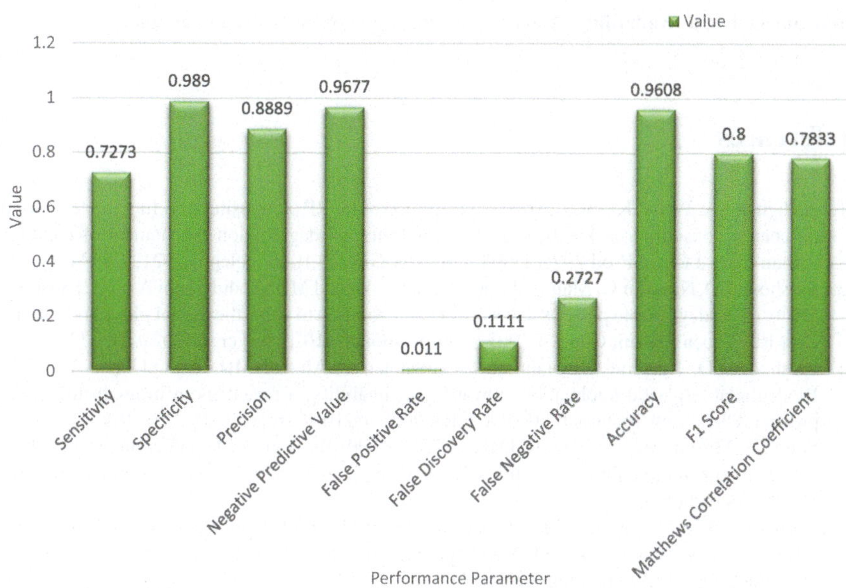

Fig. 7 Performance evaluation

6 Conclusion

Artificial intelligence and machine learning have gained extensive popularity in diverse scientific and engineering domains. These advanced tools present a novel approach to problem-solving, often outperforming conventional methods. This paper presents data-based research that utilizes a range of machine learning algorithms for the analysis of paper-reinforced composite materials. The study also explores the integration of machine learning techniques, particularly the evaluation of Support Vector Machine (SVM), Decision Tree (DT), k-Nearest Neighbor (KNN), and Random Forest (RF), demonstrates the feasibility of predicting material properties with high accuracy. The remarkable 96% accuracy achieved by the Random Forest model signifies its proficiency in this context. Additionally, the algorithms are leveraged to optimize the selection of raw materials, design and simulate various parameters, ultimately facilitating the production of optimal composite materials.

Declarations

Conflicts of Interest The authors confirm that there are no conflicts of interest associated with this work.

Ethical Compliance The authors affirm that this study adheres to all ethical guidelines.

Competing Interests No competing interests are declared by the authors.

Data and Material Availability Data and materials are not applicable to this study.

References

1. Hadigheh SA, Wei Y, Kashi S (2021) Optimisation of CFRP composite recycling process based on energy consumption, kinetic behaviour and thermal degradation mechanism of recycled carbon fibre. J Clean Prod 292:125994. https://doi.org/10.1016/j.jclepro.2021.125994
2. Moshood TD, Nawanir G, Mahmud F, Mohamad F, Ahmad MH, AbdulGhani A (2022) Sustainability of biodegradable plastics: New problem or solution to solve the global plastic pollution? Curr Res Green Sustain Chem 5:100273. https://doi.org/10.1016/j.crgsc.2022.100273
3. Moshood TD, Nawanir G, Mahmud F, Mohamad F, Ahmad MH, AbdulGhani A (2022) Biodegradable plastic applications towards sustainability: a recent innovations in the green product. Clean Eng Technol 6:100404. https://doi.org/10.1016/j.clet.2022.100404
4. Song JH, Murphy RJ, Narayan R, Davies GBH (2009) Biodegradable and compostable alternatives to conventional plastics. Philos Trans R Soc B 364(1526):2127–2139. https://doi.org/10.1098/rstb.2008.0289
5. Syduzzaman M, Al Faruque MA, Bilisik K, Naebe M (2020) Plant-Based natural fibre reinforced composites: a review on fabrication, properties and applications. Coatings 10(10):973. https://doi.org/10.3390/coatings10100973
6. Green BG, Wisnom MR, Hallett SR (2007) An experimental investigation into the tensile strength scaling of notched composites. Compos A Appl Sci Manuf 38(3):867–878. https://doi.org/10.1016/j.compositesa.2006.07.008

7. Liu J, Wang L, He Y, Zhang Y, Yuan X, Li W (2023) Experimental study of notched tensile strength of large open-hole carbon fiber reinforced polymer laminates at low temperature. Compos Commun 39:101546. https://doi.org/10.1016/j.coco.2023.101546
8. Wang W, Wang H, Zhou J, Fan H, Liu X (2021) Machine learning prediction of mechanical properties of braided-textile reinforced tubular structures. Mater Des 212:110181. https://doi.org/10.1016/j.matdes.2021.110181
9. Kumar S, Sharma SC (2023) Machine learning-based approach for energy content prediction in packaged foods. Mach Learn 1
10. Kumar S, Sharma SC (2023) Integrated model for predicting supply chain risk through machine learning algorithms. Int J Math Eng Manag Sci 8(3):353–373. https://doi.org/10.33889/IJMEMS.2023.8.3.021
11. Gholami K, Ege F, Barzegar R (2023) Prediction of composite mechanical properties: integration of deep neural network methods and finite element analysis. J Compos Sci 7(2):54. https://doi.org/10.3390/jcs7020054
12. Qiu C et al (2022) A deep learning-based composite design strategy for efficient selection of material and layup sequences from a given database. Compos Sci Technol 230:109154. https://doi.org/10.1016/j.compscitech.2021.109154
13. Balasundaram R, Sathya Devi S, Sakthi Balan G (2022) Machine learning approaches for prediction of properties of natural fiber composites: apriori algorithm. Aust J Mech Eng 1–16. https://doi.org/10.1080/14484846.2022.2030091
14. Antil SK, Antil P, Singh S, Kumar A, Pruncu CI (2020) Artificial neural network and response surface methodology based analysis on solid particle erosion behavior of polymer matrix composites. Materials 13(6):1381. https://doi.org/10.3390/ma13061381
15. Rajhi AA (2022) Mechanical characterization of hybrid nano-filled glass/epoxy composites. Polymers 14(22):4852. https://doi.org/10.3390/polym14224852
16. Pushparaj JP et al (2023) Taguchi-based artificial neural network modeling of friction process on aluminum alloy reinforced with SiC nanoparticles. J Nanomater 2023:1–6. https://doi.org/10.1155/2023/3258173
17. Vijayan D, Rajmohan T (2019) Modeling and evolutionary computation on drilling of carbon fiber-reinforced polymer nanocomposite: an integrated approach using RSM based PSO. J Braz Soc Mech Sci Eng 41(10):395. https://doi.org/10.1007/s40430-019-1892-7
18. Baseer AA, Ravi Shankar DV, Hussain MM (2020) Interfacial and tensile properties of hybrid FRP composites using dnn structure with optimization model. Surf Rev Lett 27(02):1950099. https://doi.org/10.1142/S0218625X19500999
19. Pati PR (2019) Prediction and wear performance of red brick dust filled glass–epoxy composites using neural networks. Int J Plast Technol 23(2):253–260. https://doi.org/10.1007/s12588-019-09257-0

Chapter 17
Piercing of Epoxy Laminates with Wood Drill Point Geometry

Pawan Kumar Rakesh, Rajesh Kumar, and Inderdeep Singh

1 Introduction

The structural applications (building and bridge constructions) of woven glass fiber mat reinforced epoxy (GFMRE) composites are increasing due to its chemical and thermal properties, and enhanced Young's modulus and tensile strength [1–3]. These characteristics encompass an elevated strength to mass ratio, significant stifling, minimal thermal growth, commendable deterioration resistance, and notably gradually removal piercing material resistance. Epoxy composites that are produced by basic manufacturing techniques are often in a form that is close to the near net shape. Composite products that are intricate or sophisticated are always created in sections, and then after they are finished, they are assembled. Because of this, piercing is the most significant machining operation that is commonly practiced guaranteeing that the assembly operation of complicated polymer composite products is successful. The primary aim has been to diminish both axial forces and tangential piercing forces during piercing, along with mitigating the resulting burr formation. Paul et al. [4] perfected conical, radial, and helical drill profiles to reduce axial forces while preserving fundamental profile properties, resulting in a noticeable decrease in generated forces. Tsao and Hocheng [5] compared two drill configurations (the Saw drill and Core drill systems), finding that the core saw drill offered greater benefits, and using a low feed velocity during piercing reduced induced burr formation. Velayudham and Krishnamurthy [6] explored the piercing of GFRP composites

P. K. Rakesh (✉) · R. Kumar
Department of Mechanical Engineering, National Institute of Technology Uttarakhand, Srinagar 246174, Uttarakhand, India
e-mail: pawankumarrakesh@gmail.com

I. Singh
Department of Mechanical and Industrial Engineering, Indian Institute of Technology Roorkee, Roorkee 247667, Uttarakhand, India

S. K. Sethi et al. (eds.), *Polymer Composites: From Computational to Experimental Aspects*, Materials Horizons: From Nature to Nanomaterials,
https://doi.org/10.1007/978-981-97-0888-8_17

with various drill configurations, finding that drill point configuration significantly affected axial force and burr formation, with tripod points causing the least burr formation. Abrao et al. [7] deliberated the impact of four drill geometries on high-speed piercing of GFRP composites, noting reduced burr formation with increased rotational speed, likely due to elevated piercing temperature making the matrix more pliable. Khashaba et al. [8] measured the impact of various process factors on axial forces and surface finish in GFRP composites. It was seen that an increase in required axial force with a rise in drill diameter due to the corresponding increase in shear area. The drill shape and its diameter, and feed velocity significantly influenced surface finish. In their study, Durão et al. [9] examined five drill geometries, deciding that the twist 120° shape resulted in the least amount of burr formation compared to other drill angles. Faraz et al. [10] investigated in-process piercing profiles and exit burr formation in GFRP laminates using both standard and specified uncoated carbide drill bits. Specified bits proven lower axial peaks compared to standard ones. Rakesh et al. [11] assessed the effect of drill configuration and materials on axial forces and burr formation during GFRP composite piercing. Rotational speed showed no significant impact on axial force, while increasing feed velocity led to more burr formation. According to the findings by Rajamurugan et al. [12], the size of the burr formation rises as the feed velocity and drill diameter are increased, however the fiber orientation angle does not seem to have any substantial impact on the size of the burr formation. Khashaba and El-Keran [13] showed that the axial force, tangential piercing forces, and burr formation zone, all rose when the feed was raised, and that the opposite was true for the induced temperature situation. Hocheng et al. [14] combined statistical and experimental analyses to reveal the progression of burr formation during the downward movement of the drill. Burr formation initiation and crack widening were linked to bending of the last drilled ply in response to piercing axial force. Mudhukrishnan et al. [15] conducted a study measuring piercing-induced axial force and exit burr formation, indicating that axial force increases with drill feed velocity but diminishes with rotational speed increase for all drill materials. Thakur and Singh [16] optimized process parameters (rotational speed, feed velocity, and graphene wt% fillers) in the glass fiber-reinforced epoxy laminate in piercing, enhancing interlaminate shear strength (18.49%), flexural strength (25.15%), and tensile strength (14.79%) compared to the neat epoxy composite by incorporating graphene (0.25 wt%). Geier et al. [17] investigated micro-drilling in GFRP laminates, revealing that mechanical micro-drilling of FRPs combine challenges from fibrous composite machining and downscaled machining.

An immense variety of different drill geometries have been conceived of and created to make holes in composite materials that are intact and free of harm. There has been a lot of work done to understand the piercing behavior of GFMREs, but there have only been a few studies that have been done on the study of the axial forces in relation to the piercing process of wood drill geometry. The axial piercing forces are affected by the fact that wood drills have a significantly different piercing mechanism than other types of drills. There is a possibility that the piercing behavior in laminates may be defined by the axial piercing forces. In the current research endeavor, an effort is being made to study through an experiment the relative impact

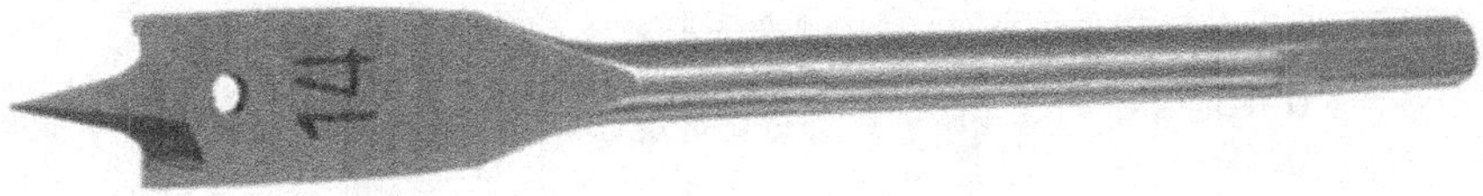

Fig. 1 Wood drill point configuration

of the wood drill configuration, the rotational speed, and the feed velocity on the axial piercing forces that are experienced while piercing GFMRE composites.

2 Laminate Fabrication and Wood Drill Geometry

The GFMRE laminates were produced using the hand-layup method, incorporating woven glass fiber mat to reinforce the epoxy. Araldite AW 106 served as the matrix, and HV 953 was employed as the hardener. For piercing holes in the GFMRE laminates, a radial piercing machine with a 1.5 kW power output and a maximum rotational speed of 4000 rpm was used. Three feed velocities of 0.12, 0.19, and 0.3 mm/rev, were chosen for rotational speed, which were set at speeds of 1120, 1800, and 2800 rpm. The wood drill configuration has a determined diameter of 14 mm (see Fig. 1). To measure axial piercing forces and tangential piercing forces, a four-component drill dynamometer (Kistler, Type 9272) with sensitivities of 3.58 ρC/N and 1.74 ρC/(Ncm), respectively, was employed. Signal amplification was achieved using a Kistler charge amplifier (type 5070).

3 Piercing Stages in Laminate

The piercing force signals throughout the piercing process of GFMRE laminates were captured using a drill dynamometer. The laminates may be divided into three main stages throughout the piercing cycle: indentation, cutting, and reaming action [18]. These steps are often known as the piercing phases. The wood drill is equipped with dual cutting edges, with the chisel edge serving as the starting point for the piercing operation. It is crucial to highlight that the piercing action begins at the midpoint of the drill and continues until the drill is fully embedded in the laminate (Fig. 2). As the laminate becomes thicker, the part of the material that hasn't been cut acts like a beam that hangs over, making it more likely to be dragged downwards by the force exerted by the wood drill tip (Fig. 3).

Figure 2 depicts the piercing force signals that arise while penetrating GFMRE laminates with wood drill tip forms. The first phase in the process involves extruding the top layer to create an indentation, causing the forefront to contact the material. Due to the first insertion of the wood drill into the laminate, there is a notable augmentation in the applied axial force. Subsequently, there is an added augmentation

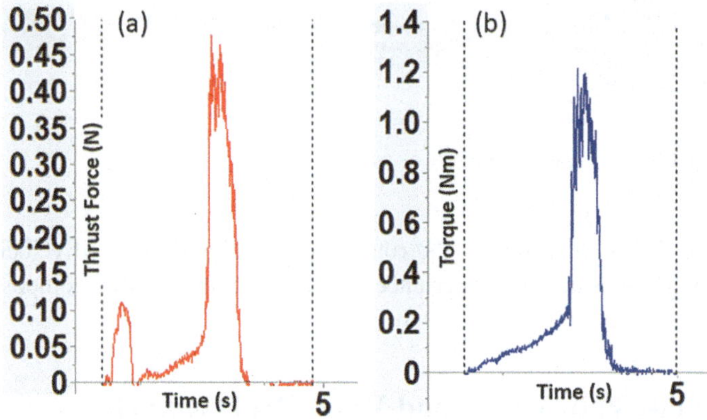

Fig. 2 Piercing of laminate with wood drill **a** Axial force, **b** Tangential piercing forces, at 1120 rpm, 0.3 mm/rev

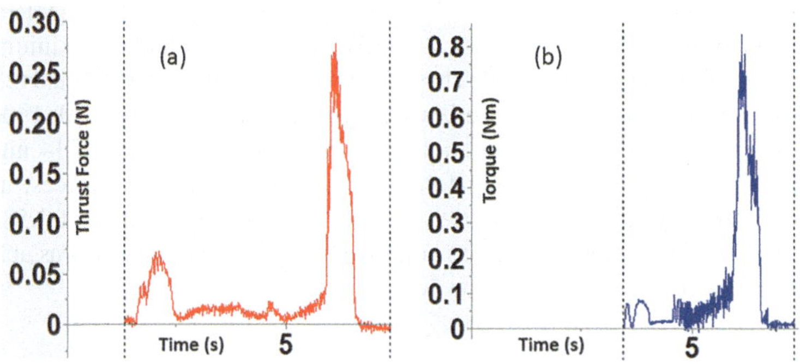

Fig. 3 Piercing of laminate with wood drill **a** Axial force, **b** Tangential piercing forces, at 2800 rpm, 0.12 mm/rev

in the propulsive force when the drill begins the process of severing the laminate to the intended width. Once the forefront reaches the bottom ply of the laminate, the decrease in axial force becomes more gradual. The reaming action, which occurs between the wood drill and the laminate, is the last stage of the piercing motion. The presence of a cyclic fluctuation in the axial power during a single rotation of the wood drill has been found, and it is essential to account for this phenomenon. Kishore et al. [19] have previously reported that the large fluctuations in axial force seen during piercing are a result of the non-uniformity within a single lamina and the presence of contact between the laminate layers. The reduction in thickness of the uncut laminate leads to a loss in its strength, which may be attributed to the consistent decrease in axial force during piercing. Ultimately, the axial force reaches a value of zero upon completion of the procedure [20–22].

The tangential piercing forces profiles of the wood drill, as seen in Fig. 2b, have a similar three-phase structure as the prior example. The first insertion of the drill into the laminate results in a fast and substantial rise in tangential piercing forces. Following this, there is an immediate increase in tangential piercing forces as the chisel edge makes complete contact with the laminate. The tangential piercing forces reach their maximum due to the drill's tip penetrating the bottom ply of the laminate. When piercing with a wood drill, the piercing process is performed in two distinct processes, as seen in Fig. 2b. The piercing action is similar in both stages, meaning that the involvement of the cutting edges leads to an increase in tangential piercing forces, which is later followed by a decline in the signal after the piercing activity is completed. As to the findings of Rakesh et al. [23], the first piercing edges create a smaller hole, which is then enlarged by the later piercing edges using the same method.

Earlier studies have shown the variation in axial force when different drill tip designs are used under the same operating conditions. Table 1 presents a comparison of the maximum axial force obtained using wood drill tip geometries, all under identical operating conditions (rotational speed: 2800 rpm; feed velocity: 0.3 mm/rev). Moreover, the axial piercing forces are directly persuaded by the piercing mechanisms of the drill tip configuration, which were discussed in the preceding section. When epoxy laminates were pierced using a wood drill with a rotational speed of 1120 rpm and a feed velocity of 0.12 mm/rev, the axial forces observed were 0.2 N and the tangential forces were 0.21 Nm. Under different piercing conditions (rotational speed of 1800 rpm and feed velocity of 0.12 mm/rev), the axial force was measured as 0.84 N and the tangential force as 0.24 Nm. By raising the feed velocity, both the axial forces and tangential forces are also rising. We may state that the feed rate exhibits a direct proportionality with the feed rate, and conversely. As the rotating speed rises, the axial and tangential pressures likewise increase as a result of the impact action of the wood drill geometry. Nevertheless, it is evident from the table that the rotational speed has a very little impact on the axial forces and tangential forces.

Table 1 Piercing forces with wood drill point configuration

S. No.	Rotational speed (rpm)	Feed velocity (mm/rev.)	Axial forces (N)	Tangential forces (Nm)
1	1120	0.12	0.20	0.21
2	1120	0.19	0.42	0.31
3	1120	0.3	0.51	0.33
4	1800	0.12	0.84	0.24
5	1800	0.19	1.22	0.43
6	1800	0.3	1.44	0.44
7	2800	0.12	0.86	0.30
8	2800	0.19	1.09	0.43
9	2800	0.3	1.56	0.61

4 Conclusions

The primary aim of this study was to investigate how the shape of the wood drill tip affects the piercing pressures exerted on GFMRE laminates. An extensive analysis was conducted on the wood drill tip configuration's piercing processes. Based on the ongoing study, the following outcomes might be inferred for consideration:

1. The wood drill tip geometries under investigation show substantial variations in their piercing mechanisms, which in turn affect the axial forces and the resulting burr formation produced by piercing.
2. The wood drill was used to find the minimum force necessary for penetration.
3. The configuration of the wood drill has a considerable impact on the axial piercing forces.

Declarations

Conflicts of Interest The authors confirm that there are no conflicts of interest associated with this work.

Ethical Compliance The authors affirm that this study adheres to all ethical guidelines.

Competing Interests No competing interests are declared by the authors.

Data and Material Availability Data and materials are not applicable to this study.

References

1. Ranakoti L, Rakesh PK, Gangil B (2021) Effect of Tasar silk waste on the mechanical properties of Jute/Grewia optiva fibers reinforced epoxy laminates. J Nat Fibers 19(15):10462–10474
2. Sharma H, Komal UK, Singh I, Misra JP, Rakesh PK (2018) Introduction to green composites, processing of green composites, pp 1–13. Springer, ISBN 978-981-13-6019-0
3. Kumar RG, Rangappa SM, Siengchin S, Zafar S (2022) A review of recent advancements in drilling of fiber-reinforced polymer composites. Compos Part C Open Access 9:100312
4. Anish P, Kapoor SG, DeVor RE (2005) Chisel edge and cutting lip shape optimization for improved twist drill point design. Int J Mach Tools Manuf 45(4–5):421–431
5. Tsao CC, Hocheng H (2005) Effects of exit back-up on delamination in drilling composite materials using a saw drill and a core drill. Int J Mach Tools Manuf 45(11):1261–1270
6. Velayudham A, Krishnamurthy R (2007) Effect of point geometry and their influence on thrust and delamination in drilling of polymeric composites. J Mater Process Technol 185(1–3):204–209
7. Abrão AM, Rubio JCC, Faria PE, Davim JP (2008) The effect of cutting tool geometry on thrust force and delamination when drilling glass fibre reinforced plastic composite. Mater Des 29(2):508–513
8. Khashaba UA, El-Sonbaty IA, Selmy AI, Megahed AA (2010) Machinability analysis in drilling woven GFR/epoxy composites: part I-effect of machining parameters. Compos A Appl Sci Manuf 41(3):391–400

9. Durão LMP, Gonçalves DJ, Tavares JMRS, de Albuquerque VHC, Vieira AA, Marques AT (2010) Drilling tool geometry evaluation for reinforced composite laminates. Compos Struct 92(7):1545–1550
10. Ali F, Heymann T, Biermann D (2011) Experimental investigations on drilling GFRP epoxy composite laminates using specialized and conventional uncoated cemented carbide drills. Mater Manuf Process 26(4):609–617. https://doi.org/10.1080/10426911003636969
11. Rakesh PK, Singh I, Kumar D (2012) Drilling of composite laminates with solid and hollow drill point geometries. J Compos Mater 46:3173–3180
12. Rajamurugan TV, Shanmugam K, Palanikumar K (2013) Analysis of delamination in drilling glass fiber reinforced polyester composites. Mater Des 45:80–87
13. Khashaba UA, El-Keran AA (2017) Drilling analysis of thin woven glass-fiber reinforced epoxy composites. J Mater Process Technol 249:415–425
14. Hocheng H, Chen CC, Tsao CC (2018) Prediction of critical thrust force for tubular composite in drilling-induced delamination by numerical and experimental analysis. Compos Struct 203:566–573
15. Mudhukrishnan M, Hariharan P, Palanikumar K (2020) Measurement and analysis of thrust force and delamination in drilling glass fiber reinforced polypropylene composites using different drills. Measurement 149:106973
16. Thakur RK, Singh KK (2023) Evaluation of drilling characteristics to explore the effect of graphene nanoplatelets on glass fiber reinforced polymer composite. Measurement 219:113233
17. Geier N, Patra K, Anand RS, Ashworth S, Balázs BZ, Lukács T, Magyar G, Tamás-Bényei P, Xu J, Davim JP (2023) A critical review on mechanical micro-drilling of glass and carbon fibre reinforced polymer (GFRP and CFRP) composites. Compos Part B Eng 254:110589
18. Kumar S, Chauhan SR, Rakesh PK, Singh I, Davim JP (2012) Drilling of glass fiber/vinyl ester composites with fillers. Mater Manuf Process 27(3):314–319
19. Kishore RA, Tiwari R, Rakesh PK, Singh I, Bhatnagar N (2011) Investigation of drilling in fiber reinforced plastics using response surface methodology. Proc Inst Mech Eng Part B J Eng Manuf 225(3):453–457
20. Kumar J, Singh Y, Rakesh PK, Singh I, Davim JP (2023) The impact of hole diameter on the molded and drilled holes in jute-fiber-reinforced epoxy composites. J Compos Sci 7(9):376
21. Rakesh PK, Singh I, Kumar D (2012) Flexural behavior of glass fiber reinforced plastic laminates with drilled hole. Proc Inst Mech Eng Part L J Mater Des Appl 226(2):149–158
22. Rakesh PK, Singh I, Kumar D (2010) Failure prediction in glass fiber reinforced plastics laminates with drilled hole under uni-axial loading. Mater Des 31(6):3002–3007
23. Rakesh PK, Singh I, Kumar D (2012) Compressive behavior of glass fiber reinforced plastic laminates with drilled hole. Adv Mater Res 410:349–352

Chapter 18
Unveiling Aquatic Plant-Animal-Microbe Interactions and Their Industrial Potential Through Mathematical Modeling

Manisha Chaudhary, Komal Kumari, Juhi Gupta, and Akarsh Verma

1 Introduction

The world's aquatic ecosystems, spanning oceans, lakes, rivers, and wetlands, constitute over two-thirds of the Earth's surface, playing an essential role in global climate stability and offering a myriad of services to meet the growing needs of the human population [1]. Within these diverse habitats, a complex web of interactions unfolds among microorganisms, plants, and animals, shaping the structural and functional dynamics of aquatic ecosystems. This comprehensive review delves into the intricate world of plant-animal-microbe interactions within aquatic environments. It investigates the biogeochemical significance of these interactions, explores the prevalence of microplastic debris and biofilm assemblages, and assesses the dominance of positive and negative interactions in shaping these ecosystems.

Aquatic biodiversity exhibits a latitudinal gradient, with the highest levels observed in tropical regions [2]. This diverse array of aquatic life is organized

M. Chaudhary
National Institute of Plant Genome Research (NIPGR), New Delhi, India

K. Kumari
Imperial Life Sciences (ILS), Gurugram 122001, India

J. Gupta (✉)
Amity Institute of Environmental Sciences, Amity University, Noida 201303, India
e-mail: jgupta2@amity.edu

A. Verma
Department of Mechanical Engineering, University of Petroleum and Energy Studies, Dehradun 248007, India

Department of Mechanical Science and Bioengineering, Osaka University, Osaka 560-8531, Japan

S. K. Sethi et al. (eds.), *Polymer Composites: From Computational to Experimental Aspects*, Materials Horizons: From Nature to Nanomaterials,
https://doi.org/10.1007/978-981-97-0888-8_18

into four distinct zones, each characterized by unique environmental conditions and ecological communities. The littoral zone, situated along the shoreline, is characterized by abundant rooted vegetation and ample sunlight penetration. Extending from the littoral zone is the limnetic zone, encompassing the open water surface away from the shore. In contrast, the profundal zone comprises the deeper waters beneath the limnetic zone, where sunlight penetration is limited. Finally, the benthic zone encompasses the sediment at the bottom of the lake or pond, providing a habitat for a diverse array of organisms adapted to low-light conditions (Fig. 1). Within the traditional ecological framework, plants, animals, and microbes occupy distinct niches, acting as primary producers, consumers, and regenerators within the complex food web. This network gives rise to a diverse range of microbial interactions, including commensalism, inhibition, food competition, predation, parasitism, and synergism [3].

Microbial diversity in aquatic ecosystems is the culmination of billions of years of evolutionary change. Aquatic microbial communities play a vital role in biogeochemical processes, influencing the transport of materials and energy, the breakdown of organic matter, and the cycling of nutrients. These processes are essential for maintaining the health and balance of aquatic ecosystems. The composition and function of these microbiomes are shaped by various environmental factors, such as light, temperature, pH, hydraulic structure, and pollutant concentrations [4].

This comprehensive review delves into the current understanding and emphasizes the significance of interactions between microbiomes, environments, and hosts, exploring the multifaceted roles of microbiomes in freshwater and marine fish and mammals. First, we review the plant-associated microbiomes, focusing

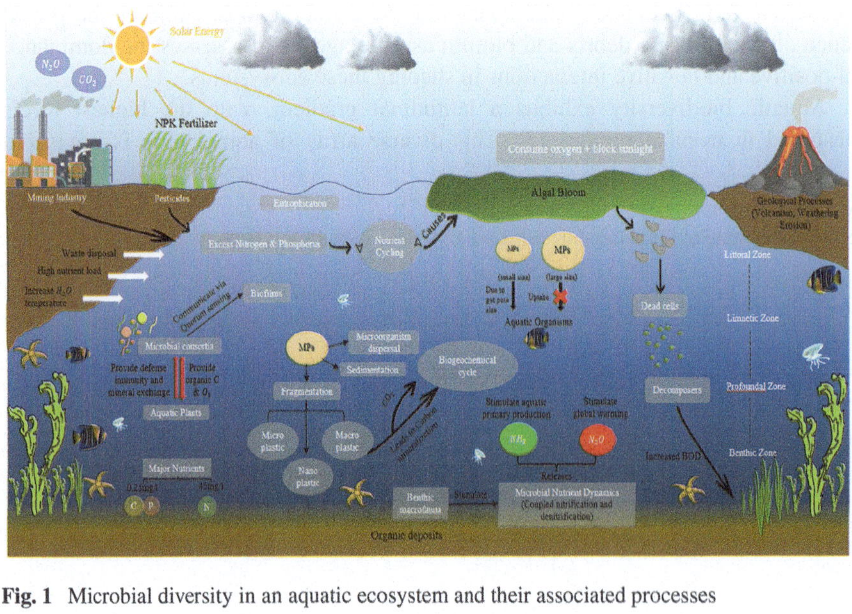

Fig. 1 Microbial diversity in an aquatic ecosystem and their associated processes

on both macrophytes and microphytes. Such associations extend beyond plants to microplastic debris, where microbial biofilms exert a profound influence on the degradation and fate of plastic waste in aquatic ecosystems. Second, we review the intricate world of animal-associated microbiomes, highlighting the symbiotic relationship between aquatic animals and their surrounding microbial milieu. The influence of the environment on host microbiomes and the bidirectional exchange of microbes between water and host guts demonstrate the nuanced interactions that shape aquatic animal health. Furthermore, the review delves into the dichotomy between positive and negative interactions in aquatic ecosystems. These positive interactions are pivotal drivers of community structure and abundance, yet they remain less studied than negative interactions.

With the advent of sophisticated computational tools, mathematical models have emerged as powerful instruments for unraveling the complexities of microbial communities within aquatic ecosystems. These models offer insights into microbial composition, functions, and behaviors, crucial for water quality prediction and understanding microbial responses to hydrodynamic conditions. Additionally, the burgeoning significance of aquatic microorganisms in various industries cannot be overlooked. The microorganisms residing within aquatic environments hold tremendous potential as sources of valuable enzymes for diverse applications in food industries, agriculture, environmental protection, and beyond. In sum, this review offers comprehensive insights into aquatic ecosystems, elucidating the complex interplay among microbiomes, organisms, and the environment, and their implications for ecosystem health and industrial applications.

2 Microbiome of Aquatic Ecosystem

Water, though composed mostly of inorganic elements and molecules, can be considered a living matrix due to the abundance of biological inhabitants it harbors. This vibrant community of microorganisms, encompassing archaea, bacteria, fungi, and protists, plays a pivotal role in maintaining the delicate balance of aquatic ecosystems. These microscopic inhabitants not only perform essential biological functions but also serve as the foundation for the microbiomes of various host animals, including invertebrates, fish, and aquatic mammals [5]. Microbes in aquatic ecosystems contribute significantly to various ecological functions. One prominent role is their involvement in primary energy production from CO_2. While photosynthetic organisms like plants, algae, and cyanobacteria contribute substantially, bacteria capable of chemosynthesis also play a role in generating complex biological molecules. This was originally attributed to extremophilic bacteria in light-deprived deep ocean regions. Bacteria's newfound ability to capture light energy without photosynthesis, enabled by proteorhodopsin, redefines their role in aquatic ecosystems, expanding their contributions to primary production and beyond [6].

The water microbiome's composition varies extensively based on geography, season, and other factors, but a core microbiome remains relatively consistent.

Temperature is a primary driver, impacting the microbiome more than geography or water depth. Rising CO_2 levels due to climate change might have a larger impact than temperature, as evidenced by studies that found CO_2's stronger influence on seawater microbiomes [7]. Disturbances in aquatic ecosystems can propagate through microbial food webs, altering interactions between primary producers and heterotrophic bacteria. A shift from symbiotic to parasitic relationships can occur under stress. Understanding the recovery of microbial communities after disturbances is crucial for ecosystem resilience. Disturbances affect not only susceptible taxa but also indirectly impact interacting species, leading to cascading effects and potential secondary extinctions. Such complex interactions can even lead to counterintuitive outcomes, where disturbances increase the abundance of certain species due to the removal of dominant competitors.

3 Functional Role of Aquatic Microbiota

Ecosystem functioning depends on multiple interactions between physical, chemical, and biological elements. Ecosystem processes like productivity and nutrient recycling stem directly from the diversity of functional traits within biotic communities, shaped by species composition and diversity driven by both biotic introductions and environmental pressures. This balance hinges on both biotic factors (e.g., species diversity, interactions) and abiotic constraints (climate, geology), as seen in the paradox of phytoplankton diversity in aquatic ecosystems [8]. Despite limitations, recent molecular techniques such as SSCP, T-RFLP, and DGGE are aiding the assessment of microbial diversity in these ecosystems, including viral communities which exert a significant impact on microbial dynamics. Within aquatic ecosystems encompassing oceans, lakes, rivers, and more, diverse microbiota consisting of bacteria, archaea, viruses, fungi, and protists play pivotal roles. They drive nutrient cycling, decomposition, primary production through photosynthesis, food web dynamics, water quality regulation, symbiotic relationships, biogeochemical cycling, disease regulation, habitat formation, and climate regulation through carbon sequestration [9]. This multifaceted functionality underscores the crucial role of aquatic microbiota in maintaining ecosystem health and balance, with implications for diverse aquatic organisms and global climate regulation.

4 Plant-Associated Microbiome

Freshwater ecosystems harbor a rich diversity of macrophytes and microphytes, including phytoplankton, diatoms, and other algae that play a dominant role in these aquatic regimes. Aquatic macrophytes encompass large plants that thrive within water bodies. The optimal growth of both macrophytes and microphytes in surface water hinges on a balance of essential nutrients, including nitrogen (N) at concentrations around 45 mg L^{-1} and phosphorus (P) at levels near 0.25 mg L^{-1}, along with organic carbon and other nutrient elements [10]. In addition to macrophytes and microphytes, diverse microbial consortia exist in various community structures, including detrital microbial mats, biofilms, and planktonic microalgal-bacterial assemblages. These consortia play a crucial role in nutrient cycling processes such as nitrification, denitrification, sulfate reduction, methanogenesis, and metal ion reduction, contributing significantly to energy flow within aquatic ecosystems. They serve as a primary food source for zooplankton, influencing water quality and environmental degradation. Chemical interactions are essential for maintaining the complex organization of aquatic ecosystems. Marine and freshwater plants, including macroalgae, employ sophisticated chemical signaling to facilitate settlement, engage in competition, deter grazing organisms, and fend off pathogens [11]. Microbial partners often play a central role in these chemical mediations, forming biofilms on host surfaces. Biofilm communities provide attachment points for macroalgae, discourage fouling by other organisms, and attract algal spores. These microbial communities synthesize bioactive compounds that suppress fouling and maintain the health of the host. Quorum-sensing inhibitors and bioactives produced by these communities prevent colonization by competitors and pathogens. Macroalgae, as crucial marine primary producers, play a significant role in shaping local ecosystems through their chemical and structural interactions [12].

Exploring the ecological roles and functional mechanisms of microbial bioactive compounds in marine and freshwater macrophytes (Fig. 2).

Seagrass and Salt Marsh Plants	Colonized by epi- and endophytic microbial communities
	Synthesize phytohormones, such as indole-3-acetic acid (IAA), to promote plant growth
	Possess a rich chemical arsenal for their defense
Mangroves	Serve as thriving primary producer habitats and nursery grounds for a diverse array of marine animals and birds
	Provide host protection against oxidative stress
	Mangrove-associated microorganisms hold immense potential for drug discovery
Freshwater Plants	Higher diversity of angiosperms
	Essential role in plant growth promotion and inhibition
	Aid in nutrient cycling

Fig. 2 Microbial bioactive compounds in aquatic macrophytes

4.1 Seagrass and Salt Marsh Plants

Seagrasses, marine flowering plants derived from terrestrial ancestors, serve as a vital food source, refuge, and nursery ground for a diverse array of marine organisms, shaping local marine ecosystems. Their roots and leaves harbor diverse epi- and endophytic microbial communities that play a complex interplay with their host, exhibiting both beneficial and occasional detrimental effects. The colonization of seagrass tissues by endophytic fungi, in particular, results in the production of bioactive compounds that safeguard the host against harmful fouling organisms and pathogens. In addition to their physical defenses, seagrasses produce a diverse array of chemical defense compounds, including rosmarinic acid, a phenolic compound, and sulfated flavonoids, to deter bacterial colonization on their surfaces. [13]. Similarly, salt marsh plants boast a rich chemical repertoire that allows them to thrive in their challenging coastal environment and defend against grazing organisms and pathogens. Studies suggest that quorum sensing (QS) and quorum sensing inhibitor (QSI) processes might play a crucial role in maintaining the delicate balance within microbe-microbe and microbe-host interactions in salt marsh ecosystems [14].

4.2 Mangroves

Mangrove ecosystems, situated at the interface of marine and terrestrial systems, serve as vital primary producers and nursery grounds for diverse bird and marine species. These ecosystems harbor a rich tapestry of microorganisms, including bacteria, archaea, fungi, and protists, which play crucial roles in biogeochemical cycles and provide their mangrove hosts with essential nutrients for growth and development, along with defensive compounds. Fungal communities, particularly notable for their diversity, predominate in the leaves, stems, flowers, roots, and rhizosphere of mangroves across diverse geographical locations. Actinomycetes, particularly those belonging to the Streptomyces genus, represent another prominent group of potentially beneficial microbial symbionts associated with mangroves [15]. The isolation of numerous antimicrobials from mangrove endophytic bacteria and fungi underscores the potential of mangrove-symbiotic microorganisms as a valuable source for drug discovery. However, further investigations are warranted to elucidate the mechanisms of action of these microorganisms within the mangrove-microbial association and in other target organisms.

4.3 Freshwater Plants

In contrast to marine environments, freshwater littoral zones harbor a higher diversity of angiosperms compared to macroalgae; however, the chemical ecology of freshwater angiosperms remains less explored relative to their marine and terrestrial counterparts [16]. Studies have shed light on the functioning of the plant holobiome in freshwater ecosystems, particularly its role in promoting and inhibiting plant growth.

While several cultured surface-colonizing bacteria from the genera Delftia, Aeromonas, and Pseudomonas have demonstrated the ability to promote the growth of Lemna minor, investigations into their phosphate solubilization, indole-3-acetic acid (IAA) production, and siderophore production did not reveal a direct correlation with the extent of plant growth promotion. The underlying mechanisms of microbial enhancement of duckweed growth remain to be elucidated. Aquitalea magnusonii, another bacterium that promotes growth, was also able to suppress the growth of inhibitory strains on the duckweed surface, demonstrating the dynamic interspecific bacterial interactions taking place on the surface of freshwater plants.

4.4 Microplastic-Associated Biofilm

Global plastic production is projected to reach an alarming 33 billion tons by 2050. Microplastics, initially documented in the 1970s, have become the dominant form of plastic pollution, sparking extensive research interest since 2004 [17].

Microplastic fragments, fibers, and granules, ranging from one micrometer to five millimeters in size, represent the most prevalent form of ocean plastic debris. Their diminutive size places them within the optimal prey range for numerous marine organisms, raising concerns about their potential to harm. The transfer of microplastics to animal tissues increases the likelihood of adverse effects, as many plastic additives and persistent waterborne chemicals are endocrine disruptors. These substances can interfere with hormone signaling pathways in target tissues, leading to disruptions in metabolic and reproductive processes. While chronic exposure to microplastics is rarely lethal, it can negatively impact individual animals, reducing feeding rates and energy reserves, with cascading effects on fecundity, growth, and behavior.

Plastic entering marine ecosystems from land-based and maritime sources undergo photooxidation due to ultraviolet (UV) radiation, leading to mechanical and biological degradation that fragments it into microplastics (>5 mm), mesoplastics (0.1 μm–5 mm), and nanoplastics (<0.1 μm) [18]. Macromolecules and microorganisms adhere to the surfaces of micro- and nanoplastics, forming an ecocorona. Microbial communities colonizing microplastic debris collected from the sea surface include heterotrophs, autotrophs, predators, and symbionts. Various policy documents and procedures, including international treaties such as the Montreal Convention, Stockholm Convention, and Minamata Convention, as well as national legal instruments, are in place to assess and restrict the release of chemical pollutants.

In aquatic ecosystems, biofilms formed by microorganisms on plastic surfaces can alter the polymer's physical and chemical properties, potentially leading to degradation. Biofilms hold promise for mitigating microplastic contamination and restoring marine ecosystems. Quorum sensing (QS) allows bacteria to detect their population density and coordinate biofilm formation using small signaling molecules

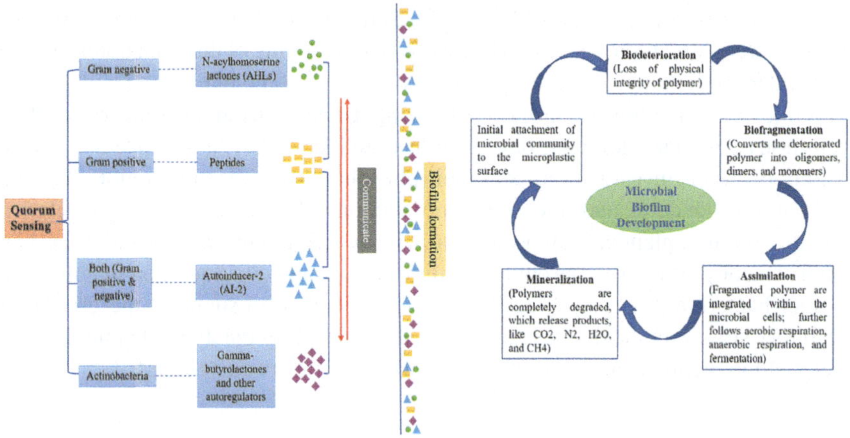

Fig. 3 Bacterial colonization through quorum sensing and biofilm development

called autoinducers or autoregulators. The main autoinducer groups include N-acylhomoserine lactones (AHLs) in gram-negative bacteria, peptides in gram-positive bacteria, autoinducer-2 (AI-2) in both gram-negative and gram-positive bacteria, and gamma-butyrolactones and other autoregulators in Actinobacteria [19] (Fig. 3).

Microplastics can create ecological niches via biofilms (plastisphere), supporting microbial colonization and growth. Microplastic-associated biofilms affect microplastic properties and pollutant adsorption. Heavy metal exposure impacts biofilm formation, altering heavy metal adsorption. Biofilm's role in microplastic-heavy metal dynamics remains unclear. Microplastic exposure adversely affects aquatic organisms' health, impacting feeding, reproduction, and immunity.

4.4.1 Biofilm Formation

Biofilms are complex communities of microorganisms, including bacteria, fungi, and algae, that thrive on submerged surfaces enveloped in a matrix of extracellular polymeric substances (EPS). These biofilms can modify and hydrolyze microplastics, induce the transformation of surface properties, leading to the degradation of additives, the release of metabolic byproducts, and ultimately influencing the fate of microplastics in marine environments [20].

The degradation of microplastics commences with the initial colonization of pioneering microbes and the subsequent formation of biofilms on the plastisphere, a distinct microhabitat surrounding microplastic particles. The hydrophobic nature of plastic surfaces facilitates microbial adhesion and biofilm development, leading to alterations in polymer buoyancy and hydrophobicity.

Environmental factors such as temperature and ionic strength can influence microbial colonization and biofilm formation on microplastics. Additionally, microbes have evolved various mechanisms to facilitate attachment, including altering cell surface charge and hydrophobicity and modifying EPS production. Substrate properties, such as crystallinity, melting temperature, and surface roughness, can also play a role in shaping the initial microbial community composition.

The maturation of biofilms into complex structures is orchestrated by cell-to-cell communication through quorum sensing (QS) mechanisms. QS signaling molecules accumulate in the extracellular environment and regulate gene expression in neighboring cells. In some bacterial species, QS also plays a role in coordinating the disassembly of biofilms.

4.4.2 Degradation of Microplastics

Microplastic degradation involves surface modification and degradation mechanisms, encompassing hydrolases and enzymes like oxidases, amidases, and laccases. Effective degradation is essential to counteract MP contamination's environmental impact. Degradation chemically reduces plastic polymer molecular weight, weakening the material, and promoting fragmentation. These fragments undergo microbial-mediated degradation, mineralizing carbon into the marine biogeochemical cycle [21].

Abiotic and biotic mechanisms drive microplastic degradation in aquatic environments. Abiotic processes encompass physical and chemical degradation due to UV radiation, wave action, additives leaching, and mechanical forces. Biotic degradation involves microbial associations on microplastic surfaces, shielding from UV radiation and impacting sedimentation (Fig. 4). Microbial activity significantly influences plastic degradation, with planktonic and epiplastic microbial communities determining microplastics' fate.

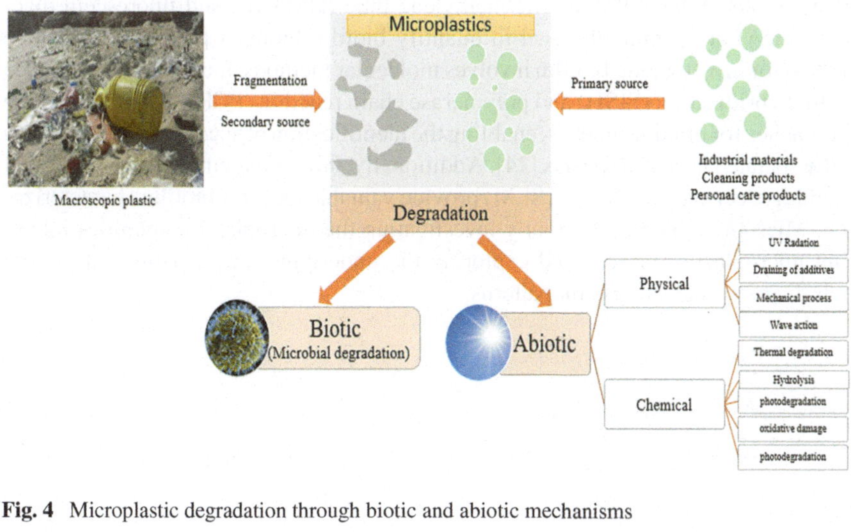

Fig. 4 Microplastic degradation through biotic and abiotic mechanisms

The interplay between plastics and abiotic factors releases dissolved organic carbon (DOC) and hydrocarbon gases, such as methane, into the surrounding marine environment [22]. These degradation byproducts serve as potential electron donors for microbial metabolism, fostering microbial colonization. Microplastics floating on ocean surfaces are predominantly colonized by aerobic microbes. However, upon settling onto the seafloor and sediments, they encounter distinct microbial communities composed primarily of anaerobic microbes that regulate degradation processes in different ways. In anaerobic environments, microbes utilize microplastics as electron donors to promote their degradation. This microbial colonization induces physicochemical alterations in polymers, including changes in crystallinity, mass loss, hydrophobicity, and surface functionalization. Through these complex interactions (actively and passively), microplastics, microbes, and their degradation products become integrated into aquatic biogeochemical cycles.

In addition to enzymes, extracellular polymeric substances (EPS) within biofilms play a pivotal role. EPS composed of polysaccharides, proteins, and lipids regulate adsorption capacity, surface tension, microbial movement, and material transport within biofilms. EPS organizes cell communities, promotes communication, gene exchange, and provides stability within the biofilm microenvironment.

4.4.3 Microbial Communities Contributing to Biofilm

The microbial communities colonizing microplastics exhibit diversity and distinct compositions. The environmental conditions of the surrounding medium, such as water columns, sediments, or coastal habitats, play a crucial role in selecting the specific composition of epiplastic microbes (Fig. 5).

The complex nature of biofilms necessitates the application of appropriate analytical techniques for their characterization. Staining assays employing dyes such as crystal violet (CV), 1,9-dimethyl methylene blue (DMMB), and fluorescein diacetate (FDA) are commonly used to quantify biofilm biomass [23]. An alternative method for investigating biofilm involves molecular methods, including fluorescence in situ hybridization (FISH) and polymerase chain reaction (PCR), offers alternative approaches for biofilm analysis, enabling the identification of specific gene sequences and associated bacterial species [24]. Additionally, flow cytometry and laser confocal scanning electron microscopy (SEM) provide valuable tools for biofilm characterization. Microbiologists are currently investigating the microbial communities inhabiting microplastic surfaces and exploring the underlying mechanisms that govern their diversity and succession patterns.

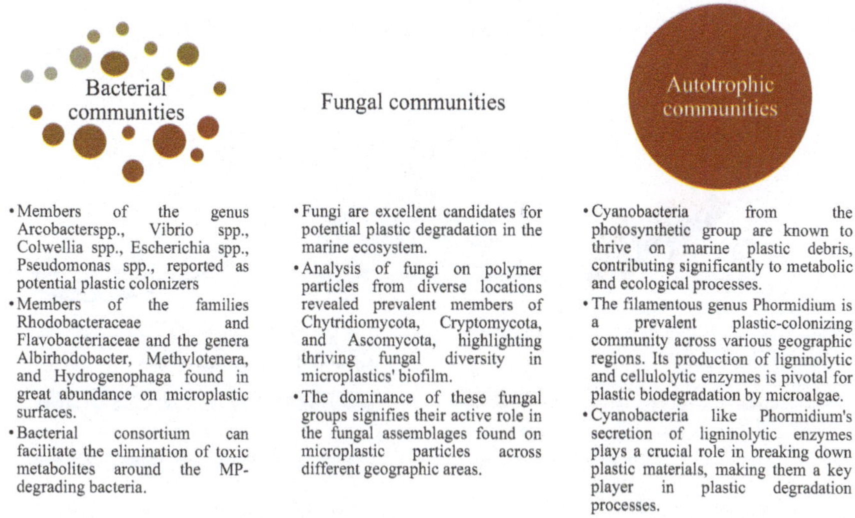

Fig. 5 Different microbial communities contributing to biofilm formation

5 Animal-Associated Microbiome

Aquatic animals are surrounded by a milieu of microbes, and environmental microbiome composition influences host microbiomes. For instance, zooplankton exposed to antibiotics experience microbiome shifts, affecting plankton population growth. The composition and function of the intestinal microbiota in aquatic animals are shaped by a complex interplay of intrinsic and extrinsic factors. Intrinsic factors, such as host genetics, developmental stage, and physiological state, establish the foundation for microbial colonization. Extrinsic factors, including diet and environmental conditions, exert a dynamic influence on the microbiota, modulating its structure and activity. The aquatic environment and the intestinal microbiota engage in a two-way interaction, influencing each other's composition and function. Waterborne bacteria shape the gut microbiota of aquatic organisms, and the gut microbiota, in turn, contributes to the bacterial community in the surrounding water, reflecting the host's adaptation to its environment.

The gastrointestinal tract (GIT) serves as the primary hub for host-microbe interactions, and its colonization and function play a critical role in aquatic animal health and well-being. The gastrointestinal tract (GIT) of fish harbors an immense bacterial community, with an estimated 108 bacterial cells per gram, encompassing a diverse range of approximately 500 predominantly aerobic or facultative anaerobic microbial species. The gut microbiota of most fish is predominantly composed of Bacteroidetes, Firmicutes, and Proteobacteria [25] with species-specific compositions reflecting the intricate balance between respiratory, skin, and gastrointestinal microbiota in aquatic mammals and fish. In aquatic vertebrates, water quality parameters and ecological

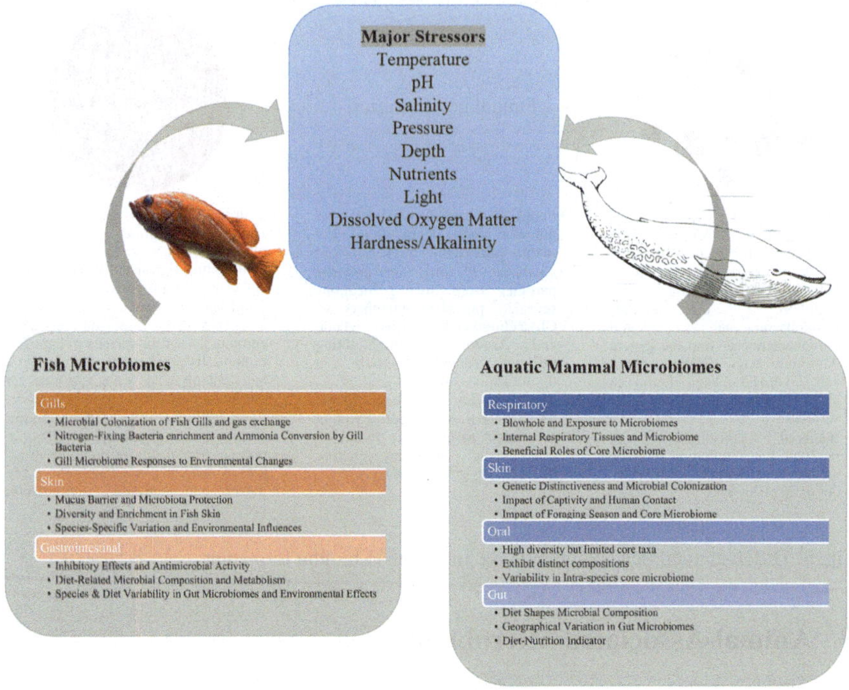

Fig. 6 Animal-associated microbiomes in an aquatic system

niches play crucial roles in shaping the composition and function of the microbiota (Fig. 6).

5.1 Fish-Associated Microbiomes

Fish and their tissue microbiomes have undergone a process of co-evolution within their aquatic microbial environment, resulting in a symbiotic relationship that benefits both partners [26]. The fish microbiota contributes to essential biological functions of the host, such as nutrient acquisition and immune defense, including the suppression of pathogenic microorganisms. In return, the host provides a nutritional environment and facilitates the colonization of both internal and external microflora.

5.1.1 Respiratory Microbiome

Microbial communities colonize gill surfaces, vital for gas exchange in fish. While Proteobacteria dominate gills across species, the proportions of Alpha, Beta, and Gammaproteobacteria, as well as genus composition, vary [27]. Nitrogen-fixing bacteria might thrive in gill mucus due to ammonia excretion. Reports showed that enrichment of Nitrosomonas, chemoautotrophic bacteria that convert ammonia to nitrite, in the gills of yellowtail kingfish (Seriola lalandi) and wild-caught red snapper [28, 29]. In another study, genes for ammonia-oxidizing and denitrifying and bacteria in the gills of common carp (Cyprinus carpio) and zebrafish (Danio rerio) and showed that these bacteria convert ammonia to nitrogen gas [30].

5.1.2 Skin Microbiome

Fish have mucus-coated epithelial surfaces acting as barriers against the aquatic environment. The skin mucus is a crucial component of their immune system, hosting diverse commensal microorganisms like bacteria, archaea, and fungi, believed to aid infection prevention [31]. Skin bacteria are enriched in Proteobacteria and Firmicutes, making reef fish skin microbiomes more diverse than seawater in terms of classes and phyla. Even among species in the same environments, skin bacteria exhibit significant variation. Dissolved oxygen weakly correlates with microbial diversity differences among fish from various sites. In the Amazon River, watercolor accounted for almost 40% of skin microbiome variance between two Amazon fish species, with oligotrophic waters decreasing abundance of Alphaproteobacteria and increasing the abundance of class Mollicutes within Firmicutes [32].

5.1.3 Gut Microbiome

The influence of the gut microbiota on immune function is profoundly shaped by factors such as species, diet, and environmental conditions. Fish harbor intricates intestinal microbiomes that play crucial roles in nutrient absorption, immune system modulation, intestinal development, energy homeostasis, and xenobiotic metabolism [33]. Among the diverse bacterial species isolated from the fish gut, anaerobic Bacillus and Aeromonas stand out for their production of metabolic enzymes, including lipases, proteases, and amylases.

Gut microbial communities exhibit substantial interspecific variability among fish species, both at the phylum and genus levels. Herbivorous fishes typically exhibit a dominant presence of Bacteroidetes and Firmicutes, particularly within the order Clostridia, such as Epulopiscium. In contrast, carnivorous fish in both freshwater and marine environments often harbor abundant Proteobacteria, especially within the orders Enterobacteriales and Vibrionales [34, 35].

5.2 *Marine Mammal-Associated Microbiomes*

5.2.1 Respiratory Microbiome

Marine mammals possess a distinctive respiratory anatomy, where the nasopharynx is separate from the oral cavity, creating a distinct environment that may harbor unique microbial communities [36]. Cetacean vents possess a unique feature in that their respiratory system directly interacts with both surface water and air microbiomes during respiration, leading to a microbial composition influenced by both external environmental factors and host biological factors [26]. Emerging evidence suggests that the internal respiratory tissues of cetaceans harbor their own distinct microbiota, a phenomenon not limited solely to aerosolized seawater. This implies that the respiratory microbiota may play a more complex and multifaceted role in cetacean health than previously thought [37]. The core microbiota of healthy animals is widely assumed to play beneficial roles in supporting the well-being of their animal hosts.

5.2.2 Skin Microbiome

The skin microbiome is a vital focus in marine mammal research due to its intimate interaction with water column microbes and the skin's role as a barrier. The foraging season significantly influences core microbiome presence in marine mammals, illustrated by divergent core microbiota in humpback whale skin [38] and blow [36] samples taken during different foraging periods. Notably, not all skin-colonizing microbes respond to environmental changes, impacting collected microbiome data. For instance, low-temperature extremophiles like Psychrobacter sp. decline with seawater temperature changes. Certain microbial groups like Moraxellaceae are common in mucosal membranes. Core genera are consistent across marine mammals from diverse environments, exemplified by Gammaproteobacteria genus Psychrobacter's prevalence in captive killer whales, bottlenose dolphins, and wild humpback whales [39].

5.2.3 Oral Microbiome

The seawater microbiome exhibited substantial compositional differences from the oral microbiome, indicating that the oral microbiota composition is host-specific. The oral microbiome displayed high diversity with a limited number of core taxa. In comparison, the oral microbial compositions of captive and wild dolphin populations differed significantly, potentially due to factors such as diet, location, and social behavior. The oral microbiome may be too variable to harbor an intra-species core microbiome, as evidenced by a study on Odontoceti cetaceans ($n = 48$), which revealed a small number of core OTUs and divergence between species [40].

5.2.4 Gut Microbiome

Diet plays a predominant role in shaping the composition of the gut microbiome. A recent study has demonstrated the crucial importance of carbohydrate fermentation in the gut microbiome of baleen whales, as evidenced by the presence of a high abundance of carbohydrate-active genes associated with both animal and plant carbohydrates. [41]. Geographic location can influence variations in gut microbiomes between marine mammal populations. The observed differences in diversity among these wild populations may stem from various factors, including proximity to human settlements and colonies of different species, exposure to pollutants and wastewater runoff, foraging location and diet, colony member dynamics, and behavior. These factors support the notion that phylogenetic relationships between the host and its associated microbiome play a significant role in shaping the gut microbial community.

5.3 Role of Benthic Macrofauna in Stimulating Microbial Nitrogen Cycle

Sediments in lakes, streams, and marine coasts serve as major reservoirs for microbial nitrogen cycling in aquatic ecosystems [42]. In these habitats, nitrifying bacteria transform ammonium into nitrite and nitrate at the sediment surface. Nitrite and nitrate then undergo reduction through various pathways, leading to dinitrogen or ammonium production, facilitated by anaerobic bacteria and archaea in anoxic sediment layers. The resulting dissolved inorganic nitrogen (DIN) diffuses into the overlying water column, fueling primary production by phytoplankton and macroalgae.

Excess nitrate and ammonium, originating from sources like fertilizer use and atmospheric deposition, contribute to harmful eutrophication effects. Benthic macrofauna, acting as intermediaries between the nitrogen cycle microbes and the water column, impact the ecosystem. By consuming microbes and microalgae, they release fixed nitrogen, driving ammonification, nitrification, denitrification, and anammox processes. Reduced DIN concentrations mitigate eutrophication, affecting primary production and curbing its negative consequences.

Macrofauna also influence nitrogen cycling by providing substrates for biofilm-forming microbes and creating oxic-anoxic interfaces within sediments [43]. Epifauna-associated biofilms contribute to nitrification and nitrous oxide emission, while infauna promote nitrification, denitrification, and nitrogen fluxes. Overall, benthic macrofauna foster ecosystem-level nitrogen removal and regeneration, influencing the aquatic ecosystem's nutrient dynamics and balance.

5.4 Phytoplankton-Derived Polysaccharides and Their Influence on Heterotrophic Bacteria

The complex interplay of zooplankton grazing, viral lysis, aging phytoplankton cell decay, and the active/passive release of photosynthetic products results in two distinct oceanic carbon reservoirs: particulate and dissolved organic matter (POM and DOM, respectively), with DOM encompassing the <0.2-μm fraction. Oceanic DOM constitutes an immense carbon pool of 662 Pg, surpassing marine biomass by over 200 times, like atmospheric CO2 levels. This renders marine DOM the principal carbon source for oceanic heterotrophic bacteria [44].

Phytoplankton sequesters excess carbon as storage polysaccharides, such as chrysolaminarin and glycogen, which dominate phytoplankton exudates. These exudation rates escalate under oxidative stress or various stressors. High molecular weight dissolved organic matter (HMW-DOM) mainly comprises polysaccharides, forming around 50% of surface water DOC and 25% in deeper samples. Conversely, dissolved amino acids and proteins are less abundant due to microbial utilization, while free monosaccharides are produced at high rates but remain in lower concentrations.

DOM's C:N and C:P ratios are higher than ambient surface water, changing with phytoplankton and nutrient conditions [45]. Bacteria actively influence phytoplankton exudates' quantity and quality, impacting interactions. Phytoplankton release polysaccharides that discourage bacterial colonization while providing a carbon source. Exopolysaccharides facilitate bacterial recruitment and retention within the algal phycosphere, where interactions range from mutualism to parasitism, driven by nutrient availability and environmental factors.

Heterotrophic bacteria acquire phytoplankton polysaccharides through exoenzymes that break down large compounds into oligo- or monomers. Fungi and microorganisms contribute enzymatic degradation products, and carbohydrate-active enzymes (CAZymes) catalyze algal polysaccharide degradation. Gammaproteobacteria and Flavobacteriia, particularly during diatom blooms, are prolific in CAZymes [46]. Bacteroidetes' polysaccharide utilization loci (PUL) house the machinery for polysaccharide detection, hydrolysis, and uptake.

6 Positive Versus Negative Interactions

Plants and their associated microorganisms are intricately intertwined, with chemical interactions mediated by natural products playing a crucial role in host health and development within aquatic ecosystems [47]. These interactions encompass a spectrum of effects, ranging from negative to beneficial for both microbial symbionts and their hosts. These symbiotic microbes exhibit metabolic synchronization with their hosts, suggesting a potential coevolutionary relationship. The metabolites synthesized by plants and microbes due to their intimate association and coaction supplement the existing pool of metabolites, thereby promoting plant growth, maintaining physiological homeostasis, and enhancing resilience against various biotic and abiotic stressors [48].

Aquatic ecosystems are characterized by a network of positive interactions, where organisms exchange vital resources such as nutrients, protection, transportation, or habitat, resulting in a benefit for at least one participant. Several mutualistic relationships have been documented, including seed-dispersing fishes that aid in mangrove expansion, crayfish and their ectosymbiotic cleaners that maintain crayfish health, and communal-spawning stream fishes that enhance fertilization success. [49].

Commensalism, where one species benefits without harming the other, is evident in phoretic relationships, like barnacles hitching rides on whales for transportation and protection. Facilitation, where one species indirectly enhances the habitat or resource availability for another, is illustrated by beavers modifying water flow and creating diverse habitats through dam construction, and salmon improving nutrient cycling and streambed structure through their digging and foraging activities [49].

Permeating the web of life, positive interspecific interactions, such as mutualism and commensalism, play a pivotal role in shaping the abundance and structure of ecological communities, yet they remain understudied compared to negative interactions like competition and predation [50].

Most mutualistic interactions involve the exchange of essential resources or services, including nutrition, transportation, and protection. However, not all exchanged resources or services fit neatly into these categories, such as gas exchange or substrate acquisition [51]. While this perspective is not exhaustive and does not account for multi-species interactions, multiple resource exchanges, or novel resources, it serves as a useful starting point for discussing positive interactions in freshwater ecosystems.

Beyond protection from predators, facilitative interactions can have positive effects by providing improved habitat. Any combination of the primary key resources can be exchanged in mutualism and commensalism, but habitat amelioration is unique to facilitation [52]. Furthermore, facilitation can be either commensalistic or mutualistic, depending on whether the facilitating species receives positive feedback from the beneficiaries.

6.1 Types of Interactions

Positive and negative interactions among plant, animal, and microbe assemblages play crucial roles in shaping aquatic ecosystems (Table 1). These interactions influence the structure, functioning, and stability of these ecosystems [49].

Climate change, a multifaceted phenomenon, exerts a profound influence on our planet, affecting us directly and indirectly through its interplay with other stressors, including chemical contaminants (Table 2). The rise in sea levels associated with climate change will lead to seawater inundation of nearshore areas, triggering

Table 1 Positive and negative interactions

Biological interactions					
	Term	Definition	Species A	Species B	Notes and examples
Positive interactions	**Mutualism**	Mutualistic interactions involve species benefiting each other	+	+	**Plant-Animal Mutualism**: Some aquatic plants provide shelter and food resources to animals (e.g., fish, crustaceans) in return for nutrient-rich waste products **Plant–Microbe Mutualism**: Aquatic plants often form symbiotic relationships with nitrogen-fixing bacteria, enhancing the plants' nutrient availability
	Facilitation	Facilitative interactions occur when one species improves the environment for another	+	+	**Plant–Microbe Facilitation**: Microbes can enhance nutrient availability by breaking down organic matter, benefiting aquatic plant growth
	Commensalism	Interaction in which one species benefits while the other has no net cost or benefit	+	0	
Negative interactions	**Competition**	Competition for resources like light, nutrients, and space can limit the growth and abundance of species	-	-	**Plant-Plant Competition**: Aquatic plants may compete for light and nutrients, leading to shifts in species composition **Animal-Animal Competition**: Fish and other aquatic animals can compete for limited prey resources, affecting their populations

(continued)

Table 1 (continued)

Biological interactions					
	Term	Definition	Species A	Species B	Notes and examples
	Predation and Herbivory	Predators feed on prey, while herbivores consume plants. These interactions can control population sizes and shape community dynamics	+	-	**Animal-Animal Predation**: Predatory interactions regulate prey populations, preventing overgrazing and maintaining ecosystem balance **Plant–Herbivore Interactions**: Herbivores can control plant growth and influence community structure
	Amensalism	Amensal interactions involve one species being negatively affected while the other is unaffected	0	-	**Plant–Microbe Amensalism**: Microbes may release toxic compounds that inhibit the growth of nearby plants
	Parasitism	Parasites derive nutrients from host organisms, often causing harm	+	-	**Animal-Animal Parasitism**: Parasites can weaken fish and other aquatic animals, affecting their health and survival

flooding and salinization. These changes could potentially disrupt the biogeochemical processes of previously contaminated soils [53]. Additionally, reduced river discharges, a consequence of climate change, will hinder the dilution of chemical contaminants from point and non-point sources, leading to a decline in water quality [54]. Carbon dioxide enrichment in aquatic systems, a hallmark of climate change, can mobilize sediment-bound metals, increasing their toxicity, particularly in acidified oceans [55].

Extensive research is still needed to fully comprehend and appreciate the prevalence and implications of positive interactions in freshwater ecosystems. Elucidating novel mechanisms and processes in freshwater systems can contribute to the development of general ecological principles, which in turn can inform our understanding of the interactions observed in freshwater environments worldwide.

Table 2 The interplay between contaminants and other stressors in aquatic ecosystems encompasses both positive and negative interactions

Positive interaction of contaminants	Negative interaction of contaminants
The rising salinity in freshwater environments can diminish the bioavailability and consequently the toxicity of various metals and other contaminants	Rising sea levels, surpassing the extent of historical soil contamination, can alter redox conditions, leading to the mobilization of redox-sensitive chemical contaminants like arsenic and mercury
Nanoparticles can safeguard organisms from ultraviolet light through scattering, sorption, and accumulation within their bodies	Climate change, encompassing global warming, can either amplify the toxicity of chemicals or, in instances where acclimation prevents toxicity increase, negatively impact energy reserves
Resident species that have evolved tolerance to chemical contaminants may hinder the establishment and proliferation of invasive species	Chemical contamination and invasive species can diminish biodiversity, disrupt ecosystem services, and alter contaminant pathways, posing elevated ecological and human health risks
Microplastics provide stable colonization surfaces for aquatic organisms, potentially driving water strider population growth in the North Pacific Subtropical Gyre	Microplastics serve as both vectors and habitats for invasive species, while also posing their own toxicity
Nutrient enrichment in aquatic systems can elevate dissolved organic carbon (DOC) levels, potentially mitigating metal contaminant bioavailability. Chemical contamination may also suppress harmful algal blooms and toxin production	The combined effects of harmful algal blooms and chemical contamination can pose a greater threat to biota than either of these factors individually

7 Mathematical Models

Mathematical models are powerful tools for understanding the composition, function, and metabolism of microbial communities [56]. They are widely used in science and engineering to analyze and predict complex systems, elucidate underlying mechanisms, and quantify processes. Mathematical models of microbial communities can be divided into two main types: ecological models and metabolic models [57]. Ecological models use sequence data to predict how different species interact with each other and how their populations change over time. Metabolic models use data on genes, enzymes, and cells to simulate how cells produce and consume energy and other molecules [58] (refer to Table 3).

These modeling approaches are used to simulate the transport and metabolism of microbial communities under different hydrodynamic conditions and to predict water quality [59]. Incorporating hydrodynamic processes into microbial modeling approaches is essential to accurately capture the dynamics of microbial communities in aquatic ecosystems.

Table 3 Comparative analysis of microbial models with pros and cons

	Models	Based on	Modeling purpose	Advantages/ Disadvantages
Ecological models (explore the associations among various species through the analysis of sequence data)	Models of microbial association	Operational taxonomic units (OTUs)	Use co-occurrence and co-exclusion patterns to infer microbial associations	Advantages: clear and concise presentation of microbial associations, flexible calculation methods, and low computational overhead Disadvantages: may not be fully representative of the real-world dynamics of aquatic environments
	Models of microbial dynamics	Operational taxonomic units (OTUs)	Simulate microbial growth and interactions to predict community dynamics	Advantages: the governing equations are adaptable to various ecosystems Disadvantage: cannot explain the mechanisms underlying variations in microbial communities
Metabolic modeling (focus on describing the mechanisms of metabolic processes and quantify the rate of reactions)	Biomass scale modeling approaches	Biomass	Simulate the temporal dynamics of microbial biomass and substrate concentration	Advantages: accurate and can capture the dynamics of a few species in aquatic ecosystems Disadvantage: require numerous model parameters limit, Intracellular processes are overlooked

(continued)

Table 3 (continued)

	Models	Based on	Modeling purpose	Advantages/ Disadvantages
	Gene-centric modeling approaches	Genetic materials	Model metabolic reactions of diverse genetic materials	Advantages: low requirements of model parameters, good accuracy, and can predict metabolic processes Disadvantage: measuring and quantifying active genetic materials in aquatic ecosystems can be challenging and expensive
	Agent-based modeling approaches	Aggregate of Bacteria	Model the metabolic behavior of microbial aggregates to simulate community dynamics	Advantages: capture intra-population variation Disadvantage: high computational cost

7.1 Ecological Models

Microbes form complex networks through diverse ecological interactions. Advanced sequencing technologies provide abundant data, such as 16S rRNA gene sequence data and metagenomic data, to enable in-depth analyses of these interactions [60]. Ecological models of microbial associations and microbial dynamics can help us to understand these networks. [61].

7.1.1 Models of Microbial Associations

Microbial association networks are built from microbial abundance matrices of operational taxonomic units (OTUs) extracted from 16S rRNA sequence data. Co-occurrence and co-exclusion relationships between OTUs are inferred using similarity measures to create networks where nodes represent OTUs and edges represent significant relationships. [62].

Emerging tools like vegan, ade4, and QIIME facilitate microbial association analysis through preprocessing raw sequence data, utilizing relative or absolute

OTU counts. Traditional methods (e.g., Pearson, Spearman, Bray–Curtis, Kullback–Leibler) and graphical models (neighborhood, inverse covariance) calculate associations [63]. Addressing challenges, SPIEC-EASI transforms compositional data, predicting microbial links from transformed data. In aquatic ecosystems, networks reveal microbe-group/environment correlations. While static networks show community state, dynamic models are vital for capturing community changes over time. Networks offer insights but lack mechanisms; dynamic methods are crucial for exploring microbial dynamics and species associations.

7.1.2 Models of Microbial Dynamics

Microbial community dynamics can be studied by fitting models to microbial sequence data [64]. A popular choice is the generalized Lotka-Volterra (gLV) model, which uses non-linear differential equations to predict how the abundance of specific community members changes over time based on their growth rates and interactions with other species.

The format is:

$$\frac{dx_i(t)}{x_i(t).dt} = b_i + \sum\nolimits_{j=1}^{N} a_{ij}.x_j(t)$$

where N represents species count, $x_i(t)$ represents the abundance of species i at a given time t, b_i represents the intrinsic growth rate of species i, and a_{ij} quantifies the interaction strength between species i and j. These parameters can be determined experimentally or inferred from time-series data through network inference techniques.

While gLV originated from lab experiments, aquatic ecosystems' open nature and strong transport processes underscore environmental factors' significant influence on microbial community assembly. To understand the impact of hydrodynamics, a mathematical model incorporating both deterministic and stochastic processes was developed. Building upon the gLV equations, this model incorporates local community dynamics, immigration from source populations, and dispersal from neighboring communities [65].

The governing equation underwent the following modifications:

$$\frac{dx_i(t)}{x_i(t).dt} = s_i + n_i + b_i + \sum\nolimits_{j=1}^{N} a_{ij}.x_j(t)$$

where s_i represents the immigration rate of species i from the source population, and n_i represents the dispersal rate of species i from the neighboring population.

Devising more sophisticated techniques to unravel observed microbial processes and growth dynamics, anchored in metabolic mechanisms, could hold the key to manipulating microbial communities for ecological restoration in aquatic ecosystems.

7.2 Metabolic Modeling

Metabolic models offer profound insights into metabolic interactions between species and facilitate dynamic analysis [66]. These models employ various non-linear rate laws to represent metabolic reactions, reflecting the diverse modeling objects and resulting in three primary categories: biomass scale modeling, gene-centric modeling, and agent-based modeling.

7.2.1 Biomass Scale Modeling Approaches

Biomass scale modeling approaches utilize a set of ordinary differential equations (ODEs) to represent the metabolic processes of specific microbial species, encompassing substrate consumption, microorganism growth, and metabolite production [67]. Biomass changes are governed by both transport processes driven by water flow and growth dynamics associated with metabolic processes, which in turn influence the concentrations of substrates and metabolites.

The Monod and Michaelis–Menten equations are among the most widely used techniques in biomass scale modeling. These equations employ substrate concentrations and specific microorganism and chemical constants to model growth rates and metabolic reaction rates, respectively. The governing equations are expressed as follows:

$$\mu = \frac{\mu_{\max}.S}{k_s + S}$$

$$v = \frac{\mu}{Y}$$

where μ represents the specific growth rate of the species biomass, μmax denotes the maximum growth rate of the species constant, S represents the substrate concentration, k_s is the half-saturation of substrate constant, v is the reaction rate of the specific metabolic process, and Y represents the biomass yield from the substrate consumption.

In complex aquatic ecosystems characterized by intricate transport processes and metabolic pathways, ordinary differential equations (ODEs) should be adapted to partial differential equations (PDEs) to effectively model the spatiotemporal variations in microorganism biomass and chemical concentrations.

Researchers employed advection–dispersion equations in conjunction with Monod equations to model nitrogen dynamics within the hyporheic zone. The modified equations governing the alterations in biomasses are presented as follows [68]:

$$\frac{\partial b_i}{\partial t} = \frac{\partial b_i}{\partial x} + \frac{\partial b_i}{\partial y} + \frac{\partial b_i}{\partial z} + \frac{\mu_{\max}.S}{k_s + S} b_i$$

where b_i denotes the biomass of species i.

While the incorporation of hydraulic modeling methods allows biomass scale models to capture the dynamics of a few species and specific metabolic pathways in aquatic ecosystems, the substantial number of model parameters required still limits their applicability in addressing questions related to the complex dynamics of aquatic ecosystems.

7.2.2 Gene-Centric Modeling Approaches

Gene-centric modeling approaches utilize genetic material as a proxy for microbial communities, simulating changes in genetic material and associated chemical compounds. These changes are governed by transport processes and dynamics processes, modeled using modified Monod equations. The core component of gene-centric models lies in the stoichiometric matrix, which comprehensively represents the stoichiometry of all biochemical reactions within the metabolic network [69].

Within the model, microbial growth is contingent upon substrate and nutrient availability, modeled using Monod equations with inhibition. The thermodynamic potential factor (FT), on the other hand, accounts for the chemical energy available to drive metabolic processes. Consequently, the equation for gene production rate is expressed as follows:

$$\frac{R_j}{F_T} = \Gamma_j \widehat{A}.\mu_j \widehat{A} \prod \frac{C_s}{K_s + C_s} \prod \frac{K_x}{K_x + C_x}$$

In this equation, Γ_j represents the abundance of gene j, μ_j denotes the specific growth rate, C_s represents the reactant or nutrient s concentration, K_s represents the half-saturation constant of reactant or nutrient s, C_x represents the inhibitor x concentration, and K_x represents the half-saturation inhibitor constant x. The variable s encompasses the set of all potentially limiting nutrients or substrates for the pathway associated with gene j, while x encompasses the set of all potential inhibitors. This gene production rate equation is then employed to model the changes in gene abundances and metabolite concentrations.

$$\frac{\partial \Gamma_j}{\partial t} = -\lambda \Gamma_j + \mathrm{R}_j + \frac{\partial}{\partial z}\left(K \frac{\partial \Gamma_j}{\partial z}\right) - v \frac{\partial \Gamma_j}{\partial z}$$

$$\frac{\partial \mathrm{C}_s}{\partial t} = \sum_j \left(\frac{\gamma_{j,s}}{|\gamma_{j,e^-}|} \widehat{A}. \frac{R_j}{n_j \widehat{A}.Y\left(\frac{\Delta \mathrm{G}_j}{|\gamma_{j,e^-}|}\right)} \right) + \frac{\partial}{\partial z}\left(K \frac{\mathrm{C}_s}{\partial z}\right) - v \frac{\partial \mathrm{C}_s}{\partial z}$$

In this equation, λ symbolizes the gene-specific mortality rate constant, $\gamma_{j,\,s}$ denotes the stoichiometric coefficient for chemical species s in the reaction associated with gene j (negative values indicate reactants and positive values indicate products), $\gamma_{j,\,e}$—signifies the stoichiometric coefficient of the electron donor in the metabolism, and Y represents the biomass production per mole of electron donor, which is a function of free energy yield [70].

Gene-centric modeling approaches hold promise in elucidating and modeling the intricate mechanisms of intracellular metabolic processes. However, current genetic material detection methods primarily indicate the presence of genetic material, falling short of providing insights into its functional state, which could potentially impact the accuracy of model outcomes.

7.2.3 Agent-Based Modeling Approaches

Agent-based models (ABMs) can represent both an entire bacterial population and individual cells. To be effective, ABMs should be implemented within a well-defined context that incorporates clear concentration gradients of substrates [71]. ABMs focus on the dynamic behavior of bacterial populations, simulating the activities of individual cells within the population and their interactions with the environment using computational methods such as game theory, evolutionary programming, and Monte Carlo techniques.

The behavior of bacterial aggregates can be categorized into growth dynamics and transport processes. Growth dynamics adhere to the central dogma of biology, encompassing transcription and translation. Transport processes encompass directional movements induced by environmental factors, such as water flow, and stochastic movements modeled using Monte Carlo methods [71].

The initial step is transcription, which can be modeled using the mRNA mass balance equation:

$$\frac{dmRNA}{dt} = \frac{k_{S,mRNA,c}}{L_{DNA}} RNAP\gamma_c - k_{D,mRNA} mRNA - \mu_g mRNA$$

where $k_{S,\ mRNA}$ denotes the *mRNA* synthesis rate constant, *C* represents the transcription activity (i.e., transcription efficiency, promoter strength), $k_{D,\ mRNA}$ denotes the *mRNA* degradation rate constant, and μ_g represents the specific growth rate of *mRNA*, with the last term in the equation accounting for growth dilution. Subsequently, the protein mass balance equation is applied to model translation:

$$\frac{dPRO}{dt} = k_{S,P} \frac{mRNA}{TxL} RPT - k_{D,PRO} PRO - \mu PRO$$

where *PRO* denotes the protein concentration, $k_{S,\ P}$ represents the protein synthesis rate constant, *mRNA* and *TxL* are the *mRNA* for *PRO* synthesis and total *mRNA* concentrations, and *RPT* signifies the concentration of gene product (the ribosomes), k_D, *PRO* shows the *PRO* degradation rate constant and μ is the specific growth rate of protein. Despite the model's intended purpose of simulating viral dynamics in an aquatic ecosystem, it overlooked the impact of environmental factors, such as ocean currents, on the simulated aggregates. This oversight prevented validation of the simulation results against real-world data.

The complex transport processes in aquatic ecosystems pose significant challenges in modeling cell-to-cell interactions, both within the overlying water column and across distinct phases. To facilitate the widespread application of ABMs in simulating large microbial communities in aquatic ecosystems, advancements, particularly in the form of simplifications, are crucial.

8 Industrial Importance

Microbes in aquatic ecosystems are increasingly recognized for their valuable byproducts and are being harnessed for the production of useful chemicals. A significant portion of aquatic bacteria are prolific sources of hydrolytic enzymes, including amylases, lipases, proteases, phospholipases, catalases, and other important industrial enzymes [72]. Among these enzymes, proteases produced by microorganisms hold particular industrial potential due to their broad biochemical applications in food processing, pharmaceutical formulations, detergents, and waste treatment [73].

Currently, alkaline proteases dominate the enzyme market, attributed to their diverse applications, with a significant portion originating from Bacillus species [74]. Additionally, lipolytic bacteria in lakes have been identified as producers of phospholipase, a crucial enzyme utilized in bakery processes, including bread making, as well as in the egg yolk industry and the refinement of vegetable oils. Likewise, an enzyme called α-amylase holds significance in numerous industrial processes and represents a quarter of the global enzyme market [75]. This enzyme is employed in the hydrolysis of starch molecules, breaking them down into their simpler form—glucose units. The selection of these industrial enzymes is based on the microbe's capacity and the expression of genes in microbial hosts or cloned versions, ensuring commercially viable quantities.

The exploration of aquatic microorganisms' genetic and biochemical diversity is still in its early stages, and the vast potential of enzymes with diverse applications remains largely untapped. Several kits are available for extracting these commercially valuable enzymes, such as the API ZYM kit, which has been used extensively by researchers to investigate the extracellular enzymatic activity of bacterial strains [73].

Various microbial applications are widely recognized for their ability to address significant agricultural challenges, including low crop yields, plant disease control, and soil health maintenance. Additionally, microbes play a crucial role in bioremediation, effectively cleaning up soil and water contaminated with organic and inorganic pollutants [76]. Enzymes, in particular, have garnered global attention due to their potential applications in the biodegradation of agricultural waste and their expanding use in various industries, including textiles, paper pulp, pharmaceuticals, and food production [73, 77]. The introduction of sophisticated biochemical and molecular approaches to microbial ecophysiology has led to groundbreaking advancements in our understanding of the role and significance of microbial processes in aquatic ecosystems.

The advent of aquatic prokaryotic metagenomic sequencing promises to provide an invaluable resource for uncovering novel metabolic capabilities of microorganisms and revolutionize our understanding of microbial diversity [73]. Numerous studies have consistently demonstrated the crucial role of these microorganisms in nutrient cycling and ecosystem structure, highlighting the urgent need to explore their potential by optimizing their growth conditions and production.

8.1 Role of Cyanobacteria in Industry

Microalgae, a dominant microbial group, exhibit immense promise as sources of natural carotenoids, along with various compounds like polyunsaturated fatty acids, oils, dyes, pigments, antioxidants, and bioactive elements, with applications ranging from pharmaceuticals to high value bioactive [78]. Cyanobacteria, a subgroup, are noteworthy for producing essential medical compounds such as gamma linolenic acid (GLA) in Spirulina platensis and Arthrospira sp., contributing to blood pressure regulation and lipid metabolism [79]. Cyanobacteria like Phormidium sp. inhibit bacterial growth, and Lyngbya majuscula synthesizes diverse chemicals with activities including antimicrobial, antifungal, and sodium-channel blocking [80]. They also generate antialgal compounds that combat algal blooms. Microalgae possess potential as biofertilizers, aiding soil quality, nutrient transfer, microbial populations, and energy efficiency.

Cyanobacteria, as industrial biotechnology platforms, facilitate bioproduct production, from biofuels (hydrogen, alcohols, isoprenoids) to bioactive and proteins [79]. Notably, microalgae are explored for biodiesel, bioethanol, and biomethane production, while carbohydrates contribute to food, cosmetic, and textile industries as stabilizers and emulsifiers. Polysaccharides from microalgae, like fucoidans and carrageenan, have medicinal uses. Industrial enzymes and recombinant proteins, from Dunaliella tertiolecta and Phaeodactylum tricornutum, offer diverse applications [81]. As new genetic engineering tools evolve, cyanobacteria's potential expands, positioning them as a vital asset in industrial biotechnology.

9 Conclusions and Future Prospectives

This review paper provides a comprehensive exploration of microbial interactions in aquatic ecosystems, shedding light on their structural, functional, and ecological significance. From plant-animal consortia to the industrial applications of aquatic microorganisms, this synthesis advances our understanding of the intricate web of life in aquatic environments and underscores their critical importance for the well-being of both ecosystems and human society. The insights presented in this review clearly demonstrate the intricate interplay between aquatic plants, animals, and microbes, where all three components derive mutual benefits from their interactions. The symbiosis and interactions among these serve more economical and commercial potential. Knowledge of Mathematical modeling and continual studying are significant in completing these complex networks. Further research in these interactions will be greatly facilitated by the introduction of novel techniques. With the advent of high throughput techniques and mathematical modeling, quantification of all possible changes occurring in ecosystem is now feasible. These techniques generate substantial datasets that require sophisticated statistical analyses and modeling. The insights gained via these analyses paved the way for finding the responsible key pathways and genes.

Declarations

Conflicts of Interest The authors confirm that there are no conflicts of interest associated with this work.

Ethical Compliance The authors affirm that this study adheres to all ethical guidelines.

Competing Interests No competing interests are declared by the authors.

Data and Material Availability Data and materials are not applicable to this study.

References

1. Häder D-P et al (2020) Anthropogenic pollution of aquatic ecosystems: Emerging problems with global implications. Sci Total Environ 713:136586. https://doi.org/10.1016/j.scitotenv.2020.136586
2. Irfan S, Alatawi AMM (2019) Aquatic ecosystem and biodiversity: a review. Open J Ecol 9(1):1–13. https://doi.org/10.4236/oje.2019.91001
3. Trombetta T et al (2020) Marine microbial food web networks during phytoplankton bloom and non-bloom periods: warming favors smaller organism interactions and intensifies trophic cascade. Front Microbiol 11. https://www.frontiersin.org/articles/10.3389/fmicb.2020.502336
4. Zhang H et al (2023) Editorial: microbial ecology and function of the aquatic systems. Front Microbiol 13. https://www.frontiersin.org/articles/10.3389/fmicb.2022.1109221
5. Wilkins LGE et al (2019) Host-associated microbiomes drive structure and function of marine ecosystems. PLOS Biol 17(11):e3000533. https://doi.org/10.1371/journal.pbio.3000533
6. Moënne-Loccoz Y et al (2014) 'Microorganisms and biotic interactions. Environ Microbiol: Fundam Appl 395–444. https://doi.org/10.1007/978-94-017-9118-2_11
7. Shilky (ORCID: 0000-0001-8292-0385) et al (2023) Role of microbes in controlling the geochemical composition of aquatic ecosystems. In: Hydrogeochemistry of aquatic ecosystems. John Wiley & Sons, Ltd., pp 265–281. https://doi.org/10.1002/9781119870562.ch12
8. Humbert J-F, Dorigo U (2005) Biodiversity and aquatic ecosystem functioning: a mini-review. Aquat Ecosyst Health & Manag 8:367–374. https://doi.org/10.1080/14634980500457773
9. Chen C-Z et al (2022) Exploring the interactions between the gut microbiome and the shifting surrounding aquatic environment in fisheries and aquaculture: a review. Environ Res 214:114202. https://doi.org/10.1016/j.envres.2022.114202

10. Srivastava JK et al (2017) Plant–microbe interaction in aquatic system and their role in the management of water quality: a review. Appl Water Sci 7(3):1079–1090. https://doi.org/10.1007/s13201-016-0415-2
11. Goecke F et al (2010) Chemical interactions between marine macroalgae and bacteria. Mar Ecol Prog Ser 409:267–300. https://doi.org/10.3354/meps08607
12. Mieszkin S, Callow M, Callow J (2013) Interactions between microbial biofilms and marine fouling algae: a mini review. Biofouling 29:1097–1113. https://doi.org/10.1080/08927014.2013.828712
13. Guan C et al (2017) Identification of rosmarinic acid and sulfated flavonoids as inhibitors of microfouling on the surface of eelgrass Zostera marina. Biofouling 33(10):867–880. https://doi.org/10.1080/08927014.2017.1383399
14. Singh VK, Mishra A, Jha B (2017) Anti-quorum sensing and anti-biofilm activity of Delftia tsuruhatensis extract by attenuating the quorum sensing-controlled virulence factor production in Pseudomonas aeruginosa. Front Cellul Infect Microbiol 7. https://www.frontiersin.org/articles/10.3389/fcimb.2017.00337
15. Law JW-F et al (2019) A review on mangrove actinobacterial diversity: the roles of streptomyces and novel species discovery. Prog Microb & Mol Biol 2(1). https://doi.org/10.36877/pmmb.a0000024
16. Gross EM, Bakker ES (2012) The role of plant secondary metabolites in freshwater macrophyte-herbivore interactions: limited or unexplored chemical defences? In: Iason GR, Dicke M (eds) The ecology of plant secondary metabolites: from genes to global processes. Cambridge University Press (Ecological Reviews), pp 154–169
17. Ma H et al (2020) Microplastics in aquatic environments: toxicity to trigger ecological consequences. Environ Pollut 261:114089. https://doi.org/10.1016/j.envpol.2020.114089
18. Galloway TS, Cole M, Lewis C (2017) Interactions of microplastic debris throughout the marine ecosystem. Nat Ecol & Evol 1(5):0116. https://doi.org/10.1038/s41559-017-0116
19. Papenfort K, Bassler B (2016) Quorum-sensing signal-response systems in gram-negative bacteria. Nat Rev Microbiol 14(9):576–588. https://doi.org/10.1038/nrmicro.2016.89
20. Debroy A, George N, Mukherjee G (2022) Role of biofilms in the degradation of microplastics in aquatic environments. J Chem Technol & Biotechnol 97(12):3271–3282. https://doi.org/10.1002/jctb.6978
21. Bacha A-U-R, Nabi I, Zhang L (2021) Mechanisms and the engineering approaches for the degradation of microplastics. ACS ES&T Eng 1(11):1481–1501. https://doi.org/10.1021/acsestengg.1c00216
22. Liu S et al (2021) Interactions between microplastics and heavy metals in aquatic environments: a review. Front Microbiol 12. https://www.frontiersin.org/articles/10.3389/fmicb.2021.652520
23. Kamimura R et al (2022) Quantitative analyses of biofilm by using crystal violet staining and optical reflection. Materials 15(19):6727. https://doi.org/10.3390/ma15196727
24. Barbosa A et al (2023) Imaging biofilms using fluorescence in situ hybridization: seeing is believing. Front Cellul Infect Microbiol 13:1195803. https://doi.org/10.3389/fcimb.2023.1195803
25. Hancz C (2022) Application of probiotics for environmentally friendly and sustainable aquaculture: a review. Sustainability 14:1–17. https://doi.org/10.3390/su142215479
26. Sehnal L et al (2021) Microbiome composition and function in aquatic vertebrates: small organisms making big impacts on aquatic animal health. Front Microbiol 12:567408. https://doi.org/10.3389/fmicb.2021.567408
27. Reverter M et al (2017) Characterisation of the gill mucosal bacterial communities of four butterflyfish species: a reservoir of bacterial diversity in coral reef ecosystems. FEMS Microbiol Ecol. 93(6):fix051. https://doi.org/10.1093/femsec/fix051
28. Legrand TPRA et al (2018) The inner workings of the outer surface: skin and gill microbiota as indicators of changing gut health in yellowtail kingfish. Front Microbiol 8:2664. https://doi.org/10.3389/fmicb.2017.02664
29. Tarnecki AM, Patterson WF, Arias CR (2016) Microbiota of wild-caught Red Snapper Lutjanus campechanus. BMC Microbiol 16(1):245. https://doi.org/10.1186/s12866-016-0864-7

30. van Kessel MAHJ et al (2016) Branchial nitrogen cycle symbionts can remove ammonia in fish gills. Environ Microbiol Reports 8(5):590–594. https://doi.org/10.1111/1758-2229.12407
31. Gomez D, Sunyer JO, Salinas I (2013) The mucosal immune system of fish: the evolution of tolerating commensals while fighting pathogens. Fish & Shellfish Immunol. 35(6):1729–1739. https://doi.org/10.1016/j.fsi.2013.09.032
32. Sylvain F-É et al (2019) Amazon fish bacterial communities show structural convergence along widespread hydrochemical gradients. Mol Ecol 28(15):3612–3626. https://doi.org/10.1111/mec.15184
33. Nayak SK (2010) Role of gastrointestinal microbiota in fish. Aquac Res 41(11):1553–1573. https://doi.org/10.1111/j.1365-2109.2010.02546.x
34. Clements KD et al (2007) Clostridia dominate 16S rRNA gene libraries prepared from the hindgut of temperate marine herbivorous fishes. Mar Biol 150(6):1431–1440. https://doi.org/10.1007/s00227-006-0443-9
35. Kim D-H, Brunt J, Austin B (2007) Microbial diversity of intestinal contents and mucus in rainbow trout (Oncorhynchus mykiss). J Appl Microbiol 102(6):1654–1664. https://doi.org/10.1111/j.1365-2672.2006.03185.x
36. Apprill A et al (2017) Extensive core microbiome in drone-captured whale blow supports a framework for health monitoring. mSystems 2(5). https://doi.org/10.1128/msystems.00119-17
37. Bik EM. et al (2016) Marine mammals harbor unique microbiotas shaped by and yet distinct from the sea. Nat Commun 7(1):10516. https://doi.org/10.1038/ncomms10516
38. Bierlich KC et al (2018) Temporal and regional variability in the skin microbiome of humpback whales along the Western Antarctic Peninsula. Appl Environ Microbiol 84(5):e02574-17. https://doi.org/10.1128/AEM.02574-17
39. Chiarello M et al (2017) Captive bottlenose dolphins and killer whales harbor a species-specific skin microbiota that varies among individuals. Sci Reports 7(1):15269. https://doi.org/10.1038/s41598-017-15220-z
40. Soares-Castro P et al (2019) Microbiota fingerprints within the oral cavity of cetaceans as indicators for population biomonitoring. Sci Reports 9(1):13679. https://doi.org/10.1038/s41598-019-50139-7
41. Sanders JG et al (2015) Baleen whales host a unique gut microbiome with similarities to both carnivores and herbivores. Nat Commun 6(1):8285. https://doi.org/10.1038/ncomms9285
42. Stief P (2013) Stimulation of microbial nitrogen cycling in aquatic ecosystems by benthic macrofauna: mechanisms and environmental implications. Biogeosciences 10(12):7829–7846. https://doi.org/10.5194/bg-10-7829-2013
43. Gutiérrez JL et al (2003) Mollusks as ecosystem engineers: the role of shell production in aquatic habitats. Oikos 101(1):79–90. https://doi.org/10.1034/j.1600-0706.2003.12322.x
44. Mühlenbruch M et al (2018) Mini-review: phytoplankton-derived polysaccharides in the marine environment and their interactions with heterotrophic bacteria. Environ Microbiol 20(8):2671–2685. https://doi.org/10.1111/1462-2920.14302
45. Thornton DCO (2014) Dissolved organic matter (DOM) release by phytoplankton in the contemporary and future ocean. Eur J Phycol. 49(1):20–46. https://doi.org/10.1080/09670262.2013.875596
46. Lombard V et al (2014) The carbohydrate-active enzymes database (CAZy) in 2013. Nucl Acids Res 42(D1):D490–D495. https://doi.org/10.1093/nar/gkt1178
47. Mathur V, Ulanova D (2023) Microbial metabolites beneficial to plant hosts across ecosystems. Microb Ecol 86(1):25–48. https://doi.org/10.1007/s00248-022-02073-x
48. Koza NA et al (2022) Microorganisms in plant growth and development: roles in abiotic stress tolerance and secondary metabolites secretion. Microorganisms 10(8):1528. https://doi.org/10.3390/microorganisms10081528
49. Silknetter S et al (2020) Positive biotic interactions in freshwaters: a review and research directive. Freshwater Biol 65(4):811–832. https://doi.org/10.1111/fwb.13476
50. Holomuzki JR, Feminella JW, Power ME (2010) Biotic interactions in freshwater benthic habitats. J North Am Benthol Soc 29(1):220–244. https://doi.org/10.1899/08-044.1

51. Hoeksema JD, Bruna EM (2015) Context-dependent outcomes of mutualistic interactions. In: Bronstein JL (ed) Mutualism. Oxford University Press. https://doi.org/10.1093/acprof:oso/9780199675654.003.0010.
52. Bronstein JL (2009) The evolution of facilitation and mutualism. J Ecol 97(6):1160–1170. https://doi.org/10.1111/j.1365-2745.2009.01566.x
53. Chapman PM (2018) Negatives and positives: contaminants and other stressors in aquatic ecosystems. Bull Environ Contam Toxicol 100(1):3–7. https://doi.org/10.1007/s00128-017-2229-9
54. Sjerps RMA, Ter Laak TL, Zwolsman GJJG (2017) Projected impact of climate change and chemical emissions on the water quality of the European rivers Rhine and Meuse: a drinking water perspective. Sci Total Environ 601–602:1682–1694. https://doi.org/10.1016/j.scitotenv.2017.05.250
55. Passarelli MC et al (2017) Assessing the influence of ocean acidification to marine amphipods: a comparative study. Sci Total Environ 595:759–768. https://doi.org/10.1016/j.scitotenv.2017.04.004
56. Hui C et al (2022) Modelling structure and dynamics of microbial community in aquatic ecosystems: the importance of hydrodynamic processes. J Hydrol 605:127351. https://doi.org/10.1016/j.jhydrol.2021.127351.
57. Coyte KZ, Schluter J, Foster KR (2015) The ecology of the microbiome: networks, competition, and stability. Science 350(6261):663–666. https://doi.org/10.1126/science.aad2602
58. Hui C et al (2021) Coupling genomics and hydraulic information to predict the nitrogen dynamics in a channel confluence. Environ Sci & Technol [Preprint]. https://doi.org/10.1021/acs.est.0c04018
59. Huang S, Zhang Y (2021) Interannual variability of air–sea exchange of mercury in the global ocean: the "seesaw effect" in the equatorial pacific and contributions to the atmosphere. Environ Sci & Technol 55(10):7145–7156. https://doi.org/10.1021/acs.est.1c00691
60. Hsu RH et al (2019) Microbial interaction network inference in microfluidic droplets. Cell Syst 9(3):229–242.e4. https://doi.org/10.1016/j.cels.2019.06.008
61. Gorter FA, Manhart M, Ackermann M (2020) Understanding the evolution of interspecies interactions in microbial communities. Philos Trans R Soc B: Biol Sci 375(1798):20190256. https://doi.org/10.1098/rstb.2019.0256
62. Deng Y et al (2012) Molecular ecological network analyses. BMC Bioinform 13(1):113. https://doi.org/10.1186/1471-2105-13-113
63. Faust K, Raes J (2012) Microbial interactions: from networks to models. Nat Rev Microbiol 10(8):538–550. https://doi.org/10.1038/nrmicro2832
64. Descheemaeker L, de Buyl S (2020) Stochastic logistic models reproduce experimental time series of microbial communities. In: Krishna S, Walczak AM (eds) eLife 9:e55650. https://doi.org/10.7554/eLife.55650
65. Gao Y et al (2021) Dams shift microbial community assembly and imprint nitrogen transformation along the Yangtze River. Water Res 189:116579. https://doi.org/10.1016/j.watres.2020.116579.
66. Chan SHJ et al (2017) Standardizing biomass reactions and ensuring complete mass balance in genome-scale metabolic models. Bioinformatics 33(22):3603–3609. https://doi.org/10.1093/bioinformatics/btx453
67. Van Wey AS et al (2016) Determination of potential metabolic pathways of human intestinal bacteria by modeling growth kinetics from cross-feeding dynamics. Food Res Int 88:207–216. https://doi.org/10.1016/j.foodres.2016.02.004
68. Marzadri A, Tonina D, Bellin A (2012) Morphodynamic controls on redox conditions and on nitrogen dynamics within the hyporheic zone: application to gravel bed rivers with alternate-bar morphology. J Geophys Res: Biogeosci 117(G3). https://doi.org/10.1029/2012JG001966
69. Louca S et al (2016) Integrating biogeochemistry with multiomic sequence information in a model oxygen minimum zone. Proc Natl Acad Sci 113(40):E5925–E5933. https://doi.org/10.1073/pnas.1602897113

70. Roden EE, Jin Q (2011) Thermodynamics of microbial growth coupled to metabolism of glucose, ethanol, short-chain organic acids, and hydrogen. Appl Environ Microbiol 77(5):1907–1909. https://doi.org/10.1128/AEM.02425-10
71. Hellweger FL et al (2016) Advancing microbial sciences by individual-based modelling. Nat Rev Microbiol 14(7):61–471. https://doi.org/10.1038/nrmicro.2016.62
72. Cheng TH et al (2020) Industrial enzymes-producing marine bacteria from marine resources. Biotechnol Reports 27:e00482. https://doi.org/10.1016/j.btre.2020.e00482
73. Microbial diversity of aquatic ecosystem and its industrial potential (2016) J Bacteriol & Mycol. Open Access 3(1). https://doi.org/10.15406/jbmoa.2016.03.00048
74. Barzkar N (2020) Marine microbial alkaline protease: an efficient and essential tool for various industrial applications. Int J Biol Macromol 161:1216–1229. https://doi.org/10.1016/j.ijbiomac.2020.06.072
75. Raveendran S et al (2018) Applications of microbial enzymes in food industry. Food Technol Biotechnol 56(1):16–30. https://doi.org/10.17113/ftb.56.01.18.5491
76. Bhatt K et al (2022) New insights into engineered plant-microbe interactions for pesticide removal. Chemosphere 309:136635. https://doi.org/10.1016/j.chemosphere.2022.136635
77. Girelli AM, Astolfi ML, Scuto FR (2020) Agro-industrial wastes as potential carriers for enzyme immobilization: a review. Chemosphere 244:125368. https://doi.org/10.1016/j.chemosphere.2019.125368
78. Galasso C, Corinaldesi C, Sansone C (2017) Carotenoids from marine organisms: biological functions and industrial applications. Antioxidants 6(4):96. https://doi.org/10.3390/antiox6040096
79. Abed RMM, Dobretsov S, Sudesh K (2009) Applications of cyanobacteria in biotechnology. J Appl Microbiol 106(1):1–12. https://doi.org/10.1111/j.1365-2672.2008.03918.x
80. Demay J et al (2019) Natural products from cyanobacteria: focus on beneficial activities. Mar Drugs 17(6):320. https://doi.org/10.3390/md17060320
81. Bilal M et al (2017) 'High-value compounds from microalgae with industrial exploitability – a review. Front Biosci 9:319–342. https://doi.org/10.2741/S490

Chapter 19
Active and Passive Solar Distillation—A Detailed Review

Vaishally Dogra, Chandra Kishore, Akarsh Verma, and Juhi Gupta

Nomenclature

A_c	Solar collector area (m^2)
A_g	Glass cover area (m^2)
A_{ss}	Sides area in solar still (m^2)
A_s	Solar still basin area (m^2)
C_w	Specific heat of water in solar still (J/kg °C)
F_R	Factor of heat removal
Gr	Grashof Number
h_b	Overall heat transfer coefficient (HTC) from basin liner to ambient (W/m^2 °C)
$h_{c,b-w}$	Convective HTC from basin liner to water (W/m^2 °C)
$h_{c,g-a}$	Convective HTC from glass cover to ambient (W/m^2 °C)
$h_{c,w-g}$	Convective HTC from water to glass cover (W/m^2 °C)
$h_{e,w-g}$	Evaporative HTC from water to glass cover (W/m^2 °C)
$h_{r,g-a}$	Radiative HTC from glass cover to ambient (W/m^2 °C)
$h_{r,w-g}$	Radiative HTC from water to glass cover (W/m^2 °C)
$h_{t,b-a}$	Total HTC from basin to ambient (W/m^2 °C)

V. Dogra · C. Kishore
Department of Mechanical Engineering, Graphic Era Deemed to Be University, Dehradun, India

A. Verma (✉)
Department of Mechanical Engineering, University of Petroleum and Energy Studies, Dehradun 248007, India
e-mail: akarshverma007@gmail.com

Department of Mechanical Science and Bioengineering, Osaka University, Osaka 560-8531, Japan

J. Gupta (✉)
Amity Institute of Environmental Sciences, Amity University, Noida 201303, India
e-mail: jgupta2@amity.edu

S. K. Sethi et al. (eds.), *Polymer Composites: From Computational to Experimental Aspects*, Materials Horizons: From Nature to Nanomaterials,
https://doi.org/10.1007/978-981-97-0888-8_19

$h_{t,g-a}$	Total HTC from glass cover to ambient (W/m^2 °C)
$h_{t,w-g}$	Total HTC from water to glass cover (W/m^2 °C)
$I(t)_c$	Intensity of solar radiation over inclined surface of solar collector (W/m^2)
$I(t)_s$	Intensity of solar radiation over inclined surface of solar still (W/m^2)
K_i	Thermal conductivity of insulating material (W/m °C)
K_g	Glass cover thermal conductivity (W/m °C)
K_w	Water thermal conductivity (W/m °C)
L	Latent heat of vaporization (J/kg)
L_i	Thickness of insulating material (m)
L_g	Thickness of glass cover (m)
m_{ew}	Hourly output from solar still (kg/m^{2}h)
M_{ew}	Daily output from solar still (kg/m^{2}day)
M_w	Mass of water in basin (kg)
Nu	Nusselt number
P_{gi}	Partial vapor pressure at inner surface glass temperature (N/m^2)
Pr	Prandtl number
P_w	Vapor pressure at water temperature (N/m^2)
q_b	Rate of heat transfer from basin liner to ambient (W/m^2)
$q_{c,b-w}$	Rate of convective heat transfer from basin liner to water (W/m^2)
$q_{c,g-a}$	Rate of convective heat transfer from glass cover to ambient (W/m^2)
$q_{c,w-g}$	Rate of convective heat transfer from water to glass cover (W/m^2)
$\dot{q}_{e,w-g}$	Flux rate of evaporative heat transfer within still from water to glass cover (W)
$q_{e,w-g}$	Evaporative heat transfer rate within still from water to glass cover (W/m^2)
$q_{r,g-a}$	Rate of radiative heat transfer from glass cover to ambient (W/m^2)
$q_{r,w-g}$	Rate of radiative heat transfer from water to glass cover (W/m^2)
$q_{t,g-a}$	Rate of radiative heat transfer from glass cover to ambient (W/m^2)
$q_{t,w-g}$	Rate of total heat transfer from water to glass cover (W/m^2)
q_w	Rate of convective heat transfer from basin liner to ambient (W/m^2)
Q_u	Useful thermal energy gain from the solar collector (W/m^2)
t	Time (sec)
T_a	Ambient temperature (°C)
T_b	Basin temperature (°C)
T_{gi}	Inner surface glass cover temperature (°C)
T_{go}	Outer surface glass cover temperature (°C)
T_{sky}	Temperature of sky (°C)
T_w	Water temperature (°C)
T_{wo}	Temperature of basin water at t = 0 (°C)
U_b	Overall bottom heat loss coefficient (W/m^2 °C)
U_{LC}	Overall HTC for solar collector (W/m^2 °C)
U_{LS}	Overall HTC for solar still (W/m^2 °C)
U_S	Overall side heat loss coefficient (W/m^2 °C)

U_t	Overall top heat loss coefficient from water surface to ambient air (W/m^2 °C)
V	Wind velocity (m/s)
X_w	Mean characteristic length of solar still between basin and water surface (m)
α'_b	Fraction of solar flux absorbed by basin liner
α'_g	Fraction of solar flux absorbed by glass
α'_w	Fraction of solar flux absorbed by water
$(\alpha\tau)_{eff}$	Effective absorptivity-transmittance product of solar flux
ε_g	Emissivity of glass
ε_w	Emissivity of water
η_{active}	Overall thermal efficiency of active solar still
$\eta_{passive}$	Overall thermal efficiency of passive still
σ	Stefan Boltzmann constant (5.67×10^{-8} (W/m^2K^4)

1 Introduction

Water is a critical component for sustaining life on Earth and is essential for both economic and social progress. In regions of Africa, access to water that meets quality standards is severely restricted. The technique of solar distillation has been utilized for many years to purify water. The first U.S. patent for a device designed for solar distillation was awarded to Norman Wheeler and Walton Evans in 1870. During the 1960s and 1970s, a number of advanced solar distillation facilities were established on the Greek islands, with capacities varying between 2,000 and 8,500 cubic meters per day. A Multi-Effect Distillation facility was erected in Abu Dhabi in 1984, boasting a capacity of 120 cubic meters per day, and remains operational. Chile was home to the first large-scale, modern solar still, constructed in 1872, which consisted of 64 basins and provided 20,000 L of water daily to a local mining community. During World War II, the U.S. Navy initiated the first large-scale production of solar stills, manufacturing 200,000 inflatable units for use in naval lifeboats.

For leading a normal life, energy is the basic need for human beings. To sustain these energy needs, solar technologies are very important for under-developed and developed countries. Solar energy is abundant, pollution-free, and everlasting renewable source of energy. Solar distillation is used for improving the quality of saline water and aids to produce potable water [1]. The heat of the sun causes water to evaporate; separating it from the impurities, these water vapors then condense on the still glass cover and are collected for further use (refer to Fig. 1).

Salt emerges as a potential byproduct in the desalination process. Contemporary focus in the field of desalination is increasingly shifting toward devising economically viable methods for generating potable water for human use. Solar distillation is recognized as an eco-friendly technology that is technically viable for the production of freshwater, with considerable potential for conserving high-grade energy.

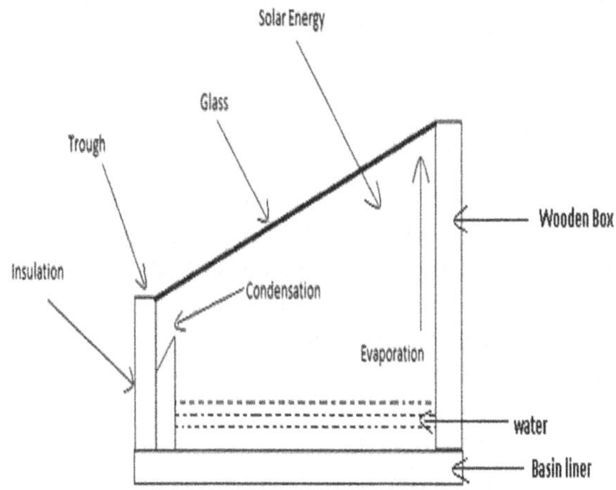

Fig. 1 Diagrammatic representation depicting a basic solar distillation unit

Solar distillation systems can be broadly categorized into two types: active and passive systems. Research by Murugavel et al. [2] suggests that the productivity of a still can be enhanced by employing various basin materials. A review by Sampathkumar et al. [3] on active stills concluded that the exergy efficiency of active solar distillation systems surpasses that of passive systems. Tiwari et al. [4] conducted a parametric analysis of solar stills, finding that the temperatures of both the external and internal glass covers influenced the daily yield in both active and passive systems. Moreover, active systems displayed higher daily efficiency compared to their passive counterparts.

The Jebel Ali Desalination facility in the United Arab Emirates holds the distinction of being the largest desalination plant globally, with a production capacity of 640,000 cubic meters of distilled water per day. Israel uses the largest percentage of desalinated water in the world and produces 40% of its freshwater for human use from solar distillation methods. This study focuses on various types of solar distillation methods for producing freshwater for human consumption and other uses.

2 Types of Solar Stills

2.1 Active Stills

Active systems collect, store, convert, and harness solar energy by using external energy sources such as blowers, pumps, photo voltaic, and thermal collectors. In an active solar still, the evaporation rate and, consequently, productivity are enhanced due to the additional thermal energy delivered to the still's basin via an external source. Classification of active solar stills is given as follows[5]:

i. High-temperature distillation:- It consists of the use of a solar collector panel to feed hot water in the basin.
ii. Pre-heated water application:- Heated water is continuously introduced into the basin at a stable flow rate.
iii. Nocturnal production:- Once in a day the hot water is added into the still basin.

2.1.1 High-Temperature Active Solar Distillation

Solar collectors transform solar radiation into heat which is transferred to water. This causes a rise in water temperature and hot water is fed to basin of still.

Solar Still Twinned with Flat Plate Collector

Solar still twinned with flat plate collector works in either of the two modes mentioned below:-

i. Forced circulation mode, where a pump is used for supplying water to the still.
ii. Natural circulation mode, where the difference in water density is the parameter for water flow.

In a solar still integrated with a flat plate collector (FPC) and operating under forced circulation, a pump facilitates the movement of water from the still's basin through the FPC and back into the basin. Solar stills functioning under natural circulation or thermosyphon modes offer simplicity, reliability, and cost-effectiveness.

Double Effect Active Solar Still:

In a forced circulation system that integrates a flat plate collector (FPC) with a solar still basin, supplementary heat is provided to the incoming water. This elevates the water temperature, accelerating the evaporation process within the basin. The vaporized water, after relinquishing its latent heat of vaporization, condenses on the interior surface of the first glass cover. The surrendered latent heat is then transferred via convection to the water flowing over this first glass layer, causing it to heat up and vaporize. This vapor, in turn, releases its latent heat upon condensing on the inner surface of the second glass cover, a phenomenon known as the second effect. The entire setup is termed as a double-effect active solar distillation system.

The condensed water on the inner faces of both glass covers is guided by gravity into collection troughs for potable and other uses. Prasad et al. [6] presented an analysis of this double-effect active solar distillation system, taking into account both climatic and design variables. Instead of an FPC, a compound parabolic concentrator (CPC) was employed, offering the benefit of manual seasonal adjustments for optimal solar radiation capture. The thermal model of the system was developed under the assumption of quasi-steady state conditions and no vapor leakage. Hourly yields for each effect were calculated based on energy balance equations for individual components. It was noted that the second effect's reutilization of latent heat contributed to an overall increase in system output. Numerical analyses were conducted, and the results were compared to those of a single-effect active solar distillation unit. With a minimal water flow rate in the upper basin, the double-effect system demonstrated a marked improvement in performance. In this double-effect setup, the basin water is heated not only by the FPC but also by solar radiation passing through the glass cover and water mass, as outlined in reference [7].

Regenerative Active Solar Still

In the system under consideration, a solar still's basin is integrated with flat plate collectors, arranged in a series configuration, via a tubular heat exchanger. One extremity of this tube is linked to the collector's inlet, while the opposite end is connected to the collector's outlet. This arrangement facilitates optimal heat transfer, thereby enhancing the latent heat of evaporation on the inner side of the glass cover. To regulate the elevated temperature of the glass, cool water is circulated over it. The vaporized water then undergoes condensation upon relinquishing its latent heat to the water flowing atop the glass cover, and this condensed water can be redirected to the basin to augment yield. Singh et al. [8] examined the thermal behavior of a regenerative active solar distillation system operating in a thermosyphon mode. The study evaluated the overall thermal efficiency based on the energy balance across various system components. Comparative analyses were also performed between active and passive solar stills, both with and without water flow over the glass cover. Numerical findings indicated a marked improvement in the system's overall efficiency when water was circulated over the glass cover. Reference [9] further illustrates that in a regenerative active solar still, a higher rate of evaporation is attained by actively supplying heat to the still basin.

Solar Still Integrated with a Parallel Flat Plate Collector

The collector features a flat plate arranged in parallel, situated above insulating material and separated by an air gap. Beneath this absorber, water is channeled. This entire setup is housed within a wooden enclosure, topped by a glass sheet positioned over the absorber, which is coated in black paint prior to the placement of the glass. An insulated conduit links the collector's outlet to the solar still, allowing for either forced or natural water circulation between the two components. Research conducted by Rai et al. [10] focused on the efficacy of a solar still integrated with a flat plate solar collector operating in forced circulation mode. Their findings revealed a 24% increase in daily distillate output compared to a standalone single-basin solar still. Similarly, Yadav et al. [11] investigated the performance of a solar still coupled with a flat plate solar collector under thermosyphon conditions. Their study concluded that the daily yield improved by 30–35% relative to an uncoupled solar still, making the integrated system a cost-effective alternative to using a flat plate collector alone.

Solar Still Coupled with Parabolic Concentrator

In natural circulation conditions, Singh et al. [12] conducted an analytical assessment of water temperature within an active solar still that was equipped with flat plate collectors as well as a parabolic concentrator. The findings demonstrated that the system with a concentrator had a better efficiency than one with a parabolic collector due to the concentrator's higher evaporative heat transfer coefficient.

This kind of solar still is connected to a parabolic-shaped concentrator, which, as shown in the reference [13], concentrates solar radiation on a vast scale and directs it to a tiny absorber or receiver region. Through pipes, the water to be distilled is delivered to the receiver using either forced or natural circulation. Sun tracking technology, which monitors the sun's position and adjusts the collectors appropriately to focus more sunrays and maximize efficiency, can improve concentrator performance.

Solar Still Coupled with Evacuated Tube Collector

Evacuated tube collectors are often favored over flat plate collectors for high-temperature distillation applications due to their ease of design, reduced heat loss, cost-effectiveness, and superior performance. Unlike flat plate collectors, where sunlight is perpendicular to the collector surface only around noon—resulting in some sunlight being reflected and thus lower efficiency—evacuated tube collectors maintain a near-perpendicular angle to sunlight for the majority of the day due to their cylindrical form, leading to increased efficiency (see reference [14] for the corresponding figure).

Shobha et al. [15] conducted an evaluation of a solar still integrated with an evacuated tube collector-based solar water heater, comparing theoretical predictions with actual test results. The experimental setup consisted of a double-sloped, single-basin solar still and an evacuated tube collector solar water heater. Theoretical analyses were performed for varying water depths and operational conditions, utilizing the design parameters of the still and relevant energy equations. The experimental data indicated that the still's effectiveness diminishes as water depth increases. Active stills demonstrated thermal efficiencies that were 32 to 48% higher than their passive counterparts. In a separate study, Ragh Vendra et al. [16] examined a solar still coupled with an evacuated tube collector operating in natural circulation mode. Their findings revealed daily energy and exergy efficiencies of 33.0% and 2.5%, respectively, with a daily output of 3.8% kg/m^2 at a water depth of 0.03 m.

Solar Still Integrated with Heat Pipe

Calle et al. [17] provided a dynamic model of a solar-aided multi-effect distillation plant. A modular and hierarchical modeling approach was used to construct the model, which was then used to develop, calibrate, and verify several specific-phenomenon sub-models using experimental data. The model serves as a valuable tool for scrutinizing plant performance under diverse conditions and operational strategies, offering insights for optimizing future control methods. This conclusion is supported by the strong correlation observed between simulation outcomes and empirical data.

The system architecture comprises a Vertical Multiple Effect Diffusion (VMED) still and a heat pipe solar collector, as depicted in reference [18]. The VMED still features parallel dividers separated by narrow air gaps, the outer layers of which are in contact with saltwater-saturated wicks. A closed circuit is established between the solar collector and the VMED still by connecting the evaporation tubes to the solar collector and the condensation tube to the initial divider of the still. This closed-circuit functions as a heat pipe and is evacuated by a vacuum pump, while being filled with a consistent quantity of distilled water to act as the working fluid. Solar radiation absorbed by the collector is transferred to the still via the latent heat of the working fluid through the heat pipes. To enhance the production of distillate, solar energy is recirculated within the still.

Solar Still Coupled with Solar Pond

A solar pond is a man-made water body designed to capture and store solar energy, resulting in a temperature increase that provides thermal energy. The pond is structured into three distinct layers: the Upper Convective Zone (UCZ), the Lower Convective Zone (LCZ), and an intermediary layer known as the Non-Convective Zone (NCZ). Solar rays traverse both the UCZ and NCZ to reach the LCZ, which contains a high concentration of salt solution. This salt solution acts as a store for the thermal energy which is further used for distillation. Velmurugan et al. [19] constructed a system comprising a settling tank, a single-basin solar still, a storage reservoir, a mini solar pond, and a stepped solar still. Industrial wastewater served as the input material, first passing through the settling tank and then the mini solar pond for preheating. In a separate study, Salata et al. [20] explored the feasibility of achieving process temperatures exceeding 100 °C using solar ponds in conjunction with heat transformers. They observed that the pond area required to produce a cubic meter of desalinated water per day ranged from 1000 to 4000 m^3, contingent on the heat flux extracted. Their research concluded that consistently achieving temperatures above 100 °C was possible only under specific, concatenated conditions and that the heat generated at these elevated temperatures could be applied to standard thermal processes in seawater desalination.

As detailed in reference [21], single-basin and stepped solar stills were individually paired with a mini solar pond, and their performance metrics were evaluated.

The studies revealed that output increased in correlation with solar intensity and the temperature differential between water and glass, while it decreased as wind speed increased. Optimal productivity, reaching 100%, was achieved when a fin-type stepped solar still was combined with sponge and pebbles.

Solar Still Coupled with Hybrid PV/T System

In solar PV cells, most of the absorbed solar radiations are not converted to electricity. The excess unabsorbed energy reduces its efficiency by increasing the temperature of the PV cell. Boubekriet al. [22] studied the numerical simulation of an active solar still coupled with a PV/T solar water heater system. Mathematical modeling was carried out for heat transfer and energy balance. The analysis observed that when still was coupled with a PV/T storage tank, the increase in production was estimated at 131.06, 47.61, and 137.50% for summer, winter, and spring, respectively. Hybrid PV system, as shown in reference [23], consists of PV modules coupled with heat extraction devices (thermal modules). The system delivers both heat and electricity by converting sunlight into electricity with the help of PV modules and collects the residual thermal energy by thermal modules.

2.1.2 Pre-Heated Water Active Solar Still

In this technique, preheated water is used to raise the temperature of water in the basin. Solar stills may use the hot water from numerous sources, including factories and power plants, to boost output. In order to increase production, hot water is provided to basins either directly or through heat exchangers. A single-basin-type solar still with a still size of 1×1 m^2 and insulation thickness of 0.05 m was built by Tiwari et al. [24]. The still was mostly utilized in places that consumed waste heat. To cool the glass cover, water is flown over the glass as illustrated in the reference [24]. According to experimental findings, the yield rose according to the rise in incoming water temperature both for mass and for water flow. A yield of 0.5 kg/m^2 was obtained.

2.1.3 Nocturnal Active Solar Still

Nocturnal distillation is the use of stored energy for solar still evaporation during the hours when there is no sunlight to create yield. This can be done by using hot water that is produced from sources other than solar energy or by using solar energy that is stored in vast water depths in typical solar stills during the hours when the sun isn't shining. Tiwari et al. [25] explored nocturnal water production utilizing waste heat to warm brine, where the day yield of the still relies on the initial brine temperature. The investigation concluded that given the same state of the still and climatic circumstances, the daily distillate production in the tubular solar still was greater than that in the traditional solar still.

2.2 Passive Stills

A passive solar still doesn't rely on mechanical apparatuses or traditional energy sources for its operation. Characterized by its straightforward design, ease of construction, and minimal maintenance needs, a passive still primarily consists of three key elements: a metal basin liner, a saline water reservoir, and a glass cover. It functions based on the fundamental processes of evaporation and condensation. Saline or contaminated water is poured into the basin, and sunlight is allowed to penetrate through the glass cover to facilitate evaporation. As the water evaporates, contaminants and microbes are left behind in the basin. The purified water vapor condenses on the inner surface of the glass cover, from where it trickles down and is collected in a trough for subsequent use. To improve output, simple alterations can be made within the basin of passive solar stills. Due to their operation at lower temperatures, the productivity of passive stills is generally lower compared to other types.

2.2.1 Single-Slope Basin Still

A unidirectional sloping basin consists of a basin coated in black paint, filled with either brackish or sea water, and entirely enclosed by a transparent covering. Solar rays penetrate this transparent layer and are absorbed by the basin, leading to the evaporation of the heated water. The vapor then condenses on the interior surface of the glass cover, flows downward along the slope, and is gathered in a collection tray, as illustrated in Fig. 2.

2.2.2 Wick Stills

After passing through the glass cover, solar energy is absorbed by the wick's surface. A portion of this absorbed energy heats the water that moves through the wick via capillary action. A significant amount of heat is retained within the still, facilitating energy transfer from the wick surface to both the glass cover and the surrounding air. The water coursing through the wick becomes heated and evaporates, subsequently condensing on the underside of the glass cover. Upon releasing its latent heat of vaporization, the condensed vapor descends by gravitational force and is collected in a drainage channel.

Tiwari et al. [26] introduced a double-condensing, multiple-wick solar still design that increased the condensing surface area by adding an additional galvanized iron sheet beneath a blackened, wet jute cloth, separated by a small gap around the

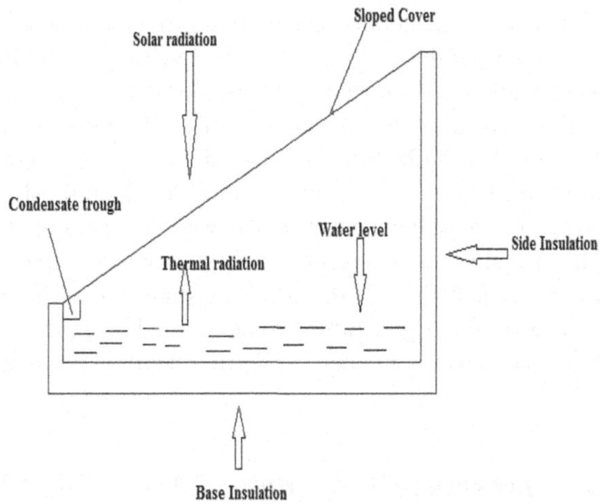

Fig. 2 Single-slope basin-type solar still

absorbing surface. Their analysis showed that this design yielded nearly 20% more distillate compared to a standard multiple-wick solar still. Tanaka [27] performed a theoretical study on a tilted wick solar still integrated with an external Flat Plate Collector, reporting that adjusting the incline of both the still and the reflector in any season could enhance daily distillate production. On average, the wick-type solar still increased distillate output by about 21% year-round compared to traditional solar stills. As indicated in reference [28], a wick-type solar still is a form of passive still featuring wick material that floats on the basin water.

2.2.3 Stepped Solar Still

In a study by El-Agouz [29], a revised version of a stepped solar still was introduced, incorporating a storage tank for saline and sea water as well as a cotton-based energy storage material. The findings indicated that the modified design exhibited superior daily efficiency compared to traditional solar stills. A separate study [30] described a stepped solar still that is architecturally and geometrically similar to a conventional solar still, with the exception that its absorber plate is constructed in a stepped manner. This design feature enhances the system's productivity. Saline or brackish water is initially directed into the first stepped tray. Overflowing water then moves to the second tray and eventually collects in a standard basin at the bottom. Water input is halted once a stable water level is achieved in the basin. The design's stepped trays facilitate increased evaporation rates, thereby improving overall performance.

Alaudeen et al. [30] developed a stepped solar still integrated with a Flat Plate Collector and assessed its performance at varying water depths. Inclined Flat Plate Collectors, covering a total area of 240 cm^2, were positioned between the trays. The system was further optimized using various packing materials such as sponges, rocks, and pebbles to enhance water yield. Experimental results were then compared with theoretical models. The highest productivity rate of 1468 kg/m^2 was achieved at a water depth of 2 cm, while the lowest rate of 1150 kg/m^2 was observed at a 4 cm depth. Economic analysis revealed that the payback period for the stepped solar still was 80 days shorter than that of a conventional still.

Kabeel et al. [31] also examined the impact of tray dimensions on the performance of stepped solar stills. Both a conventional single-sloped solar still and a modified stepped solar still were tested. The water temperature in the stepped still was manipulated using a vacuum tube solar collector. To further boost yield, a wick was added to the vertical sides of the still. The study found that reducing the water depth and increasing the tray depth up to an optimal point led to enhanced performance. The best results were obtained for a water depth of 5 mm and a tray width of 120 mm.

3 Theoretical Simulation of Active Solar Distillation System

3.1 *Theoretical Analysis*

3.1.1 Heat Transfer

In active solar stills, the process of heat transfer takes place through two primary mechanisms, as outlined in [32]. The first involves the transfer of heat from the water to the inner surface of the glass, while the second mechanism concerns the transfer of heat from the outer surface of the glass to the ambient atmosphere. The governing quantities for these heat transfers are calculated based on established laws of heat transfer.

Internal Heat Transfer

The internal heat transfer includes radiation, convection, and evaporation from the water surface to the inner surface of the glass cover within the solar still.

Radiative Heat Transfer

Radiative heat transfer between water to glass as given by Tiwari and Tiwari [33] considers the view factor as unity as follows:

$$q_{r,w-g} = h_{r,w-g}\left(T_w - T_{gi}\right) \quad (1)$$

where the coefficient of radiative heat transfer is given as:

$$h_{r,w-g} = \varepsilon_{eff}\sigma[\frac{T_w + 273)^2 + (T_{gi} + 273)^2}{T_w + T_{gi} + 546}] \quad (2)$$

and the effective emittance can be calculated by:

$$\varepsilon_{eff} = \frac{1}{\left(\frac{1}{\varepsilon_g}\right) + \left(\frac{1}{\varepsilon_w}\right) - 1} \quad (3)$$

Convective Heat Transfer

The rate of convective heat transfer due to the temperature difference between the water surface and the inner glass surface is given by:

$$q_{c,w-g} = h_{c,w-g}\left(T_w - T_{gi}\right) \quad (4)$$

The convective heat transfer coefficient as expressed by Badran et al. [34], Dimri et al. [35], and Dunkle [36] is given by:

$$h_{c,w-g} = 0.884[\left(T_w - T_{gi}\right) + \frac{\left(P_w - P_{gi}\right)\left(T_w + 273\right)}{\left(268.9 \text{ x } 10^{-3} - P_w\right)}]^{1/3} \quad (5)$$

P_w and P_{gi} are the partial pressures of water and vapor at the glass temperature and are represented as:

$$P_W = \exp[25.317 - (\frac{5144}{273 + T_w})] \quad (6)$$

$$P_{gi} = \exp[25.317 - (\frac{5144}{273 + T_{gi}})] \quad (7)$$

Convective heat transfer from basin to water given by [14] is expressed as:

$$q_{c,b-w} = h_{c,b-w}(T_b - T_w) \quad (8)$$

Convective heat transfer coefficient between basin to water is given as:

$$h_{c,b-w} = \frac{K_w}{X_w} C\ (GrPr)^n,\ C = 0.54 \text{ and } n = 1/4 \quad (9)$$

Correlation for convective heat transfer coefficient given by Malik et al. [37] is expressed as:

$$h_{c,b-w} = \frac{NuK_w}{\delta_w} \quad (10)$$

for

$$Gr < 10^{5,}\ Nu = 1 \quad (11)$$

for

$$10^5 < Gr < 2 \times 10^7, \; Nu = 0.54(GrPr)^{0.25} \quad (12)$$

for

$$Gr > 2 \times 10^7, \; Nu = 0.14(GrPr)^{0.33} \quad (13)$$

Evaporative Heat Transfer

Equation for evaporative heat transfer between water and glass is given by [38]:

$$q_{e,w-g} = h_{e,w-g}\left(T_w - T_{gi}\right) \quad (14)$$

Evaluation of evaporative heat transfer coefficient as done by [34] and [36] is as follows:

$$h_{e,w-g} = 16.273 \text{ x } 10^{-3} \text{ x } h_{c,w-g}\left[\frac{P_w - P_{gi}}{T_w - T_{gi}}\right] \quad (15)$$

Rate of total heat transfer from water to glass can be calculated as:

$$q_{t,w-g} = h_{t,w-g}\left(T_w - T_{gi}\right) \quad (16)$$

where

$$q_{t,w-g} = q_{c,w-g} + q_{e,w-g} + q_{r,w-g} \quad (17)$$

and

$$h_{t,w-g} = h_{c,w-g} + h_{e,w-g} + h_{r,w-g} \quad (18)$$

External Heat Transfer

Heat exchange between the glass surface and the surrounding air takes place through convective processes, while the interaction between the glass and the sky occurs via radiative means. The quantities of heat exchanged in these interactions are regulated by established thermal principles.

Top Loss Heat Transfer Coefficient

Heat lost from the outer glass surface to the atmosphere by convective and radiative modes of heat transfer is called total top heat loss. Total top heat loss coefficient defined in [34] is given by:

$$q_{t,g-a} = q_{r,g-a} + q_{c,g-a} \tag{19}$$

$$q_{t,g-a} = h_{t,g-a}(T_{go} - T_a) \tag{20}$$

where

$$h_{t,g-a} = h_{r,g-a} + h_{c,g-a} \tag{21}$$

Radiative heat transfer is defined as:

$$q_{r,g-a} = h_{r,g-a}(T_{go} - T_a) \tag{22}$$

Radiative heat transfer coefficient is given by:

$$h_{r,g-a} = \varepsilon_g \sigma \left[\frac{(T_{go} + 273)^4 - (T_{sky} + 273)^4}{T_{go} - T_a} \right] \tag{23}$$

T_{sky} in the above equation is six less than atmospheric temperature, i.e.,

$$T_{sky} = T_a - 6 \tag{24}$$

The convective heat transfer from glass to the atmosphere as given by [33] is defined as:

$$q_{c,g-a} = h_{c,g-a}(T_{go} - T_a) \tag{25}$$

Convective heat transfer coefficient in terms of wind velocity is:

$$h_{c,g-a} = \begin{cases} 5.7 + 3.8V & for\ V \le 5\,\mathrm{m/s} \\ 6.15V^{0.8} & for\ V > 5\,m/s \end{cases} \tag{26}$$

Total internal heat loss coefficient ($h_{t,w-g}$) and conductive heat transfer coefficient of glass (K_g/L_g) is expressed as U_{wo} : U_{wo} :

$$U_{wo} = [\frac{1}{h_{t,w-g}} + \frac{L_g}{K_g}] \tag{27}$$

The overall top loss coefficient from the water surface to the atmosphere through glass cover is given by:

$$U_t = \frac{h_{t,w-g}\, h_{t,g-a}}{(h_{t,g-a} + U_{wo})} \tag{28}$$

Bottom and Side Loss Heat Transfer Coefficient

Heat is conveyed from the water in the basin to the surrounding atmosphere via insulative mechanisms, as well as through convective and radiative processes originating from the basin's lateral and bottom surfaces. The rate at which heat is conducted from the basin liner to the atmosphere is described as [39].

$$q_b = h_b(T_b - T_a) \tag{29}$$

The heat transfer coefficient from the basin liner to the atmosphere is given as [39].

$$h_b = [\frac{L_i}{K_i} + \frac{1}{h_{t,b-a}}]^{-1} \tag{30}$$

where

$$h_{t,b-a} = h_{c,b-a} + h_{r,b-a} \tag{31}$$

The bottom loss heat transfer coefficient from water mass to ambient is.

$$U_b = \left[\frac{1}{h_w} + \frac{1}{h_b}\right]^{-1} \tag{32}$$

Equation (31) could be rearranged as.

$$U_b = \frac{h_w h_b}{h_w + h_b} \tag{33}$$

Conduction heat is lost through vertical walls and insulation of still and the total side loss heat transfer coefficient can be given as.

$$U_s = \left(\frac{A_{ss}}{A_s}\right) U_b \tag{34}$$

Side still area (A_{ss}) is very small as compared to still basin area (A_s); therefore, U_s is neglected.

The overall heat transfer coefficient including bottom, side, and top losses as expressed by [39] is.

$$U_{LS} = U_t + U_b \tag{35}$$

3.2 Thermal Modeling

In a study by Tiwari et al. [40], the energy balance equations for an active solar still were observed, focusing on different heat transfer coefficients. The equations can be formulated as follows.

3.2.1 Inner and Outer Surface of Glass Cover

Inner surface of glass cover

$$\alpha_g' I(t)_s + q_{r,w-g} + q_{c,w-g} + q_{e,w-g} = \frac{K_g}{L_g}\left(T_{gi} - T_{go}\right) \tag{36}$$

Substituting Eq. (16) in Eq. (36).

$$\alpha_g' I(t)_s + h_{t,w-g}\left(T_w - T_{gi}\right) = \frac{K_g}{L_g}\left(T_{gi} - T_{go}\right) \tag{37}$$

Equation (37) can be simplified as.

$$T_{gi} = \frac{\alpha_g' I(t)_s + h_{t,w-g} T_w + (K_g/L_g) T_{go}}{h_{t,w-g} + (K_g/L_g)} \tag{38}$$

Outer surface of glass cover

$$\frac{K_g}{L_g}\left(T_{gi} - T_{go}\right) = q_{r,g-a} + q_{c,g-a} \tag{39}$$

Substituting Eqs. (19) and (20) in Eq. (39):

$$\frac{K_g}{L_g}\left(T_{gi} - T_{go}\right) = h_{t,g-a}(T_{go} - T_a) \tag{40}$$

Substituting Eq. (38) in Eq. (40):

$$\frac{K_g}{L_g}(T_{gi} - T_{go}) = \frac{h_{t,w-g}(K_g/L_g)}{h_{t,w-g} + (K_g/L_g)}(T_w - T_{go}) + \frac{\alpha_g' I(t)_s (K_g/L_g)}{h_{t,w-g} + (K_g/L_g)} \tag{41}$$

Substituting Eq. (27) in Eq. (41):

$$\frac{K_g}{L_g}\left(T_{gi} - T_{go}\right) = U_{wo}\left(T_w - T_{go}\right) + h_k \alpha_g' I(t)_s \frac{K_g}{L_g} \tag{42}$$

where

$$h_k = \frac{K_g/L_g}{h_{t,w-g} + (K_g/L_g)} \quad (43)$$

Substituting Eq. (43) in Eq. (40):

$$\alpha'_g I(t)_s h_k + U_{wo}(T_w - T_{go}) = h_{t,g-a}(T_{go} - T_a) \quad (44)$$

Simplifying Eq. (44).

$$T_{go} = \frac{\alpha'_g I(t)_s h_k + U_{wo} T_w + h_{t,g-a} T_a}{h_{t,g-a} + U_{wo}} \quad (45)$$

3.2.2 Basin Liner

The rate at which the basin plate absorbs energy is equivalent to the sum of the rate at which energy is transferred to the water and the rate of energy loss through the bottom and sides of the basin. This relationship can be articulated as follows:

$$\alpha'_b(1 - \alpha'_g)(1 - \alpha'_w) I(t)_s = q_w + q_b \quad (46)$$

Substituting Eqs. (8 and (9) in Eq. (46).

$$\alpha'_b(1 - \alpha'_g)(1 - \alpha'_w) I(t)_s = h_w(T_b - T_w) + h_b(T_b - T_a) \quad (47)$$

After simplifying Eq. (47).

$$T_b = \frac{-\alpha'_b I(t)_s + h_w T_w + h_b T_a}{h_w + h_b} \quad (48)$$

3.2.3 Water Mass

$$\alpha'_w(1 - \alpha'_g) I(t)_s + h_w(T_b - T_w) + Q_u = M_w C_w \frac{dT_w}{dT} + [h_{t,w-g}(T_w - T_{go})] \quad (49)$$

Substituting values of T_{go} and T_b from Eqs. (36b) and (38) in (39)

$$Q_u + I(t)_s\left[\alpha'_b \frac{h_w}{(h_w + h_b)} + \alpha'_w + \alpha'_g \frac{h_{t,w-g}}{(h_{t,g-a} + U_{wo})}\right] = M_w C_w \frac{dT_w}{dT} + \left[\frac{h_{t,w-g} h_{t,g-a}}{(h_{t,g-a} + U_{wo})}(T_w - T_{go})\right] + \left[\frac{h_w h_b}{(h_w + h_b)}(T_w - T_a)\right] \quad (50)$$

The above equation can be simplified as.

$$Q_u + (\alpha\tau)_{eff} I(t)_s = M_w C_W \frac{dT_w}{dT} + (U_t + U_b)(T_w - T_a) \tag{51}$$

where

$$(\alpha\tau)_{eff} = [\alpha'_b \frac{h_w}{(h_w + h_b)} + \alpha'_w + \alpha'_s \frac{h_{t,w-g}}{\left(h_{t,g-a} + U_{wo}\right)}] \tag{52}$$

and Q_u is the additional energy which is supplied to the active still by solar collectors per unit area of the basin to increase the basin temperature and is expressed as:

$$Q_u = NF_R[(\alpha\tau)_c I(t)_c - U_{LC}(T_w - T_a)] \tag{53}$$

($N =$ number of collectors which is assumed to be 1).

Substituting Eq. (40) in (39), one gets:

$$F_R\left[(\alpha\tau)_c I(t)_c - U_{LC}(T_w - T_a)\right] + (\alpha\tau)_{eff} I(t)_s = M_w C_w \frac{dT_w}{dT} + (U_t + U_b)(T_w - T_a) \tag{54}$$

Substituting U_{LS} in the above equation:

$$F_R(\alpha\tau)_c I(t)_c + (\alpha\tau)_{eff} I(t)_s = (MC)_w \frac{dT_w}{dT} + [(U_{LS}) + (F_R U_{LC})(T_w - T_a) \tag{55}$$

Representation of the above equation can be given as:

$$I_{eff} = M_w C_w \frac{dT_w}{dT} + U_{eff}(T_w - T_a) \tag{56}$$

where

$$I_{eff} = F_R(\alpha\tau)_c I(t)_c + (\alpha\tau)_{eff} I(t)_s \tag{57}$$

$$U_{eff} = (U_{LS} + F_R U_{LC}) \tag{58}$$

Dividing Eq. (44) by $M_w C_w$, it becomes:

$$\frac{dT_w}{dT} + \left[\frac{U_{eff}}{M_w C_w}\right] T_w = \left[\frac{U_{eff}}{M_w C_w}\right] T_a + \left[\frac{I_{eff}}{M_w C_w}\right] \tag{59}$$

The aforementioned equation can be rewritten as:

$$\frac{dT_w}{dT} + aT_w = f(t) \tag{60}$$

where 'a' and 'f(t)' can be represented as:

$$a = \left[\frac{U_{eff}}{M_w C_w}\right] \tag{61}$$

$$f(t) = \left[\frac{I_{eff} + U_{eff} T_a}{M_w C_w}\right] \tag{62}$$

Solution to Eq. (48) is given as:

$$T_w = \frac{\overline{f(t)}}{a}[1 - e^{-at}] + T_{wo} e^{-at} \tag{63}$$

where T_{wo} is the temperature of basin water at $t = 0$ and f(t) is the average value of f(t) for the time interval between 0 and t.

Now hourly output is given by:

$$m_{ew} = \frac{h_{e,w-g}\left(T_w - T_{gi}\right)}{L} \times 3600 \times A_s \tag{64}$$

Overall thermal efficiency of the active solar still is expressed as:

$$\eta_{\mathbf{active}} = \frac{\sum_{i=1}^{24} m_{ew} L}{\sum (I(t)_c \times A_c \times 3600) + \sum (I(t)_s \times A_s \times 3600)} \tag{65}$$

$\sum_{i=1}^{24} m_{ew}$ in above equation expresses the total daily output and can be represented as:

$$M_{ew} = \sum_{i=1}^{24} m_{ew} \tag{66}$$

4 Thermal Modeling of Solar Passive Distillation System

The heat quantities for passive still, according to [41], are as follows:

Evaporative heat flux required to evaporate water from basin area of still is given by:

$$\dot{q}_{ew} = h_{e,w-g} \times A_s \times \left(T_w - T_g\right) \tag{67}$$

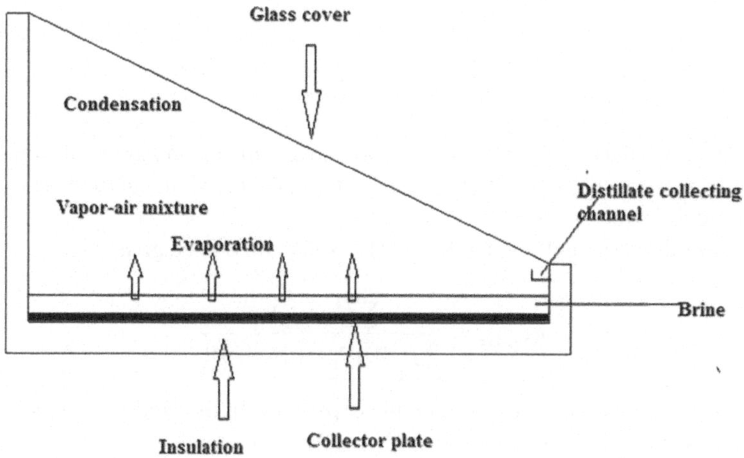

Fig. 3 Schematic diagram of a passive solar still

Water and glass cover temperatures attainable in passive still as shown in Fig. 3. for time 't' and solar radiations $I(t)_s$ can be given as:

$$T_w = \frac{(\alpha\tau)_{eff}\left(1 - e^{-at}\right)I(t)_s + U_{LS}(T_{wo} - T_a)e^{-at} + U_{LS}T_a}{U_{LS}} \tag{68}$$

$$T_g = \frac{\alpha_g' I(t)_s A_g + h_{t,w-g} T_w A_s + h_{t,g-a} T_a A_g}{h_{t,g-a} A_g + h_{t,w-g} A_s} \tag{69}$$

where

$$(\alpha\tau)_{eff} = \{\alpha_w' + \left[\frac{h_{t,w-g}}{\left(h_{t,g-a}A_g + h_{t,w-g}A_s\right)}\right]\alpha_g' A_g + \left[\frac{h_{c,b-w}}{\left(h_{t,b-a} + h_{c,b-w}\right)}\right]\alpha_b'\} \tag{70}$$

Consisting of the following:

The overall top heat loss coefficient from the water surface to ambient and overall bottom heat loss coefficient for the still are given as:

$$U_t = \frac{h_{t,g-a} h_{t,w-g}}{h_{t,g-a} A_g + h_{t,w-g} A_s} \tag{71}$$

$$U_b = \frac{h_{t,b-a} h_{bw}}{h_{t,b-a} + h_{bw}} \tag{72}$$

And overall heat transfer coefficient for still is:

$$U_{LS} = [U_t \frac{A_g}{A_s} + U_b] \tag{73}$$

where

$$A_g = A_s \, \mathrm{Sec}\, \theta \tag{74}$$

The energy balance equations for a passive solar still, encompassing the inner and outer glass covers, the basin liner, and the water mass, are identical to those used for an active solar still.

The overall thermal efficiency for passive solar still can be given as:

$$\eta_{passive} = \frac{\sum_{i=1}^{24} m_{ew} L}{\sum (I(t)_s \times A_s \times 3600)} \tag{75}$$

Lastly, the authors have a considerable experience in the field of atomistic-scale simulations and polymer composites [42–141].

5 Conclusions

Based on the research and developments done in active and solar distillation, important points can be concluded as follows:

- Active solar stills exhibit greater productivity compared to passive solar stills, primarily due to the lower operating temperatures of the latter.
- When it comes to cost-effectiveness, solar stills that are integrated with flat plate solar collectors using a thermosyphon system have an advantage over those paired with a standard flat plate collector.
- Stills that are combined with evacuated tube collectors demonstrate higher yields than those integrated with flat plate collectors.
- Production increase becomes very significant for PV/T system as it delivers both heat and electricity.
- The hot water supplied to pre-heated water active solar still basin through heat exchangers results in productivity increase.
- Double condensing multiple-wick solar stills achieve a greater yield compared to their simple multiple-wick counterparts.
- In terms of both water purity and overall performance, stepped solar stills outperform traditional solar stills.
- Improvement in still performance can be enhanced by using energy storage materials such as sponge cubes, pebbles, gravel, etc.

Declarations

Conflicts of Interest The authors confirm that there are no conflicts of interest associated with this work.

Ethical Compliance The authors affirm that this study adheres to all ethical guidelines.

Competing Interests No competing interests are declared by the authors.

Data and Material Availability Data and materials are not applicable to this study.

References

1. Mohamed A, Eltauil ZM omara (2014) Enhancing the solar still performance using solar photovoltaic, flat plate collector and hot air, Desalination 349 1–9
2. Murugavel KK, Chockalingam KK, Srithar K (2008) Progress in improving the effectiveness of the single basin passive solar still. Desalination 220:677–686
3. sampathkumar K, Arjunan TV, Pitchandi P, Senthikumar P (2010) Active solar distillation-A detailed review. Renew Sustain Energy Rev 14 1503–1526
4. Tiwari GN (2009) Vimal Dimri, Arvind Chel, Parametric study of an active and passive solar distillation system: Energy and exergy analysis. Desalination 242:1–18
5. Tiwari GN, Thomsan JM (1994) Emran Khan, Optimisation of glass cover inclination for maximum yield in a solar still. Heat Recov Sys CHP 14:447–455
6. Bhagwan Prasad, Tiwari GN (1996) Analysis of double effect active solar distillation, Energy Convers. Mgmt 37(11), pp 1647–1656
7. Kumar S (1999) Tiwari GN, Optimization of daily yield for an active double effect distillation with water flow. Energy Convers Manage 40:703–715
8. Singh AK, Tiwari GN (1993) Thermal evaluation of regenerative active solar distillation under thermosyphon mode. Energy Convers. Mgmt 34(8):697–706
9. Tiwari GN, Sinha S (1993) Parametric studies of active regenerative solar still. Energy Convers Manage 34(3):209–218
10. Rai SN, Tiwari GN (1983) Single basin solar still coupled with flat plate collector. Energy Convers Manage 23:145–149
11. Yadav YP, Prasad AS (1995) Performance analysis of a high temperature solar distillation system. Energy Convers Manage 36:365–374
12. Singh SK, Bhatnagar VP, Tiwari GN (1996) Design parameters for concentrator assisted solar distillation system. Energy Convers Manaement 37(2):247–252
13. Zeinab S (2007) Abdel Rehim, Ashraf Lasheen, Experimental and theoretical study of a solar desalination system located in Cairo, Egypt. Desalination 217:52–64
14. Tiwari GN (2007) Dimri Vimal. Singh Usha, Chel Aravind, Sarkar Bikash, Comparative thermal performance evaluation of an active solar distillation system, Int J Energy Res 31:1465–1482
15. Shobha BS, Vilastwe, Rajesh AM (2012) Performance evaluation of a solar still coupled to an evacuated tube collector type solar water heater, Int J Innov Eng Technol (IJIET), 1:2012, ISSN:2319–1058
16. Ragh Vendra Singh, Shiv Kumar, Hasan MM, Emran Khan M, Tiwari GN (2013) Performance of a solar still integrated with evacuated tube collector in natural mode, Desalination 318, 25–33
17. Alberto de la Calle, Javier Bonilla, Lidia Roca, Patricia Palenzuela, Dynamic modeling and simulation of a solar-assisted multi-effect distillation plant, Desalination 357(0) 65–76
18. Hiroshi T, Yasuhito N, Masahito T (2005) Indoor experiments of the vertical multiple effect diffusion type solar still coupled with a heat pipe solar collector. Desalination 177:291–302
19. Velmurugan V (2009) Mandlin J, Stalin B, Srithar K, Augmentation of saline streams in solar stills integrating with a mini solar pond. Desalination 249(1):143–149
20. Salata F, Coppi M (2014) A first approach study on the desalination of sea water using heat transformers powered by solar ponds. Appl Energy 136:611–618

21. Velmurugan V, Mandlin J, Stalin B, Srithar K (2009) Augmentation of saline streams in solar stills integrating with a mini solar pond. Desalination 249(1):143–149
22. Boubekri M, Chaker A, Cheknane A (2013) Modelling and simulation of the continuous production of an improved solar still coupled with a PV/T solar water heater system. Desalination 331:6–15
23. Shiv K (2009) Tiwari GN, Estimation of internal heat transfer coefficients of a hybrid (PV/T) active solar still. Sol Energy 83:1656–1667
24. Tiwari GN, Madhuri Garg HP (1985) Effect of water flow over the glass cover of a single basin solar still with an intermittent flow of waste hot water in the basin. Energy Convers Manage 25(3):315–322
25. Tiwari GN (1988) Kumar Ashok, Nocturnal water production by tubular solar stills using waste heat to preheat brine. Desalination 69:309–318
26. Tiwari GN, Sharma SB, Sodha MS (1984) Performance of a double condensing multiple wick solar still. Energy Convers. Mgmt 24(2):155–159
27. Tanaka H (2009) Tilted wick solar still with external flat plate reflector: Optimum inclination of still and reflector. Desalination 249:411–415
28. Mahdi JT, Smith BE, Sharif AO (2011) An experimental wick-type solar still system: design and construction. Desalination 267:233–238
29. El-Agouz SA (2014) Experimental investigation of stepped solar still with continuous water circulation. Energy Convers Manage 86:186–193
30. Alaudeen A, Johnson K, Ganasundar P, Syed Abuthahir A, Srithar K (2013) Study on stepped type basin in a solar still. J King Saud University—Engineering Sci
31. Kabeel AE, Khalil A, Omara ZN, Younes MM (2012) Theoretical and experimental parametric study of modified stepped solar still. Desalination 289:12–20
32. Sampathkumar K, Arjunan TV, Pitchandi P, Senthikumar P (2010) Active solar distillation—A detailed review. Renew Sustain Energy Rev 14 1503–1526
33. Tiwari GN, Tiwari AK (2008) Solar distillation practice for water desalination systems. Anamaya Publishers, New Delhi
34. Badran OO, Abu-Khader MM (2007) Evaluating thermal performance of a single slope solar still. Heat Mass Transfer 43:985–99
35. Dimri V, Sarkar B, Singh U, Tiwari GN (2008) Effect of condensing cover material on yield of active solar still: an experimental validation. Desalination 178–189
36. Dunkle RV (1961) Solar water distillation, the roof type solar still and a multi effect diffusion still. International developments in heat transfer. In: ASME: Proceedings of International Heat Transfer, University of Colorado 5:895–902
37. Malik MAS, Tiwari GN, Kumar A, Sodha MS (1985) Solar Distillation. Pergamon Press, Oxford
38. Velmurugan V, Srithar K (2007) Solar stills integrated with a mini solar pond- analytical simulation and experimental validation. Desalination 216:232–241
39. Tiwari GN (2004) Solar energy; fundamentals, design, modeling and application. New Delhi: Narosa Publishing House. pp 278–306
40. Tiwari GN (2009) Vimal Dimri, Arvind Chel, Parametric study of an active and passive solar distillation system. Energy and exergy analysis, Desalination 242:1–18
41. Syahru S, Hamdullahpur F, Dincer I (2002) Exergy analysis of fluidized bed drying of moist particles. Exergy 2:87–98
42. Verma A, Jain N, Sethi SK (2022) Modeling and simulation of graphene-based composites. In Innovations in Graphene-Based Polymer Composites (pp 167–198). Woodhead Publishing
43. Verma A, Parashar A, van Duin AC (2022) Graphene-reinforced polymeric membranes for water desalination and gas separation/barrier applications. In Innovations Graphene-Based Polym Compos (pp. 133–165). Woodhead Publishing
44. Verma A, Parashar A (2017) The effect of STW defects on the mechanical properties and fracture toughness of pristine and hydrogenated graphene. Phys Chem Chem Phys 19(24):16023–16037

45. Verma A, Parashar A (2018) Molecular dynamics based simulations to study failure morphology of hydroxyl and epoxide functionalised graphene. Comput Mater Sci 143:15–26
46. Verma A, Parashar A (2018) Molecular dynamics based simulations to study the fracture strength of monolayer graphene oxide. Nanotechnology 29(11):115706
47. Verma A, Parashar A (2018) Structural and chemical insights into thermal transport for strained functionalised graphene: a molecular dynamics study. Mater Res Express 5(11):115605
48. Verma A, Parashar A, Packirisamy M (2018) Tailoring the failure morphology of 2D bicrystalline graphene oxide. J Appl Phys 124(1):015102
49. Verma A, Parashar A (2018) Reactive force field based atomistic simulations to study fracture toughness of bicrystalline graphene functionalised with oxide groups. Diam Relat Mater 88:193–203
50. Singla V, Verma A, Parashar A (2018) A molecular dynamics based study to estimate the point defects formation energies in graphene containing STW defects. Mater Res Express 6(1):015606
51. Verma A, Parashar A, Packirisamy M (2019) Role of chemical adatoms in fracture mechanics of graphene nanolayer. In: Materials Today: Proceedings 11:920–924
52. Verma A, Parashar A, Packirisamy M (2018) Atomistic modeling of graphene/hexagonal boron nitride polymer nanocomposites: a review. Wiley Interdiscip Rev: Comput Mol Sci 8(3):e1346
53. Verma A, Kumar R, Parashar A (2019) Enhanced thermal transport across a bi-crystalline graphene–polymer interface: an atomistic approach. Phys Chem Chem Phys 21(11):6229–6237
54. Verma, Akarsh, Amit Gaur, Singh VK (2017) Mechanical properties and microstructure of starch and sisal fiber biocomposite modified with epoxy resin mechanical properties and microstructure modified with epoxy resin. 6(1), pp. 500–520
55. Verma A, Parashar A, Jain N, Singh VK, Rangappa SM, Siengchin S (2020) Surface modification techniques for the preparation of different novel biofibers for composites. In Biofibers Biopolym Biocomposites (pp 1–34). Springer, Cham
56. Verma A, Singh VK (2016) Experimental investigations on thermal properties of coconut shell particles in DAP solution for use in green composite applications. J Mater Sci Eng 5(3):1000242
57. Singh K, Jain N, Verma A, Singh VK, Chauhan S (2020) Functionalized graphite–reinforced cross-linked poly (vinyl alcohol) nanocomposites for vibration isolator application: Morphology, mechanical, and thermal assessment. Mater Perform Charact 9(1):215–230
58. Verma A, Singh VK, Arif M (2016) Study of flame retardant and mechanical properties of coconut shell particles filled composite. Res Rev: J Mater Sci 4(3):1–5
59. Verma A, Jain N, Rastogi S, Dogra V, Sanjay SM, Siengchin S, Mansour R (2020) Mechanism, anti-corrosion protection and components of anti-corrosion polymer coatings. In Polymer Coatings (pp 53–66). CRC Press
60. Chaudhary A, Sharma S, Verma A (2022) WEDM machining of heat treated ASSAB'88 tool steel: a comprehensive experimental analysis. Materials Today: Proceedings 50:946–951
61. Bisht N, Verma A, Chauhan S, Singh VK (2021) Effect of functionalized silicon carbide nanoparticles as additive in cross-linked PVA based composites for vibration damping application. J Vinyl Add Tech 27(4):920–932
62. Verma A, Jain N, Parashar A, Singh VK, Sanjay MR, Siengchin S (2020) Design and modeling of lightweight polymer composite structures. In Lightweight Polym Compos Struct (pp. 193–224). CRC Press
63. Dogra V, Kishore C, Verma A, Rana AK, Gaur A (2021) Fabrication and experimental testing of hybrid composite material having biodegradable bagasse fiber in a modified epoxy resin: evaluation of mechanical and morphological behavior. Appl Sci Eng Prog 14(4):661–667
64. Verma A, Baurai K, Sanjay MR, Siengchin S (2020) Mechanical, microstructural, and thermal characterization insights of pyrolyzed carbon black from waste tires reinforced epoxy nanocomposites for coating application. Polym Compos 41(1):338–349

65. Verma A, Budiyal L, Sanjay MR, Siengchin S (2019) Processing and characterization analysis of pyrolyzed oil rubber (from waste tires)-epoxy polymer blend composite for lightweight structures and coatings applications. Polym Eng Sci 59(10):2041–2051
66. Verma A, Negi P, Singh VK (2019) Experimental analysis on carbon residuum transformed epoxy resin: Chicken feather fiber hybrid composite. Polym Compos 40(7):2690–2699
67. Verma A, Singh VK (2018) Mechanical, microstructural and thermal characterization of epoxy-based human hair–reinforced composites. J Test Eval 47(2):1193–1215
68. Jain N, Verma A, Ogata S, Sanjay MR, Siengchin S (2022) Application of machine learning in determining the mechanical properties of materials. In Machine Learn Appl Compos Mater (pp 99–113). Springer, Singapore.
69. Verma A, Singh VK, Verma SK, Sharma A (2016) Human hair: a biodegradable composite fiber–a review. International Journal of Waste Resources 6(206):1000206
70. Kataria A, Verma A, Sanjay MR, Siengchin S, Jawaid M (2022) Physical, morphological, structural, thermal, and tensile properties of coir fibers. In Coir Fiber and its Composites (pp. 79–107). Woodhead Publishing
71. Jain N, Verma A, Singh VK (2019) Dynamic mechanical analysis and creep-recovery behaviour of polyvinyl alcohol based cross-linked biocomposite reinforced with basalt fiber. Mater Res Express 6(10):105373
72. Chaurasia A, Verma A, Parashar A, Mulik RS (2019) Experimental and computational studies to analyze the effect of h-BN nanosheets on mechanical behavior of h-BN/polyethylene nanocomposites. J Phys Chem C 123(32):20059–20070
73. Verma A, Negi P, Singh VK (2018) Physical and thermal characterization of chicken feather fiber and crumb rubber reformed epoxy resin hybrid composite. Adv Civ Eng Mater 7(1):538–557
74. Bharath KN, Madhu P, Gowda TG, Verma A, Sanjay MR, Siengchin S (2020) A novel approach for development of printed circuit board from biofiber based composites. Polym Compos 41(11):4550–4558
75. Verma A, Joshi K, Gaur A, Singh VK (2018) Starch-jute fiber hybrid biocomposite modified with an epoxy resin coating: fabrication and experimental characterization. J Mech Behav Mater 27(5–6):20182006
76. Verma A, Singh C, Singh VK, Jain N (2019) Fabrication and characterization of chitosan-coated sisal fiber–Phytagel modified soy protein-based green composite. J Compos Mater 53(18):2481–2504
77. Rastogi S, Verma A, Singh VK (2020) Experimental response of nonwoven waste cellulose fabric–reinforced epoxy composites for high toughness and coating applications. Mater Perform Charact 9(1):151–172
78. Verma A, Zhang W, Van Duin AC (2021) ReaxFF reactive molecular dynamics simulations to study the interfacial dynamics between defective h-BN nanosheets and water nanodroplets. Phys Chem Chem Phys 23(18):10822–10834
79. Verma A, Jain N, Parashar A, Gaur A, Sanjay MR, Siengchin S (2021) Lifecycle assessment of thermoplastic and thermosetting bamboo composites. In Bamboo Fiber Compos (pp. 235–246). Springer, Singapore
80. Arpitha GR, Verma A, Sanjay MR, Siengchin S (2021) Preparation and experimental investigation on mechanical and tribological performance of hemp-glass fiber reinforced laminated composites for lightweight applications. Adv. Civil Eng. Mater 10(1):427–439
81. Verma A, Samant SS (2016) Inspection of hydrodynamic lubrication in infinitely long journal bearing with oscillating journal velocity. Journal of Applied Mechanical Engineering 5(3):1–7
82. Verma A, Jain N, Rangappa SM, Siengchin S, Jawaid M (2021) Natural fibers based biophenolic composites. In Phenolic Polym Based Compos Mater (pp. 153–168). Springer, Singapore
83. Arpitha GR, Jain N, Verma A, Madhusudhan M (2022) Corncob bio-waste and boron nitride particles reinforced epoxy-based composites for lightweight applications: fabrication and characterization. Biomass Convers Biorefinery, pp 1–8. https://doi.org/10.1007/s13399-022-03717-1

84. Verma A, Singh VK. Experimental characterization of modified epoxy resin assorted with almond shell particles. ESSENCE-Int J Environ Rehabil Conserv, 7 (1), pp 36–44
85. Verma A (2022) A perspective on the potential material candidate for railway sector applications: PVA based functionalized graphene reinforced composite. Appl Sci Eng Prog 15(2):5727–5727
86. Verma A, Jain N, Singh K, Singh VK, Rangappa SM, Siengchin S (2022) PVA-based blends and composites. In Biodegradable Polymers, Blends and Composites (pp 309–326). Woodhead Publishing
87. Raja S, Verma A, Rangappa SM, Siengchin S (2022) Development and experimental analysis of polymer based composite bipolar plate using Aquila Taguchi optimization: Design of experiments. Polym Compos 43(8):5522–5533
88. Prabhakaran S, Sharma S, Verma A, Rangappa SM, Siengchin S (2022) Mechanical, thermal, and acoustical studies on natural alternative material for partition walls: A novel experimental investigation. Polym Compos 43(7):4711–4720
89. Sethi SK, Gogoi R, Verma A, Manik G (2022) How can the geometry of a rough surface affect its wettability?-A coarse-grained simulation analysis. Prog Org Coat 172:107062
90. Chaudhary A, Sharma S, Verma A (2022) Optimization of WEDM process parameters for machining of heat treated ASSAB'88 tool steel using Response surface methodology (RSM). Materials Today: Proceedings 50:917–922
91. Verma A, Jain N, Sanjay MR, Siengchin S. Viscoelastic properties of completely biodegradable Polymer-Based composites. In Vibration Damping Behav Biocomposites (pp 173–188). CRC Press
92. Verma A, Jain N, Mishra RR (2022) Applications and drawbacks of Epoxy/Natural fiber composites. In Handbook of Epoxy/Fiber Composites (pp. 1–15). Singapore: Springer Singapore
93. Lila MK, Verma A, Bhurat SS (2022) Impact behaviors of Epoxy/Synthetic fiber composites. In Handbook of Epoxy/Fiber Composites (pp 1–18). Singapore: Springer Singapore
94. Kumar G, Mishra RR, Verma A (2022) Introduction to molecular dynamics simulations. In forcefields for Atomistic-Scale Simulations: Materials and Applications (pp 1–19). Springer, Singapore.https://doi.org/10.1007/978-981-19-3092-8_1
95. Verma A, Sharma S (2022) Atomistic simulations to study thermal effects and strain rate on mechanical and fracture properties of graphene like BC3. In Forcefields AtIstic-Scale SimulS: Mater Appl (pp. 237–252). Springer, Singapore
96. Chaturvedi S, Verma A, Sethi SK, Ogata S (2022) Defect energy calculations of Nickel, copper and aluminium (and Their Alloys): molecular dynamics approach. In Forcefields AtIstic-Scale Simuls: Mater Appl (pp 157–186). Springer, Singapore
97. Shankar U, Gogoi R, Sethi SK, Verma A (2022) Introduction to materials studio software for the Atomistic-Scale simulations. In Forcefields AtIstic-Scale Simuls: Mater Appl (pp 299–313). Springer, Singapore
98. Shankar U, Sethi SK, Verma A (2022) Forcefields and modeling of polymer coatings and nanocomposites. In Forcefields AtIstic-Scale SimulS: Mater Appl (pp 81–98). Springer, Singapore
99. Verma A, Ogata S (2022) Computational modelling of deformation and failure of bone at molecular scale. In Forcefields AtIstic-Scale Simuls: Mater Appl (pp 253–268). Springer, Singapore
100. Homer ER, Verma A, Britton D, Johnson OK, Thompson GB (2022) Simulated migration behavior of metastable Σ3 (11 8 5) incoherent twin grain boundaries. In IOP Conference Series: Materials Science and Engineering 1249(1), p 012019
101. Chaturvedi S, Verma A, Sethi SK, Rangappa SM Siengchin S (2022) Stalk fibers (rice, wheat, barley, etc.) composites and applications. In Plant Fibers, Their Compos, Appl (pp 347–362). Woodhead Publishing
102. Arpitha GR, Verma A, MR, S., Gorbatyuk, S., Khan, A., Sobahi, T.R., Asiri, A.M. and Siengchin, S. (2022) Bio-composite film from corn starch based vetiver cellulose. Journal of Natural Fibers 19(16):14634–14644

103. Thimmaiah SH, Narayanappa K, Thyavihalli Girijappa Y, Gulihonenahali Rajakumara A, Hemath M, Thiagamani SMK, Verma A (2022) An artificial neural network and Taguchi prediction on wear characteristics of Kenaf-Kevlar fabric reinforced hybrid polyester composites. Polym Compos 44(1):261–273
104. Kataria A, Verma A, Sethi SK, Ogata S (2022) Introduction to interatomic Potentials/ Forcefields. In Forcefields AtIstic-Scale SimulS: Mater Appl (pp 21–49). Springer, Singapore
105. Kataria A, Verma A, Sanjay MR, Siengchin S (2022) Molecular modeling of 2D graphene grain boundaries: Mechanical and fracture aspects. Mater Today Proc 52:2404–2408
106. Verma A, Ogata S (2023) Magnesium based alloys for reinforcing biopolymer composites and coatings: A critical overview on biomedical materials. Advanced Industrial and Engineering Polymer Research. https://doi.org/10.1016/j.aiepr.2023.01.002
107. Bharath KN, Madhu P, Gowda TY, Verma A, Sanjay MR, Siengchin S (2021) Mechanical and chemical properties evaluation of sheep wool fiber–reinforced vinylester and polyester composites. Materials Performance and Characterization 10(1):99–109
108. Aditya Kataria, Swati Chaturvedi, Vaibhav Chaudhary, Akarsh Verma, Naman Jain Mavinkere Rangappa Sanjay, Suchart Siengchin, 2023. Cellulose fiber-reinforced composites—History of evolution, chemistry, and structure. In Cellulose Fibre Reinforced Composites (pp 1–22). Woodhead Publishing. DOI: https://doi.org/10.1016/B978-0-323-90125-3.00012-4
109. Swati Chaturvedi, Aditya Kataria, Vaibhav Chaudhary, Akarsh Verma, Naman Jain Mavinkere Rangappa Sanjay, Suchart Siengchin (2023) Bionanocomposites reinforced with cellulose fibers and agro-industrial wastes. In Cellulose Fibre Reinforced Composites (pp 1–22). Woodhead Publishing. https://doi.org/10.1016/B978-0-323-90125-3.00017-3
110. Arpitha GR, Mohit H, Madhu P, Verma A (2023) Effect of sugarcane bagasse and alumina reinforcements on physical, mechanical, and thermal characteristics of epoxy composites using artificial neural networks and response surface methodology. Biomass Convers Biorefinery. https://doi.org/10.1007/s13399-023-03886-7
111. Deji R, Verma A, Kaur N, Choudhary BC, Sharma RK (2022) Density functional theory study of carbon monoxide adsorption on transition metal doped armchair graphene nanoribbon. In: Materials Today: Proceedings 54:771–776
112. Deji R, Verma A, Choudhary BC, Sharma RK (2022) New insights into NO adsorption on alkali metal and transition metal doped graphene nanoribbon surface: A DFT approach. J Mol Graph Model 111:108109
113. Verma A, Jain N, Parashar A, Singh VK, Sanjay MR, Siengchin S (2020) Lightweight graphene composite materials. In Lightweight Polym Compos Struct (pp 1–20). CRC Press
114. Verma A, Parashar A (2020) Characterization of 2D nanomaterials for energy storage. In Recent Adv Theor, Appl, Comput Exp Mech (pp 221–226). Springer, Singapore
115. Deji R, Jyoti R, Verma A, Choudhary BC, Sharma RK (2022) A theoretical study of HCN adsorption and width effect on co-doped armchair graphene nanoribbon. Comput Theor Chem 1209:113592
116. Verma A, Parashar A, Packirisamy M (2019) Effect of grain boundaries on the interfacial behaviour of graphene-polyethylene nanocomposite. Appl Surf Sci 470:1085–1092
117. Deji R, Verma A, Kaur N, Choudhary BC, Sharma RK (2022) Adsorption chemistry of co-doped graphene nanoribbon and its derivatives towards carbon based gases for gas sensing applications: Quantum DFT investigation. Mater Sci Semicond Process 146:106670
118. Chaturvedi S, Verma A, Singh SK, Ogata S (2022) EAM Inter-Atomic Potential—Its implication on Nickel, copper, and aluminum (and Their Alloys). In Forcefields AtIstic-Scale SimulS: Mater Appl (pp 133–156). Springer, Singapore. https://doi.org/10.1007/978-981-19-3092-8_7
119. Verma A, Parashar A, Singh SK, Jain N, Sanjay SM, Siengchin S (2020) Modeling and simulation in polymer coatings. In Polymer Coats (pp 309–324). CRC Press
120. Dogra V, Kishore C, Mishra A, Gaur A, Verma A (2023) Sol-Gel preparation and wetting behaviour analysis of hydrophobic Zirconium based nano-coating: Implications for solar panel coating. Chem Eng J Adv 15:100507. https://doi.org/10.1016/j.ceja.2023.100507

121. TG YG, Ballupete Nagaraju S, Puttegowda M, Verma A, Rangappa SM, Siengchin S (2023) Biopolymer-Based composites: An Eco-Friendly alternative from agricultural Waste Biomass. J Compos Sci 7(6):242. https://doi.org/10.3390/jcs7060242
122. Arpitha GR, Jain N, Verma A (2023) Banana biofiber and glass fiber reinforced hybrid composite for lightweight structural applications: Mechanical, thermal, and microstructural characterization. Biomass Convers Biorefinery, pp 1–10. https://doi.org/10.1007/s13399-023-04300-y
123. Sati P, Verma A, Zindal A, Chauhan S, Singh VK (2023) PVA biopolymer-acidic functionalized graphene hybrid nano composite for vibration isolation application: An experimental approach with variable reflux and vacuum timings. Chem Phys Impact 6:100212. https://doi.org/10.1016/j.chphi.2023.100212
124. Agrawal PK, Sharma P, Verma A, Singh VK, Chaudhary AK, Chauhan S (2023) Impact of graphite particle surface modification on the strengthening of cross-linked polyvinyl alcohol composites: A comprehensive investigation. In: Proceedings of the Institution of Mechanical Engineers, Part L: Journal of Materials: Design and Applications. https://doi.org/10.1177/14644207231188601
125. Dogra V, Kishore C, Mishra A, Verma A, Gaur A (2023) Coatings: Types and synthesis techniques. In Coating Mater: Comput Asp, Appl ChallS (pp. 17–31). Singapore: Springer Nature Singapore. https://doi.org/10.1007/978-981-99-3549-9_2
126. Gupta HS, Shankar U, Verma A, Gogoi R, Sethi SK (2023) Computational aspects: Self-clean coatings, plastics and polymers in coatings. In Coating Mater: Comput Asp, Appl Challs (pp 167–184). Singapore: Springer Nature Singapore. https://doi.org/10.1007/978-981-99-3549-9_8
127. Gupta HS, Sethi SK, Verma A (2023) Applications of coating materials: a critical overview. Coat Mater: Comput Asp, Appl Challs, pp 81–109. https://doi.org/10.1007/978-981-99-3549-9_4
128. Tongco JV, Sethi SK, Kumar A, Verma A, Shankar U (2023) Introduction to coatings: types and their synthesis. Coat Mater: Comput Asp, Appl Challs (pp 1–15). Singapore: Springer Nature Singapore. https://doi.org/10.1007/978-981-99-3549-9_1
129. Kataria A, Zafar S, Verma A, Ogata S (2023) Molecular dynamics simulations in coatings. In Coating Mater: Comput Asp, Appl Challs (pp 125–156). Singapore: Springer Nature Singapore. https://doi.org/10.1007/978-981-99-3549-9_6
130. Manasa CM, Prince B, Arpitha GR, Verma A (2023) Hybrid feature selection techniques to improve the accuracy of rice yield prediction: a machine learning approach. In Coating Mater: Comput Asp, Appl Challs (pp 409–421). Singapore: Springer Nature Singapore. https://doi.org/10.1007/978-981-99-3549-9_19
131. Singh SK, Chaurasia A, Verma A (2023) Basics of density functional theory, molecular dynamics, and monte carlo simulation techniques in materials science. In Coating Mater: Comput Asp, Appl Challs (pp 111–124). Singapore: Springer Nature Singapore. https://doi.org/10.1007/978-981-99-3549-9_5
132. Zafar S, Verma A (2023) Continuum Mechanics-Based simulations in coatings. In Coating Mater: Comput Asp, Appl Challs (pp 185–216). Singapore: Springer Nature Singapore. https://doi.org/10.1007/978-981-99-3549-9_9
133. Robinson J, Verma A, Homer ER, Thompson GB (2023) Nanotwin stability in alloyed copper under ambient and cryo-temperature dependent deformation states. Mater Sci Eng, A 871:144866. https://doi.org/10.1016/j.msea.2023.144866
134. Verma A, Johnson OK, Thompson GB, Chesser I, Ogata S, Homer ER (2023) Insights into factors that affect non-Arrhenius migration of a simulated incoherent Σ3 grain boundary. Acta Mater 258:119210. https://doi.org/10.1016/j.actamat.2023.119210
135. Nagaraju SB, Sathyanarayana K, Somashekara MK, Pradeep DG, Puttegowda M, Verma A (2023) Artificial neural networks for predicting mechanical properties of Al2219-B4C-Gr composites with multireinforcements. Proc Inst Mech Eng C J Mech Eng Sci. https://doi.org/10.1177/09544062231196038

136. Sharma S, Verma A, Rangappa SM, Siengchin S, Ogata S (2023) Recent progressive developments in conductive-fillers based polymer nanocomposites (CFPNC's) and conducting polymeric nanocomposites (CPNC's) for multifaceted sensing applications. J Market Res 26:5921–5974. https://doi.org/10.1016/j.jmrt.2023.08.300
137. Dora TL, Singh SK, Mishra RR, Verma A (2023) Role of Crystal Orientation, Temperature, and Strain Rate on the Mechanical Characterization of Nickel: An Atomistic-scale investigation. Journal of Micromanufacturing. https://doi.org/10.1177/25165984231195519
138. Jain N, Sharma P, Verma A, Gupta J (2023) Enhancement of thermo-mechanical, creep-recovery, and anti-microbial properties in PVA-based biodegradable films through cross-linking with oxalic acid: implications for packaging application. Biomass Convers Biorefinery, pp 1–10. https://doi.org/10.1007/s13399-023-04748-y
139. Sharath BN, Madhu P, Verma A (2023) Enhancing tribological performance: A review of ceramic reinforced aluminium hybrid composites for high-temperature engineering applications. Hybrid Advances 4:100094
140. Shariff MM, Arpitha GR, Jain N, Shankar U, Verma A, Shivakumar ND (2023) A comparative study on the effect of reinforcing boron nitride/alumina in epoxy-based hybrid composite with Millettia pinnata leaf powder and glass sheets: Experimental fabrication, mechanical and micro-structural characterization. Hybrid Advances 4:100095
141. Mandal S, Jena PC, Gangil S, Pal S, Haydary J, Sharma RK Verma A (2023) Ni-supported pigeon pea stalk biochar as a catalyst for ex situ tar cracking in biomass gasification. Biomass Convers Biorefinery, pp 1–11. https://doi.org/10.1007/s13399-023-04974-4

SPRINGER NATURE

GPSR Compliance

The European Union's (EU) General Product Safety Regulation (GPSR) is a set of rules that requires consumer products to be safe and our obligations to ensure this.

If you have any concerns about our products, you can contact us on ProductSafety@springernature.com

In case Publisher is established outside the EU, the EU authorized representative is:

Springer Nature Customer Service Center GmbH
Europaplatz 3
69115 Heidelberg, Germany

The manufacturer's authorised representative in the EU is Springer Nature Customer Service Centre GmbH, Europaplatz 3, 69115 Heidelberg, Germany. If you have any concerns regarding our products, please contact ProductSafety@springernature.com

Printed and bound by CPI Group (UK) Ltd, Croydon, CR0 4YY
30/09/2026
02225314-0014